UAV-Based Remote Sensing
Volume 1

Special Issue Editors

Felipe Gonzalez Toro
Antonios Tsourdos

MDPI • Basel • Beijing • Wuhan • Barcelona • Belgrade

MDPI

Special Issue Editors

Felipe Gonzalez Toro
Queensland University of Technology
Australia

Antonios Tsourdos
Cranfield University
UK

Editorial Office
MDPI
St. Alban-Anlage 66
Basel, Switzerland

This edition is a reprint of the Special Issue published online in the open access journal *Sensors* (ISSN 1424-8220) from 2016–2017 (available at: http://www.mdpi.com/journal/sensors/special_issues/UAV_remote_sensing).

For citation purposes, cite each article independently as indicated on the article page online and as indicated below:

Lastname, F.M.; Lastname, F.M. Article title. *Journal Name* **Year**, *Article number, page range.*

First Edition 2018

Volume I
ISBN 978-3-03842-777-3 (Pbk)
ISBN 978-3-03842-778-0 (PDF)

Volume I–II
ISBN 978-3-03842-857-2 (Pbk)
ISBN 978-3-03842-858-9 (PDF)

Table of Contents

About the Special Issue Editors v

Preface to "UAV-Based Remote Sensing" vii

Zain Anwar Ali, Daobo Wang and Muhammad Aamir
Fuzzy-Based Hybrid Control Algorithm for the Stabilization of a Tri-Rotor UAV
doi: 10.3390/s16050652 1

Abdulla Al-Kaff, Fernando García, David Martín, Arturo De La Escalera and José María Armingol
Obstacle Detection and Avoidance System Based on Monocular Camera and Size Expansion
Algorithm for UAVs
doi: 10.3390/s17051061 19

Miguel Alvarado, Felipe Gonzalez, Peter Erskine, David Cliff and Darlene Heuff
A Methodology to Monitor Airborne PM$_{10}$ Dust Particles Using a Small Unmanned
Aerial Vehicle
doi: 10.3390/s17020343 41

Guanbing Bai, Jinghong Liu, Yueming Song and Yujia Zuo
Two-UAV Intersection Localization System Based on the Airborne Optoelectronic Platform
doi: 10.3390/s17010098 66

Fotios Balampanis, Iván Maza and Aníbal Ollero
Coastal Areas Division and Coverage with Multiple UAVs for Remote Sensing
doi: 10.3390/s17040808 84

Anthony DeMario, Pete Lopez, Eli Plewka, Ryan Wix, Hai Xia, Emily Zamora, Dan Gessler
and Azer P. Yalin
Water Plume Temperature Measurements by an Unmanned Aerial System (UAS)
doi: 10.3390/s17020306 109

Luke J. Evans, T. Hefin Jones, Keeyen Pang, Silvester Saimin and Benoit Goossens
Spatial Ecology of Estuarine Crocodile (*Crocodylus porosus*) Nesting in a Fragmented Landscape
doi: 10.3390/s16091527 119

Changhong Fu, Ran Duan, Dogan Kircali and Erdal Kayacan
Onboard Robust Visual Tracking for UAVs Using a Reliable Global-Local Object Model
doi: 10.3390/s16091406 129

Mateo Gašparović and Luka Jurjević
Gimbal Influence on the Stability of Exterior Orientation Parameters of UAV Acquired Images
doi: 10.3390/s17020401 151

Colin Greatwood, Thomas Richardson, Jim Freer, Rick Thomas, Rob MacKenzie,
Rebecca Brownlow, David Lowry, Rebecca Fisher and Euan Nisbet
Atmospheric Sampling on Ascension Island Using Multirotor UAVs
doi: 10.3390/s17061189 167

Mason Itkin, Mihui Kim and Younghee Park
Development of Cloud-Based UAV Monitoring and Management System
doi: 10.3390/s16111913 191

Michal Kedzierski and Paulina Delis
Fast Orientation of Video Images of Buildings Acquired from a UAV without Stabilization
doi: 10.3390/s16070951 . . . . **210**

**Sarantis Kyristsis, Angelos Antonopoulos, Theofilos Chanialakis, Emmanouel Stefanakis,
Christos Linardos, Achilles Tripolitsiotis and Panagiotis Partsinevelos**
Towards Autonomous Modular UAV Missions: The Detection, Geo-Location and
Landing Paradigm
doi: 10.3390/s16111844 . . . . **226**

Boyang Li, Yifan Jiang, Jingxuan Sun, Lingfeng Cai and Chih-Yung Wen
Development and Testing of a Two-UAV Communication Relay System
doi: 10.3390/s16101696 . . . . **239**

Huanyu Li, Linfeng Wu, Yingjie Li, Chunwen Li and Hangyu Li
A Novel Method for Vertical Acceleration Noise Suppression of a Thrust-Vectored VTOL UAV
doi: 10.3390/s16122054 . . . . **260**

Zhe Liu, Zulin Wang and Mai Xu
Cubature Information SMC-PHD for Multi-Target Tracking
doi: 10.3390/s16050653 . . . . **281**

Jianchen Liu, Bingxuan Guo, Wanshou Jiang, Weishu Gong and Xiongwu Xiao
Epipolar Rectification with Minimum Perspective Distortion for Oblique Images
doi: 10.3390/s16111870 . . . . **302**

Chenglong Liu, Jinghong Liu, Yueming Song and Huaidan Liang
A Novel System for Correction of Relative Angular Displacement between Airborne Platform
and UAV in Target Localization
doi: 10.3390/s17030510 . . . . **319**

Yalong Ma, Xinkai Wu, Guizhen Yu, Yongzheng Xu and Yunpeng Wang
Pedestrian Detection and Tracking from Low-Resolution Unmanned Aerial Vehicle
Thermal Imagery
doi: 10.3390/s16040446 . . . . **341**

**Douglas G. Macharet, Héctor I. A. Perez-Imaz, Paulo A. F. Rezeck, Guilherme A. Potje,
Luiz C. C. Benyosef, André Wiermann, Gustavo M. Freitas, Luis G. U. Garcia,
Mario F. M. Campos**
Autonomous Aeromagnetic Surveys Using a Fluxgate Magnetometer
doi: 10.3390/s16122169 . . . . **367**

About the Special Issue Editors

Felipe Gonzalez Toro, Associate Professor at the Science and Engineering Faculty, Queensland University of Technology (Australia), with a passion for innovation in the fields of aerial robotics and automation and remote sensing. He creates and uses aerial robots, drones or UAVs that possess a high level of cognition using efficient on-board computer algorithms and advanced optimization and game theory approaches that assist us to understand and improve our physical and natural world. Dr. Gonzalez leads the UAVs-based remote sensing research at QUT. As of 2017, he has published nearly 120 peer reviewed papers. To date, Dr. Gonzalez has been awarded $10.1M in chief investigator/partner investigator grants. This grant income represents a mixture of sole investigator funding, international, multidisciplinary collaborative grants and funding from industry. He is also a Chartered Professional Engineer, Engineers Australia—National Professional Engineers Register (NPER), a member of the Royal Aeronautical Society (RAeS), The IEEE, American Institute of Aeronautics and Astronautics (AIAA) and holder of a current Australian Private Pilot Licence (CASA PPL).

Antonios Tsourdos obtained a MEng on Electronic, Control and Systems Engineering, from the University of Sheffield (1995), an MSc on Systems Engineering from Cardiff University (1996) and a PhD on Nonlinear Robust Autopilot Design and Analysis from Cranfield University (1999). He joined the Cranfield University in 1999 as lecturer, was appointed Head of the Centre of Autonomous and Cyber-Physical Systems in 2007 and Professor of Autonomous Systems and Control in 2009 and Director of Research—Aerospace, Transport and Manufacturing in 2015. Professor Tsourdos was a member of the Team Stellar, the winning team for the UK MoD Grand Challenge (2008) and the IET Innovation Award (Category Team, 2009). Professor Tsourdos is an editorial board member of: Proceedings of the IMechE Part G Journal of Aerospace Engineering; IEEE Transactions of Aerospace and Electronic Systems; Aerospace Science & Technology; International Journal of Systems Science; Systems Science & Control Engineering; and the International Journal of Aeronautical and Space Sciences. Professor Tsourdos is Chair of the IFAC Technical Committee on Aerospace Control, a member of the IFAC Technical Committee on Networked Systems, Discrete Event and Hybrid Systems, and Intelligent Autonomous Vehicles. Professor Tsourdos is also a member of the AIAA Technical Committee on Guidance, Control and Navigation; AIAA Unmanned Systems Program Committee; IEEE Control System Society Technical Committee on Aerospace Control (TCAC) and IET Robotics & Mechatronics Executive Team.

Preface to "UAV-Based Remote Sensing"

Active technological development has fuelled rapid growth in the number of Unmanned Aerial Vehicle (UAV) platforms being deployed around the globe. Improved sensors and enhanced image processing techniques have consolidated and confirmed UAVs as the technology of choice. Many jurisdictions have regulated stringent restrictions on flying UAVs in numerous scenarios where they could be of great value. Despite the increased regulation, UAVs or drones have rapidly become the tool of choice for both the environmental science and the remote sensing communities. This is due, in no small part, to a lowering cost of entry. The last few years has seen significant technological development in UAV platforms, sensor miniaturisation, and robotic sensing. UAV technology progresses apace. The design of novel UAV systems and the use of UAV platforms integrated with RGB, multispectral, hyperspectral, thermal imaging, gas sensing and/or laser scanning sensors have now been demonstrated in both research and practical applications. Novel UAV platforms, UAV-based sensors, robotic sensing and imaging techniques, the development of processing workflows, as well as the capacity of ultra-high temporal and spatial resolution data, provide both opportunities and challenges that will allow engineers and scientists to address novel and important scientific questions in UAV and sensor design, remote sensing and environmental monitoring. This work features papers on UAV sensor design; improvements in UAV sensor technology; obstacle detection, methods for measuring optical flow; target tracking; gimbal influence on the stability of UAV images; augmented reality tools; segmentation in digital surface models for 3D reconstruction; detecting the location and grasping objects; multi-target localization; vision-based tracking in cooperative multi-UAV systems; noise suppression techniques; rectification for oblique images; two-UAV communication system; fuzzy-based hybrid control algorithms; pedestrian detection and tracking as well as a range of atmospheric, geological, agricultural, ecological, reef, wildlife, building and construction; coastal area coverage; search and rescue (SAR); water plume temperature measurements; aeromagnetic and archaeological survey applications.

Felipe Gonzalez Toro, Antonios Tsourdos
Special Issue Editors

![sensors](sensors logo)

MDPI

Article

Fuzzy-Based Hybrid Control Algorithm for the Stabilization of a Tri-Rotor UAV

Zain Anwar Ali [1,*], Daobo Wang [1] and Muhammad Aamir [2]

[1]   College of Automation Engineering, Nanjing University of Aeronautics and Astronautics, Nanjing 210016, China; dbwangpe@nuaa.edu.cn
[2]   Electronic Engineering Department, Sir Syed University of Engineering and Technology, Karachi 75300, Pakistan; muaamir5@yahoo.com
*   Correspondence: zainanwar86@hotmail.com; Tel.: +86-1301-693-1051

Academic Editor: Felipe Gonzalez Toro
Received: 3 February 2016; Accepted: 28 April 2016; Published: 9 May 2016

Abstract: In this paper, a new and novel mathematical fuzzy hybrid scheme is proposed for the stabilization of a tri-rotor unmanned aerial vehicle (UAV). The fuzzy hybrid scheme consists of a fuzzy logic controller, regulation pole-placement tracking (RST) controller with model reference adaptive control (MRAC), in which adaptive gains of the RST controller are being fine-tuned by a fuzzy logic controller. Brushless direct current (BLDC) motors are installed in the triangular frame of the tri-rotor UAV, which helps maintain control on its motion and different altitude and attitude changes, similar to rotorcrafts. MRAC-based MIT rule is proposed for system stability. Moreover, the proposed hybrid controller with nonlinear flight dynamics is shown in the presence of translational and rotational velocity components. The performance of the proposed algorithm is demonstrated via MATLAB simulations, in which the proposed fuzzy hybrid controller is compared with the existing adaptive RST controller. It shows that our proposed algorithm has better transient performance with zero steady-state error, and fast convergence towards stability.

Keywords: Unmanned Aerial Vehicle; Tri-Rotor UAV; RST controller; fuzzy hybrid controller

1. Introduction

One of the best inventions of today's era is the small flying machine commonly called a UAV. This research is dedicated to such types of UAVs, which are commonly used in the monitoring of disaster management and military operations, as well as small indoor activities [1–3]. The research on UAVs is based on the different knowledge banks of aeronautics, signal processing, and control automation. For this research, multiple hardware-based tests are performed to design the best flying machines with precise control mechanisms.

The current trend is focused on the design of advanced, lightweight, and perfect UAVs that can be operated in any disastrous situations over remote areas. UAVs are classified as either fixed-wing or rotary wing [4]. Rotor-based UAVs are multiple input and multiple output (MIMO) multivariable systems [5]. Rotorcraft have a great advantage over fixed-wing aircraft with respect to various applications, like vertical takeoff and vertical landing (VTOL) capability and payloads. Rotor-based UAVs include many types, such as bi-rotor, tri-rotor, quad-rotor and hex-rotor [6]. Moreover, a tri-rotor UAV with VTOL ability is considered in this paper.

Real-world application of UAVs require intense hardware testing. Before the experimental testing of our proposed algorithm in the real world, we have to simulate the numerical nonlinear simulations for the Euler angles, control commands, rotational velocities, and translational velocities [7]. In this research, our main concern is to rectify the error which occurs in a yaw moment due to the unpaired reaction of the rotors, thereby producing torque. Brushless direct current (BLDC) motors are installed in the triangular frame of the tri-rotor craft to nullify the tilt angle moment.

The dynamics of the UAV are highly nonlinear and multi-variable, with a lot of parameter uncertainties, many effects to which a potential controller has to be robust. The aerodynamics of the actuator blades (flapping of blade and propeller), inertial torques (angular speed of propellers), and gyroscopic effects (which change the orientation of the UAV) are found in [8]. The redundancy in the rotors of a UAV formulates them towards a set of partial collapses. Although the maneuverability and performance will probably be condensed in the case of such a collapse, it is required that a controller stabilizes the system and tolerates reduced mode functions, such as safe arrival, steady hover, *etc.* [9,10].

Previously, many control methods were used for the stabilization of UAVs, including the conventional proportional integral derivative (PID) controller, fuzzy controller, adaptive controller, and so on [11]. For controlling the parameters of a UAV an adaptive controller has a capability to give good performance in the presence of model and parametric uncertainties, while MRAC is concerned with the vibrant reaction of the controlled system to asymptotic convergence. It follows the reference system in spite of parametric model uncertainties in the system [12].

In [13] the proposed MRAC for controlling the dynamics of a quad-rotor in the presence of actuator uncertainties was considered to enhance an existing linear controller, offering autonomous waypoint following. The stability of the adaptive controller was ensured by the Lyapnauv theorem and, in a nonlinear structure, the algorithm is applied for indoor flight test.

In [14] the hybrid control scheme to fault tolerant control (FTC) for a quad-rotor aircraft in the presence of faults in their rotors during the flight have been explored and tested on the MRAC algorithm and a gain-scheduled PID (GS-PID) control. MRAC and GS-PID are used in collaboration with a linear quadratic regulator (LQR) to control the attitude of the UAV. MRAC is based on MIT rules for controlling the height and other parameters of a Qball-X4 Quad-Rotor aircraft.

Takagi-Sugeno fuzzy rules were previously used in [15,16] to control the nonlinear behavior of the vehicle. On the other hand, in [17], a twin controller approach that consists of a backstepping controller to control the nonlinear dynamics of the system and linguistic logic rules of a fuzzy logic controller (Mamdani) is used to control the attitude of tilt of a tri-rotor UAV. In [18] a dual controller approach with an adaptive fuzzy sliding mode controller is used to control the mini UAV, in which sliding mode control is utilized to control the nonlinear behavior of the UAV, and then fuzzy logic rules are implemented on it. The hybrid controller approach was also addressed in [19] in which a fuzzy-PID controller with a PSO algorithm is applied on tri-rotor dynamics.

Hwoever, in this paper, we proposed a fuzzy hybrid controller consisting of a RST with MRAC, based on MIT rules working as a main controller in the model to deal with the nonlinear system. We compare the performance of our proposed fuzzy hybrid controller with the robust adaptive RST controller of [20].

Moreover, the adaptive gains of the RST controller are (*i.e.*, regulation gain "G_R", pole-placement gain G_S, and tracking gain "G_T") tuned by a fuzzy logic controller (Mamdani technique). This means that our main controller is a RST with MRAC based on MIT rules, and for the tuning purpose we use the Mamdani fuzzy logic controller. We have to implement the gains of RST by adding fuzzy logic between uniform scales of membership functions. It shows the best results as compared to the adaptive RST controller [21,22].

In this paper, we are incorporating RST controller with our proposed system in two separate ways. First, the system is undergoes through Robust adaptive RST controller, after that we use Fuzzy-Hybrid based MIT algorithm and then conclude the results by taking the difference of robustness.

The core contributions in this research are as follows: (1) a novel fuzzy-based adaptive robust RST controller is derived by accumulating the MIT rule in the control law to remove the model disturbance and to derive the steady-state error to zero; (2) the proposed controller uses the angular responses as an input control command, which shows more accurate and practical insight in the real world; (3) in spite of the model disturbance, the close loop system error converges to zero, proved in Theorem 2; and (4) lastly, the polynomial characteristic solution is based on the Diophantine equation while least square estimation is used to check system stability and proved in Theorem 3.

The breakup of this paper is structured as follows. The system modeling, dynamic representation of a tri-rotor UAV, and main engine model is discussed in Section 2. Section 3 demonstrates the dynamic control strategies and the control algorithm of the UAV. Moreover, the simulation results and discussions are discussed in Section 4. Lastly, Section 5 states the conclusions.

2. System Model and Preliminaries

2.1. Tri-Rotor Modeling

The equation of motion of a rigid body is defined by Newton's second law of motion [23,24]. Linear and angular forces change with respect to the timeframe, called the initial reference frame, in which the UAV has a similar velocity, force components, and moments, which are used to develop the six degrees of freedom nonlinear equations of motion. The nonlinear aerodynamic forces, aerodynamic moments, rotation motion, and translational motion of a UAV are defined by using differential Equations (1)–(4).

Equations of aerodynamic force:

$$
\begin{aligned}
F_X - mg\,\sin\theta &= m\left(\dot{u} + qw - rv\right) \\
F_Y + mg\,\cos\theta\,\sin\varphi &= m\left(\dot{v} + ru - pw\right) \\
F_Z + mg\,\cos\theta\,\cos\varphi &= m\left(\dot{w} + pv - qu\right)
\end{aligned}
\tag{1}
$$

Equations of aerodynamic moments:

$$
\begin{aligned}
L &= I_x\dot{p} - I_{xz}\dot{r} + qr\left(I_z - I_y\right) - I_{xz}pq \\
M &= I_y\dot{q} + rp\left(I_x - I_z\right) + I_{xz}\left(p^2 - r^2\right) \\
N &= -I_{xz}\dot{p} + I_z\dot{r} + pq\left(I_y - I_x\right) + I_{xz}qr
\end{aligned}
\tag{2}
$$

Rotational rates:

$$
\begin{aligned}
p &= \dot{\varphi} - \dot{\psi}\,\sin\theta \\
q &= \dot{\theta}\,\cos\varphi + \dot{\psi}\,\cos\theta\,\sin\varphi \\
r &= \dot{\psi}\,\cos\theta\,\cos\varphi - \dot{\theta}\,\sin\varphi
\end{aligned}
\tag{3}
$$

Euler angles and body angular velocities:

$$
\begin{aligned}
\dot{\theta} &= q\,\cos\varphi - r\,\sin\theta \\
\dot{\varphi} &= p + q\,\sin\varphi\,\tan\theta + r\cos\varphi\,\tan\theta \\
\dot{\psi} &= \left(q\,\sin\varphi + r\,\cos\varphi\right)\sec\theta
\end{aligned}
\tag{4}
$$

The four control commands of the tri-rotor UAV are (Col, Lat, Lon, Ped), which is similar to the conventional helicopter, in which Col is Collective, Lat is Lateral, Lon is Longitudinal, and Ped is pedal. (Col, Lat) are used to control the roll rate, (Lon) control the pitch rate and (Ped) controls the yaw rate of a UAV and tilt angle by using the parameter "$\propto$" [25,26]. (p, q, r) and (U, V, W) are the rotational velocity and translational velocity of the coordinate system. (L, M, N) and (ϕ, θ, ψ) are the external moments and rotational angles of a fixed body frame.

A tri-rotor aerial vehicle exhibits many physical effects, like inertial torque, effects of aerodynamics, effects of gravity, effects of gyroscope and frictional effects, *etc.* In the presence of these physical effects it is quite difficult to design a controller which can easily handle all of the physical effects and stabilize the UAV in a fair amount of time, because it has six degree of freedom (6-DOF) with a highly-nonlinear, multivariable, under-actuated, strongly-coupled model with the rotors, as shown in Figures 1 and 2 taken from the design of Mohamed MK [27].

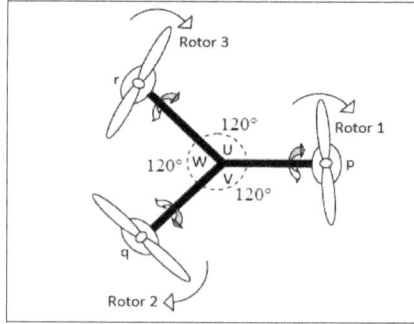

Figure 1. Top view of a tri-rotor UAV along with rotational and transitional rates.

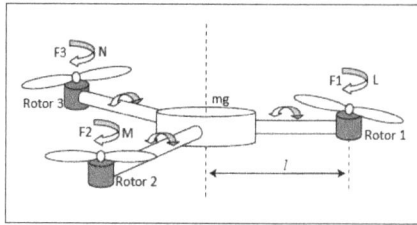

Figure 2. 3-D view of a tri-rotor UAV along with aerodynamic moment components.

The tri-rotor UAV, including translational and rotational subsystems, needs a vibrant strategy of nonlinear sequential control in the 6-DOF model. This research addresses error controlling in a tri-rotor aircraft by managing the torque produced by unpaired rotor reactions. To overcome this issue, implementation of different designs has been made with the help of BLDC motors, which actually control the nullifying angle. This method is useful for quicker turn by tilting the rotor's axis.

2.2. Dynamic Representation of a Tri-Rotor UAV

The orientation of the UAV is explained by Euler angles having altitude, roll (θ), pitch (φ), and yaw (Ψ) control and it rotates along (x, y, z) axes, respectively. The translational and rotational movement of the tri-rotor UAV into the dimensional space and dynamics of the rigid body derive from Newton's law. Moreover, moments, forces, velocity components, and aerodynamic components are described in Table 1.

Table 1. Dynamic constants of a tri-rotor UAV.

x, y, z Axis System	Roll (ϕ)	Pitch (θ)	Yaw (ψ)
Aerodynamic Force Components	X	Y	Z
Aerodynamic Moment Components	L	M	N
Translational Velocity	U	V	W
Angular Rates	p	q	r
Three-Axis Inertia	I_x	I_y	I_z

The overall system configuration is defined in Figure 2, where "L" is the distance from the center of the body frame to all three rotors, labelled as L1, L2, and L3. The rotor forces are f1, f2, f3, and the rotor torque is defined as τ_1, τ_2, and τ_3, respectively. The angular velocity of the system is "ρ".

2.3. Main Engine (Electric Motors)

Brushless Direct Current (BLDC) motors can be used as a main engine of the UAV to achieve required electric propulsion in the system. BLDC motors found their application in the field of robotics, space crafts and medical devices, due to higher torque speed features, greater performance, minimum repairs and variable degree of speed [28]. Generally BLDC is more costly than simple DC motors, due to its better efficiency and reliability [29]. The electrical and mechanical equations of BLDC are given below.

$$V = Ri + (L - M)\frac{di}{dt} + E \tag{5}$$

$$E = K_e\omega_m F(\theta_e) \tag{6}$$

$$T = K_t i_a F(\theta_e) \tag{7}$$

$$T_e = J\frac{d^2\theta_m}{dt^2} + \beta\frac{d\theta_m}{dt} \tag{8}$$

$$\theta_e = \frac{P}{2}\theta_m \tag{9}$$

$$\omega_m = \frac{d\theta_m}{dt} \tag{10}$$

In which, V is the applied voltage; R is the total resistance; E is the back electromagnetic force; L is the total inductance of the motor; M is the mutual inductance of the motor; ω_m is the angular speed of the motor; θ_e is the rotation angle of the motor; T_e is the electrical torque produced by the motor; and K_e is the back-EMF constant.

3. Designing of Controller

3.1. Tri-Rotor Dynamic Control Strategies

The flight dynamics and control strategy of a tri-rotor UAV is the same as traditional aircraft. The placement or orientation of flight dynamics control is a product of roll, pitch, and yaw. The control scheme of the UAV includes altitude, roll, pitch, yaw, and tilt angle control, having a major role for the displacement control the parameters of the system.

Altitude Control Mechanism: To achieve the desired altitude, the speed of all rotors must be same $\rho1 = \rho2 = \rho3$. Increasing the speed of all rotors constantly will eventually raise the altitude of the UAV, such that the angular velocity of motors becomes equal.

Roll Control Mechanism: Roll control is achieved by regulating the front rotors speed. Decreasing rotor 1 velocity, rolls the system to the left and rotor 2 rolls the system towards right-side. Roll control has two conditions.

i   When moving clockwise roll $\rho2 > \rho1 > \rho3$.
ii   When moving counter-clockwise roll $\rho2 < \rho1 < \rho3$.

Pitch Control Mechanism: Regulating the speed of Rotor 1 and rearward rotors will change the pitch. The system pitches downward if the speed of Rotors 1 and 2 decreases while Rotor 3 speed keeps rising. If we decrease the speed of Rotor 3 and increase the speed of Rotors 1 and 2, the UAV pitch rises and fly flight reverses. Pitch control also has two conditions:

i   When nose-up $\rho2 = \rho3 > \rho1$.
ii   When nose-down $\rho2 = \rho3 < \rho1$.

Yaw Control Mechanism: The product of reaction torque and tilt angle "$\propto$" of Rotor 3 is used to control yaw movement. The value of the tilt angle is too small and helps to maneuver the UAV quickly. The yaw control condition is: $\rho1 = \rho2 = \rho1$, with $\propto = 0$.

3.2. Control Algorithm

In this section the overall control hierarchy is defined to control the attitude and altitude of the tri-rotor UAV. In which we assume the desired attitude variables are $K_1 = \varnothing$ (roll), $K_2 = \theta$ (pitch), and $K_3 = \Psi$ (yaw), respectively, and K_T is a generalized term for the rotational angles of the system. Now the control algorithm for the attitude controlling of an actual system of the UAV is written as:

$$G_{KT_1}(j) = \frac{B(j)}{A(j)} = Y(t) \tag{11}$$

The degree of the system model numerator $B(j)$ and denominator $A(j)$ is found to be "1" and "3". Where $B = (B^+ * B^-)$, such that B^+ is *variable and* B^- *is the constant.*

Equation (12) presents the desired attitude response of our UAV model:

$$G_{KT_2}(j) = \frac{B_m(j)}{A_m(j)} = Y_m(t) \tag{12}$$

Now the gradient theory which was defined by the model reference adaptive control method is implemented:

$$\deg A_c = 2 * \deg A(j) - 1 = 5 \tag{13}$$

So, the RST controller will be second-ordered, and now $\deg A_0 = \deg A(j) - \deg B^+ - 1 = 1$.

Remark 1. *For the perfect system model* $A_m(j) = A(j)$ *and* $B_m(j) = B(j)$*. Where* A_C *and* A_{Cm} *is the characteristic polynomials of the actual and desired system models and* $A_{Cm} = A_C$ *for the constraints and will not affect the system to change the close loop poles of the model. Otherwise,* A_{Cm} *differes the system model mismatch.*

Fuzzy Logic Controller. The vibrant performance of the fuzzy logic controller is described by the set of linguistic procedures that was established by a knowledgeable acquaintance in [30–32], in which the system "error" and variation in error rate are the input constraints, and RST are the variable outputs in our proposed controller. Formerly, RST can be improved online, using the set of rules, existing error, and variation in the error. In general, the error in the angle, combined with the mechanism output, increases. Furthermore, the controller performs well whether the error rises or the rate of error difference falls. It is important that, in the minor error phase, the earmarked control output is required to influence the change in error as soon as the error falls suddenly.

With the help of Equation (13), the degree of the proposed controller is found to be 5. The fuzzy logic-based adaptive RST controller is written in Equations (14)–(16):

$$FR = q^2 + (r_0 \times q) + r_1 \tag{14}$$

$$FS = (s_0 \times q^2) + (s_1 \times q) + s_2 \tag{15}$$

$$FT = (t_0 \times q \times A_0) + (t_1 \times A_0) \tag{16}$$

Put the values of *FR, FS, FT* in the above:

$$U_{FR_{KT}}(j) = FR(e, \frac{de}{dt})G_R \times e(j) \tag{17}$$

$$U_{FS_{KT}}(j) = FS(e, \frac{de}{dt})G_S \times \Sigma e(ij) \Delta(t) \tag{18}$$

$$U_{FT_{KT}}(j) = FT(e, \frac{de}{dt})G_T \times \frac{\Delta e(j)}{\Delta t} \tag{19}$$

where G_R, G_S *and* G_T are the delayed control gains of the signal scaling factors.

Remark 2. G_{KT_1} (j) *and* G_{KT_2} (j), *are the system models. Moreover, the system diverges at a low phase angle with an unstable control signal. The proposed control algorithm is not better in this case. Previously, in [33], a better controller was proposed for this type of case, a model having a low phase angle lying in the complex plane with some damping issues. As a result, in [34], zero cancellation in the system is situated esoteric the area which will cancel.*

Remark 3. *The proposed controller design at sampling of time [NT], which will remove the difficulties in the planning stage of controller implementation.*

Theorem 1. *After designing the control system, the next step is stability analysis, which shows the robustness level of the designed controller. The stability is highly vulnerable due to modeling errors called sensitivity. Therefore, model mismatch and model sensitivity are added in the system divergence between the actual and desired response depending upon the performance of the control system and its stability which will be taken from [34] Theorem 5.4, and examine the stability of the control system by using the model disturbance.*

Lemma. *The proposed equation illustrates the close loop system model having (NT) sampling period:*

$$\frac{\left[Y^T\left(t\right)\right]^{NT}}{FR^{NT}} = \frac{\left[G_{KT1}^T\left(j\right)G_{KT2}^T\left(j\right)H^{T,NT}\right]^{NT}G_{KT1}^{NT}\left(j\right)\left[\frac{FT^{NT}}{FR^{NT}}\right]}{1+\left[G_{KT1}^T\left(j\right)G_{KT2}^T\left(j\right)H^{T,NT}\right]^{NT}G_{KT1}^{NT}\left(j\right)\left[\frac{FS^{NT}}{FR^{NT}}\right]}$$

$H^{T,NT}$ is the conversion rate and will depend on the orientation of the desired signal.

Remark 4. *The disturbance in the pole placement method is obsolete in [34]. Since the reference model, observer polynomial, and reference model disturbance act as constraints, they are proved in convergence analysis.*

Proof of Theorem 1. The output of the unvarying model is considered in [35]:

$$Y^T\left(t\right) = G_{KT1}^T\left(j\right)G_{KT2}^T\left(j\right)H^{T,NT}\left[G_{KT1}^{NT}\left(j\right)\right]^{NT}\left[\frac{FT^{NT}}{FR^{NT}}FR^{NT} - \frac{FS^{NT}}{FR^{NT}}\left[Y^T\left(t\right)\right]^{NT}\right]^T$$

Remark 5. *Figure 3, gives complete work flow of proposed system using model reference adaptive control algorithm.*

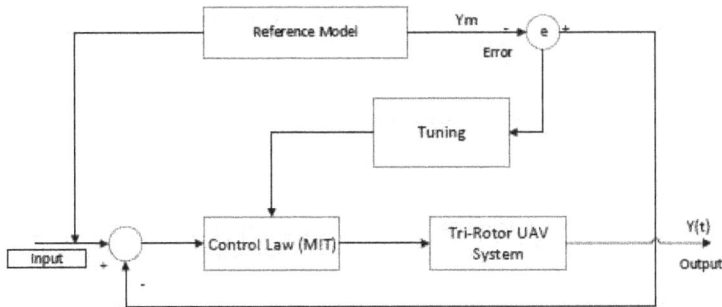

Figure 3. The model reference adaptive control system.

Convergence: Taking the input constraints of our controller signals gives the close loop error of the system model and unstable part of the disturbance. In other words, our proposed algorithm is able to stabilize the unstable part of the noise or disturbance model. The convergence at the desired value of parameters is done by an optimal control method based on the MIT rule to identify the errors.

Remark 6. *For the proposed controller the adaptive gain is in the range of 0.15 to 5 and above this range the controller performance deteriorates.*

$$e\,(j) = Y_{actual}\,(t) - Y_{m(desired)} \tag{20}$$

The sensitivity derivative is presented in Equation (21):

$$Y\,(t) = \frac{B\,(j)\,T}{(A\,(j)\,R + B\,(j)\,S)} \tag{21}$$

$$U_c\,(t) = \frac{A\,(j)\,R + B\,(j)\,S}{B\,(j)\,T} * Y\,(t) \tag{22}$$

The convergence proof contracts distinctly with the constraints, firstly identifying the sensitivity derivative of all of the parameters $(t_0,\ t_1, r_0,\ r_1, s_0,\ s_1,\ s_2)$ of the controller.

The Diophantine equation represents as $A_C = A\,(j)\,R + B\,(j)\,S$ and $A_C = A_0 A_m\,(j)$; therefore:

$$A\,(j)\,R + B\,(j)\,S = A_0 A_m\,(j) \tag{23}$$

The MIT rule-based sensitivity derivative of Equation (20) is:

$$e\,(j) = \frac{B\,(j)\,T}{A\,(j)\,R + B\,(j)\,S} * U_c\,(t) - Y_m \tag{24}$$

Theorem 2. *The model in Equation (11) with controller Equation (14–16) on the basis of system $(A\,(j), B\,(j))$, having UAV model disturbance, the error of close loop output Equation (24) goes to zero asymptotically if and only if $A_{Cm} = A_C$.*

Proof of Theorem 2. The close loop output error *Equation* (24) is settled by A_C and compared with *Equation* (23).

1. If, and only if, $A\,(j) = A_m\,(j)$ and $B\,(j) = B_m\,(j)$; therefore, the close loop output error responds to e (j) and goes to zero asymptotically because A_C is stable in Equation (13) and A_{Cm} is supposed to be stable.

2. Contradiction: if $A\,(j) \neq A_m\,(j)$ and $B\,(j) \neq B_m\,(j)$, the value is not cancelled by the close loop output error in Equation (24). The instability is in the denominator of e (j) if A_{Cm} is unstable. Hereafter, if $A_C \neq A_{Cm}$ or A_{Cm} is unstable, then e (j) close loop output error raises, unbounded, and it is necessary for A_{Cm} to be stable to make the close loop output error zero.

Now put in the value of $T = (t_0 q + t_1)A_0$ in Equation (24):

$$e\,(j) = \frac{B\,(j)\,(t_0 q + t_1)\,A_0}{A\,(j)\,R + B\,(j)\,S} * U_c\,(t) - (Y_m) \tag{25}$$

Therefore, w.r.t "t_0", the partial derivative of Equation (25) is:

$$\left\{ \begin{array}{l} \frac{\delta e(j)}{\delta(t_0)} = \frac{B(j)qA_0}{(A(j)R + B(j)S)} * U_c\,(t) \\[2mm] \frac{\delta e(j)}{\delta(t_0)} = \left(\frac{B(j)qA_0}{(A(j)R + B(j)S)} \right) * \frac{A(j)R + B(j)S}{B(j)T} * Y\,(t) \\[2mm] \frac{\delta e(j)}{\delta(t_0)} = \frac{A_0 q}{T} * Y\,(t) \end{array} \right. \tag{26}$$

The partial derivative of Equation (25) w.r.t "t_1" is:

$$
\begin{cases}
\frac{\delta e(j)}{\delta(t_1)} = \left(\frac{B(j)}{(A(j)R+B(j)S)} \right) * U_c(t) \\
\frac{\delta e(j)}{\delta(t_1)} = \left(\frac{B(j)}{(A(j)R+B(j)S)} \right) * \frac{A(j)R+B(j)S}{B(j)T} * Y(t) \\
\frac{\delta e(j)}{\delta(t_1)} = \frac{A_0}{T} * Y(t)
\end{cases}
\tag{27}
$$

From Equation (24), replace the value of R: $(R = (q)^2 + r_0 q + r_1)$

$$
e(j) = \frac{B(j)T}{A(j)(q^2 + r_0 q + r_1) + B(j)S} * U_c(t) - (Y_m)
\tag{28}
$$

Partial differentiate Equation (25) w.r.t "r_0":

$$
\begin{cases}
\frac{\delta e(j)}{\delta(r_0)} = - \left(\frac{B(j)TAS}{(A(j)R+B(j)S)^2} \right) * U_c(t) \\
\frac{\delta e(j)}{\delta(r_0)} = - \left(\frac{B(j)TAS}{(A(j)R+B(j)S)^2} \right) * \frac{A(j)R+B(j)S}{B(j)T} * Y(t) \\
\frac{\delta e(j)}{\delta(r_0)} = - \frac{AS}{A_0 A_m(j)} * Y(t)
\end{cases}
\tag{29}
$$

Partial differentiate Equation (25) w.r.t "r_1":

$$
\frac{\delta e(j)}{\delta(r_1)} = - \frac{A(q)}{A_0 A_m(j)} * Y(t)
\tag{30}
$$

Put the value of S, $S = (s_0 * q^2 + s_1 * q + s_2)$ in Equation (24):

$$
e(j) = \frac{B(j)T}{A(j)R + B(j)(s_0 q^2 + s_1 q + s_2)} * U_c(t) - (Y_m)
\tag{31}
$$

Now the Partial derivative of Equation (31) w.r.t "s_0" :

$$
\begin{cases}
\frac{\delta e(j)}{\delta(s_0)} = - \frac{\left(B(j)^2 * q^2 \right)}{(A(j)R+B(j)S)^2} * U_c(t) \\
\frac{\delta e(j)}{\delta(s_0)} = - \frac{\left(B(j)^2 * q^2 \right)}{(A(j)R+B(j)S)^2} \frac{A(j)R+B(j)S}{B(j)T} * Y(t) \\
\frac{\delta e(j)}{\delta(s_0)} = - \frac{B(j) * q^2}{A_0 A_m(j)} * Y(t)
\end{cases}
\tag{32}
$$

Now taking Partial derivative of Equation (31) w.r.t "s_1"

$$
\frac{\delta e(j)}{\delta(s_1)} = - \frac{B(j)q}{(A_m(j)A_0)} * Y(t)
\tag{33}
$$

Likewise, Partial derivative of the Equation (31) w.r.t "s_2" gives

$$
\frac{\delta e(j)}{\delta(s_2)} = - \frac{B(j)}{(A_m(j)A_0)} * Y(t)
\tag{34}
$$

By applying the MIT algorithm in the desired model of the system, where Equations (35) and (36) denote the cost function which is based on the MIT rule. Whitaker demonstrates the difference in system bounds as a function of the system error and the gradient of the system error with respect to the system constraints and takes the partial derivative of the gradient error with respect to its constraints. The constraints of the particular model with the initial estimate $J(KT)$, and the rate of change of speed among the desired and actual model is taken from [36]. To minimize the error with respect to time so that the desired response is achieved requires an optimum control with cost function.

Theorem 3. *In [20] the least square estimation J $\left(\overline{KT}\right)$ and their polynomial characteristic solution depends upon the Diophantine equation and stability of A_{Cm}.*

Proof of Theorem 3. The solution also depends upon the Diophantine equation as well as the stability of A_{Cm}. By using Theorem 2:

1. If $e(j) \to 0$ as $l \to \infty$ and leading $J\left(\overline{KT}\right) = 0$, the solution gives the smallest positive value of cost function.
2. A_{Cm} is a contradiction, if it is not stable from Theorem 2; $e(j) \nrightarrow 0$ as $l \to \infty$, thus $J\left(\overline{KT}\right) \neq 0$. This is not optimal because, as seen in step 1 of the proof, there exists a solution that makes $J\left(\overline{KT}\right) = 0$.

$$J(KT) = \frac{1}{2}e^2(KT) \tag{35}$$

$$\frac{dKT}{(dt)} = -\gamma' \frac{1}{l_o} Y_{(m)} e = -\gamma * Y_{(m)} * e(j) \tag{36}$$

After that, apply the MIT rule to derivate control variables (s_0, s_1, s_2, r_0, r_1, t_0, and t_1) and set in the controller gives

$$\frac{d(s_0)}{dt} = -\gamma_e \frac{\delta e}{\delta s_0} \frac{d(s_0)}{dt} = \frac{\gamma_e q^2 B(j) Y(t)}{A_m(j) A_0} \tag{37}$$

$$\frac{d(s_1)}{dt} = -\gamma_e \frac{\delta e}{\delta s_1} \frac{d(s_1)}{dt} = \frac{\gamma_e B(j) qY(t)}{A_m(j) A_0} \tag{38}$$

$$\frac{d(s_2)}{dt} = -\gamma_e \frac{\delta e}{\delta s_2} \frac{d(s_2)}{dt} = \frac{\gamma_e B(j) Y(t)}{A_m(j) A_0} \tag{39}$$

$$\frac{d(r_0)}{dt} = -\gamma_e \frac{\delta e}{\delta r_0} \frac{d(r_0)}{dt} = \frac{\gamma_e A(j) qY(t)}{A_m(j) A_0} \tag{40}$$

$$\frac{d(r_1)}{dt} = -\gamma_e \frac{\delta e}{\delta r_1} \frac{d(r_1)}{dt} = \frac{\gamma_e A(j) Y(t)}{A_m(j) A_0} \tag{41}$$

$$\frac{d(t_0)}{dt} = -\gamma_e \frac{\delta e}{\delta t_0} \frac{d(t_0)}{dt} = -\frac{\gamma_e B(j) q}{A_0} \tag{42}$$

$$\frac{d(t_1)}{dt} = -\gamma_e \frac{\delta e}{\delta t_1} \frac{d(t_1)}{dt} = -\frac{\gamma_e}{(A_m(j))} \tag{43}$$

Now, the main controller equation becomes:

$$U_{F-Hybrid_{\emptyset, \theta, \Psi}} = \left(\frac{(F(T_0 q + T_1) A_0)}{F(q^2 + r_0 q + r_1)}\right) * (Uc(t)) - \left(\frac{F(S_0 q^2 + S_1 q + S_2)}{F(q^2 + r_0 q + r_1)}\right) Y(t)$$

The change in UAV orientation depends upon the rate of change of the control commands, which makes the system respond quite better towards stability by using our proposed algorithm.

The linguistics levels of the fuzzy hybrid controller are assigned as (BN) below negative, (SN) small negative, (ZR) zero, (SP) small positive, and (BP) big positive. The fuzzy logic controller if-then rules are defined in Tables 2–4 such that error "e" is the rotor speed having range in −10 to +10, the derivative error range is −5 to +5, and the output range of the fuzzy hybrid controller is 0 to 1 with R = 0.667, S = 0.5, and T = 0.5, respectively.

Table 2. Fuzzy logic controller If-Then rule for "R".

$\overrightarrow{de/dt}$	BN	SN	ZR	SP	BP
error ↓					
BN	ZR	SP	MP	MP	SP
SN	SP	SP	SP	MP	MP
ZR	SP	MP	MP	MP	MP
SP	SP	SP	MP	SP	SP
BP	ZR	SP	MP	LP	S

Table 3. Fuzzy logic controller If-Then rule for "S".

$\overrightarrow{de/dt}$	BN	SN	ZR	SP	BP
error ↓					
BN	ZR	ZR	SP	BP	BP
SN	ZR	SP	SP	SP	BP
ZR	ZR	SP	SP	SP	BP
SP	ZR	SP	SP	SP	BP
BP	ZR	ZR	SP	SP	BP

Table 4. Fuzzy Logic Controller If-Then rule for "T".

$\overrightarrow{de/dt}$	BN	SN	ZR	SP	BP
error ↓					
BN	SP	SP	ZR	ZR	ZR
SN	BP	SP	SP	SP	BP
ZR	BP	SP	SP	SP	BP
SP	BP	SP	SP	SP	SP
BP	BP	BP	BP	SP	SP

The input error membership function of the fuzzy logic controller (FLC) is shown in Figure 4. Figure 5 shows the input derivative error of FLC. Moreover, Figures 6–8 show the output gains of RST-based FLC logic.

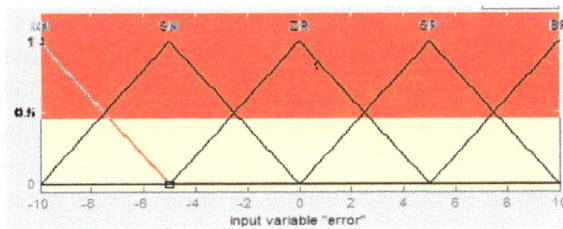

Figure 4. Input error membership function of fuzzy Logic.

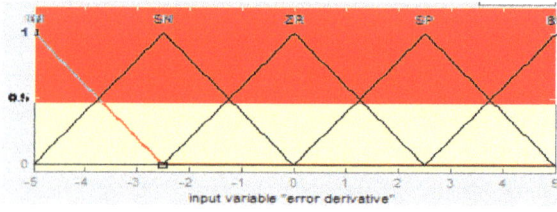

Figure 5. Input derivative error membership function of fuzzy logic.

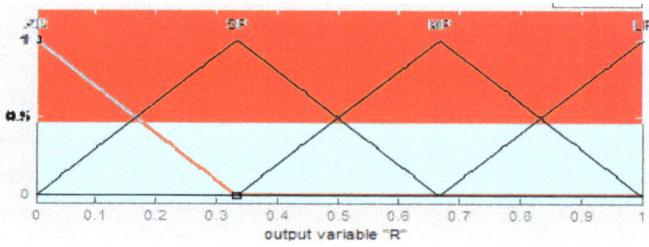

Figure 6. Regulation output gain membership function of fuzzy logic.

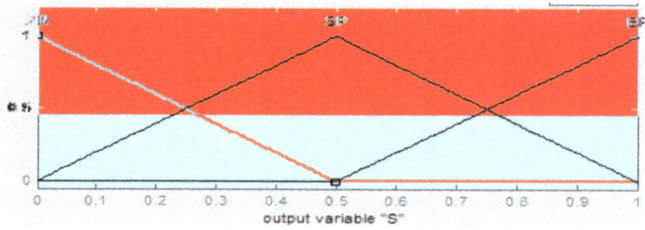

Figure 7. Pole-placement output gain membership function of fuzzy logic.

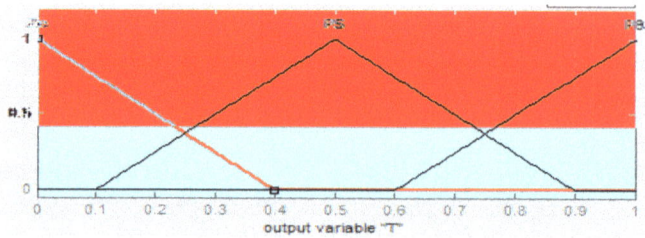

Figure 8. Tuning the output gain membership function of fuzzy logic.

4. Simulation Results and Discussions

The validity and robustness of our proposed algorithm are presented in this section. Moreover, the nonlinear simulations for the stabilization of tri-rotor parameters are shown in Table 5. All of the simulations were done via Simulink, MATLAB. In all simulations, we compared our proposed fuzzy-hybrid controller with the adaptive RST controller of [20].

Table 5. Tri-rotor parameters.

Parameters	Values	Si Units
Ix	0.3105	$kg \cdot m^2$
Iy	0.2112	$kg \cdot m^2$
Iz	0.2215	$kg \cdot m^2$
l	0.3050	m
Mass	0.785	kg

Ideally, we can say that there are two subsystems of the UAV; one is rotational, while the other is a translational velocity response. The rotational subsystem is responsible for controlling the initial errors, thereby stabilizing the attitude of UAV and converges to zero. Pitch, roll, and yaw angles are also converging to zero to realize the hovering state. The rotational subsystem and Euler angles do not depend on translational components. However, the translational components depend on Euler angles.

The four control commands, *i.e.*, altitude, lateral, longitudinal, and angular, are shown in Figure 9. The settling and rise time are within one to one and a half seconds for each input channel along with zero overshoot. By comparing the adaptive RST controller control commands, it is not perfectly linear because in the adaptive RST controller the control commands do not perfectly converge to zero and have undershoot and overshoot, initially, which causes the UAV to dislocate from the desired position due to the presence of errors. However, our proposed method completely converges to zero and exactly reaches the perfect location as shown in Figure 10, which shows the actual reached height and vertical speed response of the UAV.

The Euler angular responses are shown in Figure 11; one can observe that initial angles are non-zero, but it will converge to zero, which means that it will perform the attitude hold control by comparing with the adaptive RST controller. Figure 12 shows the rotational rate responses and, by comparing the transitional rate responses given in Figure 13, observe that there are several "bumps" but ultimately converge to the desired value. In Figures 12 and 13 our proposed algorithm converges to zero at about one second and have no settling time error, but the adaptive RST controller settled at about two to two and a half seconds.

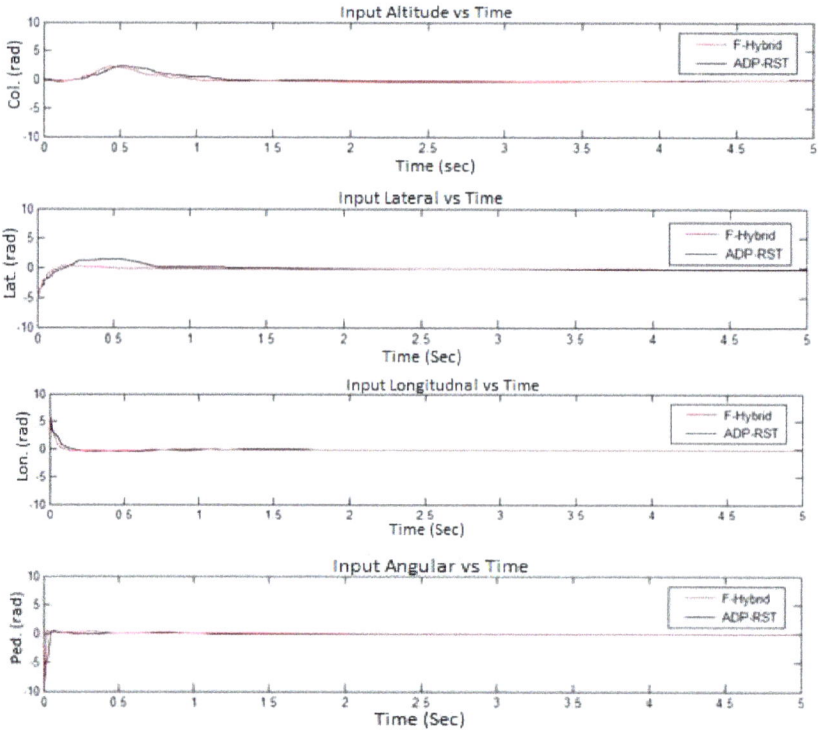

Figure 9. Comparison of F-Hybrid with adaptive RST for control commands.

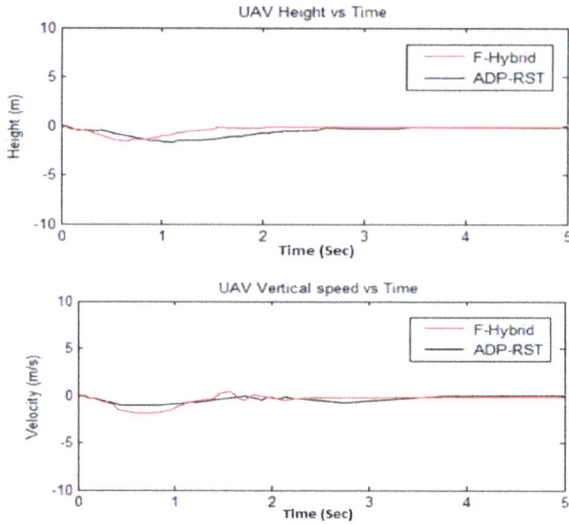

Figure 10. Comparison of F-Hybrid with adaptive RST for height and vertical speed.

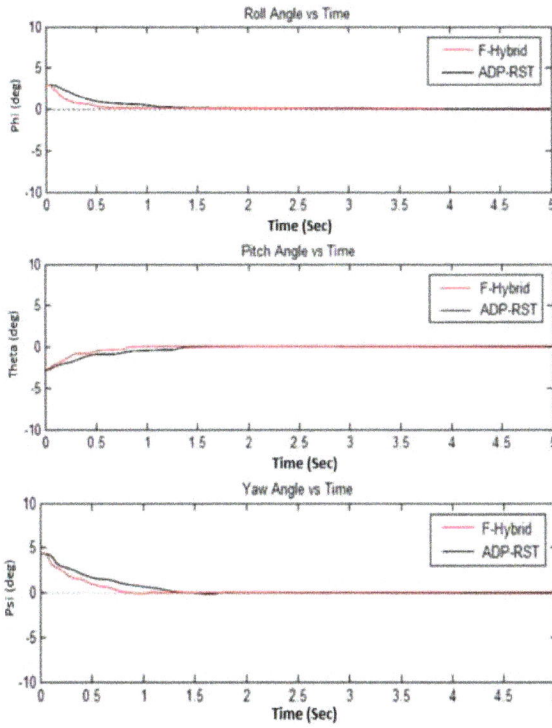

Figure 11. Comparison of F-Hybrid with adaptive RST for Euler angle response.

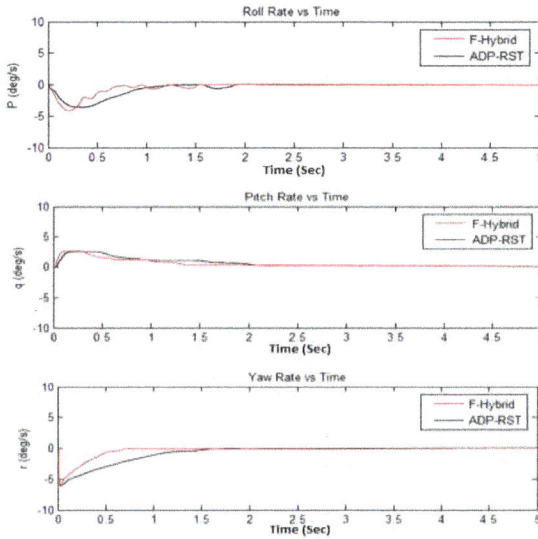

Figure 12. Comparison of F-Hybrid with adaptive RST for rotational velocity responses.

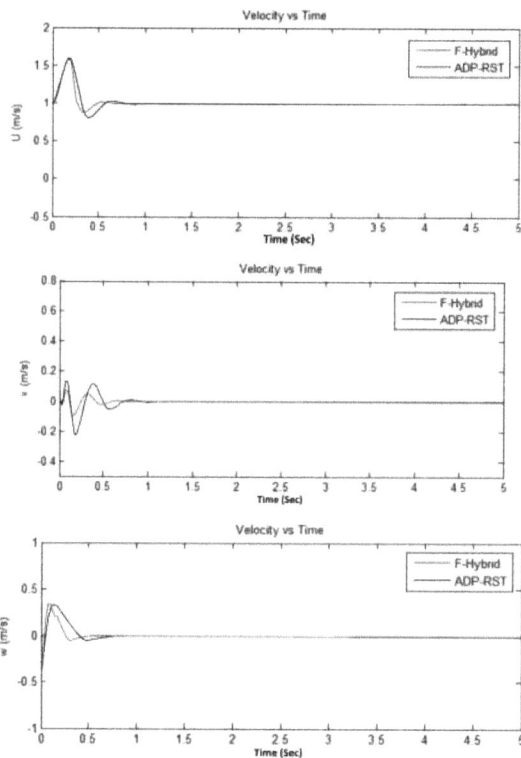

Figure 13. Comparison of F-Hybrid with adaptive RST for translational velocity responses.

The sampling time was set to 0.2 s for all simulations, whereas every simulation was done in five seconds. The simulation results from Figures 9–12 show that our proposed algorithm is quite fast and converges to zero without any settling time error. The Euler angle initial conditions are $\varnothing = 2.5$, $\theta = -2$, and $\psi = 4.7$ degrees. In Figure 13, the translational velocity components are U = 1, V = W = 0, meaning it can easily control the attitude of UAV, without any overshoot, undershoot, and transient errors. Furthermore, the settling time response also gets better than the previously-used adaptive RST controller. In a very short time of about 1 to 1.5 s, all of the simulated parameter values yield to the original trimmed conditions. The results show that our proposed controller is able to stabilize the attitude angles and altitude of the tri-rotor UAV.

5. Conclusions

In this paper, a new and novel approach of a fuzzy hybrid controller is presented for stabilizing the nonlinear behavior of a tri-rotor UAV and to achieve the desired altitude. The model of the tri-rotor UAV is based on the Newton-Euler method. The method is applied on the translational and rotational velocity subsystems of the aircraft. To observe the stability of the UAV we must stabilize the attitude and altitude responses. The RST with MRAC nonlinear controller algorithm is based on the MIT rule and the sensitivity function of the closed-loop system has shown significant results. Moreover, fuzzy logic controllers have been proposed with linguistic logic, which enhanced the performance. The effectiveness and stability of our algorithm are implemented using nonlinear simulation and it is observed that the proposed method has better transient characteristics and performance with zero steady state error on rigid environments with reasonable rise and settling time. Furthermore, the proposed method shows better robustness and very fast convergence towards stability in the presence of a system disturbance.

Acknowledgments: This work is sponsored by the National Natural Science Foundation of China (NSFC) under the grant number is 61503185 and supported by Key Laboratory of College of Automation Engineering, Nanjing University of Aeronautics and Astronautics, Nanjing, Jiangsu 210016, China.

Author Contributions: The central arguments and structure of the paper were equally devised by each of the authors through a series of discussions. Zain Anwar Ali conceived the idea and carry out experimental work under supervision of Daobo Wang and Muhammad Amir was involved in the writing of this research article and in the results.

Conflicts of Interest: The authors declare no conflict of interest.

References

1. Yoon, S.; Lee, S.J.; Lee, B.; Kim, C.J.; Lee, Y.J.; Sung, S. Design and flight test of a small Tri-rotor unmanned vehicle with a LQR based onboard attitude control system. *Int. J. Innov. Comput. Inf. Control* **2013**, *9*, 2347–2360.
2. Jones, P.; Ludington, B.; Reimann, J.; Vachtsevanos, G. Intelligent Control of Unmanned Aerial Vehicles for Improved Autonomy. *Eur. J. Control* **2007**, *13*, 320–333. [CrossRef]
3. Metni, N.; Hamel, T. A UAV for bridge inspection: Visual servoing control law with orientation limits. *Autom. Constr.* **2007**, *17*, 3–10. [CrossRef]
4. Tayebi, A.; McGilvray, S. Attitude stabilization of a four-rotor aerial robot. In Proceedings of the IEEE Conference on Decision and Control, Paradise Island, Bahamas, 14–17 December 2004; Volume 2, pp. 1216–1221.
5. Salazar-cruz, S.; Kendoul, F.; Lozano, R.; Fantoni, I. Real-Time Stabilization of a Small Three-Rotor Aircraft. *IEEE Trans. Aerosp. Electr. Syst.* **2008**, *44*, 783–794. [CrossRef]
6. Guenard, N.; Hamel, T.; Moreau, V. Dynamic modeling and intuitive control strategy for an "X4-flyer". In Proceedings of the International Conference on Control and Automation, Hungarian Academy of Science, Budapest, Hungary, 26–29 June 2005; Volume 1, pp. 141–146.
7. Yoo, D.-W.; Oh, H.-D.; Won, D.-Y.; Tahk, M.-J. Dynamic Modeling and Stabilization Techniques for Tri-Rotor Unmanned Aerial Vehicles. *Int. J. Aeronaut. Space Sci.* **2010**, *11*, 167–174. [CrossRef]
8. Benallegue, A.; Mokhtari, A.; Fridman, L. High-order sliding-mode observer for a quadrotor UAV. *Int. J. Robust Nonlinear Control* **2008**, *18*, 427–440. [CrossRef]
9. Kim, J. Model-Error Control Synthesis: A New Approach to Robust Control. Ph.D. Thesis, Texas A&M University, USA, 2002.
10. Castillo, P.; Lozano, R.; Dzul, A. *Modelling and Control of Mini-Flying Machines*; Springer-Verlag London: London, UK, 2005.
11. Desa, H.; Ahmed, S.F. Adaptive Hybrid Control Algorithm Design for Attitude Stabilization of Quadrotor (UAV). *Arch. Des Sci.* **2013**, *66*, 51–64.
12. Mohammadi, M.; Shahri, A.M. Adaptive Nonlinear Stabilization Control for a Quadrotor UAV: Theory, Simulation and Experimentation. *J. Intell. Robot. Syst.* **2013**, *72*, 105–122. [CrossRef]
13. Dydek, Z.T.; Annaswamy, A.M. Adaptive Control of Quadrotor UAVs in the Presence of Actuator Uncertainties. In Proceedings of the AIAA Guidance, Navigation, and Control Conference 2010, Atlanta, Georgia, 20–22 April 2010.
14. Sadeghzadeh, I.; Mehta, A.; Zhang, Y. Fault/Damage Tolerant Control of a Quadrotor Helicopter UAV Using Model Reference Adaptive Control and Gain-Scheduled PID. In Proceedings of the AIAA Guidance, Navigation, and Control Conference, Portland, Oregon, 8–11 August 2011.
15. Chadlia, M.; Aouaoudab, S.; Karimic, H.R.; Shid, P. Robust fault tolerant tracking controller design for a VTOL aircraft. *J. Frankl. Inst.* **2013**, *350*, 2627–2645. [CrossRef]
16. Aouaouda, S.; Chadli, M.; Karimi, H.-R. Robust static output-feedback controller design against sensor failure for vehicle dynamics. In Proceedings of the IET Control Theory & Applications, 12 June 2014; Volume 8, pp. 728–737. [CrossRef]
17. Chiou, J.S.; Tran, H.K.; Peng, S.T. Attitude control of a single tilt tri-rotor UAV SYSTEM: Dynamic modeling and each channel's nonlinear controllers design. *Math. Probl. Eng.* **2013**, *2*. [CrossRef] [PubMed]

18. Yeh, F.K.; Huang, C.W.; Huang, J.J. Adaptive fuzzy sliding-mode control for a mini-UAV with propellers. In Proceedings of the InSICE Annual Conference (SICE), Tokyo, Japan, 13–18 September 2011; pp. 645–650.

19. Chiou, J.S.; Tran, H.K.; Peng, S.T. Design Hybrid Evolutionary PSO Aiding Miniature Aerial Vehicle Controllers. *Math. Probl. Eng.* **2015**, *12*. [CrossRef]

20. Ali, Z.A.; Wang, D.; Javed, R.; Aklbar, A. Modeling & Controlling the Dynamics of Tri-rotor UAV Using Robust RST Controller with MRAC Adaptive Algorithm. *Int. J. Control Autom.* **2016**, *9*, 61–76.

21. Cuenca, A.; Salt, J. RST controller design for a non-uniform multi-rate control system. *J. Process Control* **2012**, *22*, 1865–1877. [CrossRef]

22. Landau, I.D. The R-S-T digital controller design and applications. *Control Eng. Pract.* **1998**, *6*, 155–165. [CrossRef]

23. Stevens, B.L.; Lewis, F.L. *Aircraft Control and Simulation*; Wileys: New York, NY, USA, 1992.

24. Padfield, G.D. The Theory and Application of Flying Qualities and Simulation Modeling. In *Helicopter Flight Dynamics*; AIAA: Education Series, Washington, DC, 1996.

25. Deuflhard, P. *Newton Methods for Nonlinear Problems*, 1st ed.; Springer: Berlin, Germany, 2005.

26. Weihua, Z.; Go, T.H. Robust decentralized formation flight control. *Int. J. Aerosp. Eng.* **2011**. [CrossRef]

27. Mohamed, K.M. Design and Control of UAV Systems: A Tri-Rotor UAV Case Study. Ph.D. Thesis, The University of Manchester, Manchester, UK, 31 October 2012.

28. Shanmugasundram, R.; Zakaraiah, K.M.; Yadaiah, N. Modeling, simulation and analysis of controllers for brushless direct current motor drives. *J. Vib. Control* **2012**. [CrossRef]

29. Sincero, G.; Cros, J.; Viarouge, P. Efficient simulation method for comparison of brush and brushless DC motors for light traction application. In Proceedings of the 13th European Conference on Power Electronics and Applications, Barcelona, Spain, 8–10 September 2009; pp. 1–10.

30. Mamdani, E.H. Application of fuzzy logic to approximate reasoning using linguistic synthesis. *IEEE Trans. Comput.* **1977**, *100*, 1182–1191. [CrossRef]

31. Passino, K.M.; Yurkovich, S. Fuzzy Control by Addison-Wesley: Reading, PA, USA, 1998.

32. Mamdani, E.H. Application of fuzzy algorithms for control of simple dynamic plant. *Proc. Inst. Electr. Eng.* **1974**, *121*, 1585–1588. [CrossRef]

33. Salt, J.; Sala, A.; Albertos, P. A transfer-function approach to dual-rate controller design for unstable and non-minimum-phase plants. *IEEE Trans. Control Syst. Technol.* **2011**, *19*, 1186–1194. [CrossRef]

34. Astrom, K.J.; Wittenmark, B. *Computer-Controlled Systems: Theory and Design*; Prentice Hall: Englewood Cliffs, NJ, USA, 1984.

35. Salt, J.; Albertos, P. Model-based multirate controllers design. *IEEE Trans. Control Syst. Technol.* **2005**, *13*, 988–997. [CrossRef]

36. Schmidt, P. Intelligent model reference adaptive control applied to motion control. In Proceedings of the Record of the 1995 IEEE Industry Applications Conference Thirtieth IAS Annual Meeting IAS-95, Orlando, FL, USA, 8–12 October 1995.

sensors

MDPI

Article

Obstacle Detection and Avoidance System Based on Monocular Camera and Size Expansion Algorithm for UAVs

Abdulla Al-Kaff [*,†,‡], **Fernando García** [‡], **David Martín** [‡], **Arturo De La Escalera** [‡] and **José María Armingol** [‡]

Intelligent Systems Lab, Universidad Carlos III de Madrid, Leganes, 28911 Madrid, Spain; fegarcia@ing.uc3m.es (F.G.); dmgomez@ing.uc3m.es (D.M.); escalera@ing.uc3m.es (A.D.L.E.); armingol@ing.uc3m.es (J.M.A.)
* Correspondence: akaff@ing.uc3m.es; Tel.: +34-916-246-217
† Current address: Department of System Engineering and Automation, Avenida de la Universidad, 30, Leganes, 28911 Madrid, Spain.
‡ These authors contributed equally to this work.

Academic Editors: Felipe Gonzalez Toro and Antonios Tsourdos
Received: 19 December 2016; Accepted: 4 May 2017; Published: 7 May 2017

Abstract: One of the most challenging problems in the domain of autonomous aerial vehicles is the designing of a robust real-time obstacle detection and avoidance system. This problem is complex, especially for the micro and small aerial vehicles, that is due to the Size, Weight and Power (SWaP) constraints. Therefore, using lightweight sensors (i.e., Digital camera) can be the best choice comparing with other sensors; such as laser or radar.For real-time applications, different works are based on stereo cameras in order to obtain a 3D model of the obstacles, or to estimate their depth. Instead, in this paper, a method that mimics the human behavior of detecting the collision state of the approaching obstacles using monocular camera is proposed. The key of the proposed algorithm is to analyze the size changes of the detected feature points, combined with the expansion ratios of the convex hull constructed around the detected feature points from consecutive frames. During the Aerial Vehicle (UAV) motion, the detection algorithm estimates the changes in the size of the area of the approaching obstacles. First, the method detects the feature points of the obstacles, then extracts the obstacles that have the probability of getting close toward the UAV. Secondly, by comparing the area ratio of the obstacle and the position of the UAV, the method decides if the detected obstacle may cause a collision. Finally, by estimating the obstacle 2D position in the image and combining with the tracked waypoints, the UAV performs the avoidance maneuver. The proposed algorithm was evaluated by performing real indoor and outdoor flights, and the obtained results show the accuracy of the proposed algorithm compared with other related works.

Keywords: obstacle detection; collision avoidance; size expansion; feature points; UAV; monocular vision

1. Introduction

During the last decade, with the developments in microelectronics and the increase of computing efficiency, the use of Unmanned Aerial Vehicles (UAVs) is no longer restricted to the military purposes only. Recently, with the advent of small and micro aerial vehicles (sUAVs and MAVs), the requirements of operations and applications in low altitudes are increased.

Due to their ability to operate in remote, dangerous and dull situations, sUAVs and MAVs especially helicopters and vertical take-off and landing (VTOL) rotor-craft systems are increasingly used in many applications; such as surveying and mapping, rescue operation in disasters [1], spatial information acquisition, data collection from inaccessible areas and geophysics exploration [2,3], cooperate in manipulation and transportation [4], buildings inspection [5,6] and navigation purposes [7,8].

Nowadays, with the current technology and the variety and complexity of the tasks, modern UAVs aim at higher levels of autonomy and performing flight stabilization. For the autonomous UAVs, the ability of detection and avoidance of obstacles with high level of accuracy is considered to be a challenging problem.

The difficulty appears because of the size of UAVs is getting smaller and thus, the weight is getting lighter. Therefore, taking into account these properties, sUAVs and MAVs have not the ability of carrying heavy sensors such as laser [9–11] or radar [12]. Hence, the suitable solution is to use the on-board cameras due to its advantage of lightweight and low power consumption.

In addition to its lightweight and the low power consumption, the cameras provide rich information of the environment. Therefore, they are considered as important sensors mounted on the small and micro UAVs.

In vision-based navigation systems, different approaches were presented to solve the problem of obstacle detection and avoidance. Approaches such as [13–15], built a 3D model of the obstacle in the environment. Other works calculate the depth (distance) of the obstacles, such as in [16,17]. A technique based on stereo cameras, in order to estimate the proximity of the obstacles, was introduced in [18]. At which, the system detects the size and the position of the obstacles based on the disparity images and the view angle. Furthermore, this technique calculates the relation of the size and the distance of the detected obstacle to the UAV. All these approaches have the disadvantage of the high cost in the computational time.

Whilst bio-inspired (insect, animal or human like) approaches estimate the presence of the obstacle efficiently, without calculating the 3D model, such as using optical flow [19–21] or perspective cues [7,22,23]. However, optical flow approaches cannot identify the forward movement, due to the aperture problem, thus frontal obstacles would provide only movement component normal to the detected edges in the image, not providing frontal movement information *per se*. Perspective cues approaches work well in the structured environments [24].

Detecting and avoiding frontal obstacles using monocular camera is considered a challenging problem because of the absence of the optical flow or the motion parallax. However, size expansion provides useful information for detecting the obstacles that are moving towards the UAV.

From the bio-inspired point of view, the human visual system has the ability to extract information correctly of the objects that are moving toward them [25]. In addition, Gibson illustrated the ability of the human visual system to identify the approaching of the objects related to the expansion of its size, by both eyes or even one eye [26].

From this aspect, in this paper, a bio-inspired approach using a monocular camera is presented to mimic the human behavior of obstacle detection and avoidance applied on UAVs. The system is divided into two main stages: **Vision-Based Navigation and Guidance** in which, the obstacle detection algorithm is performed based on the input images captured from the front camera. In addition, **Motion Control**, where the avoidance decision is taken and sent to the UAV. Figure 1a shows the general overview of the system, whilst Figure 1b depicts the subsystem which focuses on the detection and avoidance stages.

The novelty in this paper is based on two main lines: First, the use of the size changes of the detected feature points, in order to provide object detection. Second, the changes in the size ratio in consecutive frames of the convex hull constructed from these points allow reliable frontal obstacle detection by means of a monocular camera and the motion of the UAV. The presence of approaching obstacles is estimated from these size expansion ratios, avoiding the need of complex 3D models, as shown in Figure 2. Reducing considerably the computation cost of the detection algorithm.

The remainder of this paper is organized as follows; Section 2 introduces the state-of-the-art work related to obstacle detection and avoidance approaches, followed by the proposed obstacle detection algorithm in Section 3. Section 4 presents the avoidance algorithm, then Section 5 discusses the experimental results. Finally, in Section 6 conclusion is summarized.

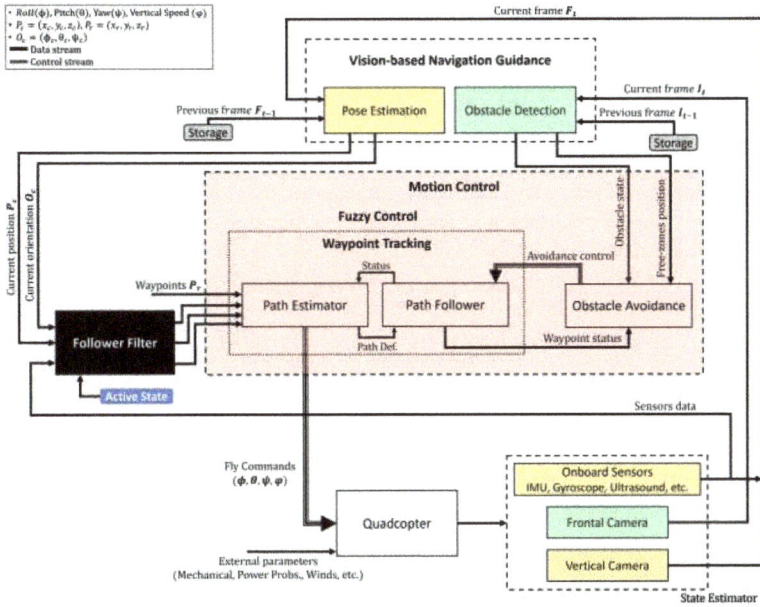

a System Overview.

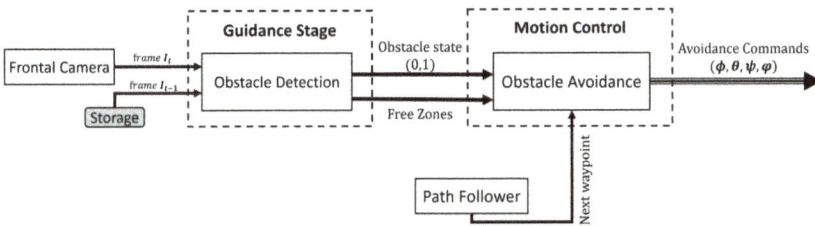

b Detection and Avoidance Subsystem.

Figure 1. General overview of the Detection and Avoidance phases.

2. Related Work

Obstacle detection and avoidance plays an important role in any autonomous navigation system. Different works were presented to solve this challenging process especially in vision-based systems.

In [27], it was presented an approach based on the texture and color variation cue to detect obstacles for indoor environments. However, this approach works with detailed textures. Furthermore, their experiments were limited to indoor environments.

Working with Hybrid MAVs, Green et al. proposed an optical flow approach, mimicking the biological flying insects, by using dual cameras mounted on a fixed-wing UAV, in order to detect and avoid lateral obstacles [28]. Besides the detection of lateral obstacles only, some limitations appeared in avoiding large obstacles like walls. In addition, from the experiments, the avoidance algorithm is insufficient if the UAV flies in a straight path.

In [29], SIFT descriptor and Multi-scale Oriented-Patches (MOPS) are combined to show 3D information of the obstacles. At which, the edges and corners of the object are extracted using MOPS by obtaining and matching the MOPS feature points of the corners, then the 3D spatial information of the MOPS points is extracted. After that, SIFT is used to detect the internal outline information. However, the presented approach has expensive computational time (577 ms).

Bills et al. [7] proposed an approach for indoor environments with a uniform structure characteristics. In this work, Hough Transform is used to detect the edges that are used to classify the essence of the scene based on a trained classifier. However, their experiments were limited to corridors and stairs areas.

A saliency method based on Discrete Cosine Transform (DCT) is presented in [30] for obstacle detection purposes. From the input images, the system assumes that the obstacle is a unique content in a repeated redundant background, then by applying amplitude spectrum suppression, the method can remove the background. Finally, by using the Inverse Discrete Cosine Transform (IDCT) and a threshold algorithm, the center of the obstacle is obtained. Furthermore, a pinhole camera model is used to estimate the relative angle between the UAV and the obstacle, this angle is used with a PD controller to control the heading of the UAV for obstacle avoidance.

In [17], the authors presented an approach for measuring the relative distance to the obstacle. At which, the camera position is estimated based on the Extended Kalman Filter (*EKF*) and the IMU data. Then the 3D position of the obstacle can be calculated by back projecting the detected features of the obstacle from its images.

An expansion segmentation method was presented in [31], in which a conditional Markov Random Field (MRF) is used to distinguish if the frontal object may represent a collision or not. Additionally, an inertial system is used to estimate the collision time. However, the experiments of this work was limited to simulations.

Another approach presented in [24] used the feature detection algorithm in conjunction with the template matching to detect the size expansions of the obstacles. However, the experiments were limited to tree-like obstacles and did not show results of other shapes.

Kim et al. presented a block-based motion estimation approach for detecting the moving obstacles (humans) [32]. In which, the input image is divided into smaller blocks, then the system compare the motion in each block through consecutive images. However, their experiments were limited in detecting large size obstacles (humans) in indoor environments.

In addition, surveys of different approaches of UAVs guidance, navigation and collision avoidance methods and technologies are presented in [33,34]. Recently, Mcfadyen et al. presented a literature review of the vision-based collision avoidance systems [35].

The work presented in this paper represents a step forward in the obstacle detection and avoidance by the use of the convex hull (area) of the obstacle identified by means of image features. The approach is able to work in real time with all different kind of obstacles and in different tested scenarios.

3. Obstacle Detection

The proposed obstacle detection algorithm mimics the human behavior of detecting the obstacles that are located in front of the UAV during motion. At which point, the collision state of the approaching obstacles is estimated instead of building 3D models, or calculating the depth of the obstacle.

The novelty and the key of this algorithm is to estimate the size ratios of the approaching obstacles from the consecutive frames during the flight as shown in Figure 2. This is achieved by estimating the change in the size property of the detected feature points (diameter), and the size of the convex hull (area) which is constructed from these points as well. When the size ratios exceed certain empirical values (explained in Section 3.2), it means that there is an obstacle detected, and can cause a danger to the UAV as shown in Algorithm 1, and Figure 3.

Algorithm 1: Obstacle Detection

Input: Image frames F

Output: Collision state Obs, Obstacle position τ

1 **Define**: Current frame F_t, Previous frame F_{t-1}, Current workspace ROI_t, Previous workspace ROI_{t-1}, Current keypoints $KP_t[\,]$, Previous keypoints $KP_{t-1}[\,]$, Number of keypoints (N, M), $n \in N, m \in M, n = m, x \in n$, Matched points pts, Distance ratio *thresh*, Object of interest *Convex*

2 **begin**

3   **while** *isFlying( )* **do**

4        $F_{t-1} \leftarrow$ getNewFrame( )

5   **if** *Obstacle Detection isActivated( )* **then**

6           $F_t \leftarrow$ getNewFrame( )

7           $(ROI_{t-1}, ROI_t) \leftarrow$ DefWrkspc(F_{t-1}, F_t)

8           $(KP_{t-1}(N), KP_t(M)) \leftarrow$ DetectKeypoints(F_{t-1}, F_t)

9           $(pts_{t-1}(n), pts_t(m)) \leftarrow$ MatchSymKeypoints$(KP_{t-1}(N), KP_t(M), thresh)$

10          $(pts_{t-1}(x), pts_t(x)) \Longleftrightarrow size(pts_t(m)) > size(pts_{t-1}(n))$

11          $(Convex_1, Convex_2) \leftarrow$ Create$(pts_{t-1}(x), pts_t(x))$

12   **if** *AreaScale$(Convex_2 : Convex_1) \geq 1.7$ and SizeScale$(pts_t(x) : pts_{t-1}(x) \geq 1.2)$* **then**

13             $Obs \leftarrow true$

14             $\tau \leftarrow$ CalcPos(Obs)

15   **goto** Algorithm 3

16   **end**

17   **end**

18        $F_{t-1} \leftarrow F_t$

19   **end**

20 **end**

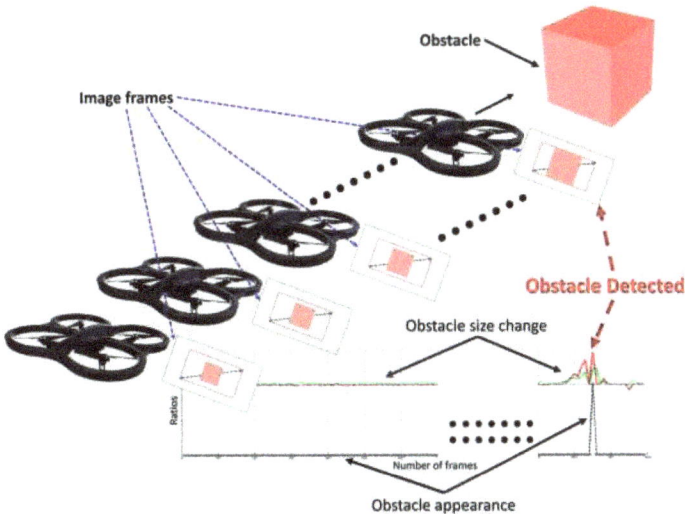

Figure 2. Concept of approaching obstacle detection.

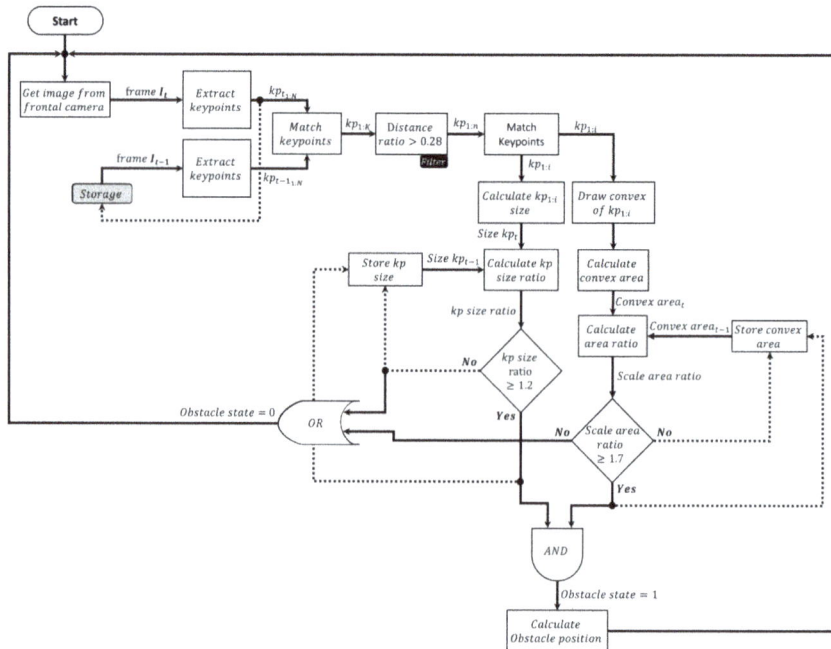

Figure 3. Obstacle detection approach flowchart.

3.1. Feature Detection and Description

In this step, an image Region Of Interest (ROI) of diagonal 62° Field of View (FOV) is taken, in order to be processed instead of the whole image, as shown in Figure 4. The selection of the diagonal 62° ROI is based on the results that are obtained from the experiments. Where, it has been found that any object detected out of the area of this ROI will not cause any danger to the UAV, and only the objects that are detected in the scope of this diagonal 62° ROI can be considered as an obstacle. Furthermore, processing the diagonal 62° ROI instead of the whole diagonal 92° image, leads to a significant minimizing in computational time. Test performed proved the viability of this approach, and the results will be discussed in following sections.

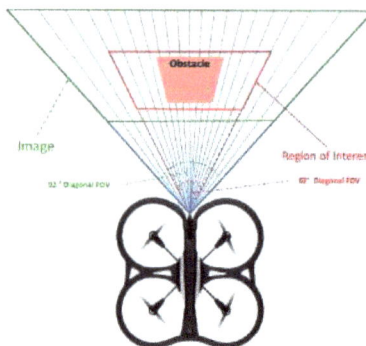

Figure 4. Define the diagonal 62° patch from the whole 92° image FOV.

Due to flying in unknown environments, the captured frames are affected by different conditions; such as illumination variation which may induce to noise and error. However, the keypoints need to be extracted accurately even under these conditions. Therefore the SIFT detector algorithm is used; because of the ability to identify and localize accurately the feature points even under different image condition specially scale and rotation properties.

According to Algorithm 1, all the keypoints are detected and its descriptors are extracted from the two consecutive frames as shown in Figure 5, then a vector of the position (x, y) and the size (s) of each keypoint is obtained.

After detection the keypoints, the Brute-Force algorithm is applied to match the keypoints from the two frames, and only the points that are found in both frames are returned. Algorithm 2 illustrates the concept of the Brute-Force method to match, and find the smallest distance of a pair of points.

Algorithm 2: Brute-Force Matcher

    Input: Array of keypoints in first image $kp_{t-1}[N]$,
            Array of keypoints in second image $kp_t[M]$
    Output: Indices of the matched points $index_1$, $index_2$

1 **Define:** Keypoints $kp_1 = (x_1, y_1), \ldots\ldots, kp_n = (x_n, y_n)$, Minimum distance d_{min}, Distance
    between keypoints d

2 **begin**

3     $d_{min} \leftarrow \infty$

4   **for** $i \leftarrow 1, N$ **do**

5   **for** $j \leftarrow i+1, N$ **do**

6             $d \leftarrow \sqrt{(x_i - x_j)^2 + (y_i - y_j)^2}$

7   **if** $d < d_{min}$ **then**

8                 $d_{min} \leftarrow d$

9                 $index_1 \leftarrow i$

10                $index_2 \leftarrow j$

11   **end**

12   **end**

13 **end**

14 **end**

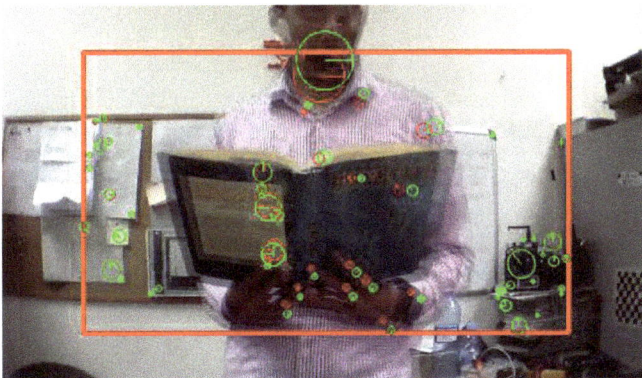

Figure 5. Keypoints extraction from two consecutive frames; keypoints extracted from frame f_{t-1} (red) and keypoints extracted from frame f_t (green).

For more accuracy, the matched keypoints are filtered, by eliminating the ones that have a minimum distance ratio bigger than an empirical threshold value (0.28). Let *mkp* is the filtered-matched keypoint which are calculated as follows:

$$mkp(n) = \begin{cases} (x,y,s), & distratio \leq 0.28 \\ 0, & otherwise \end{cases} \quad \forall n \in K \quad (1)$$

where, s is the size of the keypoint (diameter), *distratio* is the minimum distance ratio of the matched keypoints, and K is the total number of the matched keypoints.

Afterwards, the obtained keypoints by Equation (1) are compared from the second to the first frame, and then the algorithm return the matched keypoints if and only if its size is growing, as shown in Figure 6:

$$mkp(i) = \begin{cases} (x,y,s), & Size(mkp_2(i) > mkp_1(i)) \\ 0, & otherwise \end{cases} \quad \forall i \in n \quad (2)$$

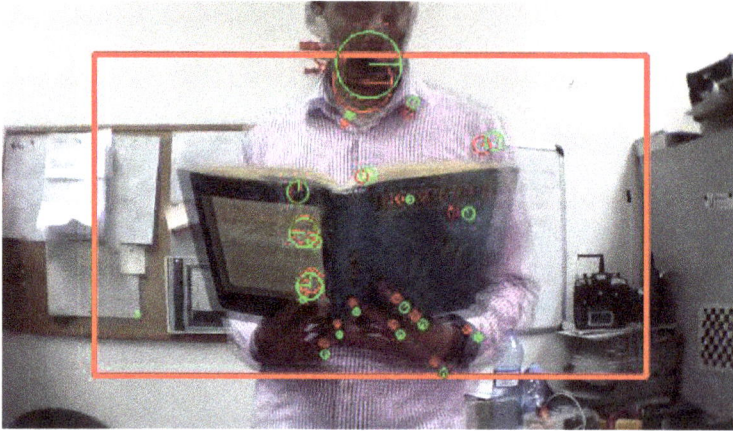

Figure 6. Filtered keypoints where the size expand from the second frame to the first frame; keypoints extracted from frame f_{t-1} (red) and keypoints extracted from frame f_t (green).

3.2. Object of Interest (OOI)

The next step of the detection algorithm is to determine the probability to detect a frontal obstacle. Hence, from the extracted and filtered keypoints by Equation (2), an Object of Interest (OOI) is created around these keypoints in both frames, by creating a convex hull of the corresponding points, as it is shown in Figure 7:

$$C = \sum_{i=1}^{N} \lambda_i mpk_i | (\forall i : \lambda_i \geq 0) \quad (3)$$

where C defines the convex hull, and λ_i is a non-negative weight assigned to the keypoints $mpk_i \in N$ and $\sum_{i=1}^{N} \lambda_i = 1$.

Next, in order to estimate the changes in the size of the area of the detected obstacles, it is considered that each convex hull as an irregular polygon. Therefore, for a given C as a convex hull, the area of C can be calculated as follows:

$$C_{area} = \frac{1}{2} \begin{vmatrix} x_1 & y_1 \\ x_2 & y_2 \\ x_3 & y_3 \\ \vdots & \vdots \\ x_n & y_n \\ x_1 & y_1 \end{vmatrix} = \frac{1}{2} \begin{bmatrix} (x_1y_2 + x_2y_3 + x_3y_4 + \cdots + x_ny_1) \\ -(y_1x_2 + y_2x_3 + y_3x_4 + \cdots + y_nx_1) \end{bmatrix} \tag{4}$$

where $x_{(1:n)}$ and $y_{(1:n)}$ are vertices, and n is the number of sides of the polygon.

Finally, the size ratio of the matched keypoints, and the area of the convex hull from the second to the first frame are calculated respectively as follows:

$$ratio(mkp) = \frac{1}{N} \sum_{i=1}^{N} \frac{Size(mkp_2(i))}{Size(mkp_1(i))} \tag{5}$$

$$ratio(C) = \frac{Size(C_2)}{Size(C_1)} \tag{6}$$

Then, the algorithm estimates the collision state, if the approaching obstacle may represent a collision or not.

$$State = \begin{cases} 1, & ratio(mkp) \geq 1.2 \wedge ratio(C) \geq 1.7 \\ 0, & otherwise \end{cases} \tag{7}$$

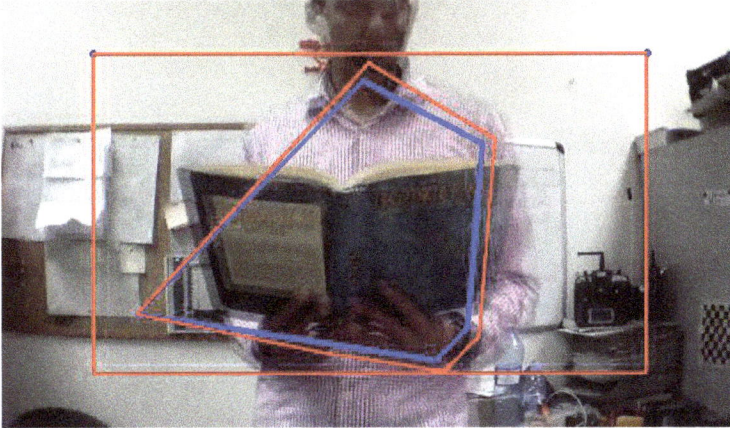

Figure 7. Convex Hull construction from detected keypoints in both frames; frame f_{t-1} (blue) and f_t (red).

Next, an empirical study about the relation of the ratios between the size of the keypoints, the area of the obstacle and the distance of the approaching obstacle has been developed and the results are illustrated in Figure 8. This relation has been estimated by performing different indoor and outdoor experiments. Assuming that the UAV is flying at a constant velocity, the best ratios are in the range of [1.2–1.5], and [1.7–2.0] for keypoints size and obstacle size area respectively, at which the obstacle can be detected in a distance of [120–50] cm.

Figure 9 shows the collision state of the detected obstacles by the monocular camera, where it provides **1** if there is an obstacle, or it provides **0** if there is no obstacle detected.

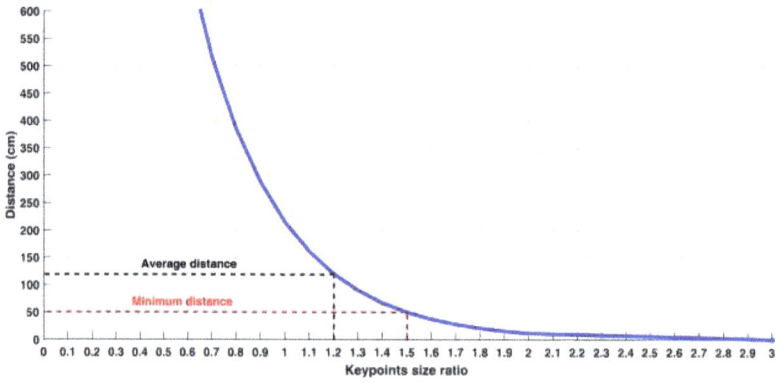

a Keypoints.

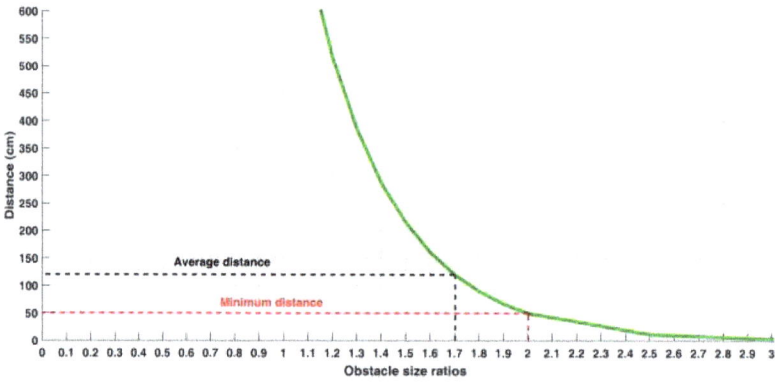

b Obstacle Area.

Figure 8. Distance and Ratios relation.

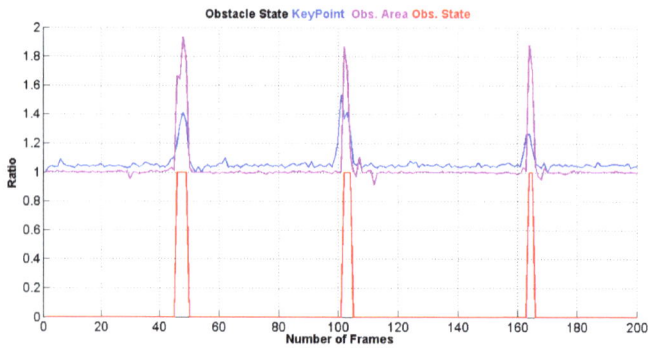

Figure 9. Obstacle State: Keypoint size ratio (Blue), Convex hull area ratio (Magenta) and Obstacle State **(0)** not fount **(1)** found (Red).

In this step, after detecting the obstacles with a collision state value 1, the algorithm estimates the position of the extremely outer points that construct the obstacle in the image (P_l, P_r, P_u, P_d), as it is shown in Figure 10, where P_l is the point the of a position that has the minimum x value, P_r has the maximum x value,and similarly, P_u and P_d have the y minimum and maximum values respectively.

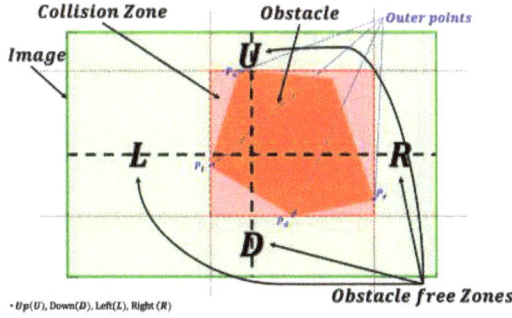

Figure 10. Estimating Obstacle outer points.

Finally, the collision-free zones *Left*, *Right*, *Up* and *Down* (in case of hanged or flying obstacles) are calculated as four rectangles surrounding the obstacle as shown en Equation (8):

$$\tau = \begin{pmatrix} \tau_l, & \tau_r, & \tau_u, & \tau_d \end{pmatrix}$$
$$= \begin{pmatrix} Zone_{L_{width}}, & Zone_{R_{width}}, & Zone_{U_{height}}, & Zone_{D_{height}} \end{pmatrix} \tag{8}$$

where, $Zone_{L_{width}}$, $Zone_{R_{width}}$, $Zone_{U_{height}}$ and $Zone_{D_{height}}$ are the width and the height of the rectangles that are created by the points (P_l, P_r, P_u, P_d), as follows:

$$Zone_L = Rectangle\left[(0, P_{u_y}), \quad (P_{l_x}, P_{u_y}), \quad (0, P_{d_y}), \quad (P_{l_x}, P_{d_y})\right]$$
$$Zone_R = Rectangle\left[(P_{r_x}, P_{u_y}), \quad (ROI_w, P_{u_y}), \quad (P_{r_x}, P_{d_y}), \quad (ROI_w, P_{d_y})\right]$$
$$Zone_U = Rectangle\left[(P_{l_x}, 0), \quad (P_{r_x}, 0), \quad (P_{l_x}, P_{u_y}), \quad (P_{r_x}, P_{u_y})\right] \tag{9}$$
$$Zone_D = Rectangle\left[(P_{l_x}, P_{d_y}), \quad (P_{r_x}, P_{d_y}), \quad (P_{l_x}, ROI_h), \quad (P_{r_x}, ROI_h)\right]$$

where, w and h are the width and the height of the ROI respectively.

4. Obstacle Avoidance

In this section, the combined mission of the waypoint tracking and the avoidance method is described. The geometrical problem is shown in Figure 11 where the avoidance technique is summarized in Algorithm 3.

To define the problem of waypoint tracking, let the UAV X flying at a velocity V, considering the UAV flies forward at a constant velocity along its *x-axis*, where:

$$X = \begin{bmatrix} x_d & y_d & z_d \end{bmatrix}^T \quad and,$$
$$V = \begin{bmatrix} u_d & v_d & w_d \end{bmatrix}^T \tag{10}$$

On the other hand, let a waypoint:

$$WP = \begin{bmatrix} x_w & y_w & z_w \end{bmatrix}^T \tag{11}$$

hence the waypoint is assumed to be tracked if x_d is achieved when both y_d and z_d are satisfied, where:

$$x_d = x_w \pm \mu_x$$
$$y_d = y_w \pm \mu_y \tag{12}$$
$$z_d = z_w \pm \mu_z$$

where, μ is the tolerance area around the waypoint position with a radius of 10 cm from the waypoint.

Let a frontal obstacle **obs** be detected by Algorithm 1, situated in the UAV path and surrounded by the collision-free zones $\tau = (\tau_l, \tau_r, \tau_u, \tau_d)$.

First, the avoidance algorithm checks all the free zones and differentiate which zone is the best to be followed. This is done by reading the position of the next waypoint and by comparing the size of the free zones, where the final maneuver will be in term of **Left-Right** or **Up-Down** motion or a combination of both. After that, a safety boundary surrounding the obstacle is assumed as shown in Figure 11, which is based on the dimensions of the UAV. This safety region is estimated to be:

$$Safety_{lr} = \left(\frac{w_{UAV}}{2} \right) + 20 \text{ [cm]} \tag{13}$$

and

$$Safety_{ud} = \left(\frac{h_{UAV}}{2} \right) + 20 \text{ [cm]} \tag{14}$$

where, w and h defines the width and the height of the UAV respectively.

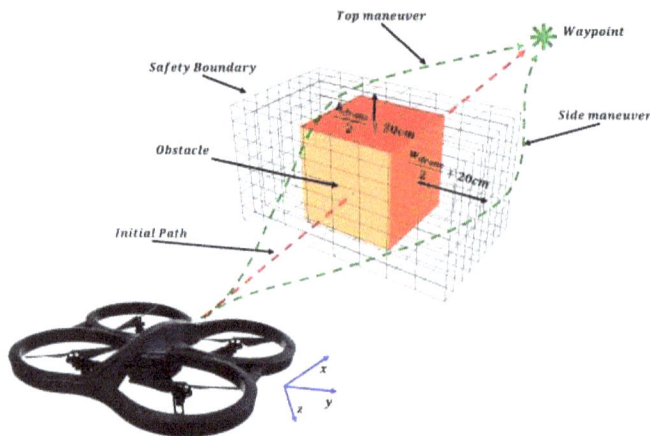

Figure 11. Obstacle Avoidance Path.

Afterwards, the algorithm reads the position of the predefined next waypoint, and calculates the new waypoint out of the path (in order to avoid the obstacle), and sends a control command (velocity control) to the UAV for maneuvering according to the waypoint position as follows:

- Horizontal maneuver (*Right* or *Left*)

$$V_{lr} = \kappa(y_d \pm Safety_{lr}) \tag{15}$$

- Vertical maneuver (*Top* or *bottom*)

$$V_{ud} = \kappa(z_d \pm Safety_{ud}) \tag{16}$$

where y_d and z_d are the UAV position in (y, z)-coordinates, and κ is a control coefficient.

Algorithm 3: Obstacle Avoidance Algorithm

    Input: Collision-free zone τ, UAV position $X = (x_d, y_d, z_d)$
    Output: Navigation command $Nav(\phi, \theta, \psi, \vartheta)$

1 **Define:** Image zones $(\tau_l, \tau_r, \tau_u, \tau_d)$, Roll ϕ, Pitch θ, Yaw ψ, Vertical speed ϑ, Next waypoint
    $WP = (x_w, y_w, z_w)$, Tolerance $(thresh_1, thresh_2)$ and Maneuver behavior (Mr_{rl}, Mr_{ud})

2 **Control Commands:**

3     $+\Delta H \leftarrow z_d + \left(\frac{h_{UAV}}{2}\right) + 20\ cm$ $-\Delta H \leftarrow z_d - \left(\frac{h_{UAV}}{2}\right) + 20\ cm$

4     $+\Delta Y \leftarrow y_d + \left(\frac{w_{UAV}}{2}\right) + 20\ cm$ $-\Delta Y \leftarrow y_d - \left(\frac{w_{UAV}}{2}\right) + 20\ cm$

5 **begin**

6   **while** *!atGoal* **do**

7   **if** *ObAreaRatio* $\leq$ **2** *and KPSizeRatio* $\leq$ **1.5** **then**

8             $WP \leftarrow Read()$

9   **if** τ_l *and* $\tau_r = 0$ **then**   `// No free-zones in Right or Left`

10               $Mr_{rl} \leftarrow 0$

11   **else if** $\tau_l = \tau_r$ **then**   `// Right and Left free-zones are equal`

12   **if** $y_w > y_d$ **then**   `// Compare y position of the UAV and next waypoint`

13                   $Mr_{rl} \leftarrow \tau_r$ `// Go Right`

14   **else**

15                   $Mr_{rl} \leftarrow \tau_l$ `// Go Left`

16   **else**

17               $Mr_{rl} \leftarrow max\ (\tau_l, \tau_r)$ `// Choose the wider free-zones`

18             $\phi \leftarrow CalcControl(Mr_{rl}, \tau_l, \tau_r)$ `// See Algorithm 4`

19   **if** τ_u *and* $\tau_d = 0$ **then**   `// No free-zones in UP or Down`

20               $Mr_{ud} \leftarrow 0$

21   **else if** $\tau_u = \tau_d$ **then**   `// Up and Down free-zones are equal`

22   **if** $z_w > z_d$ **then**   `// Compare z position of the UAV and next waypoint`

23                   $Mr_{ud} \leftarrow \tau_u$ `// Go Up`

24   **else**

25                   $Mr_{ud} \leftarrow \tau_d$ `// Go Down`

26   **else**

27               $Mr_{ud} \leftarrow max\ (\tau_u, \tau_d)$ `// Choose the wider free-zones`

28             $\vartheta \leftarrow CalcControl(Mr_{ud}, \tau_u, \tau_d)$ `// See Algorithm 4`

29             $\theta \leftarrow 1, \psi \leftarrow 0$

30             $X \leftarrow Nav(\phi, \theta, \psi, \vartheta)$

31   **else**

32             $X \leftarrow Nav(0, 0, 0, 0)$

33   **goto** Algorithm 1   `// See Detection Algorithm 1`

Algorithm 4: Calculate Avoidance Control

    Input: Image zones $(\tau_l, \tau_r, \tau_u, \tau_d)$ and Maneuver behavior (Mr_{rl}, Mr_{ud})

    Output: Roll ϕ, Pitch θ, Yaw ψ, Vertical speed ϑ

1 **Define:** Control gains $K_{ph}, K_{p\phi}$, UAV position $X = (x_d, y_d, z_d)$

2 **if** $Mr_{rl} = \tau_l$ **then**

3     $\phi \leftarrow K_{p\phi} \left(-\Delta Y - y_d \right)$

4 **else if** $Mr_{rl} = \tau_r$ **then**

5     $\phi \leftarrow K_{p\phi} \left(+\Delta Y - y_d \right)$

6 **else**

7     $\phi \leftarrow 0$

8 **if** $Mr_{ud} = \tau_u$ **then**

9     $\vartheta \leftarrow K_{ph} \left(+\Delta H - z_d \right)$

10 **else if** $Mr_{ud} = \tau_d$ **then**

11     $\vartheta \leftarrow K_{ph} \left(-\Delta H - z_d \right)$

12 **else**

13     $\vartheta \leftarrow 0$

Finally, by estimating the new UAV position after avoidance, the algorithm recalculates the new waypoints in order the UAV to be able to return back to its predefined path and activate the detection process.

In the case that the *AreaScale* is greater than 2 and the *SizeScale* of the keypoints is greater than 1.5, a "*Hover*" command is sent to the UAV. That is because if the ratios exceed these limits, this means that the obstacle is very close to the UAV (less than 50 cm), as it is shown in Figure 8.

5. Experimental Results

In order to evaluate the performance of the proposed algorithms in the previous sections, 100 different real flight experiments have been carried out, in both indoor and outdoor environments, with a total number of 1000 obstacles, taking in consideration the visual conditions (the illumination and the texture of the obstacles) which affect the accuracy of the detection.

5.1. Platform

The processing in the ground station is performed in Intel i7-3770 at 3.4 GHz CPU, with 6 GB DDR3 RAM. The connection with the UAV is established via a standard 802.11n wireless LAN card.

The experiments have been performed with a Parrot AR.Drone 2.0 quadcopter [36]. This control system is governed by the inputs (roll, pitch, yaw angles, and vertical speed), therefore the implemented controller realize the UAV actual position, orientation and velocity.

One of the most important aspects in the avoidance phase is based on the robust control system for the UAV at which it is necessary to know its dynamic model. However, to avoid the complexity in modeling, the avoidance control was applied over the internal control of the system, modifying the roll and the vertical speed in order to perform the maneuvers in y and z directions.

5.2. Scenarios

Two different scenarios have been conducted, in order to evaluate the performance of the proposed algorithms in both motion and stability, with data gathered from the experiments to test the detection and the estimation of the position of the obstacle. In each scenario, different types of obstacles (people, obstacles, pillars, trees and walls), (static and dynamic) are situated.

The first scenario is a predefined straight flight, where the UAV flies in a straight line from the starting point to the end point. Different types of obstacles with unknown previous position were situated in the UAV path. The goal of this scenario is to evaluate the accuracy, and robustness in detecting, and avoiding the obstacles in motion.

The second scenario is a hover stability flight. At which, the UAV enters to the hover flight mode, and different obstacle are approaching to it. Once an obstacle is detected, the UAV flights in the opposite direction of the obstacle (*Backward maneuver*).

5.3. Results (Obstacle Detection)

From the experiments, the obtained results demonstrate that the algorithm is able to detect the obstacles with different sizes (areas) between 8500 and 200,000 pixels, and at a distance range between 90 and 120 cm. It is shown that the minimum accuracy of the algorithm is 95.0%, and the overall accuracy is 97.4% as it is demonstrated in Table 1.

Table 1. Accuracy of Detection Algorithm.

	Indoor				Outdoor				
	People	Obstacle	Pillar	Wall	People	Obstacle	Tree	Wall	Total
Situated	200	110	80	80	200	120	140	70	1000
Detected	196	107	79	76	196	116	135	69	974
Fail	4	3	1	4	4	4	5	1	26
Accuracy (%)									
Object	98.0	97.3	98.8	95.0	98.0	96.7	96.4	98.6	97.4
Environment		97.3				97.4			

a Input frame F_t.

b Input frame F_{t+1}.

c Keypoints within the 62° FOV.

d Detected obstacle.

Figure 12. Obstacle detection: $ratio(mkp) = 1.27$, $ratio(C) = 1.76$ and $distance = 114$ cm.

Figures 12–14 illustrate the detection process of various approaching obstacles, with different size ratios. Where, Figures 12a,b, 13a,b and 14a,b are showing the two input consecutive frames to be processed. In Figures 12c, 13c or 14c it is shown the total number of the detected and matched keypoints before filtering its size expansion property. Finally, the filtered keypoints and the constructed polygon of the detected obstacle are shown in Figures 12d, 13d and 14d.

a Input frame F_t.   **b** Input frame F_{t+1}.

c Keypoints within the 62° FOV.   **d** Detected obstacle.

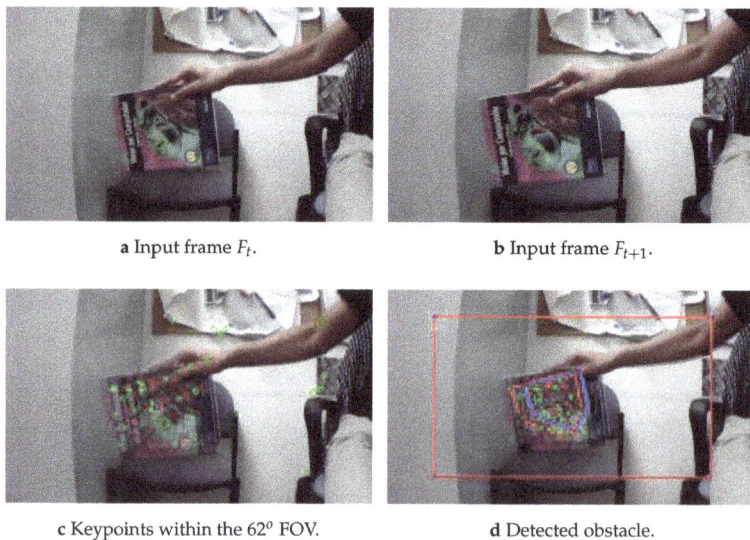

Figure 13. Obstacle detection: $ratio(mkp) = 1.25$, $ratio(C) = 1.71$ and $distance = 92$ cm.

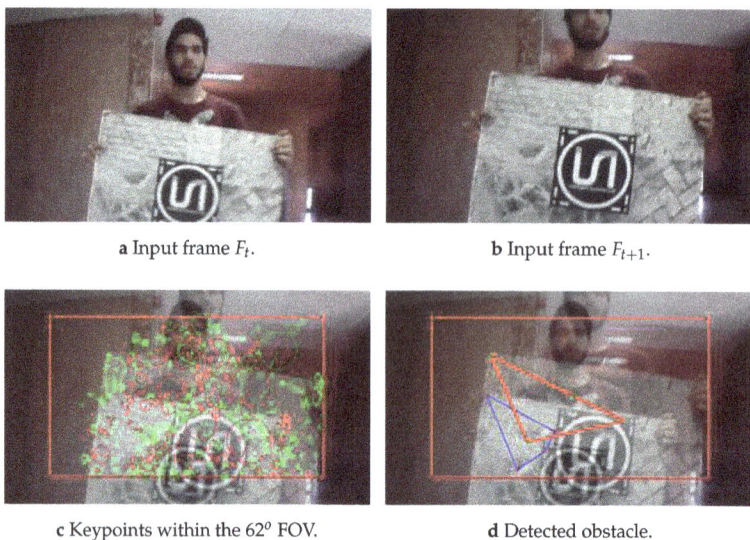

a Input frame F_t.   **b** Input frame F_{t+1}.

c Keypoints within the 62° FOV.   **d** Detected obstacle.

Figure 14. Obstacle detection: $ratio(mkp) = 1.20$, $ratio(C) = 2.15$ and $distance = 126$ cm.

Table 1 summarizes the accuracy of the detection algorithm. The table shows the total number of the obstacles that either situated in the UAV path (*first scenario*) or moving towards the UAV (*second scenario*), the number of the detected obstacles and the number of fails.

From the table, it is illustrated that the accuracy of the detection process in the indoor scenarios is better than the accuracy in outdoor environments. This is due to the constancy of the light conditions in indoors rather than outdoors, which are suffered from various lighting effects.

Two main reasons for the fail of detection; the first one is the disability of extracting sufficient number of keypoints, and that is either because of the low light conditions or because of the absence of the texture on the obstacle surface such as in the case of some pillars and walls as shown in Figure 15.

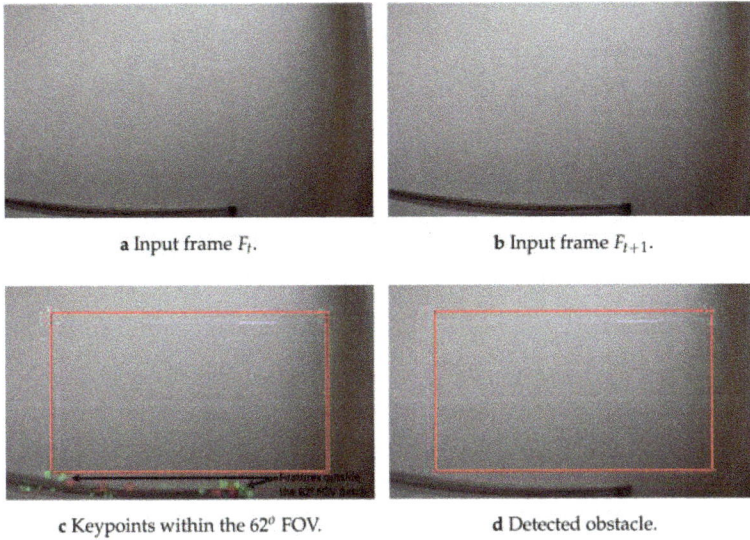

a Input frame F_t.

b Input frame F_{t+1}.

c Keypoints within the 62^0 FOV.

d Detected obstacle.

Figure 15. Obstacle detection fail (wall) (absence of texture): $ratio(mkp) = 1$ and $ratio(C) = 1$.

The second reason is the direction of the motion of the obstacle, the algorithm is able to detect the moving obstacle if the motion is towards the UAV.

Figure 16 shows an example of the second scenario, where the UAV flies in hover mode, and the object is moving, however, this movement is not in the direction of the UAV. Therefore, it does not consider as an obstacle.

However, in most cases of the moving obstacles according to Table 1, the algorithm could not detect the appearance of the obstacles if the motion is around the UAV such as in the case of the people and obstacles.

In addition, the proposed algorithm is evaluated against two related works of detecting frontal obstacles based on monocular vision. As it shown in Table 2, the proposed algorithm provides more accuracy (97.4%) comparing to *SURF + Template matching* method [24] which provides 97%, and *relative distance estimation* approach [17] that provides 97.1% of accuracy.

Furthermore, the computational time of the detection algorithm is estimated around 52.4 ms. This is due to the processing ROIs of 62° FOV, which leads to decrease the processing time up to 50% from 106.1 ms comparing to processing the whole 92° FOV images. In addition, this computational time is estimated for the detection of 800–1200 keypoints. On the one hand, if the number of detected keypoints exceeds 6000, the computational time peaks to a maximum of 100 ms, on the other hand, if it is below 300 keypoints, then the required computation time is reduced to 30 ms.

a Input frame F_t.

b Input frame F_{t+1}.

c Keypoints within the 62° FOV.

d Detected obstacle.

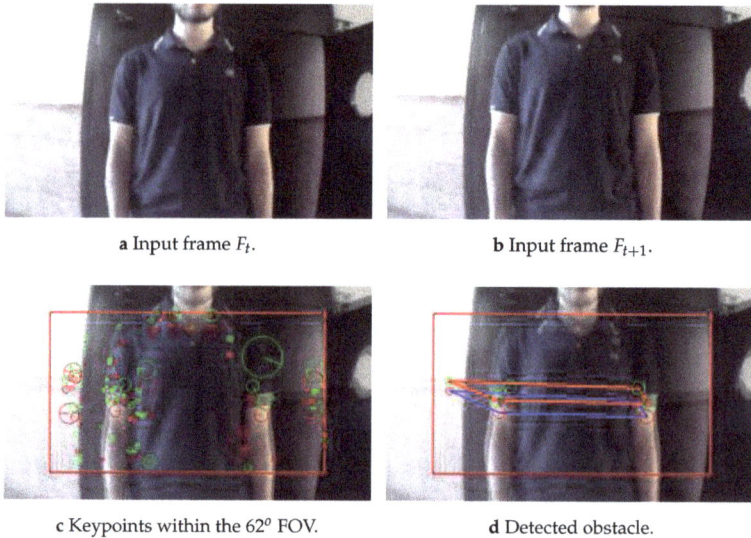

Figure 16. Obstacle detection fail (people) (motion around the UAV)-KPSizeRatio = 1.07 and ObAreaRatio = 1.03.

Table 2. Comparison of Frontal Obstacle Detection.

Algorithm	Total	Detected	Fail	Accuracy (%)
SURF + Template matching [24]	107	104	3	97
Relative distance estimation [17]	35	34	1	97.1
Proposed Algorithm	1000	974	26	97.4

5.4. Results (Obstacle Avoidance)

Figures 17 and 18 demonstrate an example of a set of experiments presenting the first scenario. In these experiments, the UAV is flying in a velocity of 2 m/s. All the started from the same *start* point, and during the the flight, an obstacle is situated in the UAV path. Figure 17 illustrates the UAV ability to perform avoidance maneuvers in the Left or Right directions of a total number of 9 experiments.

Similarly, in Figure 18, the success in avoiding hanged obstacles performing vertical maneuvers in the z direction by passing above and under the obstacle in a total number of 10 experiments is represented.

Finally, Table 3 shows a comparison of avoidance accuracy between two methods from the bibliography and the proposed algorithm. The accuracy results display that the best performance belongs to the proposed algorithm reaching 93% of accuracy.

Table 3. Comparison of Avoidance Accuracy.

Algorithm	Total	Success	Failure	Accuracy (%)
SURF + Template matching [24]	23	20	3	87
Relative Distance Estimation [17]	35	31	4	88.57
Proposed Algorithm	100	93	7	93

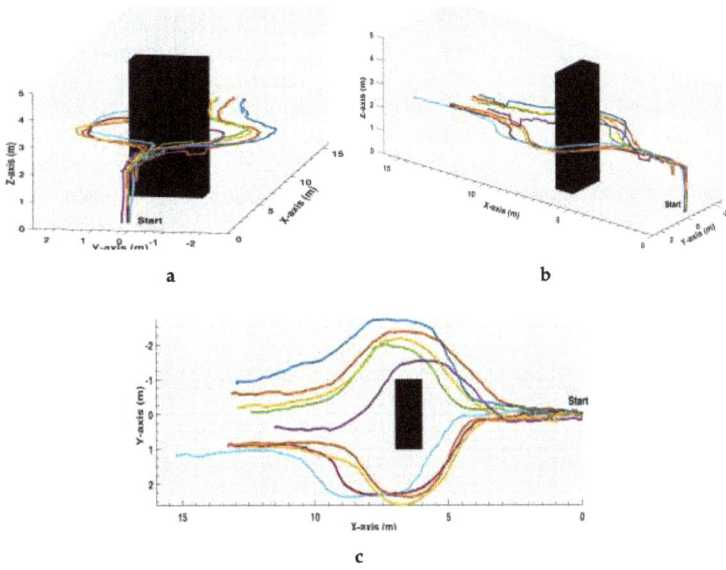

Figure 17. *Left-Right* Avoidance Maneuver, 9 experiments; (**a**): Front, (**b**): 3D, and (**c**): 2D.

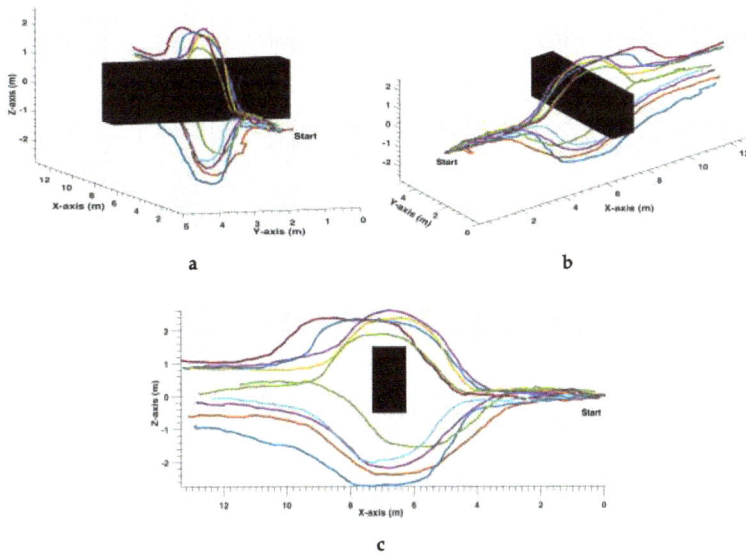

Figure 18. *Up-Down* Avoidance Maneuver, 10 experiments; (**a**): Front, (**b**): 3D, and (**c**): 2D.

6. Conclusions

In this paper, two algorithms have been presented as a framework to cope with cutting-edge UAVs technology. Real-time obstacle detection and avoidance is studied as a complex and essential task for intelligent aerial vehicles in transportation systems. The proposed algorithms take the advantages of onboard camera to accomplish complex tasks, that is, safe obstacle sensing and detection tasks.

The selected configuration of 62° ensured the capabilities of detecting the border of an object of the size of the actual drone (58.4 × 1.3 × 54.4) located in the center of the image at distances higher the 15 cm, which allows to avoid obstacles, for higher obstacles or closer distances, the drone proved to be able to stop and preform hover movement, avoiding the collision. Bigger obstacles located at longer distances were avoided due to the use a high quality camera able to detect obstacle at long distances. In case of faster speeds required, the frame rate calculation and the angle should be adjusted, to allow the drone to do the calculation at proper detection. However, the change of the field of view of the camera, would only be advisable in order to allow further maneuverability in extremely dense scenarios, with short distance detection requirements which are not common in aerial scenarios where UAVs are deployed.

Keypoints used approach is based on the use of SIFT features. The performance obtained proved to be good in both computational time and overall detection performance scheme. The nature of the approach made it possible to adapt it to different set of features beyond the use of SIFT, such as the FAST-BRIEF pair [37] and BRISK [38] which proved to provide better performance in different scenarios. Future works will try to analyze the advantages of adding these sets of features to the presented approach.

The usefulness and advantages of the presented reliable solutions are demonstrated through real results under demanding circumstances, such as, complex artificial and human obstacles. Hence, complex scenarios are evaluated and difficulties are successfully overcame by means of monocular camera processing, where the relative size expansion of the obstacles are estimated and the approaching obstacles are detected from a distance between 90 and 120 cm with 97.4% of total accuracy. The various performed tests proved both, the trustable performance of the algorithms provided and the improvements in comparison to the previous works presented in literature.

The strengths of the presented applications are clearly stated in the paper, where according to the UAV path, the obstacle detection and avoidance algorithms demonstrate hight accuracy in the detection process and send maneuver commands to the UAV based on the obstacle and UAV positions. However, the specific drawbacks that should be taken into account are mainly related with the nature of the sensing devices used, that is, the monocular camera has the drawback of the high sensitivity to lighting conditions; such as direct sun light may lead to lack of information.

Future works will focus on two main lines. The tests of the application with different features, as commented. The second line is focused on the development of advanced control system, with higher capacities. These includes the use of fuzzy logic, and other advanced AI approaches.

Acknowledgments: Research supported by the Spanish Government through the Cicyt project ADAS ROAD-EYE (TRA2013-48314-C3-1-R).

Author Contributions: A. Al-Kaff, F. Garcia and D. Martin conceived and designed the experiments; A. Al-Kaff performed the experiments; A. Al-Kaff, F. Garcia and D. Martin analyzed the data; A. de la Escalera and J.M Armingol contributed reagents/materials/analysis tools; A. Al-Kaff, F. Garcia and D. Martin wrote the paper." Authorship must be limited to those who have contributed substantially to the work reported.

Conflicts of Interest: The authors declare no conflict of interest.

References

1. Erdos, D.; Erdos, A.; Watkins, S. An experimental UAV system for search and rescue challenge. *IEEE Aerosp. Electron. Syst. Mag.* **2013**, *28*, 32–37.

2. Kamel, B.; Santana, M.C.S.; De Almeida, T.C. Position estimation of autonomous aerial navigation based on Hough transform and Harris corners detection. In Proceedings of the 9th WSEAS International Conference On Circuits, Systems, Electronics, Control & Signal Processing, Athens, Greece, 29–31 December 2010; pp. 148–153.

3. Fraundorfer, F.; Heng, L.; Honegger, D.; Lee, G.H.; Meier, L.; Tanskanen, P.; Pollefeys, M. Vision-based autonomous mapping and exploration using a quadrotor MAV. In Proceedings of the 2012 IEEE/RSJ International Conference on Intelligent Robots and Systems (IROS), Vilamoura, Algarve, Portugal, 7–12 October 2012; pp. 4557–4564.
4. Michael, N.; Fink, J.; Kumar, V. Cooperative manipulation and transportation with aerial robots. *Auton. Robots* **2011**, *30*, 73–86.
5. Eschmann, C.; Kuo, C.M.; Kuo, C.H.; Boller, C. Unmanned aircraft systems for remote building inspection and monitoring. In Proceedings of the 6th European Workshop on Structural Health Monitoring, Dresden, Germany, 2–6 July 2012.
6. Choi, S.; Kim, E. Image Acquisition System for Construction Inspection Based on Small Unmanned Aerial Vehicle. In *Advanced Multimedia and Ubiquitous Engineering*; Springer: Heidelberg, Germany, 2015; pp. 273–280.
7. Bills, C.; Chen, J.; Saxena, A. Autonomous MAV flight in indoor environments using single image perspective cues. In Proceedings of the 2011 IEEE International Conference on Robotics and Automation (ICRA), Shanghai, China, 9–13 May 2011; pp. 5776–5783.
8. Blosch, M.; Weiss, S.; Scaramuzza, D.; Siegwart, R. Vision based MAV navigation in unknown and unstructured environments. In Proceedings of the 2010 IEEE International Conference on Robotics and Automation (ICRA), Anchorage, AK, USA, 3–8 May 2010; pp. 21–28.
9. Shim, D.; Chung, H.; Sastry, S. Conflict-free navigation in unknown urban environments. *IEEE Robot. Autom. Mag.* **2006**, *13*, 27–33.
10. Luo, D.; Wang, F.; Wang, B.; Chen, B. Implementation of obstacle avoidance technique for indoor coaxial rotorcraft with Scanning Laser Range Finder. In Proceedings of the 2012 31st Chinese Control Conference (CCC), Hefei, China, 25–27 July 2012; pp. 5135–5140.
11. Shang, E.; An, X.; Li, J.; He, H. A novel setup method of 3D LIDAR for negative obstacle detection in field environment. In Proceedings of the 2014 IEEE 17th International Conference on Intelligent Transportation Systems (ITSC), Qingdao, China, 8–11 October 2014; pp. 1436–1441.
12. Ariyur, K.; Lommel, P.; Enns, D. Reactive inflight obstacle avoidance via radar feedback. In Proceedings of the American Control Conference, Portland, OR, USA, 8–10 June 2005; Volume 4, pp. 2978–2982.
13. Broggi, A.; Cattani, S.; Patander, M.; Sabbatelli, M.; Zani, P. A full-3D voxel-based dynamic obstacle detection for urban scenario using stereo vision. In Proceedings of the 2013 16th International IEEE Conference on Intelligent Transportation Systems (ITSC), The Hague, Netherlands, 6–9 October 2013; pp. 71–76.
14. Gao, Y.; Ai, X.; Rarity, J.; Dahnoun, N. Obstacle detection with 3D camera using U-V-Disparity. In Proceedings of the 2011 7th International Workshop on Systems, Signal Processing and their Applications (WOSSPA), Tipaza, Algeria, 9–11 May 2011; pp. 239–242.
15. Na, I.; Han, S.H.; Jeong, H. Stereo-based road obstacle detection and tracking. In Proceedings of the 2011 13th International Conference on Advanced Communication Technology (ICACT), PyeongChang, South Korea, 19–22 February 2011; pp. 1181–1184.
16. Li, J.; Li, X.-M. Vision-based navigation and obstacle detection for UAV. In Proceedings of the 2011 International Conference on Electronics, Communications and Control (ICECC), Ningbo, China, 9–11 September 2011; pp. 1771–1774.
17. Saha, S.; Natraj, A.; Waharte, S. A real-time monocular vision-based frontal obstacle detection and avoidance for low cost UAVs in GPS denied environment. In Proceedings of the 2014 IEEE International Conference on Aerospace Electronics and Remote Sensing Technology (ICARES), Yogyakarta, Indonesia, 13–14 November 2014; pp. 189–195.
18. Majumder, S.; Shankar, R.; Prasad, M. Obstacle size and proximity detection using stereo images for agile aerial robots. In Proceedings of the 2015 2nd International Conference on Signal Processing and Integrated Networks (SPIN), Noida, Delhi-NCR, India, 19–20 February 2015; pp. 437–442.
19. Merrell, P.C.; Lee, D.J.; Beard, R.W. Obstacle avoidance for unmanned air vehicles using optical flow probability distributions. *Mob. Robots XVII* **2004**, *5609*, 13–22.
20. Hrabar, S.; Sukhatme, G.; Corke, P.; Usher, K.; Roberts, J. Combined optic-flow and stereo-based navigation of urban canyons for a UAV. In Proceedings of the 2005 IEEE/RSJ International Conference on Intelligent Robots and Systems (IROS 2005), Edmonton, AB, Canada, 2–6 August 2005; pp. 3309–3316.

21. Beyeler, A.; Zufferey, J.C.; Floreano, D. 3D Vision-based Navigation for Indoor Microflyers. In Proceedings of the 2007 IEEE International Conference on Robotics and Automation, Rome, Italy, 10–14 April 2007; pp. 1336–1341.

22. Celik, K.; Chung, S.J.; Clausman, M.; Somani, A. Monocular vision SLAM for indoor aerial vehicles. In Proceedings of the IEEE/RSJ International Conference on Intelligent Robots and Systems, St. Louis, MO, USA, 10–15 October 2009; pp. 1566–1573.

23. Chavez, A.; Gustafson, D. Vision-based obstacle avoidance using SIFT features. In Proceedings of the 5th International Symposium on Advances in Visual Computing: Part II, ISVC'09, Las Vegas, NV, USA, 30 November–2 December 2009; Springer: Berlin, Heidelberg, 2009; pp. 550–557.

24. Mori, T.; Scherer, S. First results in detecting and avoiding frontal obstacles from a monocular camera for micro unmanned aerial vehicles. In Proceedings of the 2013 IEEE International Conference on Robotics and Automation (ICRA), Karlsruhe, Germany, 6–10 May 2013; pp. 1750–1757.

25. Shirai, N.; Yamaguchi, M.K. Asymmetry in the perception of motion-in-depth. *Vis. Res.* **2004**, *44*, 1003–1011.

26. Gibson, J.J. *The Ecological Approach to Visual Perception: Classic Edition*; Psychology Press: Boston, MA, USA, 2014.

27. De Croon, G.; de Weerdt, E.; De Wagter, C.; Remes, B. The appearance variation cue for obstacle avoidance. In Proceedings of the 2010 IEEE International Conference on Robotics and Biomimetics (ROBIO), Tianjin, China, 14–18 December 2010; pp. 1606–1611.

28. Green, W.; Oh, P. Optic-Flow-Based Collision Avoidance. *IEEE Robot. Autom. Mag.* **2008**, *15*, 96–103.

29. Lee, J.O.; Lee, K.H.; Park, S.H.; Im, S.G.; Park, J. Obstacle avoidance for small UAVs using monocular vision. *Aircr. Eng. Aerosp. Technol.* **2011**, *83*, 397–406.

30. Ma, Z.; Hu, T.; Shen, L.; Kong, W.; Zhao, B. A detection and relative direction estimation method for UAV in sense-and-avoid. In Proceedings of the 2015 IEEE International Conference on Information and Automation, Gothenburg, Sweden, 24–28 August 2015; pp. 2677–2682.

31. Byrne, J.; Taylor, C.J. Expansion segmentation for visual collision detection and estimation. In Proceedings of the International Conference on Robotics and Automation (ICRA'09), Kobe, Japan, 6–7 August 2009; pp. 875–882.

32. Kim, J.; Do, Y. Moving Obstacle Avoidance of a Mobile Robot Using a Single Camera. *Procedia Eng.* **2012**, *41*, 911–916.

33. Kendoul, F. A Survey of Advances in Guidance, Navigation and Control of Unmanned Rotorcraft Systems. *J. Field Rob.* **2012**, *29*, 315–378.

34. Yu, X.; Zhang, Y. Sense and avoid technologies with applications to unmanned aircraft systems: Review and prospects. *Prog. Aerosp. Sci.* **2015**, *74*, 152–166.

35. Mcfadyen, A.; Mejias, L. A survey of autonomous vision-based See and Avoid for Unmanned Aircraft Systems. *Prog. Aerosp. Sci.* **2016**, *80*, 1–17.

36. Krajnik, T.; Vonasek, V.; Fiser, D.; Faigl, J. AR-drone as a platform for robotic research and education. In *Research and Education in Robotics-EUROBOT*; Springer: Berlin/Heidelberg, Germany, 2011; pp. 172–186.

37. Schmidt, A.; Kraft, M.; Fularz, M.; Domagala, Z. The comparison of point feature detectors and descriptors in the context of robot navigation. *J. Autom. Mob. Robot. Intell. Syst.* **2013**, *7*, 11–20.

38. Jeong, C.Y.; Choi, S. A comparison of keypoint detectors in the context of pedestrian counting. In Proceedings of the 2016 International Conference on Information and Communication Technology Convergence (ICTC), Jeju Island, Korea, 31 July–26 August 2016; pp. 1179–1181.

sensors

MDPI

Article

A Methodology to Monitor Airborne PM$_{10}$ Dust Particles Using a Small Unmanned Aerial Vehicle

Miguel Alvarado [1,*], Felipe Gonzalez [2], Peter Erskine [1], David Cliff [3] and Darlene Heuff [4]

[1] Environment Centre, Sustainable Mineral Institute, The University of Queensland, 4072 Brisbane, Australia; p.erskine@uq.edu.au

[2] Science and Engineering Faculty, Queensland University of Technology (QUT), 4000 Brisbane, Australia; felipe.gonzalez@qut.edu.au

[3] People Centre, Sustainable Mineral Institute, The University of Queensland, 4072 Brisbane, Australia; d.cliff@mishc.uq.edu.au

[4] Advanced Environmental Dynamics Pty Ltd., Ferny Hills, 4055 Queensland, Australia; darlene.heuff@aedconsultants.com.au

* Correspondence: m.alvaradomolina@uq.edu.au; Tel.: +61-7-3346-4065 or +61-7-3346-4086

Academic Editor: Assefa M. Melesse
Received: 15 December 2016; Accepted: 4 February 2017; Published: 14 February 2017

Abstract: Throughout the process of coal extraction from surface mines, gases and particles are emitted in the form of fugitive emissions by activities such as hauling, blasting and transportation. As these emissions are diffuse in nature, estimations based upon emission factors and dispersion/advection equations need to be measured directly from the atmosphere. This paper expands upon previous research undertaken to develop a relative methodology to monitor PM$_{10}$ dust particles produced by mining activities making use of small unmanned aerial vehicles (UAVs). A module sensor using a laser particle counter (OPC-N2 from Alphasense, Great Notley, Essex, UK) was tested. An aerodynamic flow experiment was undertaken to determine the position and length of a sampling probe of the sensing module. Flight tests were conducted in order to demonstrate that the sensor provided data which could be used to calculate the emission rate of a source. Emission rates are a critical variable for further predictive dispersion estimates. First, data collected by the airborne module was verified using a 5.0 m tower in which a TSI DRX 8533 (reference dust monitoring device, TSI, Shoreview, MN, USA) and a duplicate of the module sensor were installed. Second, concentration values collected by the monitoring module attached to the UAV (airborne module) obtaining a percentage error of 1.1%. Finally, emission rates from the source were calculated, with airborne data, obtaining errors as low as 1.2%. These errors are low and indicate that the readings collected with the airborne module are comparable to the TSI DRX and could be used to obtain specific emission factors from fugitive emissions for industrial activities.

Keywords: PM$_{10}$; monitoring; blasting; unmanned aerial vehicle (UAV); multi-rotor UAV; optical sensor

1. Introduction

The use of Unmanned Aerial Vehicles (UAVs) to collect environmental data has increased exponentially in only a few years. Their capacity to reach places where humans could be at risk or it could become too expensive for small operations, makes them very useful for gas and particulate monitoring in the mining industry. Several authors have published articles where progress on the use of UAVs for industrial use or investigation has been made by using two UAVs simultaneously [1], combining multiple gas sensors and imagery [2,3], using gas tracking algorithms [4–6], using solar energy [7], and making use of microwave sensors [8].

The methodology applied and tests undertaken for this investigation are a continuation of the work published by Alvarado et al. [9,10]. Improvements to the methodology include: the development of a new modular sensor, focusing only on the use of a multi-rotor UAV with a tailored sampling probe, and the use of a 5 m tower to validate the PM_{10} readings obtained by the UAV. The emission rates obtained from a point source created for the flight tests, were evaluated to determine the capability of the methodology to be used at open pit mine sites to estimate emission factors. In the following sections these experiments are described and their results discussed.

2. Experimental Development

2.1. Dust Monitoring Module

Following Alvarado et al. [9,10] a sampling module and data logger transmission system were developed using the OPC-N2 particle counter from Alphasense (Great Notley, Essex, UK) [11] with a minimum response time of 0.2 s. In addition to the particle counter the dust monitoring module is integrated with a:

- Barometer, temperature and humidity sensor, BME280 (Bosch, Reutlingen, Germany);
- Raspberry Pi 3 board (Raspberry Pi Foundation, Cambridge, UK);
- uBlox LEA6 GPS (uBlox, Thalwil, Switzerland);
- XBee 2.4 GHz Radio (XBee, Minnetonka, MN, USA);
- 3DR Airspeed Sensor (3DR, Berkeley, CA, USA);
- Lipo battery (3.7 V) (3DR, Berkeley, CA, USA).

The architecture of the dust monitoring module is presented in Figure 1. Information from each sensor (e.g., dust, pressure, temperature and humidity) was sent to a ground station via the XBee 2.4 GHz radio for monitoring, and recorded in a micro-SD card used by the Raspberry Pi. The dust monitoring module was programmed to have a response time of 0.2 s. Time synchronisation was mainly achieved using the GPS time stamp of each reading. The name of each log file was also recorded with the time and date synchronised with the internet using a portable WiFi hot spot.

Dust monitoring module

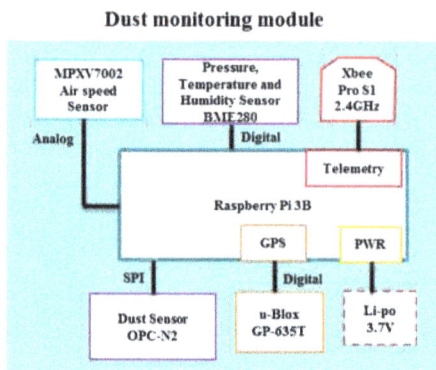

Figure 1. Architecture of dust monitoring module.

The total weight of the dust monitoring module was of 382.0 g and had the following dimensions: 14.0 cm × 9.0 cm × 10.0 cm. The small multi-rotor UAV used for the investigation was an IRIS+ from 3DR (Berkeley, CA, USA), with an endurance of approximately 10 min. The IRIS+ multi-rotor UAV was the selected flying platform due to robustness, user-friendly controls, versatility and ability to be modified and serviced. The IRIS+ was adapted to carry an air sampling probe and an enclosure for the dust monitoring module as shown in Figure 2.

Figure 2. Physical configuration of the multi-rotor UAV with the sampling probe and dust monitoring module attached.

2.2. Aerodynamics Downwash/Upwash Experiment and Analysis

When using a multi-rotor UAV two main issues are identified when obtaining a representative air sample from the atmosphere: difficulty of producing isokinetic flow, and designing an air sampling probe and determining the best position in the multi-rotor UAV.

To understand the disturbance produced by the propellers in the volume of air that will be sampled, an aerodynamic experiment was designed. Through this experiment, it was possible to observe the behavior of the volume of air surrounding the multi-rotor UAV and determine if isokinetic flow would be possible. Isokinetic flow is an important consideration when monitoring or sampling for particles below 10 μm in diameter, otherwise the resulting value could be under or over-predicted [12]. However, being able to obtain isokinetic flow in non-static conditions can be extremely difficult [13,14]. Air flow intake needs to be controlled so the air streamlines are unaltered during the transportation of particles from the atmosphere past the probe's nozzle. To achieve these conditions several factors have to be considered: air velocity has to be equal inside and outside the sampling nozzle, air intake has to have the same direction (iso-axial), and has to be placed out of the air mixing zone [14–16]. Other variables to consider for isokinetic measurements with aircrafts are: diffusion, sedimentation, and turbulent inertial deposition. However, the variables considered will vary depending on the type of aircraft [15–17]. Related literature on the topic mainly focuses on fixed wing UAVs which, due to their flight characteristics, can provide better conditions for isokinetic sampling [16,17]. Factors such as angled flight, static flight or hovering, and constant change in wind direction, make achieving isokinetic conditions very challenging for multi-rotor UAVs. Von der Weiden et al. [12], used a different approach due to the difficulty of sampling in isokinetic conditions. They created a correction calculator for air sampling that does not meet the criteria necessary to be isokinetic and iso-axial. The calculator considers variation in sedimentation, deposition due to inlets, bends, contractions, and diffusion factors. However, their approach was only used and tested for the fixed wing UAV specified in their study.

For the aerodynamics experiment, the UAV was placed indoors and mounted on a 2.5 m pole. Measurements were taken with a Kestrel 2500 anemometer (Minneapolis, MN, USA) which was positioned around the UAV following a grid pattern at different x, y and z positions (Figure 3). With the readings collected, a 3D visualization map was generated to observe the airflow produced by the propeller downwash and upwash (Figure 4).

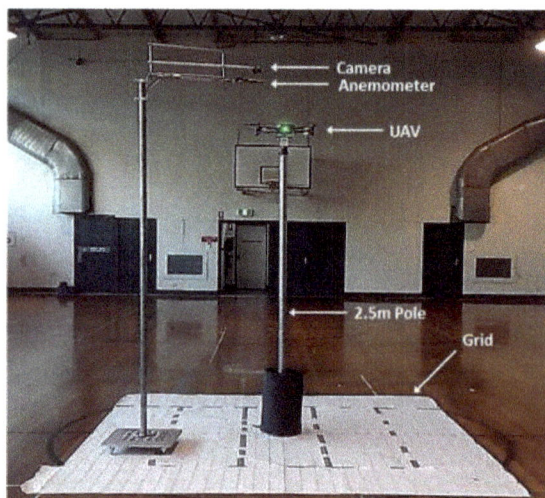

Figure 3. Set up of the multi-rotor UAV and airspeed measuring pole for aerodynamics test.

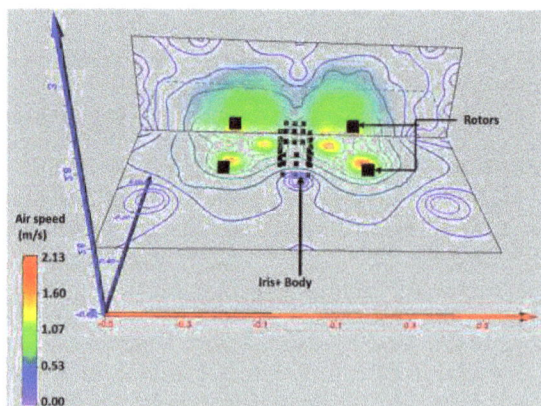

Figure 4. 3D visualization of the different airspeed regions produced by the propellers upwash at the top and sides of the multi-rotor UAV. Axis units in meters.

It was determined that the downwash produced by the multi-rotor UAV does not generate a uniform pattern by observing the readings and visualisations. The downwash varies by forming a cone which increases its base as distance from the center of the UAV increases (Figure 4). The upper side of the UAV on the other hand has a constant air speed flux which drops after a distance of approximately 40.0–45.0 cm. This constant air flux is also observed to the sides of the UAV up to a distance of 40.0–45.0 cm. The constant mixed air boundaries around the multi-rotor UAV helped define the position of the probe on the side of the UAV (horizontally) or on the top of it (vertically). A horizontal probe would increase the capacity of the UAV to collect an isokinetic sample when the wind direction is in line with the direction of the probe. However, when conducting filed tests, wind direction varied up to 68° during the testing period (10 min approximately) for three out of four flights conducted. Such ample variation in the intake angle is a considerable change of the ideal conditions required for isokinetic sampling. To overcome this issue, it was decided to use the vertical probe on top of the UAV at a distance of 47.5 cm from the center of the UAV (Figure 2). The use of a canopy on the top also increases its capacity to retain small volumes of the air targeted. This can be useful in high wind speeds. A vertical probe has the same exposure when taking air samples regardless of the orientation of the multi-rotor UAV or the wind flow.

2.3. Calcined Alumina Correction Factor

A correction factor was calculated for the dust source so the data collected by the sensing module could be used for further estimates. The source of dust chosen for the test was calcined alumina, which is less hygroscopic than talcum powder, used in previous tests [9]. Being less hygroscopic helped keep the optical sensors free of dust adhering to the components and accumulating inside the protective casing. In addition, according to the particle size distribution of the calcined alumina used for the experiment, the dust had a PM_{10} content of approximately 30%. This value is ten times higher than the average PM_{10} content found in talcum powder. A higher PM_{10} content allowed testing the sensor with small amounts of dust. The highest concentration of PM_{10} measured during the experiment was of 17.32 mg/m^3 (19.42 mg/m^3 of TSP). An OPC-N2 optical particle counter and a TSI DRX 8533 [18] were collocated with a gravimetric sampler, an AirChek2000 (Eighty Four, PA, USA), to calculate their correction factor (Figures 5 and 6). Concentrations from the OPC-N2 and the TSI DRX were obtained directly from the logging files produced by the interface software provided by the manufacturers. Tests were done in a dust chamber of approximately 0.104 m^3.

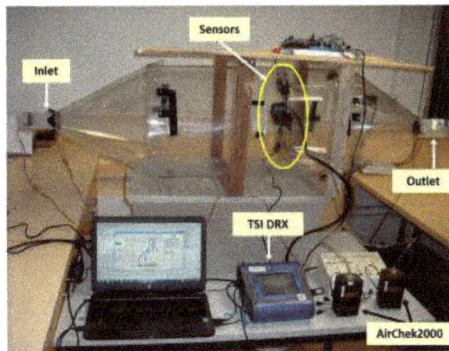

Figure 5. Set up of monitoring devices and dust chamber for calcined alumina calibration tests.

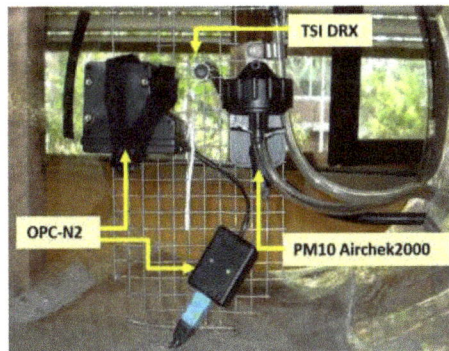

Figure 6. Layout of all monitoring equipment and sensors used to obtain the particle correction factor.

The tests consisted of inducing dust into the chamber with the aid of an axial fan. Three tests were conducted using a regular supply of dust at intervals lasting 3 min for the first, 6 min for the second and 9 min for the third test. In addition to the tests where dust was supplied, two blank tests were conducted to know background dust levels, resulting in an average concentration of 4.82 µg/m^3 of PM$_{10}$. Once all the data was collected, lineal regression analysis was used to calculate the correction factors between devices.

The samples collected with the AirChek2000 were used as the main reference. The filters used by the sampler were sent to a certified laboratory (ALS Environmental, New Castle, Australia) for analysis and determination of calcined alumina PM$_{10}$ concentrations. The correction factors calculated are shown in Table 1. Figure 7 shows the resulting predicted values for the AirChek2000 making use of the particle correction factor calculated with the correlation between the TSI DRX and the air sampler. A coefficient of determination (R^2) of 0.92 for the correlation indicates a good fit between the optical sensor and the air sampler.

Table 1. Correction factors calculated with the analysis of the data from the dust chamber.

Device	TSI DRX		AirChek2000	
	Correction Factor	R^2	Correction Factor	R^2
OPC N2 (airborne module)	0.342	0.62	1.362	0.73
TSI DRX	NA	NA	1.312	0.92

NA = Not Applicable.

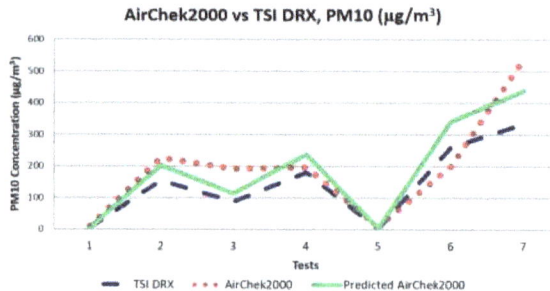

Figure 7. Correlation obtained to calculate the correction factor for the TSI DRX against the gravimetric sampler.

In Figure 8, the process to obtain the correction factors is presented as a flow diagram. This diagram also summarises the other two experiments conducted to validate the readings of the dust monitoring sensors. These experiments are described in the following sections.

2.4. Variable Wind Speed Modeling

In addition to the previous test a variable air speed experiment was designed to study the response of the OPC-N2 with the probe attached using different wind speeds. For this experiment a wind tunnel was built to produce a laminar flow and have control of the air speed changes (Figure 9). The test area of the wind tunnel was 40.0 cm cross section, in which the probe (attached to the OPC-N2), a secondary OPC-N2 without a probe, and the TSI DRX (making use of the conductive tubing), were placed at the same height (15.0 cm from base of test area) for uniformity.

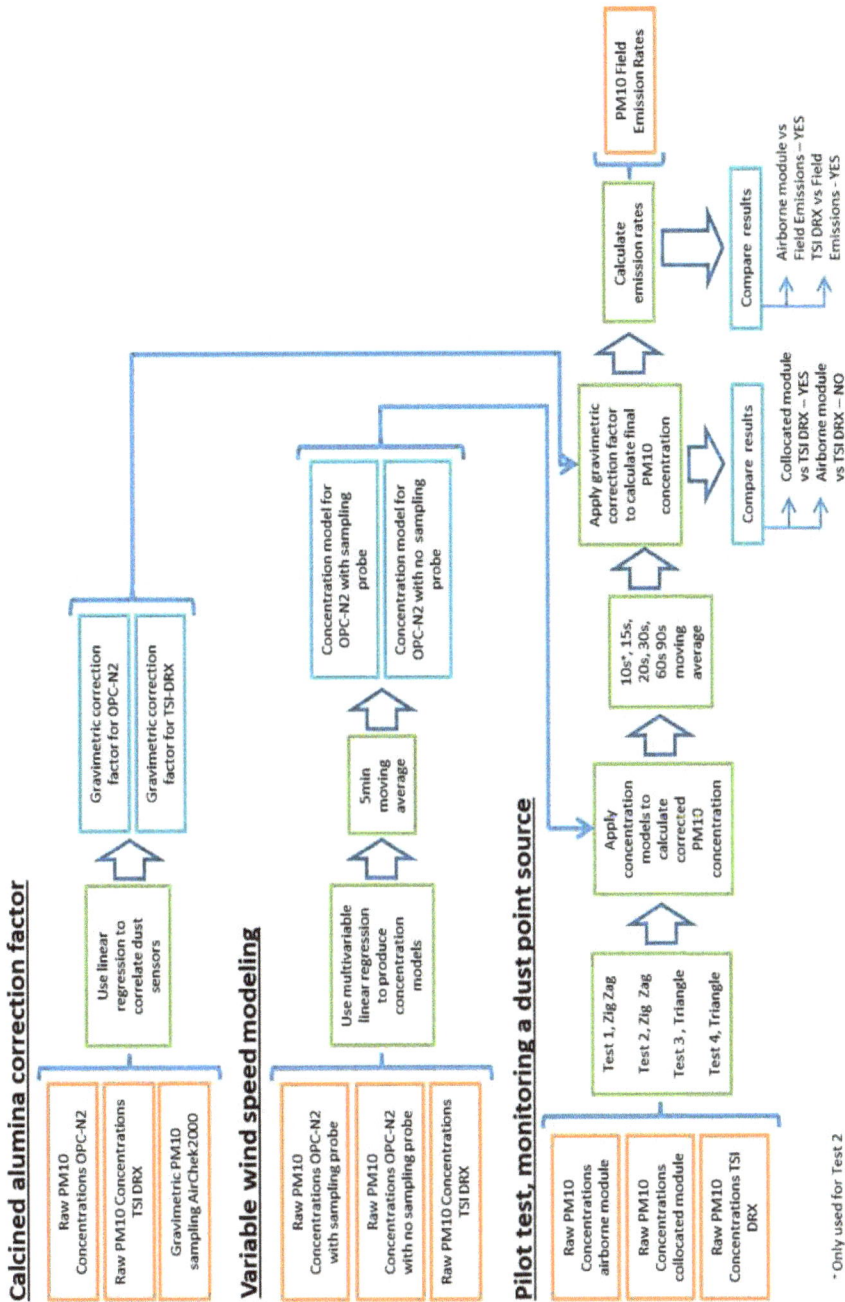

Figure 8. Flow diagram summarizing tests undertaken to validate readings collected with the dust monitoring sensors.

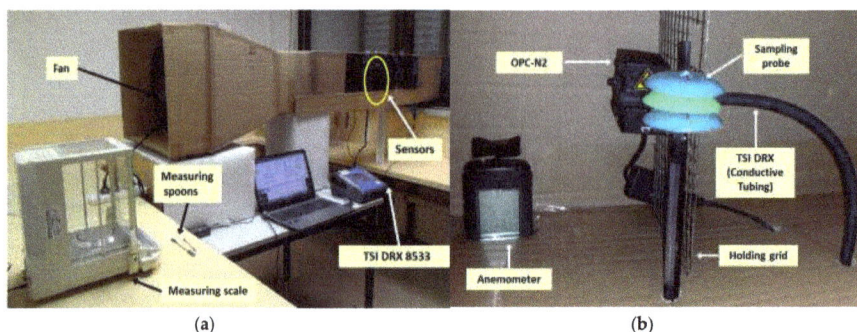

Figure 9. Wind tunnel and equipment used for the variable wind speed tests, (**a**) overall view of the wind tunnel and equipment used; and (**b**) test area set with the anemometer, probe, OPC-N2 and TSI DRX.

Four individual experiments were conducted to generate a variable wind speed model. Each experiment used a different wind speed: 1.11 m/s, 1.67 m/s, 3.06 m/s and 3.89 m/s. The air flow was produced with fans and measured with a SkyWatch ATMOS anemometer (Sudbury, UK). Dust particles were supplied using calcined alumina at an approximate rate of 0.015 g every 10 min. The concentration read by the OPC-N2 was logged and reported by the software interface provided by Alphasense Ltd. (Great Notley, Essex, UK).

Two equations were generated out of this experiment, one for the OPC-N2 connected to the probe and one equation for the OPC-N2 which was collocated with the TSI DRX during the flight tests. The resulting coefficients for the airborne module system equation and for the collocated module are shown in Equations (1) and (2) and Table 2:

$$C_{UAV} = (2.118 \times C_R) - (1.175 \times R.H.) - (1.261 \times T) + (1.822 \times U) + 102.082 \qquad (1)$$

$$C_{Coll} = (0.197 \times C_R) - (0.623 \times R.H.) - (0.641 \times T) + (1.125 \times U) + 57.292 \qquad (2)$$

Table 2. Resulting coefficients for variables of equations for dust monitoring sensors with their *p*-value and confidence level.

Sensor	Variable	Coefficients	*p*-Value
Airborne module (C_{UAV})	Intercept	102.082	2.5×10^{-223}
	Raw Conc. PM_{10} (C_R, µg/m³)	2.118	~0.0
	Relative Humidity (R.H. %)	−1.175	4.2×10^{-235}
	Temperature (T, °C)	−1.261	2.1×10^{-231}
	Air speed (U, m/s)	1.822	2.8×10^{-159}
	R^2	0.55	NA
Collocated module sensor (C_{Coll})	Intercept	57.292	2.7×10^{-84}
	Raw Conc. PM_{10} (C_R, µg/m³)	0.197	~0.0
	Relative Humidity (R.H. %)	−0.623	5.6×10^{-79}
	Temperature (T, °C)	−0.641	1.7×10^{-72}
	Air speed (U, m/s)	1.125	2.0×10^{-74}
	R^2	0.66	NA

NA = Not Applicable.

The R^2 for both equations using a multivariate linear regression was above 0.5. The airborne module had a coefficient of 0.55 and the collocated module sensor of 0.66. Coefficients with a *p*-value < 0.05 (95% confidence) is regarded as statistically significant. Coefficients within a 95% confidence level interval indicate they have a strong influence in the model outcome [19]. For Equations (1) and (2), all variables have a confidence level of approximately 100% (calculated as 1-*P* and expressed as a percentage). The model produced reliable results when used to correct raw data collected from the dust monitoring modules. However, both equations were calculated without the use of the feeding dust rate, due to resulting overestimated values, with errors up to 1000% when applying the model to field data collected during the flight tests. A possible explanation to the high error produced could be that the dust was supplied manually and this could produce less precise measurements than using an automated system. Due to this limitation, the models generated did not include the dust supply variable (W). Nevertheless, results of the calculations were the models were used produced estimates with low percentage errors.

2.5. Particle Size Distribution for Variable Wind Speed Experiment

Research has demonstrated that factors like wind speed, wind direction, air intake angle and shape of the nozzle, among others, affect the proportion of particles that are measured by monitoring devices [12]. Due to these considerations, dust particles should be measured under isokinetic and iso-axial conditions. In order to assess how the particle size distribution of the air sampled would be affected by the probe and different wind speeds, the data was normalised based on a percentage calculated with the average quantity of particles counted in determined size ranges.

Figure 10 shows the normalized data for all wind speeds classified by particle size diameter (using the bins set by the OPC-N2). It is observed that the particles ranging from 0.38–0.54 μm are dominant when using the probe with a percentage difference of 45.5%. These are followed by particles ranging from 0.54–1.59 μm with an average percentage difference of 9.6%. This data indicates that up to a diameter of 2.0 μm, particles are expected to be in higher proportion that particles with larger diameter. Particles with a diameter from 1.59 μm to 10 μm presented a lower difference of 1.0%. Finally, the total difference estimated was of 7.6%. Through this analysis, it is possible to observe that particles up to approximately 2.0 μm are dominant when measuring with the sampling probe suggested in the methodology. However, results of the modeling estimates suggest that the effect of the probe can be corrected to obtain comparable results to field values.

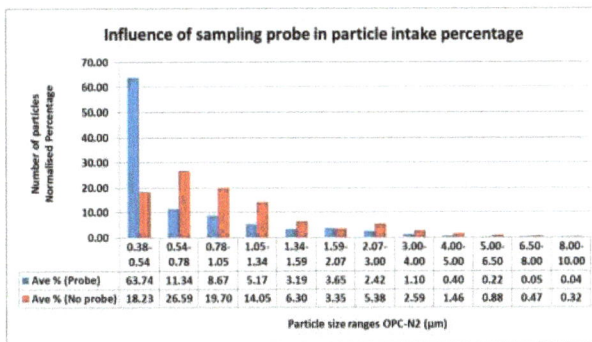

Influence of sampling probe in particle intake percentage

Particle size ranges OPC-N2 (μm)	0.38-0.54	0.54-0.78	0.78-1.05	1.05-1.34	1.34-1.59	1.59-2.07	2.07-3.00	3.00-4.00	4.00-5.00	5.00-6.50	6.50-8.00	8.00-10.00
Ave % (Probe)	63.74	11.34	8.67	5.17	3.19	3.65	2.42	1.10	0.40	0.22	0.05	0.04
Ave % (No probe)	18.23	26.59	19.70	14.05	6.30	3.35	5.38	2.59	1.46	0.88	0.47	0.32

Figure 10. Influence of the sampling probe used with the OPC-N2 over particle counting readings per particle size range.

Figure 11a,b shows the impact that variable wind speeds produce in the OPC-N2 by itself and with the sample probe attached. The distribution of particles counted by the sensor follows the same pattern that was observed in Figure 10. The average difference between wind speeds per bin is higher when not using the sampling probe, with difference of 10.3% in particles within 0.38–0.54 μm. The highest difference was also observed in the particle range from 0.38–0.54 μm when using the sampling probe.

Impacts by the probe were expected, nevertheless as stated previously, by using the models it was possible to correct these effects in the sample intake and obtain results with low errors.

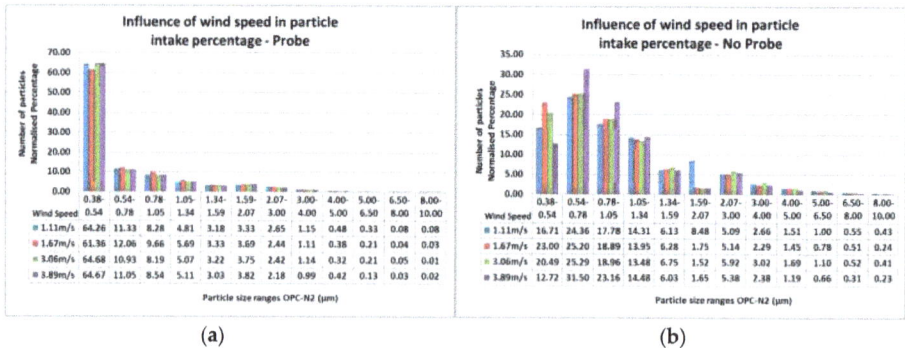

Influence of wind speed in particle intake percentage - Probe

Wind Speed	0.38- 0.54	0.54- 0.78	0.78- 1.05	1.05- 1.34	1.34- 1.59	1.59- 2.07	2.07- 3.00	3.00- 4.00	4.00- 5.00	5.00- 6.50	6.50- 8.00	8.00- 10.00
1.11m/s	64.26	11.33	8.28	4.81	3.18	3.33	2.65	1.15	0.48	0.33	0.08	0.08
1.67m/s	61.36	12.06	9.66	5.69	3.33	3.69	2.44	1.11	0.38	0.21	0.04	0.03
3.06m/s	64.68	10.93	8.19	5.07	3.22	3.75	2.42	1.14	0.32	0.21	0.05	0.01
3.89m/s	64.67	11.05	8.54	5.11	3.03	3.82	2.18	0.99	0.42	0.13	0.03	0.02

Particle size ranges OPC-N2 (µm)

(a)

Influence of wind speed in particle intake percentage - No Probe

Wind Speed	0.38- 0.54	0.54- 0.78	0.78- 1.05	1.05- 1.34	1.34- 1.59	1.59- 2.07	2.07- 3.00	3.00- 4.00	4.00- 5.00	5.00- 6.50	6.50- 8.00	8.00- 10.00
1.11m/s	16.71	24.36	17.78	14.31	6.13	8.48	5.09	2.66	1.51	1.00	0.55	0.43
1.67m/s	23.00	25.20	18.89	13.95	6.28	1.75	5.14	2.29	1.45	0.78	0.51	0.24
3.06m/s	20.49	25.29	18.96	13.48	6.75	1.52	5.92	3.02	1.60	1.10	0.52	0.41
3.89m/s	12.72	31.50	23.16	14.48	6.03	1.65	5.38	2.38	1.19	0.66	0.31	0.23

Particle size ranges OPC-N2 (µm)

(b)

Figure 11. Influence of wind speed variation over particle counting readings per particle size range for (a) OPC-N2 with a sampling probe attached; and (b) OPC-N2 without a sampling probe.

For the TSI DRX, the concentration for different particle size fractions was analysed. Particle fractions considered for the analysis were: $PM_{1.0}$, $PM_{2.5}$, $PM_{4.0}$ (respirable according to ISO 12103-1, A1), and PM_{10}. Figure 12 shows the averaged raw concentration per wind speed used. It is observed that the greatest difference between readings of a particle size with different wind speeds is produced by $PM_{1.0}$ and PM_{10} fractions with values approximately 50.0% greater than the minimum average reading. Lower average concentrations were constant for wind speeds of 1.67 m/s and 3.89 m/s, and greater for wind speeds of 1.11 m/s and 3.89 m/s. These constant patterns observed in the different particle size fractions, produced a better outcome when generating the models.

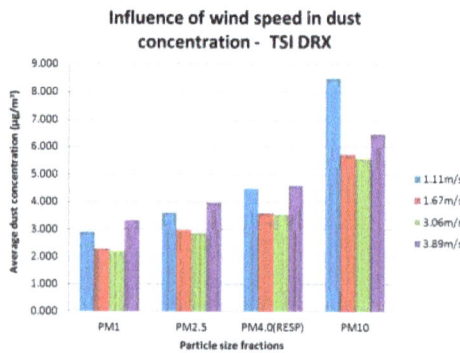

Influence of wind speed in dust concentration - TSI DRX

Legend: 1.11m/s, 1.67m/s, 3.06m/s, 3.89m/s

Particle size fractions: PM1, PM2.5, PM4.0(RESP), PM10

Figure 12. Influence of wind speed variation over dust concentration (raw) for different particle size fractions reported by the TSI DRX.

3. Flight Tests, Monitoring a Dust Point Source

An experiment was designed to collect experimental data similar to that obtained at open pit mine sites from activities such as hauling, stockpiling, blasting, etc. These activities are area sources or fugitive emissions, however to create an experimental site in which variables could be easily monitored and controlled, a stack emission was the most feasible option. The experiment consisted of using a 5.0 m stack (5.0 cm diameter), through which calcined alumina was supplied and expelled with the use of an electric leaf blower. Due to the dimensions of the stack, it was not possible to sample PM_{10}

concentration readings from inside the stack isokinetically. However, this contributed in making the scenario similar to a real fugitive emission, measuring directly with the multi-rotor UAV from the plume generated. A 5.0 m tall "monitoring tower" was installed at a distance of 10.0 m downwind from the stack. At the top of the tower, a TSI DRX and a duplicate of the dust monitoring module were installed. This tower was used to compare dust concentration readings from stationary devices against airborne data (Figure 13).

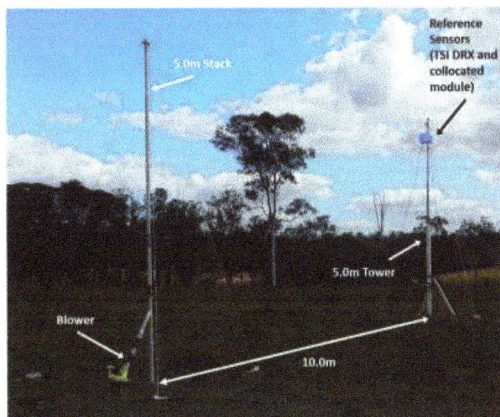

Figure 13. Experimental set up for the flight tests with the source stack and the monitoring tower.

A variable quantity of calcined alumina was blown out of the stack at approximately 30 s intervals. The total weight of powder expelled per interval was measured using an electronic balance (ELB600, Shimadzu, Kyoto, Japan). In-situ meteorological data was collected with an Oregon Scientific WMR200 weather station placed at the test site. Status and readings from the two modular sensors were followed via the radio link and controlled using SSH connection. In order to use the data produced by all the monitoring devices (dust monitoring modules, TSI DRX, and meteorological station) it was necessary to synchronise their datasets by programming the weather station and the TSI DRX. However, the dust monitoring modules had to be adjusted by internet clock link using WiFi and a portable hot-spot connection.

3.1. Procedure for Flight Tests 1 and 2

In Alvarado et al. (2015), a circular flight pattern was used to scan the area and create a grid to characterize the dust plume. However, for this flight test a different grid was designed for this test due to the small size of the plume and the endurance of the UAV, which was of approximately 10 min. Two grids were designed for the first two tests. The grids were oriented northwest, downwind from the stack (source), and followed a zig-zag path around the monitoring tower (Figure 14).

For Test 1 the multi-rotor UAV was programmed to follow a predetermined path using the "Tower" mobile application from 3DR, hovering at waypoints for 15 s. The pattern was repeated at heights of 7.0 m and 9.0 m covering an area of 225.0 m^2 approximately. The hovering periods were estimated on 15 s lapses according to the analysis of data produced in the laboratory during the calibration tests. This was estimated by averaging concentrations at different time periods and obtaining their correlation (linear association) using the Pearson product-moment correlation coefficient (r). Making a plot of the coefficients, it was possible to choose a time period that could reduce noise in the readings produced by the OPC-N2, and also allowed enough readings to be collected during the 10 min flights (Figure 15).

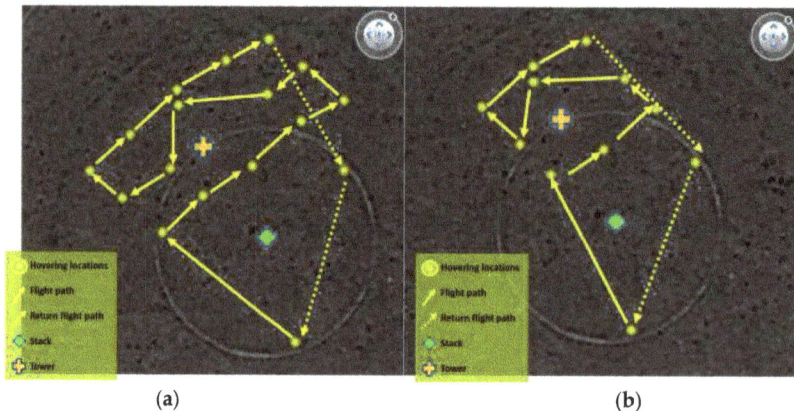

Figure 14. Characterizing zig-zag grids designed for (**a**) Test 1 and (**b**) Test 2.

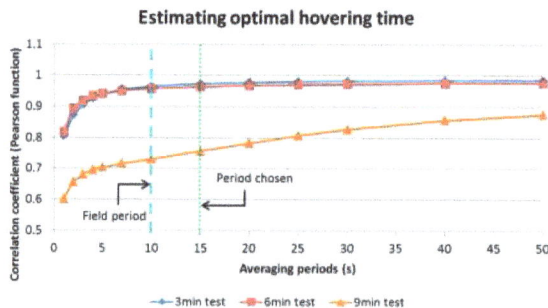

Figure 15. Calculated correlation coefficients (*r*) for different averaging periods of PM_{10} concentrations for three sets of tests.

However, once in the field it was observed that the battery did not last long enough to cover the flight programmed. A second grid was designed for Test 2. For this test, the multi-rotor UAV hovered at the waypoints for 10 s, repeating the pattern at heights of 7.0 m, 9.0 m and 11.0 m. The grid covered an approximate area of 130.0 m^2. For this experiment, the OPC-N2 software interface was not used to log the resulting concentration. A custom made program was written to record the data from all sensors that integrated the dust monitoring module. The OPC-N2 provided the particles counted for 16 predetermined bins, from which only 12 where considered due to the particle size targeted (PM_{10}) (see Table 3).

To calculate the concentration for each reading produced by OPC-N2, the total bin volume of particles was estimated and multiplied by the particle density with a value of 1.65×10^{12} µg/m^3 [11] to determine the total dust mass. The total mass was divided by the total volume of air sampled to calculate the final concentration. The total volume of air sampled was estimated with the sampling period and the sample flow rate (data provided by the sensor). Sampling period was set to 1 s as all data was averaged to match the sampling period of the TSI DRX.

Table 3. Bins considered to calculate PM_{10} concentrations and their boundaries according to particle diametre.

Bins	Bin Low Boundary, Particle Diametre (μm)	Bin High Boundary, Particle Diametre (μm)
0	0.38	0.54
1	0.54	0.78
2	0.78	1.05
3	1.05	1.34
4	1.34	1.59
5	1.59	2.07
6	2.07	3.00
7	3.00	4.00
8	4.00	5.00
9	5.00	6.50
10	6.50	8.00
11	8.00	10.00

3.2. Procedure for Flight Tests 3 and 4

Tests 3 and 4 followed a different flight strategy with the objective to only collect data flying next to the monitoring tower to determine the best way to validate readings from the dust monitoring module. These tests consisted of hovering close to the sides and behind the collocated devices in a triangular pattern. The flight was made in hover (loiter) mode and controlled manually. In total 4 flights were undertaken, two with a zig-zag pattern and two with a triangle pattern, each one of them with a total duration of 10 min approximately. Table 4 summarizes the characteristics of the four tests conducted for this section.

Table 4. Characteristics of the four flight tests conducted.

Test	Pattern	Flight Mode	Hovering Periods (s)	Approximate Area Covered (m²)	Heights Flown (m)
1	Zig-Zag	Auto (programmed)	10	225.0	7, 9
2	Zig-Zag	Auto (programmed)	15	130.0	7, 9, 11
3	Triangle	Manual	Variable	10.6	5 m approx.
4	Triangle	Manual	Variable	10.2	5 m approx.

4. Flight Test Results

4.1. Results and Analysis of Data Collected with the Collocated Module

Linear multivariate regression was used to observe the relation between variables and obtain the correlation between the TSI DRX and the two dust monitoring modules. The variables included in the analysis were: air speed (m/s), relative humidity (%), temperature (C), raw concentration ($\mu g/m^3$), and emission rate ($\mu g/s$). When the raw data was used as an input no correlation was observed. However, a better correlation was found by using the moving average of the concentrations. This was done in a similar way to how the OPC-N2 reports readings in its standalone interface (provided by Alphasense Ltd.) using a 5 min moving average.

For all tests, moving averages of 15 s, 20 s and 30 s were used to determine the best fit for the linear multivariate regression (Section 2.3). This resulted in having 20 s–30 s periods with the highest correlations with R^2 of up to 0.8 for the collocated module. Tables 5 and 6 presents a summary of the different approaches used to correlate data collected with the TSI DRX and the collocated module in the four tests analysed.

Table 5. Models generated from data collected with the TSI DRX and the collocated module using corrected concentrations and field data for Tests 1 and 2.

	Corrected Concentration Test 1			Raw Concentration Test 1			Corrected Concentration Test 2			Raw Concentration Test 2		
						Coefficients for Different Moving Averages						
Variables	15	20	30	15	20	30	15	20	30	15	20	30
Intercept	3527.905	3806.125	4339.276	2916.909	1225.689	−32.446	−734.98	−670.73	−413.48	−311.39	−279.21	−194.93
Air speed (m/s)	−28.921	−15.235	8.638	−32.403	−20.484	−11.931	−5.052	−3.76	−1.56	0.53	0.74	0.84
Rel. Humidity (%)	−53.060	−65.410	−84.957	−17.782	0.900	14.214	6.08	4.89	1.12	2.74	2.43	1.07
Temperature (°C)	−63.276	−55.941	−46.391	−78.939	−46.455	−21.134	15.65	15.79	13.55	7.37	6.60	5.58
Conc. PM_{10} ($\mu g/m^3$)	25.568	25.720	23.828	5.347	6.200	6.584	4.85	3.69	1.40	1.77	1.88	1.88
Emission Rate ($\mu g/s$)	0.002	0.002	0.003	1.0×10^{-4}	-9.5×10^{-5}	-2.0×10^{-4}	3.0×10^{-4}	3.55×10^{-5}	4.0×10^{-4}	1.1×10^{-4}	9.96×10^{-5}	1.5×10^{-4}
R^2	0.61	0.64	0.61	0.67	0.81	0.89	0.46	0.54	0.59	0.63	0.74	0.80

Note: Numbers in *italics* indicate a confidence level lower than 95%.

Table 6. Models generated from data collected with the TSI DRX and the collocated module using corrected concentrations and field data for Tests 3 and 4.

	Corrected Concentration Test 3			Raw Concentration Test 3			Corrected Concentration Test 4			Raw Concentration Test 4		
						Coefficients for Different Moving Averages (s)						
Variables	15	20	30	15	20	30	15	20	30	15	20	30
Intercept	−3205.983	−2758.832	−2363.600	−2471.678	−2031.929	−1727.493	−2603.777	−2453.5	−1930.551	−2244.417	−2164.951	−1773.315
Air speed (m/s)	−70.601	−60.273	−48.545	−58.775	−49.292	−39.101	−9.773	2.415	−8.908	3.379	3.002	2.615
Rel. Humidity (%)	26.778	24.405	24.722	17.740	15.217	16.573	14.641	13.608	10.115	9.471	8.975	6.509
Temperature (°C)	83.074	67.556	52.224	80.139	64.912	48.573	69.263	65.052	50.422	71.857	69.719	58.798
Conc. PM_{10} ($\mu g/m^3$)	16.947	16.412	13.282	80.139	3.383	2.839	14.059	14.349	14.679	2.813	2.916	3.058
Emission Rate ($\mu g/s$)	−0.003	−0.002	−0.001	3.396	−0.002	−0.002	3.0×10^{-4}	3.0×10^{-4}	8.66×10^{-5}	4.0×10^{-4}	3.0×10^{-4}	2.0×10^{-4}
R^2	0.41	0.46	0.52	0.41	0.47	0.54	0.42	0.47	0.54	0.44	0.49	0.57

Note: Numbers in *italics* indicate a confidence level lower than 95%.

Equations (1) and (2) were used to correct the raw concentrations. It is observed that the R^2 values for correlations made with corrected concentrations, when compared against R^2 values from regressions made with raw concentrations, were very close. This indicated that a model can be generated without making the correction. However, the use of corrected values was necessary when calculating emission rates. Figure 16 presents the values predicted using the model generated for Test 2, compared with the readings from the TSI DRX. An R^2 of 0.74 (moving average of 20 s) was obtained with linear regression. By using a moving average of 30 s the R^2 can be improved to 0.8, reducing greatly the noise observed in the corrected values. Such R^2 indicates that the sensor module design can obtain reliable PM_{10} readings comparable to a TSI DRX in the field when placed at the monitoring tower.

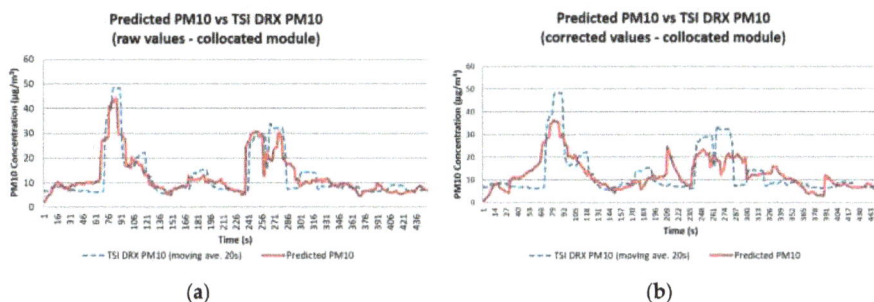

Figure 16. Comparison between PM_{10} concentration readings obtained with the TSI DRX and the predicted values calculated using the models generated for Test 2 with (**a**) raw values; and (**b**) corrected data.

From Tables 4 and 5 and comparing R^2 from Tests 1 and 2 against Tests 3 and 4, it was observed that Tests 1 and 2 have higher values. This could indicate that the multi-rotor UAV was flown closer than necessary to the tower, affecting the natural flow of the plume across the reference devices.

4.2. Comparison of Average Concentrations, Airborne Module Results Tests 2

Table 7 presents the correlation between the data collected with the airborne module for Test 2 and the TSI DRX. A linear multivariate regression was also performed for this dataset. The data is presented as an example to show that models created for the airborne module whilst traveling to different waypoints during the test do not have a high correlation with data collected with the TSI DRX. A statistical analysis was conducted to investigate if the model could be improved by including other variables that affect the dispersion of dust, such as wind direction and UTM coordinates of the UAV. However, when the model was generated and used, the resulting predicted values were not correlated and produced a R^2 of 0.1–0.2.

Figure 17 presents the results of the concentrations calculated using the corrected concentration and raw field values. The comparison between TSI DRX values and the predicted TSI DRX values presents noise, and their R^2 indicates that 40%–60% of the data is represented by the model created. For Figure 17a with corrected values, errors were identified due to their negative value, and removed. The regression was recalculated obtaining an R^2 of 0.53. This value is closer to the coefficient obtained with the regression generated with the raw concentration values.

Table 7. Models generated with data collected with the TSI DRX and experimental modules using corrected concentrations and field data for Test 2.

	Corrected Concentration			Raw Concentration		
	Coefficients for Different Moving Average (s)					
Variable	15	20 *	30	15	20 *	30
Airborne module						
Intercept	296.29	319.96	214.72	182.79	221.04	182.19
Air speed (m/s)	−0.46	−0.77	−1.39	−1.91	−1.47	−0.67
Rel. Humidity (%)	−3.30	−3.58	−2.62	−2.35	−2.81	−2.44
Temperature (°C)	−3.66	−4.30	−3.01	−2.32	−3.06	−2.47
Conc. PM_{10} ($\mu g/m^3$)	−1.4	−1.11	−0.04	−4.39	−3.04	−1.72
Dist. to source (m)	*	*	*	0.63	0.70	0.61
Emission Rate ($\mu g/s$)	0.00018	0.0002	0.0002	0.0002	0.0002	0.0002
R^2	0.43	0.49 [†]/0.54	0.45	0.45	0.52 [†]/0.56	0.51

Note: Numbers in *italics* indicate a confidence level lower than 95%. * Distance value not used to obtain the best fit of the equation. [†] R^2 value before errors were eliminated.

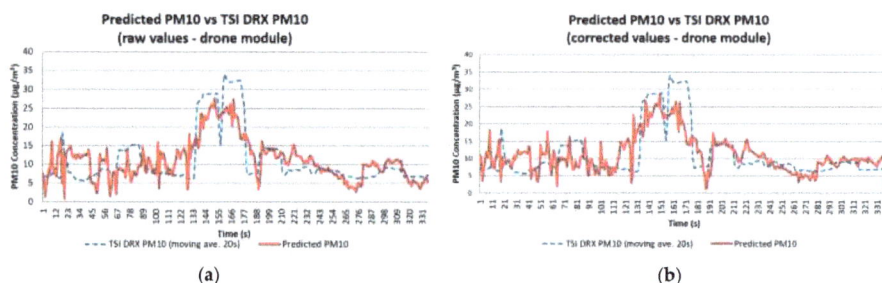

Figure 17. Comparison between PM_{10} concentration readings obtained with the TSI DRX and the predicted values calculated for Test 2 using the models generated with (**a**) raw values; and (**b**) corrected data.

The confidence level for all the variables used in the correlation of the raw data, was above 95%, indicating their high significance in the model to calculate dust emission concentrations. It was observed that for the emission rate field values, even though they had a value greater than 95%, by excluding this value from the model equation the resulting corrected values (used to calculate concentrations or emission rates) had a better fit. As mentioned before, this variation could be due to manual recording of the feeding rate.

4.3. Comparison of Average Emission Rates, Airborne Module Results Tests 1 and 2

Analysis of the zig zag dataset collected with the airborne module required isolating groups of waypoints corresponding to the 10 s and 15 s hovering periods programmed, and then averaging them by 10 s and 15 s, respectively. Once again, the moving average of the TSI DRX values together with the corrected values of the airborne module were calculated for periods of 10 s (only Test 2), 15 s, 20 s and 30 s. The percentage error between the reference device and the airborne module was calculated, to determine if the corrected concentration would be representative for the analysis. A total of 23 averaged hovering positions were used for Test 1 (Figure 18). Two thirds of 24 hovering positions collected were used from Test 2 (Figure 19). The hovering positions not considered for analysis were left out due to errors in the logging of data with the airborne module.

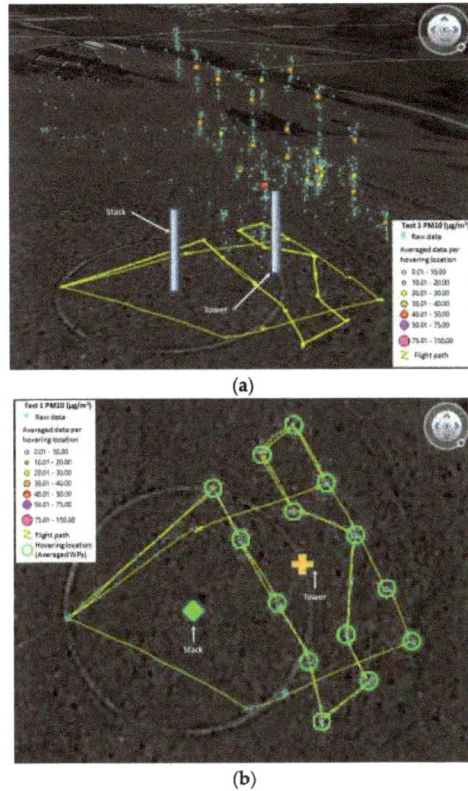

(a)

(b)

Figure 18. Field data collected for Test 1 with the airborne module before and after processing, (**a**) 3D view; and (**b**) top view with averaged locations indicated (repeated per height programmed).

As discussed in Section 4.2, concentrations of the airborne module are not considered comparable to the TSI DRX in Tests 1 and 2. Therefore, to validate the field values obtained, the average emission rate for the TSI DRX and the airborne module were calculated. Table 8 presents the results of the emission rates obtained with the readings of the TSI DRX and the corrected values of the airborne module. The emission rate was calculated using the general Gaussian equation for sources at ground level [20]:

$$Q = C \times 2\pi \times \sigma y \times \sigma z \times U, \tag{3}$$

With Q being the emission rate in $\mu g/s$, C is the concentration in $\mu g/m^3$, σy and σz the dispersion factors for a puff emission, and U the wind speed in m/s. A Gaussian model was used for this investigation due to its simplicity and extensive use for different modeling scenarios [21–23]. Background levels were estimated by reviewing the data from all flight tests and checking the minimum concentrations registered, which ranged from 0.0 $\mu m/m^3$ to 5.6 $\mu m/m^3$.

The values collected with the TSI DRX in Test 1 presented errors ranging from 8.9% to 192.9% when compared against an estimated average field emission rate of 20,449.15 $\mu g/s$.

The lower value corresponded to a moving average of 20 s. The percentage errors produced by comparing the airborne module against the field emission rate have smaller differences ranging from 8.9% to 9.8%. The lowest percentage error corresponds to the emission rate calculated with a moving average of 90s. A high variability in the emission rates was observed between values calculated with the TSI DRX and airborne module.

(a)

(b)

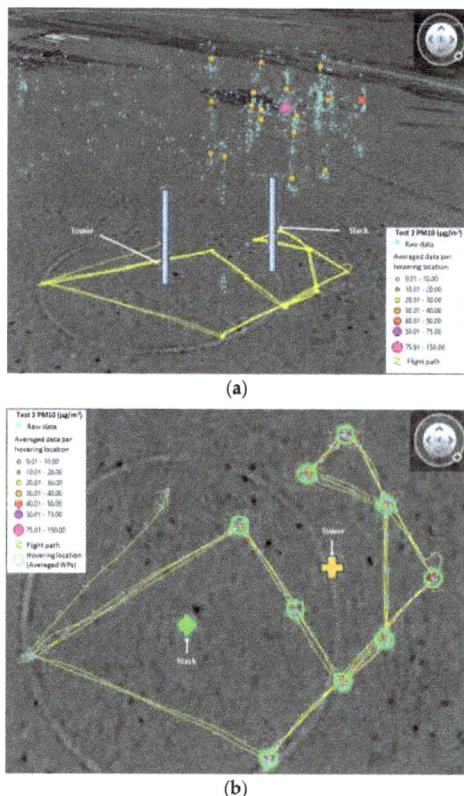

Figure 19. Field data collected for Test 2 with the airborne module before and after processing, (**a**) 3D view; and (**b**) top view with averaged locations indicated (repeated per height programmed).

Calculations of the emission rate for Test 1 were performed considering an atmospheric stability class D. The average wind speed of 3.0 m/s observed in this Test and this should correspond to a stability class B. However, wind speed values were measured at a height of 3.0 m (instead of using the standard height of 10.0 m), and friction with the ground would slow the wind down resulting in a different atmospheric stability. Considering this situation, a stability class D, better described the atmospheric conditions. Emission rates calculated with stability class B parameters produced errors ranging from 150% to 1333%, hence being considered not applicable for the analysis and interpretation of data.

Emission rates were calculated using stability class D parameters, having an average wind speed of 5.3 m/s for Test 2. Error values varied from 144.6% to 181.7% between emission rates calculated with the concentrations from the TSI DRX and the airborne module (Table 8). However, when comparing the emission rates against the average field emission rate of 15,383.67 $\mu g/m^3$, the airborne module had percentage errors ranging from 18.5% to 28.4%. Even though these values overestimated the field value, they are closer than the emissions calculated using the values obtained by the TSI DRX located in the tower, ranging from 47.8% to 56.1%.

The raw values of the airborne module, for Tests 1 and 2, presented higher errors in their emission rates. This indicated that the correction made to the readings with the model developed in the laboratory, produced a better fit, having a higher correlation between the real value and the value obtained with the airborne module.

Table 8. Summary of concentrations and emission rates calculated for Tests 1 and 2 with their percentage error.

Measurement	Average Concentration (µg/m³)	Conc. Error (%), TSI DRX vs. Airborne Module	Emission Rate (µg/s)	Emission Error (%), Calculated Value vs. Field Value	Average Concentration (µg/m³)	Conc. Error (%), TSI DRX vs. Airborne Module	Emission Rate (µg/s)	Emission Error (%), Calculated Value vs. Field Value
	Test 1				Test 2			
TSI DRX								
Raw data	18.579	NA	10,646.214	47.9	14.339	NA	7213.353	53.1
10 s Mov. Ave.	NA *	NA	NA *	NA *	13.766	NA	6924.893	54.9
15 s Mov. Ave.	29.165	NA	16,712.084	18.27	13.874	NA	6979.260	54.6
20 s Mov. Ave.	32.500	NA	18,623.284	8.93	13.504	NA	6793.132	55.8
30 s Mov. Ave.	39.484	NA	22,625.078	10.64	13.414	NA	6747.759	56.1
60 s Mov. Ave.	93.055	NA	53,322.700	160.76	14.377	NA	7232.394	52.9
90 s Mov. Ave.	104.552	NA	59,911.108	192.98	15.952	NA	8024.461	47.8
Airborne module								
Raw data	3.201	82.7	1834.189	91.0	1.549	89.29	779.523	94.9
10 s Mov. Ave. Corr.	NA *	NA *	NA *	NA *	36.245	163.39	18,232.334	18.5
15 s Mov. Ave. Corr.	32.336	10.9	18,529.561	9.4	36.373	162.2	18,296.784	18.9
20 s Mov. Ave. Corr.	32.129	1.1	18,410.932	9.8	36.559	170.7	18,390.435	19.6
30 s Mov. Ave. Corr.	32.357	18.8	18,541.318	9.3	37.782	181.7	19,005.532	23.5
60 s Mov. Ave. Corr.	32.382	65.2	18,555.661	9.3	39.258	173.1	19,748.159	28.4
90 s Mov. Ave. Corr.	32.487	68.9	18,615.632	8.9	39.012	144.7	19,624.319	27.6

NA = Not Applicable. Mov. Ave. = Moving Average. Mov. Ave. Corr. = Moving Average of Corrected data from the airborne module. * A Moving Average of 10s was calculated only for Test 3 as the hovering period for each way point programmed was of 10 s, for Test 1 the hovering period was of 15 s.

4.4. Comparison of Emission Rates, Airborne Module Results for Tests 3 and 4

The triangle flight pattern tests required different approach for their analysis. The main objective was to collect as much data close to the TSI DRX and the collocated module by flying to the sides and behind the monitoring devices. Once the data was collected, it was grouped into high density areas. This resulted in six areas for Test 3 and five areas for Test 4 (Figures 20 and 21). The moving average of all values was calculated considering intervals of 15 s, 20 s, 30 s, 60 s and 90 s. Then, each area was isolated and averaged per hovering period. This was done to determine the corrections with least error against the TSI DRX data. After observing the data in Test 4, it was considered necessary to eliminate an area located west of the devices and another located south of them, due to a high amount of blank readings (Figure 21). The other two areas located west and south of the devices presented very few blank logging records. All other areas selected for Tests 3 and 4 did not present drop-outs of sensor packets. The number of waypoints averaged per area defined in both tests ranged from 25 to 175, each waypoint representing a reading taken in one second.

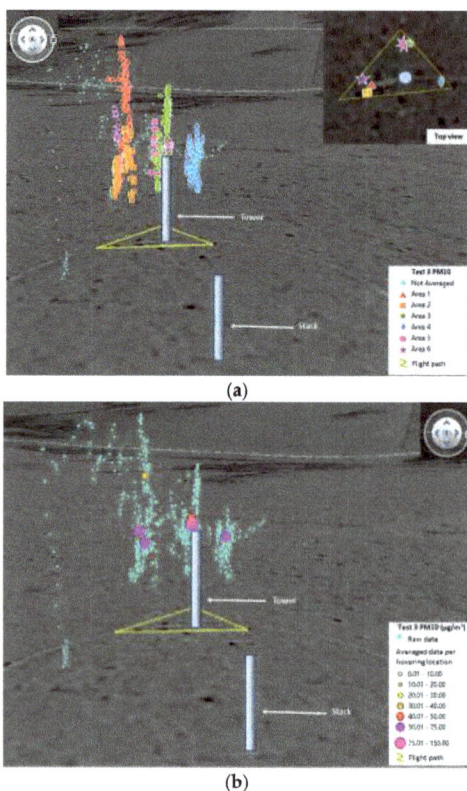

(a)

(b)

Figure 20. 3D visualization for Test 3, showing (**a**) the waypoints grouped per high density areas; and (**b**) the raw data and waypoints used once averaged per area.

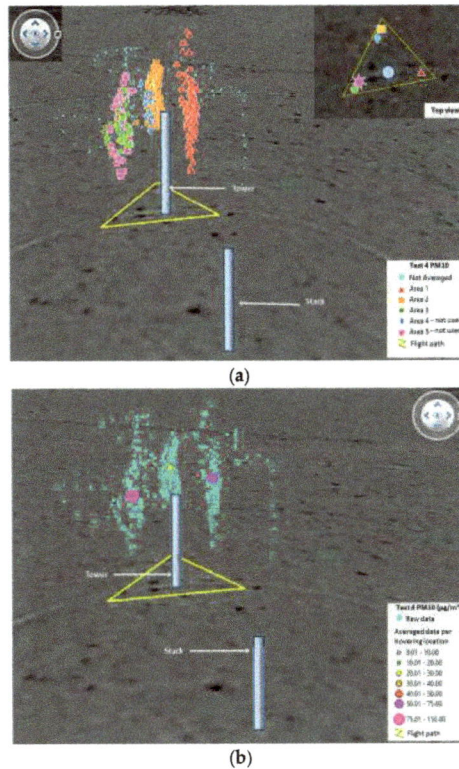

Figure 21. 3D visualization for Test 4, showing (**a**) the waypoints grouped per high density areas; and (**b**) the raw data and way points used once averaged per area.

Table 9 shows the resulting concentrations and emission rates obtained with the TSI DRX and calculated with the airborne module, as well as the calculated emission rate and their corresponding percentage error. The average estimated field value against which the airborne module values are compared was of 22,589.09 µg/s for Test 3 and of 50,864.86 µg/s for Test 4. For both tests, atmospheric stability class D was used in the calculation of emission rates, with average wind speeds of 5.2 m/s for Test 3 and of 5.1 m/s for Test 4. From Table 9 can be determined, in a similar way to Tests 1 and 2, that the values that produce the highest errors in the data are the raw values, indicating that the use of the moving average and the correction equation improved the precision of the readings considerably.

Emission rates calculated with the airborne module obtained errors as low as 1.2% in Test 3, and had the highest error in Test 4 at 43.3%. Percentage errors observed in Test 4 are considerably higher than errors presented by Test 3. This could be due mainly to the amount of data available to calculate a final emission rate with Test 3 having a complete dataset, and Test 4 having many points not considered due to errors in the log files. However, even with less data available the airborne module had a minimum error of 40.8% which is similar to errors obtained in the other tests with the TSI DRX, indicating data can still be considered useful.

Average concentrations obtained with the TSI DRX and the airborne module for both tests, have similar errors for moving averages of 15 s to 30 s, with higher variability in the 60 s and 90 s periods. This could be due to the lower amount of information in Test 4. The same behavior was also observed in Tests 1 and 2. This can lead us to the conclusion that 15 s to 30 s periods can be chosen as the preferred time periods.

Table 9. Summary of concentrations and emission rates calculated for the triangle test with their percentage error.

Measurement	Average Concentration (µg/m³)	Conc. error (%), TSI DRX vs. Airborne Module	Emission Rate (µg/s)	Emission error (%), Calculated Value vs. Field Value	Average Concentration (µg/m³)	Conc. Error (%), TSI DRX vs. Airborne Module	Emission Rate (µg/s)	Emission Error (%), Calculated Value vs. Field Value
	Test 3				Test 4			
TSI DRX								
Raw data	67.263	NA	25,328.05	12.1	76.610	NA	37,452.015	26.4
15s Mov. Ave.	57.832	NA	21,777.01	3.6	55.746	NA	27,252.411	46.4
20s Mov. Ave.	50.777	NA	19,120.22	15.4	54.135	NA	26,464.745	47.9
30s Mov. Ave.	43.630	NA	16,428.88	27.3	51.831	NA	25,338.413	50.2
60s Mov. Ave.	36.639	NA	13,796.32	38.9	51.197	NA	25,028.419	50.8
90s Mov. Ave.	34.733	NA	13,078.78	42.1	48.600	NA	23,758.769	53.3
Airborne module								
Raw data	9.2702	86.2	3490.715	84.5	8.464	88.95	4137.911	91.9
15s Mov. Ave. Corr.	61.399	06.2	23119.81	2.3	60.952	9.34	29,797.558	41.4
20s Mov. Ave. Corr.	59.276	16.7	22320.61	1.2	61.017	12.71	29,829.075	41.4
30s Mov. Ave. Corr.	55.987	28.3	21081.93	6.7	61.204	18.08	29,920.592	41.2
60s Mov. Ave. Corr.	55.455	51.4	20881.7	7.6	61.548	20.22	30,088.789	40.8
90s Mov. Ave. Corr.	54.424	56.7	20,493.45	9.3	59.052	21.51	28,868.344	43.3

NA = Not Applicable. Mov. Ave. = Moving Average. Mov. Ave. Corr. = Moving Average of Corrected data from the airborne module.

5. Discussion

Analysis of the four available datasets demonstrated that very good estimations could be calculated, with percentage errors as low as 1.1% when compared to field emission rates. The use of different grids, hovering periods, moving average periods and the collection of environmental parameters, as well as UAV telemetry data from several sources, produced a very rich dataset.

Analysis of the particle size distribution during the variable wind experiment demonstrated that particles with a diameter up to approximately 2.0 μm dominated the sample when using the sampling probe. An average difference of 7.6% was observed to affect PM_{10} readings when the sampling probe was used and similar behavior was observed when analyzing the effect of variable wind speed in the OPC-N2 with and without the sampling probe. The particles with diameters ranging from 0.38–0.54 μm were also dominant. However, results from the models generated suggest that these impacts where corrected to obtain comparable results with field values.

Having observed that correlations were only possible if a moving average function analysis was used, different time periods were investigated to balance noise reduction, flight time and area covered. Moving averages of 15 s to 30 s produce consistent results and had some of the lowest percentage errors. On the other hand, it was observed that periods of 60 s to 90 s can produce noticeable variability due to the constant movement of the multi-rotor UAV which increased the difference between data collected as the sampling period increased.

It was expected that a moving average of 15 s or 10 s would produce a better correlation for Test 1 and Test 2 respectively, as these were the hovering sampling periods for which they were programmed. However, it was observed that only Test 2, with the airborne module, produced a reasonable correlation. This could be due to the influence of other environmental variables such as wind speed, wind direction, dust supply rate, etc. For both Tests 1 and 2, the raw values of the airborne module presented the higher errors in the calculation of emission rates. This indicated that the correction made to the readings with the model developed in the laboratory does increase the correlation between the real value and the value obtained with the airborne module. Lower percentage errors resulted when correcting the values calculated with the airborne module with Equations (1) and (2), due to an overestimation observed in all values once processed, when compared with the values of the TSI DRX.

The lowest errors produced when comparing calculated emission rates from the airborne module and the field values, were produced by Test 1 and Test 4. These results indicate that by hovering for short periods of time (10 s to 15 s) it is possible to produce estimations very close to the field values. Another advantage of hovering for short periods of time is that the battery can be used more efficiently, and more hovering locations can be covered within its flying endurance. Having the UAV hovering close to the devices in the tower can assist in concluding that the concentrations corrected with the model developed in the laboratory and the field values were comparable, with a low error between them.

It was possible to identify sources of error, such as: the synchronization of sensors and devices, calibration of sensors, calculation of correction factors and models, individual specifications of sensors (response time, precision, and resistance to exposure in the environment) through the analysis of all data. However, errors obtained for the tests analysed were low and it was important to consider that the methodology tested was a relative monitoring procedure; and not an absolute method such as a gravimetric test.

6. Conclusions

Four datasets were analysed for this investigation. With the current findings, it can be concluded that calculation of emission factors for specific activities in the industrial sector is possible making use of UAV technology. By using the proposed methodology, it is possible to generate emission factors per mine site and activity targeted.

The results of the four tests indicate that the airborne module can be used to obtain PM_{10} emission rates comparable to a monitoring system located in a tower with the major advantage that the airborne module can be transported to any location, at any time, weather conditions permitting.

Work will continue to include other variables in the modeling analysis. This includes micro and local meteorological data, more telemetry data, and a different modeling approach using artificial neural networks to include more variables. Artificial neural networks will increase the reliability of the model not only by linking a greater number of significant variables but also by providing feedback from new datasets generated.

Acknowledgments: The authors wish to thank Andrew Fletcher, Nathan Unwin, Ashray Doshi, Karl O'Neil, Sam Downs, Armando Navas, Juan Jose Frausto, and Vinod Nath. The authors would like to acknowledge Japan Coal Development Australia Grant, People Centre and Environment Centre at the University of Queensland Sustainable Mineral Institute.

Author Contributions: Miguel Angel Alvarado Molina was responsible for conducting the literature review, design and coordination of experimental procedures. He had the main role in the writing of this paper. Felipe Gonzalez provided advice on experimental design and data interpretation. He also provided guidance on the structure and editing of the research article. David Cliff and Peter Erskine provided overall project guidance and advice on experimental design and data interpretation. They also provided guidance regarding the formatting and editing of the research article. Darlene Heuff provided advice on experimental design and data interpretation.

Conflicts of Interest: The authors declare no conflict of interest.

References

1. Hening, S.; Baumgartner, J.; Walden, C.; Kirmayer, R.; Teodorescu, M.; Nguyen, N.; Ippolito, C. Distributed sampling using small unmanned aerial vehicles (UAVS) for scientific missions. In Proceedings of the AIAA Infotech@Aerospace 2013 Conference, Boston, MA, USA, 19–22 August 2013.
2. Danilov, A.; Smirnov, U.D.; Pashkevich, M. The system of the ecological monitoring of environment which is based on the usage of UAV. *Russ. J. Ecol.* **2015**, *46*, 14–19. [CrossRef]
3. Haas, P.; Balistreri, C.; Pontelandolfo, P.; Triscone, G.; Pekoz, H.; Pignatiello, A. Development of an unmanned aerial vehicle UAV for air quality measurements in urban areas. In Proceedings of the 32nd AIAA Applied Aerodynamics Conference, Atlanta, GA, USA, 16–20 June 2014.
4. Gonzalez, L.F.; Lee, D.; Walker, R.A.; Periaux, J. Optimal mission path planning (MPP) for an air sampling unmanned aerial system. In Proceedings of the Austral-Asian Conference on Robotics and Automation, Sydney, Australia, 2–4 December 2009.
5. Gonzalez, L.F.; Castro, M.P.; Tamagnone, F.F. Multidisciplinary design and flight testing of a remote gas/particle airborne sensor system. In Proceedings of the 28th International Congress of the Aeronautical Sciences, Brisbane, Australia, 23–28 September 2012; Optimage Ltd.: Brisbane, Australia, 2012; pp. 1–13.
6. Malaver, A.; Gonzalez, F.; Motta, N.; Depari, A.; Corke, P. Towards the Development of a Gas Sensor System for Monitoring Pollutant Gases in the Low Troposphere Using Small Unmanned Aerial Vehicles. In Proceedings of the Workshop on Robotics for Environmental Monitoring, Sydney, Australia, 11 July 2012.
7. Malaver Rojas, A.J.; Gonzalez, L.F.; Motta, N.; Villa, T.F. Design and flight testing of an integrated solar powered UAV and WSN for remote gas sensing. In Proceedings of the 2015 IEEE Aerospace Conference, Big Sky, MT, USA, 7–14 March 2015.
8. Krüll, W.; Tobera, R.; Willms, I.; Essen, H.; von Wahl, N. Early Forest Fire Detection and Verification using Optical Smoke, Gas and Microwave Sensors. *Procedia Eng.* **2012**, *45*, 584–594. [CrossRef]
9. Alvarado, M.; Gonzalez, F.; Fletcher, A.; Doshi, A. Towards the development of a low cost airborne sensing system to monitor dust particles after blasting at open-pit mine sites. *Sensors* **2015**, *15*, 19667–19687. [CrossRef] [PubMed]
10. Alvarado, M.; Gonzalez, F.; Fletcher, A.; Doshi, A. Correction: Alvarado, M., et al. Towards the Development of a Low Cost Airborne Sensing System to Monitor Dust Particles after Blasting at Open-Pit Mine Sites. *Sensors* **2015**, *15*, 19667–19687. *Sensors* **2016**, *16*, 1028. [CrossRef] [PubMed]
11. Alphasense. Alphasense User Manual. In *OPC-N2 Optical Particle Counter*, 5th ed.; Alphasense Ltd.: Essex, UK, 2015.

12. Von der Weiden, S.L.; Drewnick, F.; Borrmann, S. Particle Loss Calculator—A new software tool for the assessment of the performance of aerosol inlet systems. *Atmos. Meas. Tech.* **2009**, *2*, 479–494. [CrossRef]

13. Lodge, J.P. *Methods of Air Sampling and Analysis*; Taylor & Francis: London, UK, 1988.

14. Pena, J.; Norman, J.; Thomson, D. Isokinetic sampler for continuous airborne aerosol measurements. *J. Air Pollut. Control Assoc.* **1977**, *27*, 337–341. [CrossRef]

15. Wilcox, J.D. Isokinetic Flow and Sampling. *J. Air Pollut. Control Assoc.* **1956**, *5*, 226–245. [CrossRef]

16. Huebert, B.; Lee, G.; Warren, W. Airborne aerosol inlet passing efficiency measurement. *J. Geophys. Res. Atmos.* **1990**, *95*, 16369–16381. [CrossRef]

17. Irshad, H.; McFarland, A.R.; Landis, M.S.; Stevens, R.K. Wind Tunnel Evaluation of an Aircraft-Borne Sampling System. *Aerosol Sci. Technol.* **2004**, *38*, 311–321. [CrossRef]

18. TSI. Operation and Service Manual, Dusttrak™ DRX. Available online: http://www.tsi.com/uploadedFiles/_Site_Root/Products/Literature/Manuals/8533--8534-DustTrak_DRX-6001898-web.pdf (accessed on 1 January 2017).

19. Napier-Munn, T. *Statistical Methods for Mineral Engineers—How to Design Experiments and Analyse Data*; Julius Kruttschnitt Mineral Research Centre: Brisbane, Australia, 2014; Volume 5.

20. Barrat, R. *Atmospheric Dispersion Modelling, an Introduction to Practical Applications*; Earthscan Publications Ltd.: Sterling, VA, USA, 2001; p. 151.

21. Roddis, D.; Laing, G.; Boulter, P.; Cox, J. *ACARP Project C22027—Development of Australia-Specific PM$_{10}$ Emission Factors for Cal Mines*; 06961P; Australian Coal Association Research Program (ACARP): Brisbane, Australia, 2015.

22. Visscher, A.D. Gaussian Dispersion Modeling. In *Air Dispersion Modeling*; John Wiley & Sons, Inc.: New York, NY, USA, 2013; pp. 141–200.

23. Chakraborty, M.K.; Ahmad, M.; Singh, R.S.; Pal, D.; Bandopadhyay, C.; Chaulya, S.K. Determination of the emission rate from various opencast mining operations. *Environ. Model. Softw.* **2002**, *17*, 467–480. [CrossRef]

    MDPI

Article

Two-UAV Intersection Localization System Based on the Airborne Optoelectronic Platform

Guanbing Bai [1,2], Jinghong Liu [1,*], Yueming Song [1] and Yujia Zuo [1,2]

[1]   Chinese Academy of Science, Changchun Institute of Optics Fine Mechanics and Physics, Key Laboratory of Airborne Optical Imaging and Measurement, #3888 Dongnanhu Road, Changchun 130033, China; baigb@ciomp.ac.cn (G.B.); himyf0319@126.com (Y.S.); mzyj0617@126.com (Y.Z.)

[2]   University of Chinese Academy of Sciences, #19 Yuquan Road, Beijing 100049, China

*   Correspondence: liu1577@126.com; Tel.: +86-177-4301-7276

Academic Editors: Felipe Gonzalez Toro and Antonios Tsourdos
Received: 24 October 2016; Accepted: 22 December 2016; Published: 6 January 2017

Abstract: To address the limitation of the existing UAV (unmanned aerial vehicles) photoelectric localization method used for moving objects, this paper proposes an improved two-UAV intersection localization system based on airborne optoelectronic platforms by using the crossed-angle localization method of photoelectric theodolites for reference. This paper introduces the makeup and operating principle of intersection localization system, creates auxiliary coordinate systems, transforms the LOS (line of sight, from the UAV to the target) vectors into homogeneous coordinates, and establishes a two-UAV intersection localization model. In this paper, the influence of the positional relationship between UAVs and the target on localization accuracy has been studied in detail to obtain an ideal measuring position and the optimal localization position where the optimal intersection angle is 72.6318°. The result shows that, given the optimal position, the localization root mean square error (RMS) will be 25.0235 m when the target is 5 km away from UAV baselines. Finally, the influence of modified adaptive Kalman filtering on localization results is analyzed, and an appropriate filtering model is established to reduce the localization RMS error to 15.7983 m. Finally, An outfield experiment was carried out and obtained the optimal results: $\sigma_B = 1.63 \times 10^{-4}$ (°), $\sigma_L = 1.35 \times 10^{-4}$ (°), $\sigma_H = 15.8$ (m), $\sigma_{sum} = 27.6$ (m), where σ_B represents the longitude error, σ_L represents the latitude error, σ_H represents the altitude error, and σ_{sum} represents the error radius.

Keywords: UAV (unmanned aerial vehicles); airborne optoelectronic platform; intersection localization; coordinate transformation; accuracy analysis; adaptive Kalman filtering

1. Introduction

As an important tool for localization, photoelectric measuring equipment is playing an important role in military and civilian applications [1–4]. According to the base variety, modern photoelectric measuring equipment is mainly divided into ground-based photoelectric theodolites, surveying vessels and airborne photoelectric platforms. However, when being used in actual reconnaissance and localization, photoelectric theodolites and surveying vessels are often affected by operating range and other factors so that they cannot track and locate the targets all of the way. In this context, owing to the high maneuverability of UAVs, airborne photoelectric platforms are playing a more and more important role in reconnaissance and localization [5–9]. In recent years, with rapid development of UAV (unmanned aerial vehicles) technology, UAV-borne optoelectronic positioning devices have found wider and wider use in reconnaissance and monitoring, and have received more and more researchers' attention [10–13]. Traditionally, an airborne photoelectric platform works in a single-station REA (range, pitch, and azimuth) localization manner, that is, to locate the target by using the distance R (of the target in relation to the platform), pitch angle E, and azimuth angle A, measured by the platform [14], as well as the position and attitude measured by airborne GPS (Global Positioning System)/INS (inertial navigation system). However, this method has a limited positioning accuracy, and as the laser range finder (LRF) it relies on has a limited operating range, this method has a narrower use. Therefore, a new localization method is needed to meet the requirement of high-precision localization.

Hosseinpoor et al. [15,16] used a UAV with RTK (real-time kinematic)-GPS for estimation and localization, and processed the localization results through extended Kalman filtering. Their method is characterized by smooth localization results, simple and achievable equipment, but with limited positioning accuracy. Conte et al. [17] achieved the target localization through a micro aerial vehicle (MAV), a vehicle applicable to ground targets within a short range. Frew [18] located the ground targets through two-UAV cooperative localization. Ross et al. [19] identified and located the ground targets through a real-time algorithm and demonstrated, through testing, a significant influence of GPS accuracy on localization accuracy. Sohn et al. [20] proposed the use of triangulation for target localization. Cheng et al. [21] located a static ground target through two-point intersection localization, and proposed the use of least-squares iteration to improve the localization accuracy. Sharma et al. [22] located moving targets on the ground by assuming zero target altitude, and improved the localization accuracy through expanded Kalman filtering. This method, however, applies to the localization only on a flat ground area. The above methods are more for near-distance ground targets and less for remote airborne targets.

To deal with the above issues, this paper proposes an improved intersection localization method. Learning from the intersection measurement method of ground-based photoelectric theodolites [4], two optoelectronic platforms with angulation function are used simultaneously to measure the pitch and azimuth angles of the target in relation to platform coordinate system. Then the angle information is integrated with the information on UAV position and attitude to determine the azimuth/pitch angles of the LOS (line of sight, from the UAV to the target) in relation to terrestrial rectangular coordinates [23]. Finally, the target location can be obtained through intersection measurement. Independent of the movement style of targets, this system can locate airborne and ground-based moving objects in a real-time manner. Since the distance between the target and the platform is not needed in localization resolving, the installation of LRF in the optoelectronic platform is not necessary. This can effectively reduce the load on UAVs, escape from the LRF range restriction, and expand the localization applicability.

2. Introduction of the Traditional Single-Station REA Localization Method

Over the past decade the most popular algorithm used in airborne positioning is the single-station REA positioning method [9,16,17,22], which is shown in Figure 1. In Figure 1, f is the focal length, T represents the actual target point, and T_i represents the position of the target point in the image whose coordinates in the image are expressed as (u_x, u_y). According to the method illustrated in [16], the three-dimensional coordinates of the target in the camera coordinate can be obtained:

$$\begin{bmatrix} x_t \\ y_t \\ z_t \end{bmatrix} = \frac{R}{F} \begin{bmatrix} u_x \\ u_y \\ f \end{bmatrix} \tag{1}$$

where R is the distance from the camera to the target point, and F represents the distance from the origin of the camera coordinate to the target image point, $F = \sqrt{u_x^2 + u_y^2 + f^2}$. The main problem of location is how to get the value of R with other known parameters. If the measured region is a flat ground surface, the value of R can be estimated by the method described in [22]. The distance between the camera and the ground plane R is estimated from the direction vector of LOS in the ground coordinate system. This method can effectively control the cost of the system and reduce the weight and volume of the platform. However, the precision of this algorithm is limited, thus it cannot meet the high-precision target location requirement in the UAV reconnaissance field. By using a LRF mounted inside the optoelectronic platform, which is used to achieve the distance between the measurement object and the platform, the value of R can also be obtained. Although this method is simple and direct, and has a high accuracy, adding an LRF in the platform will obviously increase the weight and volume of the photoelectric platform, which requires the load capacity of the UAV to be large enough.

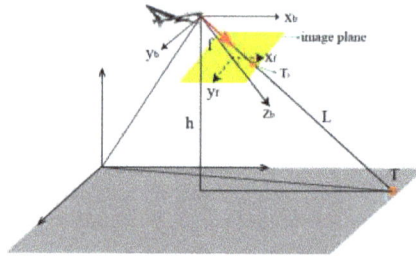

Figure 1. Schematic diagram of single-station localization.

This paper proposes a two-UAV intersection localization method to solve the problem of single-station positioning, which uses two devices locating the target at the same time to solve the problem of the localization that is needed to estimate the distance R. Compared with the single-station positioning algorithm, the two-UAV intersection localization algorithm has a higher positioning accuracy under the same measurement environment, which is verified in Section 5.3.

In summary, compared with the commonly used single-station localization method, the localization system proposed in this paper has higher location accuracy. Moreover, it can be applied in more types of UAVs since the platform used is lighter. However, it also has its own shortcomings, that is, more technical difficulty and higher cost of flight.

3. Makeup and Operating Principle of Two-UAV Intersection Localization System

This system is mainly made up of two UAVs and their onboard optoelectronic platforms, GPS, and INS. As shown in the Figure 2, there are a stabilized platform, photoelectric rotary encoder, infrared photography sensor, and infrared photography sensor in the onboard optoelectronic platform. Where the photoelectric rotary encoders are used to measure the pitch angle and azimuth angle of the photography sensors, the photography sensors are used to obtain image data and provide information on the miss distance. The stabilized platform is made up of servo motors and gyroscopes, which can keep the photography sensors steady and control the platform to rotate according to the orders. The platform has two degrees of freedom, i.e., pitch and azimuth rotation, with the functions of reconnaissance, tracking, and localization.

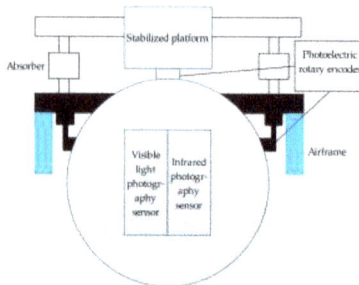

Figure 2. Structure chart of the onboard optoelectronic platform.

As shown in Figure 3, once the target is detected, the optoelectronic platform locks the target to the FOV (field of view) center and keeps tracking. During the tracking process, once the image target deviates from the center of vision due to the UAV and target relative position changes, the system can measure the miss distance (u_x, u_y) and transmit the miss distance to the control computer in real-time. According to the miss distance data, the platform servo control system can adjust the pitch and azimuth angles of the platform immediately to relock the target to the center of the FOV, in this case the target image has been ensured to be located in the center of the FOV. The optoelectronic platform locks the target to the FOV center and simultaneously measures the pitch and azimuth angles of LOS in relation to attitude measurement system. At the same time, the GPS/INS positioning system outputs the information on position and attitude of the two UAVs. Through the homogeneous-coordinates transformation, the output information is fused into a uniform coordinate system, where the target coordinates are solved through the intersection algorithm.

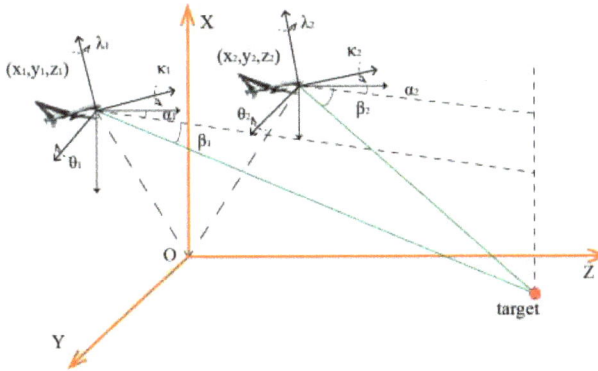

Figure 3. Schematic diagram of two-UAV intersection localization.

4. Key Technologies of Two-UAV Intersection Localization System

4.1. Establishment of Space Coordinates

The system described in this paper has six sets of space coordinates, as described below:

1. Geodetic coordinate system $C(\lambda_s, \alpha_s, h_s)$: based on international terrestrial reference system WGS-84. The position of any spatial point is expressed by longitude, latitude, and geodetic height $(\lambda_s, \alpha_s, h_s)$ [24].
2. Terrestrial rectangular coordinate system $G(O_g - X_g Y_g Z_g)$: an inertial coordinate system, as shown in the Figure 4a, where any spatial position is described by (x, y, z). The origin is the center of Earth's mass. The axis Z_g points to the North Pole, and the axis X_g is directed to the intersection point of the Greenwich meridian plane and equator. The axis Y_g is normal to the plane $X_g O_g Z_g$ and constitutes, along with the axes Z_g and X_g, a Cartesian coordinate system.
3. Geographic coordinate system of UAV $S(O_s - X_s Y_s Z_s)$: as shown in the Figure 4a, the origin is the position $(\lambda_s, \alpha_s, h_s)$ of a UAV at a certain moment, the Z_s points to true north, the X_s points to zenith, and the Y_s, along with Z_s and X_s, constitutes a right-handed coordinate system.
4. UAV coordinate system $A(O_a - X_a Y_a Z_a)$: as shown in the Figure 4b, the origin of this coordinate system coincides with that of UAV geographic coordinate system, the X_a points to the direction right above the aircraft, the Z_a points to the aircraft nose, and the Y_a, along with X_a and Z_a, forms a right-handed coordinate system. The relationship between the UAV coordinate system and geographic coordinate system is shown in the Figure 4b. The tri-axial attitude angles are λ, θ, κ measured by the inertial navigation system.

5. Camera coordinate system $T(O_t - X_tY_tZ_t)$: the origin is the intersection of the LOS and horizontal platform axis, and axis Z_t is the telescope's optic axis pointing to the target. When the axis Z_t is in the initial (or horizontal) position, the axis X_t will be directed to zenith, and the axis Y_t, along with Z_t and X_t, will constitute a right-handed coordinate system. Figure 4c shows the relationship between camera coordinate system and UAV coordinate system.
6. Reference coordinate system $R(O_r - X_rY_rZ_r)$: an auxiliary coordinate system built to facilitate intersection resolution. The origin is a definite point on the Earth ellipsoid, and the tri-axial directions are the same as in the Earth-rectangular coordinate system.

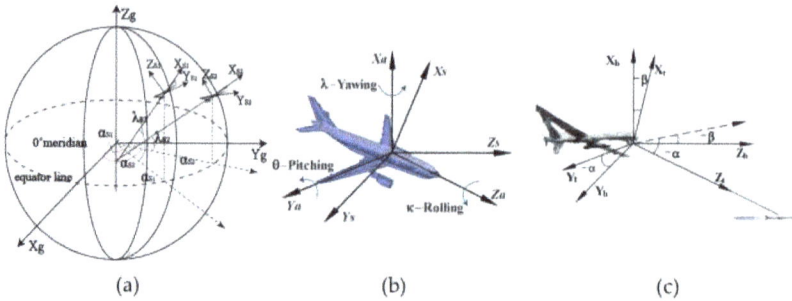

Figure 4. Definition of coordinate systems and their relations. (**a**) correlation diagram of terrestrial rectangular coordinate and geographic coordinate; (**b**) correlation diagram of geographic coordinateand UAV coordinate; (**c**) correlation diagram of UAV coordinate and camera coordinate.

The geodetic coordinate system C can be converted into the terrestrial rectangular coordinate system G in accordance with the following equation:

$$\begin{cases} x_g = (N + h_s) \cos \alpha_s \cos \lambda_s \\ y_g = (N + h_s) \cos \alpha_s \sin \lambda_s \\ z_g = [N(1 - e^2) + h_s] \sin \alpha_s \end{cases} \tag{2}$$

The terrestrial rectangular coordinate system G can be converted into the geodetic coordinate system C in accordance with the following equation:

$$\begin{aligned} \lambda_s &= a \tan\left(\frac{y_g}{x_g}\right) \\ \alpha_s &= a \tan\left(\frac{z_g + e_1^2 b \sin^3 \lambda_s}{\sqrt{x_g^2 + y_g^2} - e^2 a \cos^3 \lambda_s}\right) \\ h_s &= \frac{\sqrt{x_g^2 + y_g^2}}{\cos \alpha_s} - N \end{aligned} \tag{3}$$

where a is the length of the semi-major axis of the reference spheroid, and b is the length of semi-minor axis of the reference spheroid.

The first eccentricity: $e = \sqrt{\frac{a^2 - b^2}{a^2}}$

The second eccentricity: $e_1 = \sqrt{\frac{a^2 - b^2}{b^2}}$

Radius of curvature in the prime vertical: $N = \frac{a}{\sqrt{1 - e^2 \sin^2 \alpha_s}}$.

As shown in the Figure 4c, the camera coordinate system T revolves around Y_t for $-\beta$ and around X_t for $-\alpha$ to become the UAV coordinate system A in accordance with the principle of homogeneous-coordinates conversion:

$$\mathbf{R}_t^a(\alpha,\beta) = \mathbf{R}(\alpha)\mathbf{R}(\beta) = \begin{bmatrix} C_\beta & 0 & S_\beta & 0 \\ S_\alpha S_\beta & C_\alpha & -S_\alpha C_\beta & 0 \\ -C_\alpha S_\beta & S_\alpha & C_\alpha C_\beta & 0 \\ 0 & 0 & 0 & 1 \end{bmatrix} \tag{4}$$

where $C_\alpha = \cos\alpha$, $S_\alpha = \sin\alpha$, and both α and β are the position angles in the UAV coordinate system. As shown in Figure 4b, the UAV coordinate system A revolves around Z_a for κ, around Y_a for θ, and around X_a for λ to become the geographic coordinate system S:

$$\mathbf{R}_a^S(\lambda,\theta,\kappa) = \mathbf{R}(\lambda)\mathbf{R}(\theta)\mathbf{R}(\kappa) = \begin{bmatrix} C_\theta C_\kappa & C_\theta S_\kappa & -S_\theta & 0 \\ S_\lambda S_\theta C_\kappa - C_\lambda S_\kappa & S_\lambda S_\theta S_\kappa + C_\lambda C_\kappa & S_\lambda C_\theta & 0 \\ C_\lambda S_\theta C_\kappa + S_\lambda S_\kappa & C_\lambda S_\theta S_\kappa - S_\lambda C_\kappa & C_\lambda C_\theta & 0 \\ 0 & 0 & 0 & 1 \end{bmatrix} \tag{5}$$

where λ, θ, κ are tri-axial attitude angles of UAV coordinate system in relation to geographic coordinate system [25].

As shown in Figure 4a, the geographic coordinate system S can be converted into the terrestrial rectangular coordinate system G through a shift of h_s along the axis X_s, a rotation of λ_s around Y_s, a rotation of $-\alpha_s$ around Z_s, and a shift of $-Ne^2 \sin\lambda_s$ along the axis Z_s:

$$\mathbf{R}_S^g(h_s,\lambda_s,\alpha_s,-Ne^2\sin\lambda_s) = \begin{bmatrix} C_{\alpha_s}C_{\lambda_s} & -S_{\alpha_s} & -C_{\alpha_s}S_{\lambda_s} & h_s \\ S_{\alpha_s}C_{\lambda_s} & C_{\alpha_s} & -S_{\alpha_s}S_{\lambda_s} & 0 \\ S_{\lambda_s} & 0 & C_{\lambda_s} & -Ne^2\sin\lambda_s \\ 0 & 0 & 0 & 1 \end{bmatrix} \tag{6}$$

The terrestrial rectangular coordinate system G can be converted into the reference coordinate system R through a shift of x_r along the axis X_g, a shift of y_r along the axis Y_g, and a shift of z_r along the axis Z_g:

$$\mathbf{R}_g^r(x_r,y_r,v_r) = \begin{bmatrix} 1 & 0 & 0 & x_r \\ 0 & 1 & 0 & y_r \\ 0 & 0 & 1 & z_r \\ 0 & 0 & 0 & 1 \end{bmatrix} \tag{7}$$

4.2. Establishment of the Two-UAV Intersection Localization Model

As shown in Figure 4c, when tracking a target, the airborne optoelectronic platform will adjust the camera angle and lock the target to the FOV center. At this moment, both the LOS and the axis Z_t of the camera coordinate system are directed to the target. In the case of localization, the two UAVs will output and convert the measured data simultaneously into a uniform coordinate system for processing. In the camera coordinate system, the unit vector of LOS is expressed as $L_i = [0, 0, f, 1]^T$, where f is camera focus. Through the coordinate transformation, the expression of the LOS vector in the reference coordinate system can be obtained. Then the expression of the target position can be determined through the intersection algorithm. The coordinate transformation process is shown in the Figure 5.

T	→A	→ S	→ G	→ R
ratate(Y_t) -β	rotate(Z_a) κ	shift(X_s) -hs	shift(X_g) xr	
rotate(X_t) -α	ratate(Y_a) θ	rotate(Y_s) λs	shift(Y_g) yr	
	ratate(X_a) λ	rotate(Z_s) -αs	shift(Z_g) zr	
		shift(Z_s) Ne²sinλs		

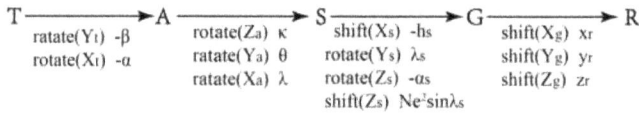

Figure 5. Coordinate transformation process.

Where α and β are the azimuth and pitch angles of camera in relation to UAV and can be measured by a photoelectric encoder; λ, θ, κ are the attitude angles of UAV in relation to geographic coordinate system and can be measured by the inertial navigation system; $(\lambda_s, \alpha_s, h_s)$ can be measured by GPS; N is the radius of curvature in the prime vertical; e is eccentricity; and the coordinates (x_r, y_r, z_r) are the expression of reference coordinate system in the terrestrial rectangular coordinate system. The data $(L, B, H) = (\lambda_s, \alpha_s, h_s)$ measured by GPS are the coordinates in the geodetic coordinate system. However, in the actual localization resolution, they shall be converted into the coordinates in the reference coordinate system for ease of calculation by determining, at first, their values in the terrestrial rectangular coordinate system and then converting these values into the coordinates in the reference coordinate system through the transformation process shown in Figure 5.

For the convenience of expression, the parameters measured by the two UAVs are marked by i ($i = 1, 2$). Through the process in Figure 5, the value of the LOS vector L_{gi} in the reference coordinate system can be determined:

$$L_{gi} = \begin{pmatrix} l_{gi} \\ m_{gi} \\ n_{gi} \\ 1 \end{pmatrix} = R_g^r R_s^g R_a^s R_t^a L_i \tag{8}$$

In fact, there exist some measurement errors in the positioning calculation process leading to the deviation between the solution of the visual axis vector direction and the actual measurement of the visual axis. Thus, the visual axis LOS1, LOS2 in the reference coordinate system may be rendezvous (shown in Figure 6a) or non-uniplannar intersections (shown in Figure 6b). As shown in Figure 6b, the dotted lines represent the actual visual axis, the solid lines represent the visual axis obtained from calculating Equation (8) and τ_1, τ_2 express the deviations of the dotted lines and the solid lines. In order to solve the problem of locating the target, we introduce a point $M(x_m, y_m, z_m)$ in the space according to the estimated target position, and the distance of M to the two visual axes LOS1, LOS2 is a minimum based on the spatial straight line principle.

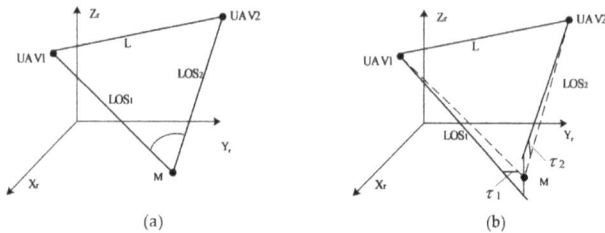

Figure 6. Positional relationship of the two visual axes. (**a**) rendezvous; (**b**) non-uniplannar intersection.

$E(x_m, y_m, z_m)$ can be calculated according to spatial geometry knowledge:

$$E(x_m, y_m, z_m) = \sqrt{\sum_{i=1}^{2} \left[(x_m - x_i^F)^2 + (y_m - y_i^F)^2 + (z_m - z_i^F)^2 \right]} \tag{9}$$

where the coordinate (x_i, y_i, z_i) shows the position of the UAV in the reference coordinate system at a certain time and $\left(x_i^F, y_i^F, z_i^F\right)$ are the foot coordinates of the pedal of the point M to LOSi, which can be obtained according to the linear parameter equation:

$$
\begin{aligned}
x_i^F &= x_i + l_{gi}\left[l_{gi}(x_m - x_i) + m_{gi}(y_m - y_i) + n_{gi}(z_m - z_i)\right] \\
y_i^F &= y_i + m_{gi}\left[l_{gi}(x_m - x_i) + m_{gi}(y_m - y_i) + n_{gi}(z_m - z_i)\right] \\
z_i^F &= z_i + n_{gi}\left[l_{gi}(x_m - x_i) + m_{gi}(y_m - y_i) + n_{gi}(z_m - z_i)\right]
\end{aligned}
\tag{10}
$$

In this case, the problem of the two-UAV intersection localization can be simplified to find out the coordinates (x_m, y_m, z_m) to make the value of E the smallest. According to the principle of least squares, the partial derivatives of x_m, y_m, z_m for E can be found and assigned to 0:

$$
\begin{aligned}
\sum_{i=1}^{2}\left[\left(1 - l_{gi}^2\right)(x_m - x_i) - l_{gi}m_{gi}(y_m - y_i) - l_{gi}n_{ni}(z_m - z_i)\right] &= 0 \\
\sum_{i=1}^{2}\left[-l_{gi}m_{gi}(x_m - x_i) + \left(1 - m_{gi}^2\right)(y_m - y_i) - m_{gi}n_{ni}(z_m - z_i)\right] &= 0 \\
\sum_{i=1}^{2}\left[-l_{gi}n_{gi}(x_m - x_i) + m_{gi}n_{ni}(y_m - y_i) + \left(1 - n_{gi}^2\right)(z_m - z_i)\right] &= 0
\end{aligned}
\tag{11}
$$

According to the formula, the linear equations of x_m, y_m, z_m can be rewritten in the form of matrix $AM = b$, where:

$$
A = \begin{pmatrix}
\sum_{i=1}^{2}\left(1 - l_{gi}^2\right) & -\sum_{i=1}^{2}l_{gi}m_{gi} & -\sum_{i=1}^{2}l_{gi}n_{gi} \\
-\sum_{i=1}^{2}l_{gi}m_{gi} & \sum_{i=1}^{2}\left(1 - m_{gi}^2\right) & -\sum_{i=1}^{2}m_{gi}n_{gi} \\
-\sum_{i=1}^{2}l_{gi}n_{gi} & -\sum_{i=1}^{2}m_{gi}n_{gi} & \sum_{i=1}^{2}\left(1 - n_{gi}^2\right)
\end{pmatrix},
\ M = \begin{pmatrix} x_m \\ y_m \\ z_m \end{pmatrix},
\ b = \begin{pmatrix}
\sum_{i=1}^{2}\left[\left(1 - l_{gi}^2\right)x_i - l_{gi}m_{gi}y_i - l_{gi}n_{ni}z_i\right] \\
\sum_{i=1}^{2}\left[-l_{gi}m_{gi}x_i + \left(1 - m_{gi}^2\right)y_i - m_{gi}n_{ni}z_i\right] \\
\sum_{i=1}^{2}\left[-l_{gi}n_{gi}x_i - m_{gi}n_{ni}y_i + \left(1 - n_{gi}^2\right)z_i\right]
\end{pmatrix}
$$

Since A is a nonsingular matrix, the solution of the system of linear equations can be obtained: $M = A^{-1}b$. Through the homogeneous-coordinates transformation expression R_i^g, the coordinates of the target in the terrestrial rectangular coordinate system can be obtained. From Equation (3), the coordinates (B_m, L_m, H_m) of the target M in the geodetic coordinate system can be derived.

5. Accuracy Analysis and Simulation Experiment

Localization accuracy analysis is an important step to judge whether a localization algorithm is good or not. There are mainly two factors influencing the accuracy of a localization algorithm. The first factor is the error of a measurement parameter. It can be learned from Equation (8) that various parameters will be integrated into the solution process of the target and their errors will undoubtedly influence the final localization accuracy. The second factor is the measuring position of a UAV in relation to the target, which, during the intersection measurement, has an important influence on measurement accuracy. In reality, the measurement accuracy can only reach a certain level due to limited modern technological and design development and equipment production costs. In this case, the positional relationship between the UAV and the target is important to localization accuracy.

5.1. Influence of UAV Position on Localization Accuracy

UAV position is of great implication to localization accuracy. When locating a target during actual military reconnaissance, the UAV needs to keep enough distance from the target to ensure stealthiness. In view of the above background, this paper analyzed the influence of different measurement positions on localization accuracy from the following aspects and obtained the relevant data for engineering reference. The measurement errors of various sensors used in the test are determined according to the maximum nominal errors given by the equipment specifications. This paper assumes that the same measuring equipment are adopted by the two UAVs, which means the parameter errors are governed by the same criteria, as shown in the Table 1.

Table 1. Distribution of random errors.

Name of Error Variable	Random Distribution	Error σ
Miss distance x	Normal distribution	4.8×10^{-5} (m)
Miss distance y	Normal distribution	4.8×10^{-5} (m)
UAV longitude	Normal distribution	1×10^{-4} (°)
UAV latitude	Normal distribution	1×10^{-4} (°)
UAV altitude	Normal distribution	10 (m)
UAV pitch	Normal distribution	0.01 (°)
UAV roll	Normal distribution	0.01 (°)
UAV yaw	Normal distribution	0.05 (°)
Camera pitch	Uniform distribution	0.01 (°)
Camera azimuth	Uniform distribution	0.01 (°)

The simulation described in this paper is carried out in the reference coordinate system, where the x axis indicates altitude and the YOZ plane represents the horizontal plane. Since the following simulation test is mainly to demonstrate the influence of UAV position in relation to the target on localization accuracy, the UAV attitude angle will not affect the test results. To facilitate the observation of test results, the three axes of the UAV coordinate system in the test are parallel to those of the reference coordinate system.

5.1.1. Influence of Baseline Length on Localization Accuracy

The target is tracked by two UAVs at the same distance (namely, the two UAVs and the target form an isosceles triangle), as shown in the Figure 7.

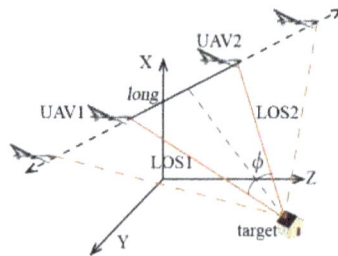

Figure 7. Positional relationship between UAVs and the target.

The distances between the target and UAV baselines are 5 km and remain unchanged. By changing the baseline *long*, the analysis result is obtained, as shown in the Figure 8.

In the Figure 8, the curves *x*, *y*, and *z* represent the localization errors $\Delta x, \Delta y, \Delta z$ (root mean square (RMS) errors rooted in 100 simulation times at one point) in the three axes. The curve *sum* indicates the error radius of the actual target position and measurement point, namely $sum = \sqrt{\Delta x^2 + \Delta y^2 + \Delta z^2}$. The test results show that, when the target is tracked by two UAVs at the same distance and altitude, and the distances from the target to UAV baselines are set as 5 km, if *long* = 7.35 km, the intersection angle ϕ will be 72.6318° and the localization accuracy will be the highest: $\Delta x = 14.83$ m, $\Delta y = 11.31$ m, $\Delta z = 16.68$ m, *sum* (min) = 25.0235 m. When the baseline *long* is 2.5–16.3 km, the intersection angle will vary from 28.1° to 116.8°, and both the total error and every error component will be 32 m. The localization result should be rejected because of the excessive error once the intersection angle is out of the range.

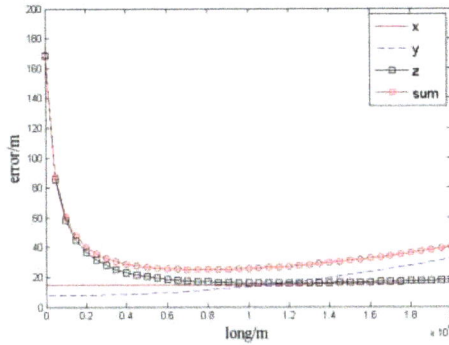

Figure 8. Error curves.

5.1.2. Localization Accuracy of Two Tracking UAVs at the Same Altitudes but Different Distances with the Target

It is known from the above that, when the distances from the target to UAV baselines are still kept as 5 km and the baseline *long* is 7.35 km, the localization accuracy will be the highest. Under these conditions, the two simulated UAVs move along the baseline extension, as shown in the Figure 9.

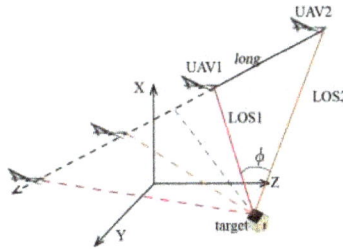

Figure 9. Positional relationship between the UAVs and the target.

As shown in the Figure 9, when the coordinates of simulated target are (0, 3675, 4000) and UAV 1 moves linearly at a constant speed from (3000, −10,000, 0) to (3000, 10,000, 0) and UAV 2 from (3000, −2650, 0) to (3000, 17,350, 0) at the same speed, the localization accuracy will be determined as shown in the Figure 10.

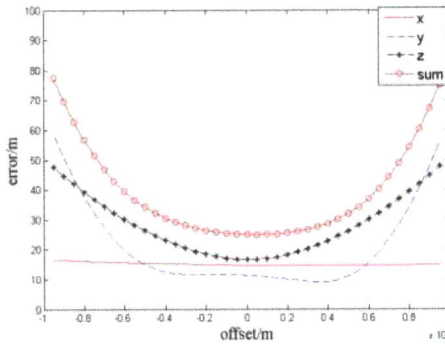

Figure 10. Error curves.

The simulation results show that the localization error increases with the *offset*. When the offset is 0 (the distances between the two UAVs and the target are the same), the total localization error (*sum*) is 25.0235 m, or the minimum. When the *offset* is 2.5 km, the *sum* is 27.1464 m. When the *offset* is 5 km, the *sum* is 34.1459 m. Once the *offset* exceeds 5 km, the localization result should be rejected because of the excessive error.

From the above simulation information, the following conclusions can be drawn:

1. According to the localization algorithm proposed by this paper, the optimal position for the two-UAV intersection system to locate the target is when the two UAVs and the target are on the same horizontal line and the azimuth of UAV 1 in relation to the target is just the opposite of that of UAV 2, namely the positions of the two UAVs in relation to the target are the same, with an intersection angle of 72.6318°. In this paper, the distance from the simulated target to a UAV baseline is 5 km and, accordingly, the baseline length in the optimal measuring position is *long* = 7.35 km. In this case, the x-axis error Δx is 14.83 m, the y-axis error Δy is 11.31 m, the z-axis error Δz is 16.68 m, and the error radius *sum* is 25.0235 m.

2. In the real world, both the target and the UAVs are moving continuously along unpredictable tracks, so it is very difficult for the UAVs to remain in the optimal measuring positions for target localization. When the two UAVs are kept parallel to each other and at the same horizontal plane as the target, an intersection angle of 28.1°–116.8° can achieve a desirable localization result.

5.2. Modified Adaptive Kalman Filtering during Two-UAV Intersection Localization

To improve the localization accuracy, the observations for the target shall be filtered. The target aimed by our system is moving in the air or on the ground. A dynamic localization model can be established by adopting the modified adaptive Kalman filtering algorithm and taking the target's triaxial movement positions and speeds as state variables, and its target position as measurement variables. In this paper, the modified Saga adaptive filtering method is used to solve the problem that the statistical characteristics of system noise and observation noise of the dynamical target are uncertain. The basic process is to calculate the system noise at the current time and the estimated value of observation noise, and compute the state estimation values by employing the new information based on each measured value $Y(k)$.

5.2.1. Modified Adaptive Kalman Filtering Modeling

An airborne optoelectronic platform often locates the targets thousands of meters away. Considering that both the target speed and the UAV speed-to-altitude ratio are not large, the target motion in the tri-axial directions can be approximated as uniform motion. Suppose the sampling time is T. The state equation of target motion will be:

$$X(k) = AX(k-1) + BU(k-1) + W(k-1) \tag{12}$$

where $X(k)$ is the target state variable at the time k; $X(k) = [x(k), v_x(k), y(k), v_y(k), z(k), v_z(k)]^T$, where $x(k), v_x(k), y(k), v_y(k), z(k), v_z(k)$ are the target's positions and speeds on the three axes. The system has no controlled variables. If $B(k) = 0$, the state-transition matrix will be:

$$A = \begin{bmatrix} 1 & T & 0 & 0 & 0 & 0 \\ 0 & 1 & 0 & 0 & 0 & 0 \\ 0 & 0 & 1 & T & 0 & 0 \\ 0 & 0 & 0 & 1 & 0 & 0 \\ 0 & 0 & 0 & 0 & 1 & T \\ 0 & 0 & 0 & 0 & 0 & 1 \end{bmatrix}$$

$W(k)$ is a white Gaussian noise sequence for system state noise, with the expectation of $q(k)$ and the covariance of $Q(k)$.

The system measurement equation is as follows:

$$Y(k) = \mathbf{H}X(k) + \mathbf{V}(k) \tag{13}$$

where $Y(k)$ is the system measurement, $Y(k) = [y_x(k), y_y(k), y_z(k)]^T$, $\mathbf{H}$ is system observation matrix,

$$\mathbf{H} = \begin{bmatrix} 1 & 0 & 0 & 0 & 0 & 0 \\ 0 & 0 & 0 & 0 & 0 & 0 \\ 0 & 0 & 1 & 0 & 0 & 0 \\ 0 & 0 & 0 & 0 & 0 & 0 \\ 0 & 0 & 0 & 0 & 1 & 0 \\ 0 & 0 & 0 & 0 & 0 & 0 \end{bmatrix}$$

$V(k)$ is a white Gaussian noise sequence for observation noise, with the expectation of $v(k)$ and the covariance of $R(k)$.

The recurrence equation of adaptive Kalman filter can be obtained:

$$\mathbf{X}(k/(k-1)) = \mathbf{A}(k)\mathbf{X}(k-1) + \mathbf{q}(k-1) \tag{14}$$

$$\mathbf{P}(k/(k-1)) = \mathbf{A}(k)\mathbf{P}(k-1)\mathbf{A}^T(k) + \mathbf{Q} \tag{15}$$

$$\mathbf{K}(k) = \frac{\mathbf{P}(k/(k-1))\mathbf{H}^T(k)}{(\mathbf{H}(k)\mathbf{P}(k/(k-1))\mathbf{H}^T(k) + \mathbf{R}(k))} \tag{16}$$

$$\mathbf{X}(k) = \mathbf{X}(k/k-1) + \mathbf{K}(k)(\mathbf{Y}(k) - \mathbf{H}\mathbf{X}(k/k-1) - \mathbf{v}(k-1)) \tag{17}$$

$$\mathbf{P}(k) = (1 - \mathbf{K}(k)\mathbf{H}(k))\mathbf{P}(k/(k-1)) \tag{18}$$

This paper uses a considerable number of error arithmetic mean values to approximate the mathematical expectation of the errors, and then uses these errors and mathematical expectations to estimate the variance of the errors:

$$\begin{aligned} \mathbf{q}(k) &= \tfrac{1}{k}\sum_{j=1}^{k}(\mathbf{X}(k) - \mathbf{A}(k)\mathbf{X}(k-1)) \\ \mathbf{q}_w(k) &= \tfrac{1}{k}\sum_{j=1}^{k}(\mathbf{X}(k) - \mathbf{A}(k)\mathbf{X}(k-1) - \mathbf{q}(k)) \\ \mathbf{r}(k) &= \tfrac{1}{k}\sum_{j=1}^{k}(\mathbf{X}(k) - \mathbf{H}(k)\mathbf{X}(k/k-1)) \\ \mathbf{r}_w(k) &= \tfrac{1}{k}\sum_{j=1}^{k}(\mathbf{X}(k) - \mathbf{H}(k)\mathbf{X}(k/k-1) - \mathbf{r}(k)) \end{aligned} \tag{19}$$

This estimation method is an unbiased estimate in which $\mathbf{q}_w(k)$, $\mathbf{r}_w(k)$ represent the standard deviation of the estimate. The square of the elements in $\mathbf{q}_w(k)$ are taken as the diagonal elements of $Q(k)$, the other elements of $Q(k)$ are assigned 0, the square of the elements in $\mathbf{r}_w(k)$ are taken as the diagonal elements of $R(k)$, and the other elements of $R(k)$ are assigned 0. The noise statistics are obtained in this way. For the time-varying system, the noise changes with time, and the old data needs to be removed, so this article uses 100 before the current time of the unbiased data, which can guarantee the accuracy and real-time of the data.

5.2.2. Filter Initialization

An adaptive Kalman filtering algorithm is a recurrence algorithm and, thus, must be initialized. This paper uses the optimal localization position for testing, so the observation noise $V(k)$ is a white Gaussian noise whose covariance $v(k)$ is a constant, namely $R(1) = \text{diag}(r1, r2, r3) = [14.83^2, 0, 0; 0, 11.31^2, 0; 0, 0, 16.68^2]$. The initial state variable $X(1)$ can be obtained from initial observations. Suppose the target moves at a constant speed in the tri-axial directions, then $X(1) = [y_x(1), (y_x(2) - y_x(1))/T,$ $y_y(1), (y_y(2) - y_y(1))/T, y_z(1), (y_z(2) - y_z(1))/T]$. The initial covariance matrix can be written as:

$$P(1) = \begin{bmatrix} r11 & \frac{r11}{T} & 0 & 0 & 0 & 0 \\ \frac{r11}{T} & \frac{2r11}{T^2} & 0 & 0 & 0 & 0 \\ 0 & 0 & r22 & \frac{r22}{T} & 0 & 0 \\ 0 & 0 & \frac{r22}{T} & \frac{2r22}{T^2} & 0 & 0 \\ 0 & 0 & 0 & 0 & r33 & \frac{r11}{T} \\ 0 & 0 & 0 & 0 & \frac{r11}{T} & \frac{2r33}{T^2} \end{bmatrix}$$

The sampling time is $T = 1$ s. The filter starts to work when $k = 2$.

5.2.3. Test Results

The filtering results from the above method are shown in the Figures 11 and 12.

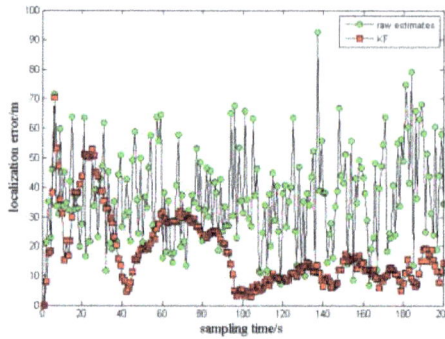

Figure 11. Localization errors after modified adaptive Kalman filtering.

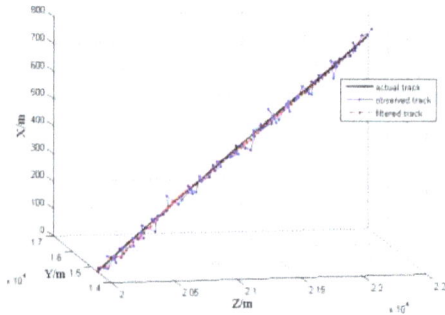

Figure 12. Localization track.

It is observed from the error distribution in Figure 11 that, after modified adaptive Kalman filtering, the localization error becomes 15.7983 m, much smaller than the original measurement of 25.0235 m. Thus, the localization accuracy has been improved significantly. Judging from the sampling time, the first 100 s constitute a data accumulation course, so the localization results are relatively divergent. However, after 100 s, the data begin to converge rapidly, and the error values are smoother than before filtering.

The Figure 12 shows the track curves of the simulated target in motion. Among them, the black curve is the actual track, the blue curve with solid dots is the observed track, and the red dashed curve is the filtered track. It can be seen that, the observed track is divergent, whereas the filtered track is smoother and fits the actual track better.

5.3. Flight Data Results

In this section, the approach previously presented is now validated using flight test data. A ground receiving station is arranged to receive the measurement data and video sequences from the two UAVs and optoelectronic platforms, and the target position is calculated in real-time using the received data. In order to ensure each set of images and the parameters from the two UAVs that the ground receiving station used in the calculation process are acquired at the same time, time synchronization calibration for the two unmanned aerial vehicles and the photoelectric platforms is necessary to be taken prior to flight. The corresponding shooting time and serial number information of the image of each frame of the video captured should be noted in order to avoid framing errors in the solution. By using these methods, we can ensure that the image data and parameters used during the resolution are captured at the same time by the two UAVs. The video images are shot by a CMOS photo-detector with the resolution ratio 1024×768, a pixel size of 5.5 µm, and a frame rate of 25 frames per second. The clock error after the time calibration of two UAVs is 5 ms, so the time difference of arrival of the two images and measured parameters used in the solution process is up to 45 ms. For non-high-speed moving objects, the error is within the acceptable range. The terrestrial solution unit uses the high-speed DSP chip TMS320F28335 (made by the TI (Texas Instruments) company, Dallas, TX, USA) to receive the data and calculate the target position with processing time of 500 µs, which can meet the real-time requirements.

In order to verify the feasibility of the algorithm, the flight test is performed in an outfield environment. The two UAVs fly with a steady rate of 100–110 m/s, flight height of 5400 m, and the camera has a focal length of 200 mm. An outfield experiment of positioning a fixed building with the algorithm of two-UAV intersection localization is carried out. The exact positioning of the target is known. With the UAVs' flight, the positional relationship between the UAVs and target is in flux. In this section several feature positions were selected to measure the localization error used to verify the simulation experiment.

Feature Position 1: As shown in Figure 13a,b, the two images' rooted in the outfield experimental video data are taken at the same time and their sizes are 1024×768 pixels. The positional relationship graph between the two UAVs and the target can be described as in Figure 13c, that is, the target is tracked by two UAVs at the same distance with the baselines *long* = 7332 m and the intersection angle ϕ 73.2°. In this case the localization error can be: $\sigma_B = 1.63 \times 10^{-4}$ (°), $\sigma_L = 1.35 \times 10^{-4}$ (°), $\sigma_H = 15.8$ (*m*), $\sigma_{sum} = 27.6$ (*m*), where σ_B represents the longitude error, σ_L represents the latitude error, σ_H represents the altitude error, and σ_{sum} represents the error radius.

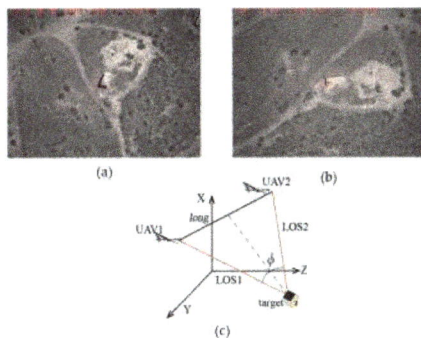

Figure 13. Feature position 1: (**a**) target detection by UAV1; (**b**) target detection by UAV2; and (**c**) positional relationship graph between the two UAVs and the target.

Feature Position 2: The positional relationship graph between the two UAVs and the target can be described as in Figure 14c, that is, the target is tracked by two UAVs at the same distance with the baselines *long* = 9337 m and the intersection angle ϕ 101.3°. In this case the localization error can be: $\sigma_B = 2.83 \times 10^{-4}$ (°), $\sigma_L = 2.71 \times 10^{-4}$ (°), $\sigma_H = 21.3$ (*m*), $\sigma_{sum} = 40.6$ (*m*).

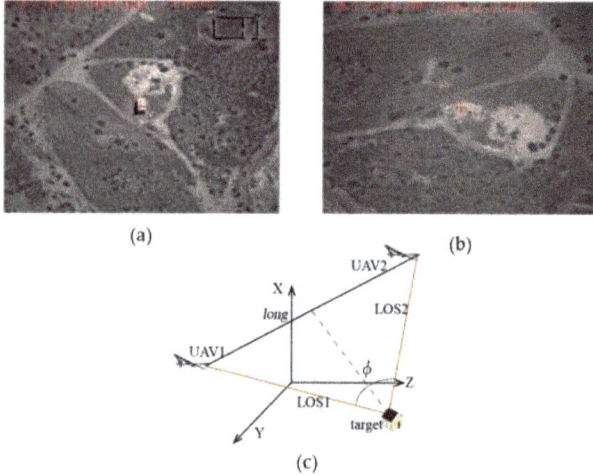

(a)   (b)

(c)

Figure 14. Feature position 2: (**a**) target detection by UAV1; (**b**) target detection by UAV2; and (**c**) positional relationship graph between the two UAVs and the target.

Feature Position 3: The positional relationship graph between the two UAVs and the target can be described as in Figure 15c, that is, the target is tracked by two UAVs at the same distance with the baselines *long* = 2749 m and the intersection angle ϕ 24.8°. In this case the localization error can be: $\sigma_B = 1.45 \times 10^{-3}$ (°), $\sigma_L = 9.46 \times 10^{-4}$ (°), $\sigma_H = 118.6$ (*m*), $\sigma_{sum} = 198.3$ (*m*).

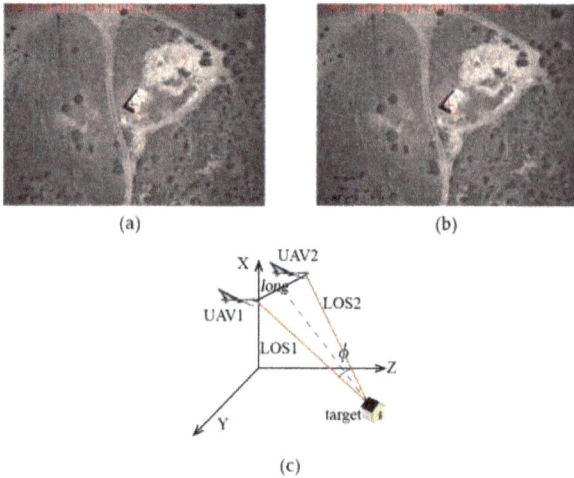

(a)   (b)

(c)

Figure 15. Feature position 3: (**a**) target detection by UAV1; (**b**) target detection by UAV2; and (**c**) positional relationship graph between the two UAVs and the target.

Feature Position 4: The positional relationship graph between the two UAVs and the target can be described as in Figure 16c, that is, the target is tracked by two UAVs at the same distance with the baselines *long* = 7332 m and the *offset* = 4200 m. In this case the localization error can be: $\sigma_B = 2.38 \times 10^{-4}$ (°), $\sigma_L = 1.58 \times 10^{-4}$ (°), $\sigma_H = 19.8$ (m), $\sigma_{sum} = 36.2$ (m).

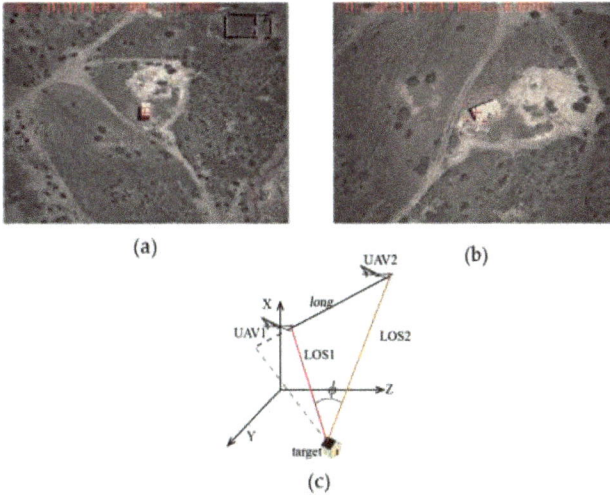

(a)

(b)

(c)

Figure 16. Feature position 4: (**a**) target detection by UAV1; (**b**) target detection by UAV2; and (**c**) positional relationship graph between the two UAVs and the target.

As shown in Table 2, the results of flight experiments show that the positioning accuracy of the two-UAV intersection localization system is highest when the two UAVs and the target are at the same distance, and the localization precision is related to the intersection angle ϕ, which is basically the same as the simulation result. The actual localization accuracy is lower than the simulation result because there exist some unknown factors in the outfield actual flight test, and future work is to take them into consideration. The experiment proves the feasibility of the system and the accuracy of the simulation analysis.

Table 2. Localization error.

Positional Relationship	Localization Algorithm	σ_B (°)	σ_L (°)	σ_H (m)	σ_{sum} (m)
Feature Position 1	Two-UAV localization	1.63×10^{-4}	1.35×10^{-4}	15.8	27.6
	single-station localization	2.69×10^{-4}	1.51×10^{-4}	19.4	37.4
Feature Position 2	Two-UAV localization	2.83×10^{-4}	2.71×10^{-4}	21.3	40.6
	single-station localization	3.81×10^{-4}	4.23×10^{-4}	45.2	72.8
Feature Position 3	Two-UAV localization	1.45×10^{-3}	9.46×10^{-4}	118.6	198.3
	single-station localization	2.71×10^{-4}	1.84×10^{-4}	21.3	38.5
Feature Position 4	Two-UAV localization	2.38×10^{-4}	1.58×10^{-4}	19.8	36.2
	single-station localization	2.46×10^{-4}	3.35×10^{-4}	24.6	51.8

In order to compare the positioning accuracy of this algorithm with the traditional single-station REA localization manner, the single-station localization experiment is added in this section. The experimental conditions are the same as above. The laser range-finder is installed inside the photoelectric platform. With laser ranging accuracy of 5 m and other equipment errors the same as the experiment before, the comparison results are shown in Table 2. When the positional relationship graph between the two UAVs and the target meets the requirements of the simulation results, the accuracy of the proposed method is obviously higher than that of the traditional stand-alone positioning method. On the contrary, as shown in Feature Position 3, when the intersection angle ϕ is beyond the 28.1°–116.8° scope of requirement, the accuracy of the proposed method is lower than that of the traditional stand-alone positioning method.

6. Conclusions

To address the limitation of the existing airborne optoelectronic localization method, this paper proposes an improved two-UAV intersection localization algorithm based on the conventional ground intersection localization method. This paper establishes a two-UAV intersection localization model, studies, in detail, the influence of UAV position on localization accuracy in order to find the optimal localization position, and quantifies the localization accuracy in different positions, thus providing a basis for the planning of the UAV track during localization. When the target is 5 km away from UAV baselines, the localization accuracy in the optimal localization position can reach 25.0235 m. For a target whose track is quite smooth, this paper introduces a modified adaptive Kalman filtering method to improve the localization accuracy to 15.7983 m. Finally, an outfield experiment was carried out to validate the two-UAV intersection localization algorithm. The localization accuracy in the optimal localization positions: $\sigma_B = 1.63 \times 10^{-4}$ (°), $\sigma_L = 1.35 \times 10^{-4}$ (°), $\sigma_H = 15.8$ (m), $\sigma_{sum} = 27.6$ (m), which is basically the same as the simulation result. Next, we will continue to study how to plan the UAV track and build a more accurate Kalman filtering model.

Author Contributions: Guanbing Bai designed the algorithm and wrote the source code and manuscript; Jinghong Liu and Yueming Song made contribution to experiments design and paper written; Yujia Zuo analyzed the experiment results and revised paper.

Conflicts of Interest: The authors declare no conflict of interest.

References

1. Eric, J.S. Geo-Pointing and Threat Location Techniques for Airborne Border Surveillance. In Proceedings of the IEEE International Conference on Technologies for Homeland Security (HST), Waltham, MA, USA, 12–14 November 2013; pp. 136–140.
2. Gao, F.; Ma, X.; Gu, J.; Li, Y. An Active Target Localization with Moncular Vision. In Proceedings of the IEEE International Conference on Control & Automation (ICCA), Taichung, Taiwan, 18–21 June 2014; pp. 1381–1386.
3. Su, L.; Hao, Q. Study on Intersection Measurement and Error Analysis. In Proceedings of the IEEE International Conference on Computer Application and System Modeling (ICCASM), Taiyuan, China, 22–24 October 2010.
4. Liu, L.; Liu, L. Intersection Measuring System of Trajectory Camera with Long Narrow Photosensitive Surface. In Proceedings of the Society of Photo-Optical Instrumentation Engineers (SPIE), Beijing, China, 26 January 2006; pp. 1006–1011.
5. Hu, T. Double UAV Cooperative Localization and Remote Location Error Analysis. In Proceedings of the 5th International Conference on Advanced Design and Manufacturing Engineering, Shenzhen, China, 19–20 September 2015; pp. 76–81.
6. Lee, W.; Bang, H.; Leeghim, H. Cooperative localization between small UAVs using a combination of heterogeneous sensors. *Aerosp. Sci. Technol.* **2013**, *27*, 105–111. [CrossRef]
7. Campbell, M.E.; Wheeler, M. Vision-Based Geolocation Tracking System for Uninhabited Aerial Vehicles. *J. Guid. Control Dyn.* **2010**, *33*, 521–531. [CrossRef]
8. Pachter, M.; Ceccarelli, N.; Chandler, P.R. Vision-Based Target Geo-location Using Camera Equipped MAVs. In Proceedings of the 46th IEEE Conference on Decision and Control, New Orleans, LA, USA, 12–14 December 2007; pp. 2333–2338.
9. Barber, D.B.; Redding, J.D.; McLain, T.W.; BeardEmail, R.W.; Taylor, C.N. Vision-based Target Geo-location using a Fixed-wing Miniature Air Vehicle. *J. Intell. Robot. Syst.* **2006**, *47*, 361–382. [CrossRef]

10. Whitacre, W.W.; Campbell, M.E. Decentralized Geolocation and Bias Estimation for Uninhabited Aerial Vehicles with Articulating Cameras. *J. Guid. Control Dyn.* **2011**, *34*, 564–573. [CrossRef]
11. William, W. Cooperative Geolocation Using UAVs with Gimballing Camera Sensors with Extensions for Communication Loss and Sensor Bias Estimation. Ph.D. Thesis, Cornell University, New York, NY, USA, 2010.
12. Campbell, M.; Whitacre, W. Cooperative Geolocation and Sensor Bias Estimation for UAVs with Articulating Cameras. In Proceedings of the AIAA Guidance, Navigation, and Control Conference, Chicago, IL, USA, 10–13 August 2009.
13. Pachter, M.; Ceccarelli, N.; Chandler, P.R. Vision-Based Target Geolocation Using Micro Air Vehicles. *J. Guid. Control Dyn.* **2008**, *31*, 297–615. [CrossRef]
14. Liu, F.; Du, R.; Jia, H. An Effective Algorithm for Location and Trackingthe Ground Target Based on Near Space Vehicle. In Proceedings of the 12th International Conference on Fuzzy Systems and Knowledge Discovery (FSKD), Zhangjiajie, China, 15–17 August 2015; pp. 2480–2485.
15. Hosseinpoor, H.R.; Samadzadegan, F.; DadrasJavan, F. Pricise Target Geolocation Based on Integeration of Thermal Video Imagery and RTK GPS in UAVs. *Int. Arch. Photogramm. Remote Sens. Spat. Inf. Sci.* **2015**, *41*, 333–338. [CrossRef]
16. Hosseinpoor, H.R.; Samadzadegan, F.; DadrasJavan, F. Pricise Target Geolocation and Tracking Based on UAV Video Imagery. *Int. Arch. Photogramm. Remote Sens. Spat. Inf. Sci.* **2016**, *XLI-B6*, 243–249. [CrossRef]
17. Conte, G.; Hempel, M.; Rudol, P.; Lundstrom, D.; Duranti, S.; Wzorek, M.; Doherty, P. High Accuracy Ground Target Geo-location Using Autonomous Micro Aerial Vehicle Platforms. In Proceedings of the AIAA Guidance, Navigation and Control Conference and Exhibit, Honolulu, HI, USA, 18–21 August 2008.
18. Frew, E.W. Sensitivity of Cooperative Target Geolocalization to Orbit Coordination. *J. Guid. Control Dyn.* **2008**, *31*, 1028–1040. [CrossRef]
19. Ross, J.; Geiger, B.; Sinsley, G.; Horn, J.; Long, L.; Niessner, A. Vision-based Target Geolocation and Optimal Surveillance on an Unmanned Aerial Vehicle. In Proceedings of the AIAA Guidance, Navigation, and Control Conference, Honolulu, HI, USA, 18–21 August 2008.
20. Sohn, S.; Lee, B.; Kim, J.; Kee, C. Vision-Based Real-Time Target Localization for Single-Antenna GPS-Guided UAV. *IEEE Trans. Aerosp. Electron. Syst.* **2008**, *44*, 1391–1401. [CrossRef]
21. Cheng, X.; Daqing, H.; Wei, H. High Precision Passive Target Localization Based on Airborne Electro-optical Payload. In Proceedings of the 14th International Conference on Optical Communications and Networks (ICOCN), Nanjing, China, 3–5 July 2015.
22. Sharma, R.; Yoder, J.; Kwon, H.; Pack, D. Vision Based Mobile Target Geo-localization and Target Discrimination Using Bayes Detection Theory. *Distrib. Auton. Robot. Syst.* **2014**, *104*, 59–71.
23. Sharma, R.; Yoder, J.; Kwon, H.; Pack, D. Active Cooperative Observation of a 3D Moving Target Using Two Dynamical Monocular Vision Sensors. *Asian J. Control* **2014**, *16*, 657–668.
24. Deng, B.; Xiong, J.; Xia, C. The Observability Analysis of Aerial Moving Target Location Based on Dual-Satellite Geolocation System. In Proceedings of the International Conference on Computer Science and Information Processing, Xi'an, China, 24–26 August 2012.
25. Choi, J.H.; Lee, D.; Bang, H. Tracking an Unknown Moving Target from UAV. In Proceedings of the 5th International Conference on Automation, Robotics and Applications, Wellington, New Zealand, 6–8 December 2011; pp. 384–389.

sensors

MDPI

Article

Coastal Areas Division and Coverage with Multiple UAVs for Remote Sensing

Fotios Balampanis *, Iván Maza and Aníbal Ollero

Robotics, Vision and Control Group, Universidad de Sevilla, Avda. de los Descubrimientos s/n, 41092 Seville, Spain; imaza@us.es (I.M.); aollero@us.es (A.O)
* Correspondence: fbalampanis@us.es

Academic Editors: Felipe Gonzalez Toro and Antonios Tsourdos
Received: 23 December 2016; Accepted: 6 April 2017; Published: 9 April 2017

Abstract: This paper tackles the problems of exact cell decomposition and partitioning of a coastal region for a team of heterogeneous Unmanned Aerial Vehicles (UAVs) with an approach that takes into account the field of view or sensing radius of the sensors on-board. An initial sensor-based exact cell decomposition of the area aids in the partitioning process, which is performed in two steps. In the first step, a growing regions algorithm performs an isotropic partitioning of the area based on the initial locations of the UAVs and their relative capabilities. Then, two novel algorithms are applied to compute an adjustment of this partitioning process, in order to solve deadlock situations that generate non-allocated regions and sub-areas above or below the relative capabilities of the UAVs. Finally, realistic simulations have been conducted for the evaluation of the proposed solution, and the obtained results show that these algorithms can compute valid and sound solutions in complex coastal region scenarios under different setups for the UAVs.

Keywords: remote sensors; Unmanned Aerial Vehicles; area partition; cell decomposition

1. Introduction

The extensive interest in the use of Unmanned Aerial Vehicles (UAVs) has led a large scientific, commercial and amateur hobbyist community to actively contribute to a broad spectrum of activities and applications. Some of these applications imply the deployment of a distributed swarm of UAVs as a sensor network, with or without the presence of other types of unmanned vehicles or static sensors.

For coastal areas, the complex geographical attributes and the increasing interest for activities in or near remote off-shore locations have raised challenges for marine environment protection and for sustainable management. European countries have vast coasts and economic zones that go far into the Atlantic and Arctic oceans and are challenging to monitor and manage. In addition, the European Strategy for Marine and Maritime Research states the need to protect the vulnerable natural environment and marine resources in a sustainable manner. The use of UAVs in coastal areas can provide increased endurance and flexibility, whereas they can reduce the environmental impact, the risk for humans and the cost of operations.

The study presented in this paper has been carried out in the framework of MarineUAS (http://marineuas.eu), a European Union-funded doctoral program to strategically strengthen research training on Autonomous Unmanned Aerial Systems for Marine and Coastal Monitoring. In particular, this study tackles the problem of complex area partitioning for a team of heterogeneous UAVs and the associated sensor-driven cell decomposition based on their on-board sensing capabilities. The proposed solution is a combination of computational geometry algorithms along with graph search strategies, which manage to partition an area regardless of the number of UAVs or their relative capabilities. Each UAV sub-area is decomposed into a sum of sensor-projection -sized cells, and a coverage strategy is computed in parallel for each of the produced configuration spaces. The strategy has been implemented as a network of ROS nodes [1], where the initial partitioning process is executed on the ground station and the cell decomposition and coverage planning are computed on-board each UAV.

The rest of this paper is organized as follows: Section 2 provides a review of the literature, presenting the current state of the art on cell decomposition and partitioning strategies for multiple vehicles. Section 3 describes the problem statement and the assumptions considered in the paper, whereas Section 4 describes the model adopted for the on-board sensors. Section 5 presents the two-step approach developed for the cell decomposition and partition of a complex coastal area considering the capabilities of a team with multiple UAVs. Results from the simulations are presented in Section 6 including a strategy for the generation of spiral inward paths for coverage. Section 7 closes the paper with a discussion on the results and future steps.

2. Related Work

In the aforementioned context, the literature provides many studies for an autonomous sensor network to achieve the task of complete area coverage. In an extensive survey for online and offline decomposition and coverage path planning algorithms, the authors in [2] provide a categorization of the techniques in the literature, showing that grid decomposition strategies and algorithms are mostly used in coverage tasks.

Several partitioning algorithms and strategies can be found for distributing known or unknown areas for a team of autonomous vehicles. In [3], the task of a safe and uniform distribution of many robotic agents has been researched, by using a generalized Voronoi diagram, creating a Voronoi partitioning of a complex area. This study also creates an initial grid decomposition, which converges to the computed Voronoi partitioning. Even though this study does not consider coverage path planning, the use of growing functions along with a modified version of the Dijkstra algorithm manages to successfully partition a complex area for multiple robots, while maintaining safe and fair distances between them. In a recent work presented in [4], the authors provide an algorithm for fair area division and partition for a team of robots, with respect to their initial positions. Their solution produces promising results regarding the algorithm's computational complexity and guarantees full area coverage without backtracking paths. While their approach is feasible and manages to successfully tackle the problems of fair partition and coverage path planning, the strategy accounts only for the fair division case and fixed cell sizes, while in some cases, the sub-areas produced do not have a uniform geometric distribution.

Regarding multi-robot task allocation and distribution, the authors in [5] provide an extensive literature review and propose a novel taxonomy called iTax. This taxonomy categorizes the problems by complexity, where the first category includes problems that can be solved linearly, whereas the other three include NP-hard problems. The problem in our study is a general multi-robot task allocation problem, belonging in the latter category of the taxonomy, that of Complex Dependencies (CD). The ability to gradually reduce the complexity of the problem and to reach a lower level of complexity in each step can be exploited in the same manner that it is analyzed in the taxonomy.

In [6], the authors provide an optimal decomposition and path planning solution using an Integer Linear Programming (ILP) solver, by taking into account a camera-sized grid decomposition. Their solution manages to obtain the desired optical samples, although they do not provide the computational time needed for the ILP to find a solution. In the same context, the authors in [7] use an enhanced exact cellular decomposition method for an area and provide a coverage path consistent with the on-board camera of the UAV. Although their solution manages to produce smooth paths with minimal turns, their algorithms are tested only over convex polygon areas. The work presented in [8] deals with the same problem of area coverage for photogrammetric sensing. The authors include energy, speed and image resolution constraints in their proposed algorithms, such as an energy fail-safe mechanism for the safe return to the landing point. However, the provided solution and experiments do not account for complex, non-concave polygonal areas.

Regarding coverage algorithms for a single or a team of vehicles in a known area, the authors in [9] address the problem by creating an evaluation framework of path length, visiting period and balance workload metrics. In order to solve the problem, they generate a point cloud in which each point serves as guard in the art gallery problem, trying to maximize the visibility. The collection of these points, along with a Constrained Delaunay Triangulation (CDT) of the area, produces a graph where these points are the nodes. Thus, the final point cloud serves as a waypoint list, and coverage is achieved by using cluster-based or cyclic coverage methods. The work presented in [10] decomposes an area by using a convex decomposition and produces parallel lines in the decomposed parts, which are used as straight line paths. The algorithm presented tries to minimize the total amount of turns and provides a complete coverage plan. This strategy produces promising results, but complete coverage is not always achieved. Moreover, in some cases, repeated coverage is performed in order for the vehicle to visit the initial position of the next decomposed region. In [11], the authors tackle the problem of complete coverage by trying to minimize the completion time for the robots. In their strategy, turns imply the decrease in speed and eventually the acceleration of the vehicles. In that manner, their algorithm tries also to minimize the number of turns. This work also uses a grid-like decomposition strategy, based on disks. Once again, the areas considered for the experimental setups are convex rectilinear polygons. Finally, the authors in [12], provide a 3D coverage path planning strategy for underwater robots, and the analysis of the probabilistic completeness of the sampling-based coverage algorithm is shown. In their study, an underwater vehicle equipped with a sonar, a Doppler Velocity Log (DVL) and a camera obtains a 3D model of the ship to be inspected, while building and smoothing a roadmap for coverage. While the results are impressive, the amount, weight and energy requirements of the sensors are not compatible with the payloads of small or medium-sized UAVs.

Then, in general, the algorithms for cell decomposition, area partition and coverage planning do not take into consideration complex area characteristics or do not assume different sensor capabilities for the UAVs. In this paper, an exact cell decomposition strategy is applied in a novel algorithmic approach for area partition in a multi-UAV coverage mission context for coastal areas.

3. Problem Statement, Assumptions and Metrics Considered

The following scenario will be considered in this paper: an extensive oil leakage has been reported on an underwater pipe near a populated coastal region R, with particular aerial restrictions due to reserved airspace, nearby airports and domestic regions. A team of UAVs can be used for remote sensing purposes in order to localize the leakage sites and the extent of the oil spill around the reported sightings. The UAVs are heterogeneous with different autonomy capabilities and different on-board sensors.

Let $U_1, U_2, \ldots, U_n$ be the team of n UAVs with initial locations $\mathbf{p}_1, \mathbf{p}_2, \ldots, \mathbf{p}_n$. These locations are relevant to the whole procedure since they are the places of the initially-reported oil sightings, and they have the highest probability of finding the location of the actual pipe leakage. Moreover, the dispersion probability of the oil spill by these locations is decreasing uniformly in all directions on the sea. As the UAVs are heterogeneous, each UAV may be in charge of the coverage of different percentages of the whole area R. Then, let Z_i be the surface in square meters to be covered by U_i.

Let us consider an exact decomposition of the shape of region R into a set $S = C_j, j = 1, \ldots, M$ of cells regardless of its complexity or the existence of no-fly areas, without any cells being outside or partially inside R. This requirement is crucial since the safe integration of UAVs in non segregated airspace is a key requirement of Single Sky European Research (SESAR) (http://www.sesarju.eu/newsroom/all-news/sesar-takes-next-steps-rpas); an exact decomposition of an area and the avoidance of flight over residential, commercial or restricted areas are ways to mitigate critical damage in case of system failures. In addition, the size of the cells should be consistent with the field of view of the sensors on-board the UAVs, as will be discussed later in the paper. The cell decomposition allows one to discretize the space in order to treat the complex geometry of R.

The goal is to compute the geometry of a set of sub-areas $A = A_i, i = 1, \ldots, n$ based on the cells C_j taking into account the following metrics:

- The closeness of the cells within A_i to the initial location $\mathbf{p}_i$ of the UAV in charge of searching that sub-area should be minimized. This can be achieved by minimizing the sum of distances between each center of cell c_{ij} from the set S and the initial locations $\mathbf{p}_i$:

$$\min_S F(S) = \min_S \sum_{i=1}^{n} \sum_{j=1}^{M_i(S)} \| \mathbf{p}_i - c_{ij} \|, \tag{1}$$

where $M_i(S)$ is the number of cells of S inside A_i.

- The size of A_i should be as close as possible to Z_i for all of the UAVs. This can be achieved by minimizing the sum of differences:

$$\min_S G(S) = \min_S \sum_{i=1}^{n} \left| \sum_{j=1}^{M_i(S)} \text{area}(C_j) - Z_i \right|, \tag{2}$$

where $\text{area}(C_j)$ represents the area of the cells inside A_i.

The former metric that takes into consideration the probability of localization led us to design an algorithm (see Section 5) where each sub-area is generated by a uniform growth from the starting locations $p_1, p_2, \ldots, p_n$. This process has another positive side effect: since this growing region process is performed in every direction, it creates "symmetric" areas that are suitable to be covered by energy-efficient spiral-like patterns [13].

In addition, A_i by construction cannot be disjointed or intersected by another sub-area, neither by a no-fly zone. This restriction guarantees that the resulting sub-areas prevent the existence of overlapping coverage paths or collisions. In case additional safety requirements were present, it would be possible to define different flight altitudes for the UAVs in adjacent sub-areas. Figure 1 shows an example of a region R partitioned among three UAVs.

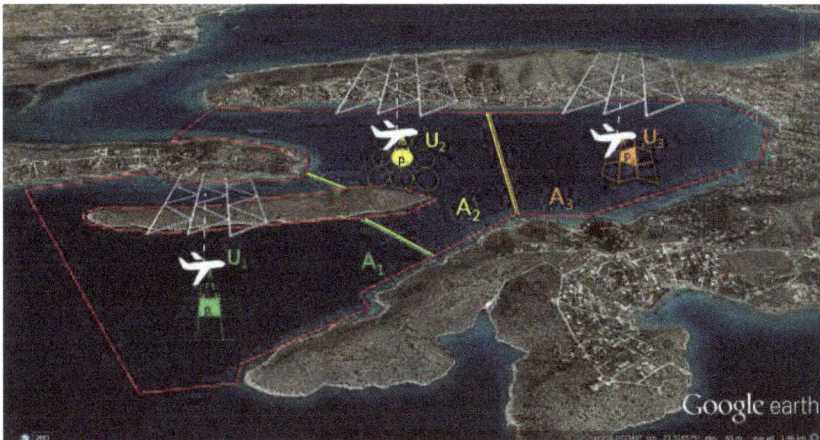

Figure 1. An example with three UAVs, each one with its allocated sub-area. The scheme is composed by two levels: the bottom layer shows the different on-board sensors' field of view projection on the sea, whereas the upper shows the cell decomposition denoted as a triangular grid on top of each UAV. U_1, U_2 and U_3 denote the UAVs, and A_1, A_2 and A_3 denote the the sub-areas of the total region R, which is constrained by the red borders. The initial positions of the UAVs are $\mathbf{p}_1$, $\mathbf{p}_2$ and $\mathbf{p}_3$.

The next section describes the model considered for the on-board sensors since the cell decomposition should be consistent with the features of the sensors.

4. Model Considered for the On-Board Sensors

Regarding the use of on-board sensors, the literature mostly refers to cameras [14]. In those cases, by knowing the length, width and focal length of the camera, as well as the altitude from the sea, the shape of the projection of the Field of View (FoV) of the camera can be calculated based on the attitude of the UAV. This is not the case for point or side scan beam sensors, which have a wide width, but a really narrow length scanning profile. Then, in the following sections, the term FoV will be used to refer to the projection of the FoV of a generic camera on the sea.

Figure 2 represents the situation considered in this paper, with the on-board sensor inside a gimbal and pointing downwards a given angle with respect to the fuselage of the UAV. Thus, pitch and roll angles of the UAV with respect to the horizontal plane are not relevant to the FoV projection, since the gimbal compensates for these angles.

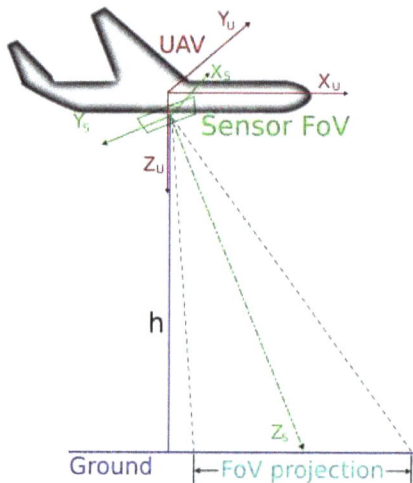

Figure 2. FoV projection calculation is relative to the coordinate frames of the system. For missions in coastal regions, the ground can be considered as flat.

The sample rate is another sensor characteristic to consider for the cell decomposition of a region. It is necessary to decompose the configuration space in a manner that a sensor can obtain at least one sample of each of the resulting cells in a unit of time (see Figure 3). Then, the projected footprint area F must guarantee that its size is proportional to the sample rate T and the UAV speed V. This can be achieved by either reducing speed or increasing altitude in order to grow the projection of the FoV. For most of the sensors, the sample rate is not an issue; however, this aspect has been pointed out for the sake of generality.

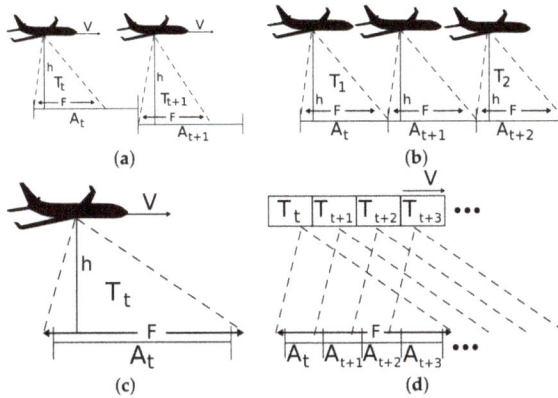

Figure 3. The appropriate cell decomposition is proportional to the velocity V, the sample rate T and the FoV projection footprint F. In (**a**), in every time step $t_0, t_1, \ldots, t_n$, F is not large enough for the sensor to take a complete sample, whereas in (**b**), T is not fast enough to obtain a sample from each area. In both cases, the problem could be solved be either reducing the speed, increasing the sample rate if possible or increasing the altitude for increasing the projection of the FoV. In (**c**), the ideal solution in the limit is shown, whereas in (**d**), the most usual case of the same portion of the sea being present in many samples is presented.

5. Area Decomposition and Partition in a Multi-UAV Context

This section describes the framework adopted by introducing the computational geometry tools for cell decomposition, as well as the algorithms for partitioning, based on the aforementioned considerations. These novel algorithms treat the segregated configuration spaces as topological graphs, allowing one to extract roadmaps for coverage planning after partitioning.

5.1. Exact Cell Decomposition

In the example shown in Figure 1, the coastal area outlines a complex shape, similar to the one in Figure 4. In these cases, the surroundings are rarely the only area restriction, since several residential or industrial areas are no-fly zones inside this complex, non-convex polygon.

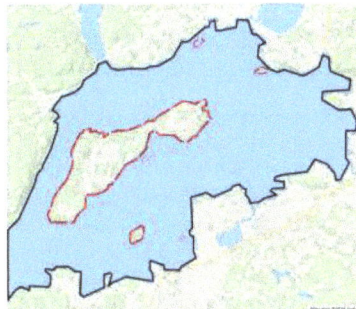

Figure 4. Trondheim fjord area (Norway) with Ytterøya island. The complex coastal area of interest is denoted by the black outer polygon, whereas the red dashed areas indicate regions that are no-fly zones.

In order to decompose these kind of areas, the decomposition strategy in [15] has been followed, applying the Constrained Delaunay Triangulation (CDT [16]). This is performed by introducing forced edge constraints that define the area and the holes as part of the input. By using a Lloyd optimization [17] on the resulting triangulation, we manage to obtain even more homogeneous triangles as this optimization improves the angles of each cell, making each one of the triangle's angles as close as possible to 60 degrees, depending on the selected iterations. By having more equilateral triangles, thus their angles closer to 60 degrees, a larger amount of area is covered in each step, and the overlapping during coverage is smaller.

The use of triangular cells for the decomposition is consistent with the complex shapes considered in the paper. The cells generated by the CDT are adapted to the shape of the borders, and this is very relevant since the center of the triangular cells is used for coverage planning. Hence, the paths computed based on this cell decomposition are initially consistent with the complex borders.

As has been previously mentioned in Section 4, each of the UAVs has a FoV projection, which guarantees that the speed along with the sample rate will manage to provide an adequate number of samples. This FoV size constraint is used as an input in the CDT method, being the maximum triangle side size. In order to guarantee that the FoV of each UAV will cover every triangular cell, regardless of its current orientation or sample rate, we need to provide a triangle side upper limit to be used as the CDT edge constraint. Then, if the centroids of the triangles are considered as waypoints, complete coverage could be achieved with the on-board sensor if a UAV follows that list of waypoints.

Considering a generic camera as the on-board sensor, the FoV projection on the ground is a trapezoid T, as can be seen in Figure 5a. The projection shape depends on the pitch angle β of the UAV, the θ FoV angle of the sensor and their relative rotation matrices. The trapezoid has bases a and b, where $a < b$, and equal sides c. We inscribe a circle having the centroid G of the trapezoid as its center. Since we want the CDT to produce the minimum amount of homogeneous triangles, an equilateral triangle W is the largest triangle that can be inscribed inside the inscribed circle of the trapezoid. In that manner, we guarantee that a UAV will always cover each produced triangle since all of the triangles of that CDT are at most as large as W (see Figure 5).

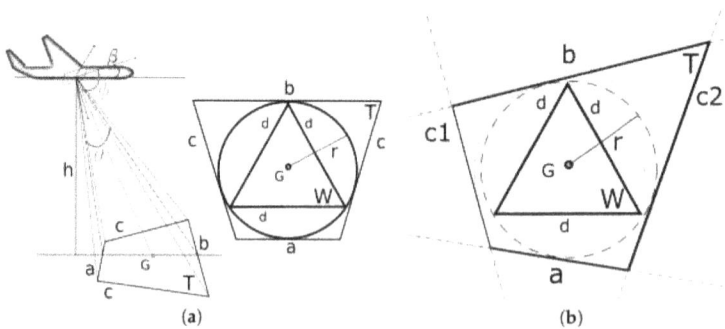

(a)   (b)

Figure 5. The FoV footprint, which is used in the test case of this paper, can be seen in (**a**) and forms a trapezoid T. The pitch angle β, along with the sensor's view angle θ, the angle ψ that pitch β and the bisector of θ form and the altitude h from the ground are used for the calculation. The CDT constraint is defined by side d of the inscribed equilateral triangle W. Since G is the centroid of T, as well as W, the inscribed circle of T always coincides with the circumscribed circle of W. Then, the orientation of the two shapes is irrelevant. In addition, since W is the largest triangle that can be produced from the triangulation, any smaller triangles will always be inside the inscribed circle of T. In (**b**), the general calculation case is shown for no normal or tangential quadrilateral FoVs. In order to draw the maximum incircle of the quadrilateral, all four expanded triangles (ac_2b, c_2bc_1, bc_1a, c_1ac_2) must be drawn and their incircles found. Afterwards, the largest circle that is also an incircle of the quadrilateral is chosen, in this case the inscribed circle that is adjacent to sides b, c_2 and a, and it is used for extracting the CDT constraint.

From basic geometry, it is known that the radius of an inscribed circle in a trapezoid with bases of a and b is $r = \frac{1}{2}\sqrt{ab}$, and the side of an inscribed equilateral triangle in a circle is calculated as a chord of that circle and is given by $d = r\sqrt{3}$. Then, the upper limit side constraint of the CDT for that FoV can be easily computed. By using the aforementioned information, an initial cell decomposition is obtained based on FoV-sized triangles (see Figure 6).

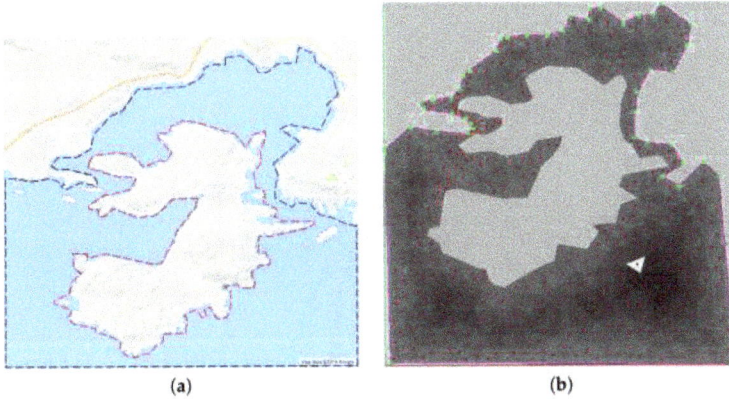

| (a) | (b) |

Figure 6. A CDT for a coastal region. In (**a**), the outer region constraints in black form a complex non-concave polygon. Several no-fly zones inside the constrained area are denoted in red. In (**b**) is depicted the CDT triangulation. The red dots represent the centroids of each cell. The shades of gray denote the Reverse Watershed Schema (RWS) formulation described later in Section 5.3.

5.2. Baseline Area Partitioning Algorithm

Let us consider an undirected graph $G = (V, E)$, where the set V of vertices represents the triangular cells of the CDT and E is the set of edges such that there is an edge from v_i to v_j if the corresponding triangles are neighbors. Two triangular cells are neighbors if a UAV can move freely between them. This graph is intended to be used also to compute roadmaps for coverage path planning after the area partition is obtained. It should be mentioned that the CDT is computed based on the largest FoV among the available UAVs. Later, once the sub-areas A_i are computed, another CDT is performed inside to fit the particular FoV of each UAV.

By treating the CDT as a graph, a baseline area partition algorithm can be designed based on two attributes for each vertex v_i of the graph: $C(v_i)$ as a unitary transition cost; and $A(v_i)$ is the identifier of the UAV that will visit v_i. These attributes are computed as an isotropic cost attribution function by a step transition algorithm, starting from the initial position of each UAV, propagating towards the other UAVs or the borders of the area. Due to the fact that this algorithm expands in waves from each of the UAV and since each agent cannot overtake triangular cells of another agent and it progresses in a breadth-first manner [18], the strategy is called Antagonizing Wavefront Propagation (AWP).

This strategy is presented in Algorithm 1 and works as follows. Let us consider each of the initial positions of the UAVs as the root node of a tree; each root node is given an initial step cost of one. In every recursion step, each vertex that has an edge connected to the parent vertex is given that cost plus one. In addition, vertex v_i gets the same $A(v_i)$ attribute of its parent vertex v_j, propagating the identifier of the UAV in that way. In case the number of vertices for the U_k UAV meets its autonomy limit, denoted by the total area Z_k it should cover, the algorithm for that UAV stops. Please note that these steps are not performed if a triangular cell already has any of these attributes.

Algorithm 1: Antagonizing wavefront propagation algorithm that computes the baseline area partition. Q is a queue list managed as an FIFO by functions *insert* and *getFirst*.

---

v_{Ik}: initial vertices/triangular cells for each UAV U_k,

S_v: area size of triangular cells v,

$N(v)$: the set of neighbors of vertex v,

$A(v)$: the UAV identifier allocated to triangular cell v,

Z_k: area coverage capability of UAV U_k in square meters

S_{vMin}: area size of the smallest triangular cell in CDT

foreach $U_k \in CDT$ **do**

$\quad$ Q.insert(v_{Ik}) and mark v_{Ik} as visited;

$\quad$ $Z_k \leftarrow Z_k - S_{v_{Ik}}$;

end

while *Q not empty* **do**

$\quad$ $v \leftarrow$ Q.getFirst();

$\quad$ **foreach** $v_i \in N(v)$ **do**

$\quad\quad$ $k \leftarrow A(v)$;

$\quad\quad$ **if** v_i *not visited AND* $Z_k > S_{vMin}$ **then**

$\quad\quad\quad$ $C(v_i) \leftarrow C(v) + 1$;

$\quad\quad\quad$ $A(v_i) \leftarrow A(v)$;

$\quad\quad\quad$ mark v_i as visited;

$\quad\quad\quad$ Q.insert(v_i);

$\quad\quad\quad$ $Z_k \leftarrow Z_k - S_{v_i}$;

$\quad\quad$ **end**

$\quad$ **end**

end

Since array Q is accessed once for every i-th cell, the *while* iteration has a complexity of $\mathcal{O}(n)$, where n is the number of vertices. The complexity of getting the first element is $\mathcal{O}(1)$; then, it inserts new elements according to the restrictions. The insertion in a stack has a complexity of $\mathcal{O}(1)$. Hence, the complexity of Algorithm 1 is $\mathcal{O}(n)$. The area partition computed is not sufficient for complex cases where a deadlock occurs after applying Algorithm 1. A further adjustment step is needed in order to assign regions where a deadlock happened, as will be described in Section 5.4.

5.3. Reverse Watershed Schema

By performing the previous step, each configuration space is either adjacent to another configuration space or to the borders of the whole region. In that manner, a second algorithm (see Algorithm 2) assigns to each vertex that already has an UAV identifier a unitary border-to-center cost attribute $D(v_i)$ of proximity from the borders to the center of the configuration space. The triangular cells that are adjacent to a border with another configuration space or to the whole area are given a high $D(v_i)$ cost and are considered as the root nodes of a tree. In each step of the algorithm, this cost is decreased and propagated to the adjacent triangular cells of these nodes. This function manages to create a border-to-center pattern resembling a watershed algorithm, and then, it is called the Reverse Watershed Schema (RWS) algorithm.

Algorithm 2: RWS algorithm for the generation of the border-to-center cost $D(v_i)$ attribute. Q is a queue list managed as an FIFO by functions *insert* and *getFirst*.

$N(v)$: the set of neighbors of vertex/triangular cell v,
$A(v)$: the UAV identifier for triangular cell v
foreach $v \in CDT$ **do**
    foreach $v_i \in N(v)$ **do**
        if $A(v) \neq A(v_i)$ **then**
            mark v as visited;
            $D(v) \leftarrow \infty$;
            Q.insert(v);
        end
    end
end
while Q *not empty* **do**
    $v \leftarrow$ Q.getFirst();
    foreach $v_i \in N(v)$ **do**
        if v_i *not visited* **then**
            $D(v_i) \leftarrow D(v) - 1$;
            mark v_i as visited;
            Q.insert(v_i);
        end
    end
end

Here, we have to note that the complexity of this algorithm is similar to the previous one. The first loop has a complexity of $\mathcal{O}(n)$, $\mathcal{O}(n)$ for the initial *foreach* loop and $\mathcal{O}(1)$ for each insertion to the stack, since the second inner *foreach* has a maximum of three iterations. For the same reasons, the second *while* loop has also a $\mathcal{O}(n)$ complexity.

5.4. Adjustment Function for Deadlock Scenarios

The previous baseline partitioning algorithm is able to perform well in most cases where the area is simple or where the initial positions of the UAVs are evenly distributed in the area. Nevertheless, it may lead to several deadlock scenarios as the growing sub-areas meet each other, as can be seen in the example shown in Figure 7. Hence, a Deadlock Handling (DLH) algorithm that adjusts the initial partitioning in the non-allocated areas is needed, by exchanging UAV identifiers or assigning UAVs to the empty areas. Two different approaches have been tested, by applying the two algorithms presented before.

As was stated before, each UAV U_k should cover an area of Z_k. In a test case, after the initial partitioning, let us consider that $Y_k \neq Z_k$ space has been allocated to U_k. In the deadlock scenarios (see Figure 7), there are areas that do not belong to any UAV. These areas are allocated to a virtual UAV U_{-1} with area $Z_{-1} = 0$. Let us consider a list L_U that contains the results of $Z_k - Y_k$ for each UAV and the area size of the smallest triangle in the CDT S_{vMin}. Thus, each UAV can have an area surplus if $Z_k - Y_k > S_{vMin}$ or a shortfall if $Z_k - Y_k < S_{vMin}$. The latter case always happens in the deadlock scenarios for U_{-1}. In each recursion of Algorithm 3, a pair of UAVs, one having an area surplus and another with a shortfall, is chosen from the list in order to gradually exchange triangular cells between them to reach the desired area size. In order to do so, a feasible transition sequence must be found, as can be seen in Figure 8.

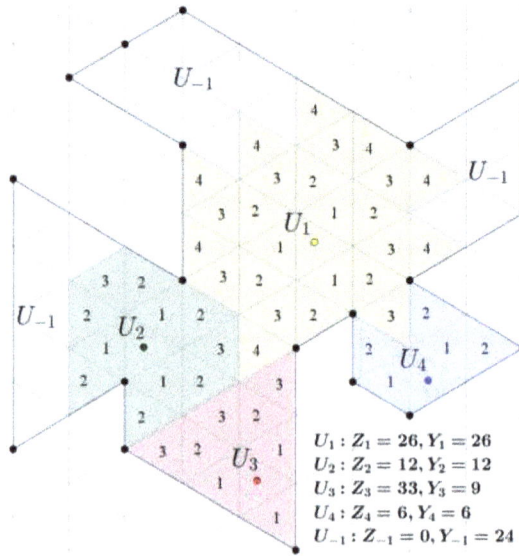

Figure 7. A deadlock scenario. Four UAVs U_1, U_2, U_3 and U_4 after the baseline partition Algorithm 1. UAVs U_1, U_2 and U_4 have met their autonomy capability of Z_k by covering Y_k area. Nevertheless, U_3 was not able to overtake any more area, being "blocked" by the other UAVs and the borders of the whole region. Colored areas indicate the configuration space of each UAV, while the numbers inside the cells indicate the isotropic cost, as has been assigned by Algorithm 1. The free or non-allocated areas belong to virtual UAV U_{-1}.

Algorithm 3: Multi-UAV partitioning Deadlock Handling (DLH) algorithm. Baseline partitioning is performed by Algorithm 1, whereas this method is for the sub-area size adjustment (if needed). Function $getSurplusUAV(L)$ gets a UAV identifier from list L that has an area surplus, whereas function $getShortfallUAV(L)$ gets the identifier of a UAV that has an area shortfall after Algorithm 1. Function $findSequence$ finds a feasible transition sequence P_{ij} between UAV U_i and U_j, whereas the *move* function performs the transfer between triangular cells.

S_{vMin}: area size of the smallest triangular cells in CDT
while $\exists U \in L_U < S_{vMin}$ **do**
$\quad$ $i \leftarrow$ getSurplusUAV(L);
$\quad$ $j \leftarrow$ getShortfallUAV(L);
$\quad$ $P_{ij} =$ findSequence(U_i, U_j);
$\quad$ **if** $Z_i - Y_i > Z_j - Y_j$ **then**
$\quad\quad$ move($Z_j - Y_j$, P_{ij});
$\quad$ **end**
$\quad$ **else**
$\quad\quad$ move($Z_i - Y_i$, P_{ij});
$\quad$ **end**
end

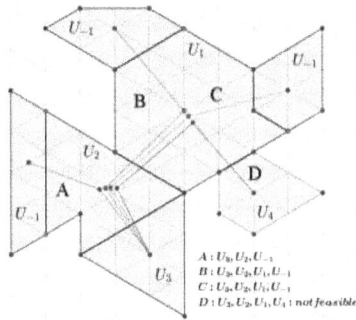

Figure 8. Transition sequence selection. After the initial partition process, U_1, U_2 and U_4 have met their sub-area size constraint and blocked the growth of U_3. As a result, three areas are not allocated (U_{-1}). The feasible transition sequences A(green), B(red) and C(blue) are used in order for U_3 to obtain the requested total area, by gradually exchanging cells in every pair of the sequence. Sequence D(black) does not lead to a partition that has an area shortfall, and thus, it is a not feasible sequence.

The complexity of this algorithm is calculated as $\mathcal{O}(Un^2)$ due to the *findSequence* function, which is actually a tree sort; U is the number of UAVs and n the number of cells. The complexity of the *move* functions is displayed below, in each of the following transposition algorithms.

Two algorithms called moveAWP and moveRWS have been implemented for the move function, which is used in Algorithm 3. In the former, the farthest vertex in the A_i sub-area of UAV U_i in sequence P_{ij} is chosen, by using the information from Algorithm 1. This vertex has also to be adjacent to the second area in the transition sequence P_{ij}. Starting from that vertex, Algorithm 1 is applied again, overtaking the requested area size in the means of exchanging UAV identifiers between those triangular cells. Recursively, this operation is performed for every item of the sequence (see Algorithm 4). The complexity is $\mathcal{O}(n)$, where n is the number of areas that are in the transition sequence. Then, since Algorithm 1 is used for the transposition function, the whole complexity is $\mathcal{O}(n * m^2)$, where m^2 is the *findSequence* algorithm complexity.

Algorithm 4: MoveAWP algorithm. C_v is the transition cost from the AWP algorithm (see Algorithm 1). Then, function $FindBiggestC_v(P[i], P[i+1])$ finds the largest transition cost value triangular cell of UAV $P[i]$ that is adjacent to UAV $P[i+1]$ in the sequence. Then, function Awp takes as variables an initial cell v, the area size that needs to be exchanged and the UAV identifier that needs to be exchanged from. The growing function is similar to Algorithm 1.

P_{ij} the transition sequence between U_i and U_j for triangular cells exchange, treated as a list
v_{init}: initial triangular cell for identifier exchange
S: area size to be moved
foreach $U_i \in P_{ij}$ **do**
$\quad v_{init} = FindBiggestC_v(P[i], P[i+1]);$
$\quad Awp(v_{init}, S, P[i+1])$
end

In the second approach, we apply the RWS algorithm in order to get a depth schema of the adjacent areas, as was described in Section 5.3. In each recursion of the algorithm (see Algorithm 5), the amount of triangular cells that are in the borders of the first pair of the transition sequence exchanges their UAV identifiers in order to change from UAV_{P_i} to $UAV_{P_{i+1}}$. If the area of these border triangular cells sum up less than the requested area, then the area size and the total amount of border triangular cells exchange their UAV identifiers. If not, then only the triangular cells in the front (in the borders) exchange their UAV identifiers. This amount of triangular cells is then exchanged to the next UAV in the sequence and so on, maintaining the aforementioned restriction, until all of the requested area and associated triangular cells are transposed from the initial UAV in the sequence to the last.

Algorithm 5: MoveRWS algorithm. Function *FindSequence* finds a valid transition sequence, as can be seen in Figure 8. This function is also called before the initial recursion of the MoveRWS algorithm. Function *ExchangeIdentifiers* makes use of the information of the RWS algorithm (see Algorithm 2), and it exchanges agent identifiers on two adjacent configuration spaces, by exchanging the amount of triangular cells that have the lowest coverage cost, but are adjacent. It also propagates and extends this cost. Function *RestOfSequence* returns the remaining sequence for the specific $P[i] \rightarrow P[i+1]$ transition, in order to initially transfer only the amount of triangular cells that are adjacent between i and $i+1$ until the final U_j UAV. In case this happens, the requested area has not been exchanged yet, so the algorithm runs recursively, and the last line takes a step back in sequence traversal.

S area size to be moved
$S_{adj(kl)}$ the area size of adjacent triangular cells between UAV k and l
P_{ij} the transition sequence between U_i and U_j for triangular cell exchange, treated as a list
foreach $U_i \in P_{ij}$ **do**
    $P_{ij} \leftarrow$ FindSequence(U_i, U_j);
    if $S_{adj(P[i],P[i+1])} > S$ **then**
        ExchangeIdentifiers(P[i], P[i+1], S);
    end
    else
        P_{rest} = RestOfSequence(U_i);
        MoveRWS($S_{adj(P[i],P[i+1])}, P_{rest}$);
        $S = S - S_{adj(P[i],P[i+1])}$;
        $i = i - 1$;
    end
end

This algorithm's complexity is $\mathcal{O}(Un^2)$ due to the use of the *findSequence* function, as has been described in Algorithm 3.

There are two main differences in these approaches. In the first approach, we have a wavefront pattern from a single triangular cell, whereas in the second, the exchange progresses as a kind of width sweep Morse function [2]. The second difference is that in the first approach, all of the triangular cells to be exchanged are the transposed UAV first, and in the second approach, only the amount of triangular cells that are in the adjacent borders are transposed in each step. In that manner, the triangular cells of the area are propagated respecting the total amount of cells that each UAV has each time, resolving overlapping UAV issues, as will be discussed in Section 6.

6. Simulation Results

The proposed algorithms are implemented in C++ using the CGAL library [19] for the constrained Delaunay triangulation and ROS (Robotic Operating System) [1] for the integration framework. In order to test the behavior of the UAVs, we have used a Software In The Loop (SITL) [20] simulation setup on a single computer, which is described later in Section 6.2.

Three coastal areas in Greece have been selected for the experiments (Figure 9). The first is a broad and populated shore near the harbor of Piraeus, Salamina. The second and third are remote islands in the Aegean archipelago, Astipalea and Sxoinousa. The first area was used for evaluating and comparing the partitioning algorithms, the second for evaluating the proposed strategy in various setups, whereas the third was used for computing narrow coverage trajectories.

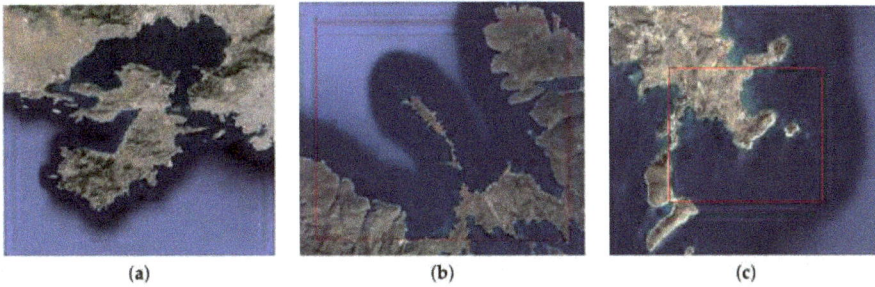

(a)   (b)   (c)

Figure 9. Selected areas for testing: (**a**) Salamina area having narrow passages and complex shapes in shores; (**b**) the Astipalea area is used for testing the suitability of the proposed algorithm; (**c**) Sxoinousa area used for coverage planning. The red square shows the region where the simulated flights occurred.

The coordinate frames chosen for the UAVs and on-board sensors and the software architecture used in the simulations are described in the following.

6.1. Coordinate Frames

In the simulations, we have considered two reference frames, one for the UAV $\{U\}$ and one for its on-board sensor $\{S\}$. For the UAV, the reference frame has its x-axis pointing forwards in relation with movement; the y-axis is given by the right-hand rule; while the z-axis points downwards. For the on-board sensor and its relation with the vehicle, the coordinate frame is shown in Figure 10.

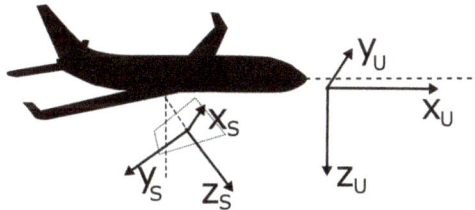

Figure 10. UAV and sensor coordinate frames and their relation. Depending on the roll (γ), pitch (β) and yaw (α) angles of the sensor with respect to the UAV fuselage, a rotation of the projected field of view occurs.

The roll (γ), pitch (β) and yaw (α) angles are used to define the orientation of the sensor. In the $\gamma = \beta = \alpha = 0$ case, the y_S sensor axis coincides with the z_U axis of the UAV while the z_S coincides with x_U. In that case, in order to translate a sensor vector to the UAV coordination frame, the rotation matrix used is:

$$\mathbf{R}_U^S(\alpha = 0, \beta = 0, \gamma = 0) = \begin{bmatrix} 0 & 0 & 1 \\ 1 & 0 & 0 \\ 0 & 1 & 0 \end{bmatrix}. \tag{3}$$

Regarding the rotation movements along the three axes, the usual convention in aviation is used, where counterclockwise rotation movements of yaw, pitch and roll are considered. Yaw is the rotation of α about the z-axis; pitch is the rotation of β about the y-axis; and roll is the rotation of γ about the x-axis. These angles change the orientation of a given frame by applying the rotation matrix:

$$\mathbf{R}_{U}^{S} = \mathbf{R}(\alpha, \beta, \gamma) = \mathbf{R}_z(\alpha)\mathbf{R}_y(\beta)\mathbf{R}_x(\gamma) =$$

$$= \begin{bmatrix} \cos\alpha\cos\beta & \cos\alpha\sin\beta\sin\gamma - \sin\alpha\cos\gamma & \cos\alpha\sin\beta\cos\gamma + \sin\alpha\sin\gamma \\ \sin\alpha\cos\beta & \sin\alpha\sin\beta\sin\gamma + \cos\alpha\cos\gamma & \sin\alpha\sin\beta\cos\gamma - \cos\alpha\sin\gamma \\ -\sin\beta & \cos\beta\sin\gamma & \cos\beta\cos\gamma \end{bmatrix}. \tag{4}$$

The on-board sensor orientation is derived by multiplying (3) by (4) and gives the full rotation matrix that allows one to transform a vector expressed in the on-board sensor frame to the UAV reference frame as:

$$\mathbf{R}_{U}^{S} = \mathbf{R}(\alpha, \beta, \gamma) = \mathbf{R}_z(\alpha)\mathbf{R}_y(\beta)\mathbf{R}_x(\gamma) =$$

$$= \begin{bmatrix} -\sin\beta & \cos\beta\sin\gamma & \cos\beta\cos\gamma \\ \cos\alpha\cos\beta & \cos\alpha\sin\beta\sin\gamma - \sin\alpha\cos\gamma & \cos\alpha\sin\beta\cos\gamma + \sin\alpha\sin\gamma \\ \sin\alpha\cos\beta & \sin\alpha\sin\beta\sin\gamma + \cos\alpha\cos\gamma & \sin\alpha\sin\beta\cos\gamma - \cos\alpha\sin\gamma \end{bmatrix}. \tag{5}$$

However, as was mentioned in Section 4, pitch and roll angles of the UAV with respect to the horizontal plane are not relevant to the FoV projection, since we are considering a gimbal on board, which compensates for these angles.

6.2. Simulation Architecture and Configuration

The simulations have been performed on computers with an Intel Core i5-5200U@2.20-GHz CPU with 8 GB of RAM and the kUbuntu 14.04 distribution of the Linux OS. The software architecture adopted is shown in Figure 11.

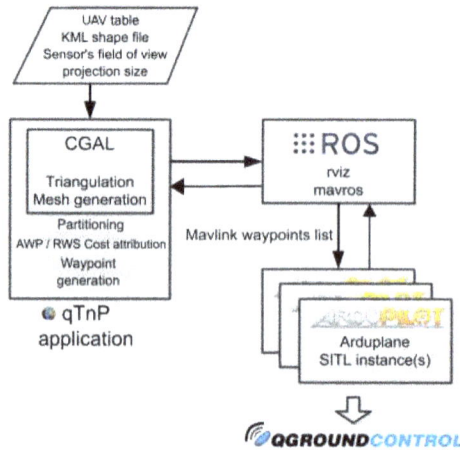

Figure 11. Software architecture with different libraries and components: the latest CGAL library (4.8.1) [19], ROS Indigo [1] components (the rviz package [21] for visualization and the mavros node [22] for the mavlink interface with the simulated UAV), an Arduplane instance [23] of the Ardupilot SITL [20], which uses the JSBSim flight dynamics model [24], and the qgroundcontrol control station [25].

The main application is based on the Qt (https://www.qt.io/) cross-platform software development framework. The setup consists of a configuration window (Figure 12) where the number of the UAVs along with their attributes can be set. These attributes are the sensor type, the FoV size referring to the maximum triangular side size, as had been defined in Section 5, a percentage of the whole region to be used in the partition step, initial positions and tasks. The configuration application sets the type of visualization that will be performed in rviz: showing the borders of each sub-area, coloring it depending on different parameters and showing the produced waypoints for coverage. Regarding the CDT, its constraints of minimum angle and initial triangulation maximum edge can be also defined, and the user can define the area of interest by uploading a KML file, including obstacles. Finally, each step of the simulation can be performed separately; performing the triangulation, extracting the partition for each UAV based on its percentage of the total region and computing coverage waypoint plans for each UAV.

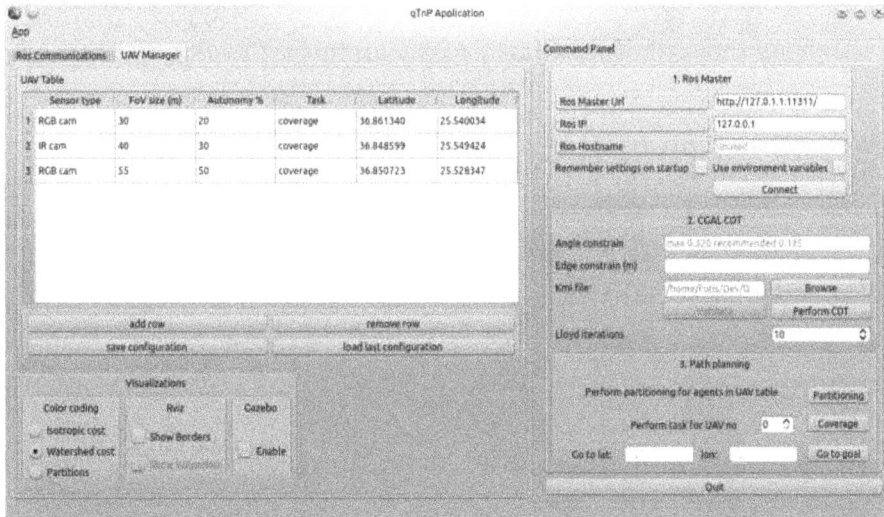

Figure 12. The qTnP main application. The first tab echoes the ROS communication messages and logs. The main "UAV Manager" tab of the application includes the UAV management table, indicating the sensor type, the cell (FoV) size, autonomy percentages and initial positions. It also includes the visualization options for rviz, showing the cost values of each of the proposed algorithms, visualizing the partitioned configuration space, showing the borders of each UAV and the produced waypoints for coverage. Finally, the command panel on the right includes connection settings, CDT-specific configuration, the KML file of the area, as well as several command buttons for the different stages of the experiments.

The implemented algorithms are part of an ROS node named qTnP (Qt Triangulation and Planning, Figure 12). This node performs all calculations and manages the communication with the rest of the ROS nodes of the configuration. Visualization of the mesh of the area, partitioned areas, cost attribution, waypoints and produced paths is handled by the rviz node, whereas the produced waypoint stacks are sent to mavros node. This node has a dual purpose. It maintains the connection with the simulated vehicles, sending waypoint list plans when the main application produces them. It also listens to the simulated UAVs, which report the mavros node on each cycle for their current position and telemetry data.

Regarding the UAV model used in the simulations and its on-board controller, the open source autopilot Ardupilot has been used. Its arduplane instance for fixed wing model aircraft has been combined with the JSBSim flight dynamics model simulator. In our setup, the system simulates the dynamics of the Rascal110 model airplane. The Arduplane controller used is the Pixhawk Flight

Management System [26]. The behavior of the vehicle during the simulated flight, as well as the produced trajectories were monitored live using the open source ground station qgroundcontrol [25].

6.3. Partitioning Algorithms Comparison in Simulation

The partitioning strategies called MoveAWP and MoveRWS described in Section 5.4 have been compared. The former uses the transition cost of the AWP algorithm, and the latter applies the RWS algorithm for adjusting the baseline partition computed by Algorithm 1. In both cases, two FoV sizes have been used, in order to show the impact in the behavior of the algorithms of small and large values. The FoV size values in the simulations refer to the maximum triangular cell side, as has been defined in Section 5. Three test scenarios were simulated with different relative capabilities for the UAVs, and the results are shown in Figure 13.

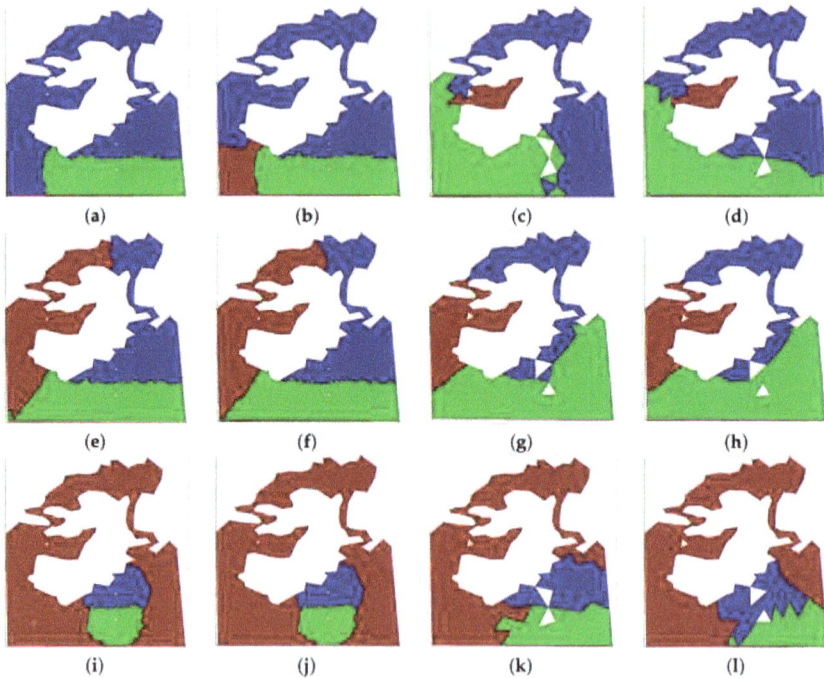

Figure 13. Partitioned area for three UAVs (indicated by the white cells) and visualized by using the ROS rviz node. Each row represents the results for different relative capabilities: the first row is the 10% (red), 60% (blue), 30% (green) case; the second row depicts the 33% (red), 33% (blue), 34% (green) case; whereas the last row shows the 80% (red), 10% (blue), 10% (green) case. In each row, each pair of images indicates the comparison of the two algorithms. (**a**,**b**) show how the MoveAWP and MoveRWS algorithms have performed with the small (250 m) FoV, whereas (**c**,**d**) show the results for the large (2 km) FoV case. (**a**) MoveAWP 250 m FoV; (**b**) MoveRWS 250 m FoV; (**c**) MoveAWP 2 km FoV; (**d**) MoveRWS 2 km FoV; (**e**) MoveAWP 250 m FoV; (**f**) MoveRWS 250 m FoV; (**g**) MoveAWP 2 km FoV; (**h**) MoveRWS 2 km FoV; (**i**) MoveAWP 250 m FoV; (**j**) MoveRWS 250 m FoV; (**k**) MoveAWP 2 km FoV; (**l**) MoveRWS 2 km FoV.

The complexity of the area has managed to highlight some issues that were not evident for the majority of simple areas. The main problem occurs during cell exchange when the initial position of the UAV is close to the borders, because a sub-area could overtake the initial position of the UAV (see Figure 13a).

Additional simulations have been performed to measure the performance of the different algorithms with respect to the metrics F and G explained in Section 3. In particular, the simulation environment shown in the second area of Figure 9 has been used for the metric F. Some results are detailed in Figure 14 for three and five UAVs and a FoV size of 30 m. In general, simulations have been executed for three and five UAVs, with initial locations evenly or randomly distributed in the area and different FoV sizes. The results for the sum of distances between each center of the triangular cell inside a sub-area and the initial location of the UAV inside that sub-area (metric F) are shown in Tables 1 and 2. In both cases, it can be seen that the moveRWS algorithm has a better performance than moveAWP, since the metric is lower.

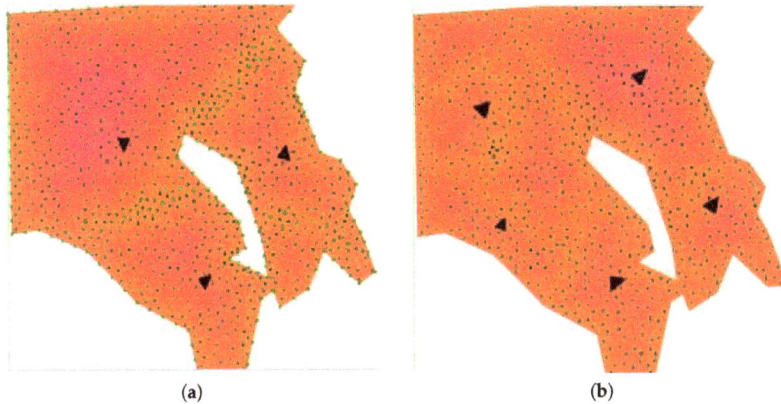

(a)   (b)

Figure 14. Area of Figure 9b selected for the comparison of the two partitioning algorithms. (**a**) is partitioned for three UAVs, whereas (**b**) for five UAVs. The depicted FoV size is 30 m in both cases. Both figures are computed with the deadlock moveRWS handling of Algorithm 5.

Table 1. An even distribution of initial locations for three and five UAVs, with different relative capabilities.

		FoV (15 m)		FoV (30 m)	
	#UAVs	moveRWS	moveAWP	moveRWS	moveAWP
Metric *F* (m)	3	333,861.84	333,909.67	82,768.44	84,979.76
Metric *F* (m)	5	437,988.74	439,642.85	129,879.24	131,516.96

Table 2. Random initial position distribution for three and five UAVs, with different relative capabilities. Like before, Algorithm 5 has performed better than Algorithm 4.

		FoV (15 m)		FoV (30 m)	
	#UAVs	moveRWS	moveAWP	moveRWS	moveAWP
Metric *F* (m)	3	508,801.74	508,751.513	211,395	214,945.82
Metric *F* (m)	5	566,971.55	568,819.45	151,389.621	155,269.99

Regarding the other metric G considered in Section 3, simulations have been performed also in the second scenario of Figure 9 with three and six UAVs with different relative capabilities and initial locations evenly and randomly distributed (see Tables 3 and 4, respectively). The goal is to compare the results computed with the baseline algorithm and the improvement achieved with the moveRWS deadlock handling algorithm, which had the better performance in the previous scenarios. Figure 15 shows the results for two particular setups with four UAVs.

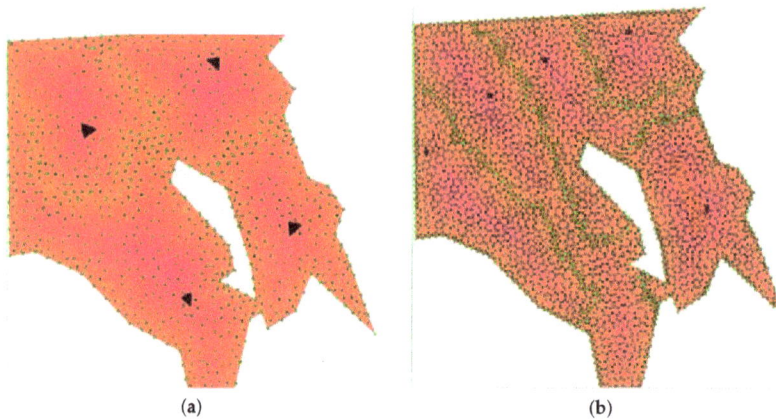

(a)   (b)

Figure 15. Area partition after applying the baseline and the deadlock moveRWS handling algorithm. (a) shows the area partitioned for four UAVs evenly distributed in the area and a FoV projection of 30 m. (b) shows the results for a FoV projection of 15 m and four UAVs randomly located. The black triangles depict the initial positions in all of the cases.

Finally, different numbers of Lloyd iterations on the resulting mesh have been tested, ranging between 20 and 60 iterations. In the simulations, different numbers of UAVs with even and random distributions for the initial locations (see Tables 3 and 4, respectively), different relative capabilities and FoV projections have been used. The results show the suitability of the proposed solution, as the average difference from the targeted relative capability of the UAVs has not exceeded a value of 1% on average for the even distribution and 1.33% for the random distribution of the initial locations; Figure 16 shows this comparison of the average difference in the even and random distribution scenarios. Moreover, the algorithm manages to properly overcome deadlock scenarios as expected, as can be seen in the various setups of Table 4, where the initial baseline algorithm has up to 30% difference from the targeted relative capabilities of the UAVs.

Table 3. For each UAV, the difference from its given capability is shown after the initial baseline algorithm and after the moveRWS deadlock treatment algorithm. An average for all UAVs, as well as the total difference is shown below each experimental setup, where G is the metric defined in Equation 2 and area(R) is the area in m^2 of the whole region R. The UAVs have evenly distributed initial positions. Setups for 3, 4, 5 and 6 UAVs have been tested, with different relative capabilities and FoV values. Different Lloyd iterations on the mesh have been tested, ranging between 20 and 60.

UAV Capability %	FoV (15 m)			FoV (30 m)		
	Lloyd Iterations			Lloyd Iterations		
	20	30	60	20	30	60
50%	0.5/0.92	0/0.62	0/0.92	0.35/0.3	0.35/0.18	0.05/0.3
30%	6.74/0.14	6.74/0.89	6.88/1.21	6.95/0.21	7.19/0.65	6.86/0.25
20%	0.01/1.04	0.01/0.26	0.01/0.28	0/0.49	0.03/0.82	0.05/0.54
Average%	2.41/0.69	2.25/0.59	2.29/0.80	2.43/0.32	2.52/0.55	2.32/0.36
G/area(R)%	7.25/2.1	6.75/1.78	6.89/2.42	7.3/0.99	7.57/1.66	6.96/1.09
20%	0.02/0.02	0.02/0.08	0.02/0.06	0/0.41	0.05/0.12	0.05/0.45
40%	0/0.06	0/0.11	0/0.05	3.89/0.28	3.91/0.59	3.6/0.49
20%	0.96/0	1.11/0	0.78/0.06	0.48/0.17	0.48/0	0.62/0.12
20%	0/0.02	0/0.17	0/0.03	0/0.29	0/0.48	0/0.07
Average%	0.25/0.03	0.28/0.09	0.2/0.05	1.09/0.37	1.11/0.3	1.07/0.28
G/area(R)%	0.98/0.12	1.13/0.35	0.8/0.2	4.37/1.11	4.44/1.18	4.27/1.13

Table 3. *Cont.*

UAV Capability %	FoV (15 m)			FoV (30 m)		
	Lloyd Iterations			Lloyd Iterations		
	20	30	60	20	30	60
20%	0.15/0.38	0.15/0.6	0.18/0.01	0.15/0.11	0.11/0.59	0.11/1.36
30%	13.73/0.34	13.75/0.31	13.72/0.15	13.1/0.24	12.98/0.76	12.63/1.64
20%	0/0.31	0/0.42	0/0.08	0/1.6	0.3/0.81	0/0.66
10%	0/0.41	0/0.86	0/0.41	0/0.71	0/0.3	0/1.21
20%	0.21/0.65	0.05/0.14	0.21/0.36	0/1.01	0/0.36	0/1.61
Average%	2.81/0.42	2.79/0.47	2.28/0.35	2.65/0.92	2.68/0.56	2.55/0.9
G/**area**(*R*)%	14.09/2.1	13.95/2.33	14.11/1.01	13.25/3.69	13.39/2.81	12.74/4.5
10%	0.15/0.21	0.15/0.73	0.15/0.41	0.11/0.72	0/0.18	0.11/0.29
20%	0.3/1.04	0.3/1.37	0.3/1.35	0/1.83	0/1.37	0/0.43
10%	0.15/0.23	0.15/0.16	0.15/0.23	0.11/0.14	0/0.97	0.11/0.39
30%	11.26/0.38	11.45/0.69	11.2/0.53	11.94/0.84	12.64/0.83	10.75/0.96
20%	0.3/0.51	0.3/0.36	0.3/0.33	0.24/0.8	0/0.98	0.24/0.55
10%	0.15/0.19	0.15/0.47	0.15/0.15	0.11/1.04	0.11/1.25	0.11/0.98
Average%	2.05/0.41	2.08/0.63	2.04/0.5	2.09/0.9	2.13/0.93	1.89/0.6
G/**area**(*R*)%	12.32/2.47	12.5/3.78	12.25/3.01	12.51/5.36	12.75/5.58	11.32/3.59

Table 4. For each UAV, the difference from its given capability is shown after the initial baseline algorithm and after the moveRWS deadlock treatment algorithm. An average for all UAVs, as well as the total difference is shown below each experimental setup, where G is the metric defined in Equation 2 and area(R) is the area in m^2 of the whole region R. The UAVs have randomly distributed initial positions. Setups for 3, 4, 5 and 6 UAVs have been tested, with different relative capabilities and FoV values. Different Lloyd iterations on the mesh have been tested, ranging between 20 and 60.

UAV Capability %	FoV (15 m)			FoV (30 m)		
	Lloyd Iterations			Lloyd Iterations		
	20	30	60	20	30	60
50%	0/0.1	0/0	30.8/0.27	0/0	0/1.49	29.81/0.5
30%	1.19/0.02	0.08/0.02	0.01/0.16	0.024/0.024	20.9/0.65	0.01/0.57
20%	0.01/0.09	0.01/0.01	0.01/0.11	0.024/0.024	0.024/0.84	0.05/1.1
Average%	0.4/0.07	0.03/0.01	10.27/0.18	0.02/0.02	6.97/0.99	9.95/0.72
G/**area**(*R*)%	1.2/0.21	0.09/0.03	30.82/0.54	0.05/0.05	20.92/2.98	29.87/2.17
20%	0/0.43	15.37/0.22	2.19/0.49	0/0.730	14.2/1.66	0.62/0.05
40%	13.1/0.33	0/0.53	0/0.22	14/0.43	0/1.89	0.07/0.14
20%	6.48/0.31	0/0.06	2.64/0.35	6.24/0.56	0/0.12	1.8/0.16
20%	0/0.42	0/0.69	0/0.35	0/0.26	0/0.12	0.02/0.07
Average%	4.9/0.37	3.84/0.38	1.21/0.36	5.06/0.5	3.55/0.95	0.67/0.11
G/**area**(*R*)%	19.58/1.48	15.37/1.5	4.83/1.43	20.24/1.98	14.2/3.79	2.67/0.42
20%	0.01/0.03	0.01/0.21	11.23/0.77	0/1.8	0.02/1.43	10.78/0.4
30%	0.53/0.02	12.85/0.24	0.02/0.68	0/0.27	13.45/0.42	0.02/1.4
20%	0.01/0.03	0.01/0.11	11.59/2.65	6.24/0.57	6.88/0.44	10.78/0.64
10%	0.01/0.07	0.01/0.19	3.382/1.19	0/0.46	0.04/0.21	3.97/0.04
20%	0.01/0.14	0.01/0.14	0.01/1.36	0/1.03	0.02/0.37	0.02/0.4
Average%	0.11/0.05	2.58/0.18	5.25/1.33	1.25/0.82	4.12/0.57	5.11/0.58
G/**area**(*R*)%	0.57/0.27	12.89/0.89	26.24/6.64	6.24/4.12	20.59/2.87	25.57/2.88
10%	0.02/0.46	4.36/0.05	4.2/0.26	0.04/0.36	0.04/0.18	0.04/0.09
20%	0.02/1.02	4.23/0.26	0.01/0.06	0.05/0.3	0.05/0.36	0.05/1.49
10%	0.02/0.42	0.01/0.47	0.01/0.01	0.04/1.2	0/0.41	0.04/0.25
30%	19.04/0.5	0.01/1.22	0.01/0.01	20.32/0.74	20/1.48	19.85/0.31
20%	0.01/0.9	0.02/0.04	0.01/0.02	10.32/0.49	10.32/0.24	10.08/0.65
10%	0.02/0.66	0.01/0.58	0.01/0.2	0.04/0.028	0/1.01	0.04/0.22
Average%	3.18/0.66	1.44/0.43	0.71/0.09	5.13/0.52	5.07/0.61	5.02/0.5
G/**area**(*R*)%	19.13/3.96	8.64/2.62	4.25/0.56	30.81/3.12	30.41/3.68	30.1/3.01

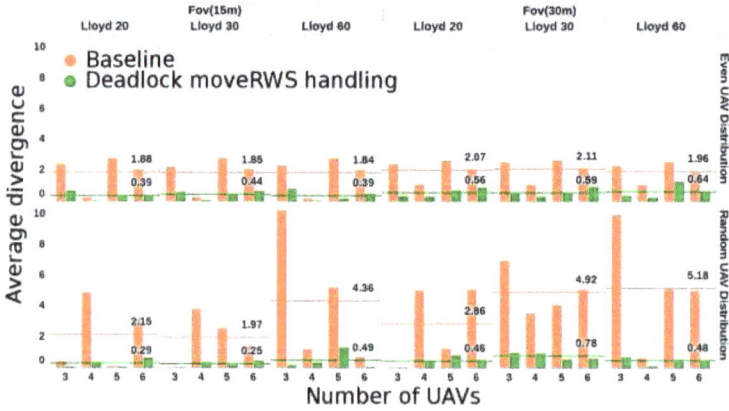

Figure 16. A graphical representation of Tables 3 and 4. Average difference after the baseline algorithm and after the deadlock moveRWS treatment algorithm. As expected, random initial positioning of UAVs creates more often deadlock scenarios for the baseline algorithm. The algorithm has been tested for 3–6 UAVs, evenly or randomly distributed in the area. FoV projections of 15 and 30 m have been tested, and in each case, a different Lloyd iteration setting (20, 30 and 60) has been set. The horizontal lines show the average difference.

6.4. Coverage Path Planning Simulation Results

The framework presented in the previous sections, and in particular Algorithm 2, can be also applied to generate waypoint lists for the UAV to achieve complete coverage of a complex coastal sub-area. By using the border-to-center cost described in Section 5.3, inward spiral-like waypoint lists W can be generated. Algorithm 6 performs a selection of vertices by initiating from the vertex that has the highest $D(v)$ cost and is closer to the starting position of the UAV. In every recursion, the closest adjacent cell v_j that has the same cost $(D(v_j) = D(v))$ is inserted in the list. In case all of the selected vertices have the same cost, the algorithm reduces the visiting cost and chooses the cell that is closer to the previous step. It should be mentioned that the complexity of this algorithm is $O(n^2)$.

Algorithm 6: Waypoint list computation for coverage. D_c is an auxiliary variable with the current border-to-center cost in each step, whereas v_{I_k} is the starting position of the UAV U_k. Function $findClosest$ finds the closest vertex to the current one that has its same border-to-center cost. CDT_k is the sub-CDT for UAV U_k. W is the produced waypoint list of vertices.

$D_c \leftarrow \infty$;
$v \leftarrow findClosest(v_{I_k}, D_c)$;
$W.insert(v)$;
foreach $v \in CDT_k$ **do**
    if $\exists v, D(v) = D_c$ **then**
        $v_j \leftarrow findClosest(v, D_c)$;
        $W.insert(v_j)$;
        $v \leftarrow v_j$;
    end
    else
        $D_c \leftarrow D_c - 1$;
    end
end

In order to show the coverage trajectories computed, tests have been performed in a particular sector of the area considered in Figure 9c. The area was partitioned into sub-areas respecting the different UAV capability considerations and coverage waypoint lists have been produced, as can be seen in Figure 17a, whereas Figure 17b shows the coverage waypoint trajectories for each UAV. Even though the number of turns is higher in comparison with a square grid decomposition strategy, the area was fully covered fulfilling the constraints considered in this paper. Table 5 shows the values of the parameters considered for the UAVs in this simulation.

Table 5. Values of the parameters considered for the UAVs in the simulations. The FoV projection size is the maximum cell side size of the triangulation. Angle γ is the on-board sensor pitch angle with respect to the horizontal plane. The relative capability percentages represent the capability of each UAV related to the whole area for covering purposes.

UAV	FoV Projection Size (m)	γ (deg)	Altitude (m)	Relative Capability
UAV 1	30	−45	100	20%
UAV 2	40	−45	80	30%
UAV 3	55	−45	120	50%

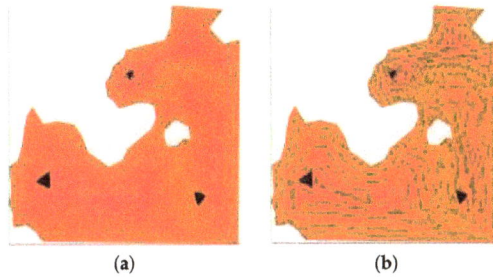

(a)   (b)

Figure 17. Area partitioning for three UAVs on the same location as in Figure 9c. White areas indicate the no-fly zones, whereas the black triangles show the initial positions of the UAVs. In (**a**), the FoV sized cell distribution is shown along with the centers of the triangles. Regarding waypoint generation for coverage, (**b**) shows the produced coverage paths for all of the UAVs. The different shades of orange indicate the border-to-center cost computed by Algorithm 2.

The detailed results for UAV 3 are depicted in Figure 18. It is shown how the sub-area is fully covered with the sensor on board, even considering a very slow data acquisition rate of 1 Hz.

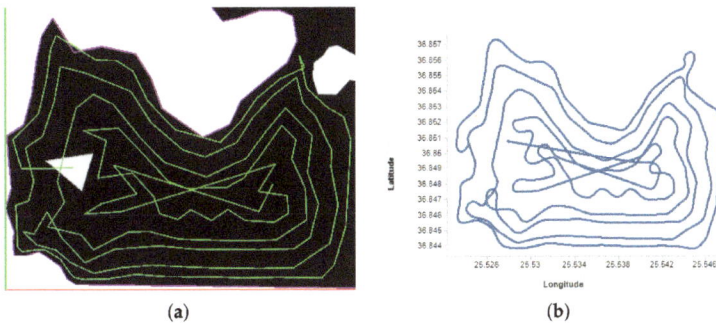

(a)   (b)

Figure 18. *Cont.*

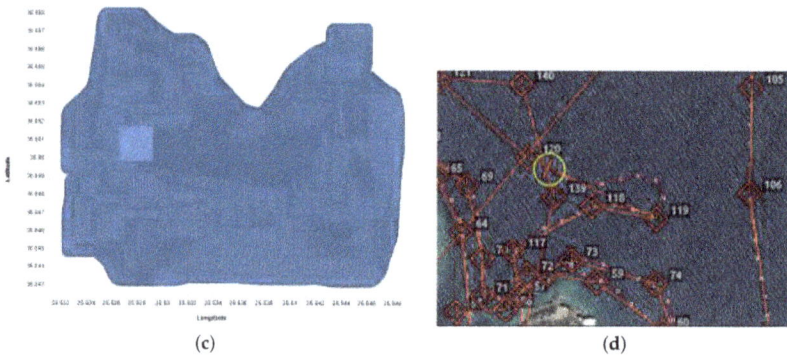

(c)

(d)

Figure 18. Coverage trajectory computed in the simulation. (**a**) shows a detailed view of the UAV 3 trajectory in Figure 17. Latitude and longitude information received during the simulated flight of that UAV is shown in (**b**,**c**) with the total sensor coverage considering a sensor working at a very slow rate of 1 Hz. Finally, (**d**) shows a screenshot of the ground station visualization during the simulations.

7. Conclusions and Future Work

This paper has presented an algorithmic approach that allows one to tackle in a common framework the problems of area decomposition, partition and coverage in a multi-UAV remote sensing context. The produced mesh and associated graph manage to be consistent with the area properties and the capabilities of the UAVs and their on-board sensors. Two novel algorithms have been proposed to solve deadlock scenarios that can be usually found when performing area partition into sub-areas taking into account the relative capabilities of the UAVs.

The current framework is actually a generic waypoint planner that is consistent with the attributes and attitude of the on-board sensor. However, it does not take into consideration the UAV platform dynamics, even though the pitch and roll upper boundary turn rates are used to calculate the maximum triangle cell. Nevertheless, a platform might, in the case of a multirotor, or might not, in the case of a fast moving fixed wing, be able to follow sharp turns that are produced. Hence, this solution does not account for waypoint to waypoint flight trajectories. These issues are usually addressed by the flight controller, for example, by assuming that a waypoint has been visited if the UAV passes close by. This metric is task specific and, in real-world applications, user defined.

Regarding future work, a comparative study has to be performed regarding mesh generation optimization using Lloyd's algorithm. In addition, removing vertices that the UAS has not managed to visit and replacing them online with the current position has to be tested in order to get information on the computational time versus the optimality of the trajectory analysis. On the other hand, uncertainties in the perception of the environment will be encoded in each of the graph vertices for the on-board computation of the trajectory, compensating for changes in the scenario and tasks. Finally, our final goal is to develop a complete system architecture for a team of heterogeneous UAS that will be able to perform in complex coastal areas, having minimal supervision during real flights.

Acknowledgments: This work is partially supported by the MarineUAS Project funded by the European Union's Horizon 2020 research and innovation program, under the Marie Sklodowska-Curie Grant Agreement No. 642153 and the AEROMAIN ProjectDPI2014-C2-1-R, funded by the Science and Innovation Ministry of the Spanish Government. The funds for covering the costs of publishing in open access have been covered by the MarineUAS Project.

Author Contributions: Fotios Balampanis has designed the algorithms, performed the experiments, wrote and edited the paper. Iván Maza has analyzed the data, revised and edited the paper. Aníbal Ollero has revised and edited the paper.

Sensors **2017**, *17*, 808

Conflicts of Interest: The authors declare no conflict of interest. The founding sponsors had no role in the design of the study; in the collection, analyses or interpretation of data; in the writing of the manuscript; nor in the decision to publish the results.

Abbreviations

The following abbreviations are used in this manuscript:

UAV Unmanned Aerial Vehicle
FoV Field of View
CDT Constrained Delaunay Triangulation
AWP Antagonizing Wavefront Propagation
RWS Reverse Watershed Schema
DLH Deadlock Handling
ROS Robotic Operating System
SITL Software In The Loop

References

1. Quigley, M.; Conley, K.; Gerkey, B.P.; Faust, J.; Foote, T.; Leibs, J.; Wheeler, R.; Ng, A.Y. ROS: An open-source Robot Operating System. In Proceeddings of the ICRA Workshop on Open Source Software, Kobe, Japan, 12–13 May 2009.
2. Galceran, E.; Carreras, M. A survey on coverage path planning for robotics. *Robot. Auton. Syst.* **2013**, *61*, 1258–1276.
3. Alitappeh, R.J.; Pimenta, L.C.A. Distributed Safe Deployment of Networked Robots. In *Springer Tracts in Advanced Robotics*; Chong, N.Y., Cho, Y.J., Eds.; Springer: Tokyo, Japan, 2016; Volume 112, pp. 65–77.
4. Kapoutsis, A.C.; Chatzichristofis, S.A.; Kosmatopoulos, E.B. DARP: Divide Areas Algorithm for Optimal Multi-Robot Coverage Path Planning. *J. Intell. Robot. Syst.* **2017**, doi:10.1007/s10846-016-0461-x.
5. Korsah, G.A.; Stentz, A.; Dias, M.B. A comprehensive taxonomy for multi-robot task allocation. *Int. J. Robot. Res.* **2013**, *32*, 1495–1512.
6. Quaritsch, M.; Kruggl, K.; Wischounig-Strucl, D.; Bhattacharya, S.; Shah, M.; Rinner, B. Networked UAVs as aerial sensor network for disaster management applications. *e & i Elektrotech. Inf.* **2010**, *127*, 56–63.
7. Li, Y.; Chen, H.; Joo Er, M.; Wang, X. Coverage path planning for UAVs based on enhanced exact cellular decomposition method. *Mechatronics* **2011**, *21*, 876–885.
8. Di Franco, C.; Buttazzo, G. Coverage Path Planning for UAVs Photogrammetry with Energy and Resolution Constraints. *J. Intell. Robot. Syst.* **2016**, *83*, 445–462.
9. Fazli, P.; Davoodi, A.; Mackworth, A.K. Multi-robot repeated area coverage. *Auton. Robots* **2013**, *34*, 251–276.
10. Bochkarev, S.; Smith, S.L. On minimizing turns in robot coverage path planning. In Proceedings of the 2016 IEEE International Conference on Automation Science and Engineering (CASE), Fort Worth, TX, USA, 21–24 August 2016; pp. 1237–1242.
11. Kapanoglu, M.; Alikalfa, M.; Ozkan, M.; Parlaktuna, O. A pattern-based genetic algorithm for multi-robot coverage path planning minimizing completion time. *J. Intell. Manufact.* **2012**, *23*, 1035–1045.
12. Englot, B.; Hover, F.S. Three-dimensional coverage planning for an underwater inspection robot. *Int. J. Robot. Res.* **2013**, *32*, 1048–1073.
13. Mei, Y.; Lu, Y.H.; Hu, Y.C.; Lee, C.G. Energy-efficient motion planning for mobile robots. In Proceedings of the IEEE International Conference on Robotics and Automation (ICRA'04), Barcelona, Spain, 18–22 April 2004; Volume 5, pp. 4344–4349.
14. Barber, D.B.; Redding, J.D.; McLain, T.W.; Beard, R.W.; Taylor, C.N. Vision-based target geo-location using a fixed-wing miniature air vehicle. *J. Intell. Robot. Syst.* **2006**, *47*, 361–382.
15. Balampanis, F.; Maza, I.; Ollero, A. Area decomposition, partition and coverage with multiple remotely piloted aircraft systems operating in coastal regions. In Proceedings of the IEEE 2016 International Conference on Unmanned Aircraft Systems (ICUAS), Arlington, VA, USA, 7–10 June 2016; pp. 275–283.
16. Boissonnat, J.D.; Devillers, O.; Pion, S.; Teillaud, M.; Yvinec, M. Triangulations in CGAL. *Comput. Geom.* **2002**, *22*, 5–19.

17. CGAL—2D Conforming Triangulations and Meshes—2.5 Optimization of Meshes with Lloyd. Available online: https://doc.cgal.org/latest/Mesh_2/index.html#secMesh_2_optimization (accessed on 9 November 2016).
18. LaValle, S.M. *Planning Algorithms*; Cambridge University Press: Cambridge, UK, 2006.
19. The CGAL Project. *CGAL User and Reference Manual*, 4.8.1 edition, CGAL Editorial Board, 2015.
20. Autopilot Software in the Loop Simulation. Available online: http://ardupilot.org/dev/docs/sitl-simulator-software-in-the-loop.html (accessed on 20 December 2016).
21. RVIZ—3D Visualization Tool for ROS. Available online: http://wiki.ros.org/rviz (accessed on 20 December 2016).
22. MAVROS—MAVLink Extendable Communication Node for ROS with Proxy for Ground Control Station. Available online: http://wiki.ros.org/mavros (accessed on 9 November 2016).
23. Fixed Wing Ardupilot Instance for Autopilot Hardware. Available online: http://ardupilot.org/plane/index.html (accessed on 9 November 2016).
24. Jsbsim, the Open Source Flight Dynamics Model in C++. Available online: http://jsbsim.sourceforge.net/ (accessed on 19 December 2016).
25. Qgroundcontrol - Ground Control Station for Small Air–Land–Water Autonomous Unmanned Systems. Available online: http://qgroundcontrol.org/ (accessed on 22 December 2016).
26. Meier, L.; Honegger, D.; Pollefeys, M. PX4: A Node-Based Multithreaded Open Source Robotics Framework for Deeply Embedded Platforms. In Proceedings of the 2015 IEEE International Conference on Robotics and Automation (ICRA), Seattle, WA, USA, 25–30 May 2015.

![sensors logo] *sensors*

MDPI

Article

Water Plume Temperature Measurements by an Unmanned Aerial System (UAS)

Anthony DeMario [1], Pete Lopez [1], Eli Plewka [1], Ryan Wix [1], Hai Xia [1], Emily Zamora [1], Dan Gessler [2,*] and Azer P. Yalin [1]

[1] Department of Mechanical Engineering, Colorado State University, Fort Collins, CO 80523, USA; ademario09@gmail.com (A.D.); palopez76@gmail.com (P.L.); eeplewka@gmail.com (E.P.); ryanwix11@gmail.com (R.W.); xia323229@gmail.com (H.X.); emqzamora@gmail.com (E.Z.); azer.yalin@colostate.edu (A.P.Y.)

[2] Alden Research Laboratory, Inc. 2000 S. College Ave Suite 300, Fort Collins, CO 80525, USA

* Correspondence: dgessler@aldenlab.com; Tel.: +1-508-829-6000

Academic Editor: Felipe Gonzalez Toro
Received: 20 December 2016; Accepted: 1 February 2017; Published: 7 February 2017

Abstract: We report on the development and testing of a proof of principle water temperature measurement system deployed on an unmanned aerial system (UAS), for field measurements of thermal discharges into water. The primary elements of the system include a quad-copter UAS to which has been integrated, for the first time, both a thermal imaging infrared (IR) camera and an immersible probe that can be dipped below the water surface to obtain vertical water temperature profiles. The IR camera is used to take images of the overall water surface to geo-locate the plume, while the immersible probe provides quantitative temperature depth profiles at specific locations. The full system has been tested including the navigation of the UAS, its ability to safely carry the sensor payload, and the performance of both the IR camera and the temperature probe. Finally, the UAS sensor system was successfully deployed in a pilot field study at a coal burning power plant, and obtained images and temperature profiles of the thermal effluent.

Keywords: thermal plume; unmanned aerial system; unmanned aerial vehicle; infrared; infrared imaging; temperature profile

1. Introduction

Many industrial processes generate waste heat as a byproduct. A typical method for dissipating the heat is known as once through water-cooling, where water is drawn in from a nearby water source (lake, rive, etc.), passed through a heat exchanger, and discharged back into the water source. This process produces a thermal discharge into the water, creating a thermal plume which can be several degrees Kelvin warmer than the original water temperature. The impact of these plumes on the environment was a factor in the development of regulations for waste heat discharges by the Environmental Protection Agency (EPA). In 1972, the Federal Water Pollution Control Act of 1948 became what is currently referred to as the Clean Water Act (CWA). The CWA revisions made it unlawful to discharge any pollutant from a point source into navigable US waters without a permit [1]. Similar effluent limitations were also placed on thermal discharges from all industrial activities with the goal of protecting aquatic environments.

To address the aforementioned effluent regulations, power plant operators or environmental consultant companies periodically characterize the effluent thermal plumes and examine their impact. Currently, the typical procedure is to pilot small boats, e.g., with two-person crews, through the discharge area. The boats tow thermistors behind them to determine the extent of the plume and establish locations for full-depth temperature readings. At desired locations, full-depth readings are taken by dropping specialized sensors (e.g., Conductivity, Temperature, Depth (CTD) sensors) downward through the water column and recording the data as a function of depth. Data collection with such methods can take a day or more to complete and can be time and labor intensive. More fundamentally, because some of the discharges are near moving water or tidal areas, the plume can change shape, location, and temperature during the measurement period, thereby giving results that may be time dependent. Practical impediments, such as the lack of boat ramps at suitable locations, can also complicate the use of a boat. The use of autonomous boats or submarines may be of interest in the future but these technologies are still under development (e.g., [2]), do not readily allow large field of view imaging (as is useful in this application), and their use would still require shoreline access for launch and recovery of the craft.

The limitations of the existing methods along with the rapid emergence (and regulatory acceptance) of unmanned aerial systems (UAS) are opening the possibility of new measurement paradigms. A UAS water temperature measurement system was conceptually developed by Alden Research Laboratory, Inc. (Fort Collins, CO, USA). Working with a team of Colorado State University mechanical engineering students and their faculty advisor, the concept for field measurements of thermal discharges into water was developed into an operational prototype system. The UAS platforms have the potential to access thermal plumes with temperature sensors without requiring nearby shoreline or boat access. Furthermore, the UAS platforms can be equipped with various imaging modalities, and with global positions systems (GPS), such that data can be geo-referenced. Potential advantages of the UAS approach include the ability to make measurements more quickly (lower cost), with better spatial and/or temporal resolution, with greater reliability (e.g., if a plume can be characterized over a shorter time scale), and in a manner that provides access to more diverse bodies of water and flow types (e.g., those where boat access is very challenging). Where boat access is readily available, the UAS data can augment the understanding of the boat collected data with a surface image of the plume (and additional depth profiles).

The present contribution represents the first simultaneous integration of both an infrared (IR) camera and an immersible temperature probe to a UAS, thereby providing a powerful new capability for measuring and visualizing the temperature of water plumes over relatively large spatial scales (~100 m–kms). Several other groups are developing UAS systems for water resource monitoring (e.g., [3]) but without this simultaneous capability. The use of a UAS to image water surface temperature, but not depth profiles, has been demonstrated by several researchers [4–6]; while these approaches are useful, they do not provide the spatial temperature (depth) information needed for the study of thermal effluent streams. One research team has shown the ability to obtain water samples from a UAS platform for subsequent laboratory analysis [7]. Returned samples are useful for chemical or biological analyses but not for in situ temperature determination. The authors are aware of only one other study where an immersible probe has been used to record temperature profiles, which is the recent work of Chung et al. [8]. The system from Chung et al. is oriented towards characterizing temperature fields over smaller spatial scales (~10 m) in lakes or other bodies of water. In contrast to past published efforts, our simultaneous use of both large field of view (FOV) IR imaging and an immersible temperature probe, both integrated to a higher capacity UAS, allows the geo-location of thermal plumes and the collection of temperature profiles over large spatial scales.

The present contribution summarizes the development, validation, and testing of a UAS based temperature imaging system for thermal plumes. The layout of the remainder of the paper is as follows. Section 2 describes the UAS platform and sensor subsystem designs and integration, Section 3 describes the component test results as well as the initial field measurements in the thermal effluent of a power plant, and Section 4 presents the research conclusions.

2. UAS Sensor System and Methodology

2.1. Overall Design Concept

The overall design concept is shown in Figure 1 and comprises three integrated functional systems: A quad-copter UAS platform to carry the sensors and camera to custom waypoints with minimal human input, an IR camera for thermal imaging of the water surface to locate the overall plume, and an immersible probe that can be lowered and raised to record the temperature depth profiles below the water surface. The use of an immersible probe, whose depth within the water is controlled by raising and lowering the full UAS system, is found to be a simple and lightweight approach with high versatility. (The use of winch based systems, while possible in principle, tend to be more complex and massive with little benefit in this case.) Ancillary subsystems include a global positioning system (GPS) sensor, data acquisition (DAQ) and storage unit, and mounting and structural elements.

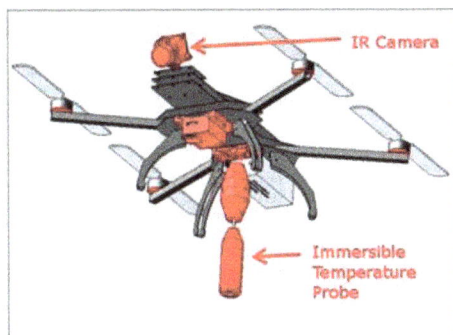

Figure 1. Overall design concept. The UAS flies over the thermal water plume using its navigational (auto-pilot) capabilities. Key components carried by the UAS include an IR camera for thermal imaging of the water surface, an immersible probe that can be lowered and raised to measure the vertical temperature profiles below the water surface, and a controller that stores the data.

2.2. UAS Platform

The system employs the Matrix-I quadcopter from Turbo Ace as a cost-effective aerial platform for proof of concept testing. Favorable characteristics include the ability to safely carry the sensor payload mass (~1.5 kg) with an "all up" (total) mass of about 4.5 kg, thereby allowing the "small UAS" designation by the Federal Aviation Administration (capped at 25 kg (55 pounds)). The Matrix-I has flight parameters consistent with plume mapping requirements, in particular, navigational accuracy better than ±1 m and a range of greater than 300 m. The endurance depends on the payload configuration and choice of batteries. The current system employs a 6 cell, 8000 mAh, 25.2 V LiPo battery with a mass of about 1.1 kg which allows up to 30 min of endurance under optimal flight conditions. Battery sets can be rotated between flights to allow time for drained batteries to be charged while continuing operation with fully charged batteries. The Matrix-I is a four-bladed platform with a carbon fiber frame and high performance propellers (38 cm length) powered by brushless motors. Advanced multi-rotor algorithms offer better performance and reliability than generic hobby quad-copter algorithms and increase motor efficiency. Custom sensor mounting components were 3-D printed from acrylonitrile butadiene styrene (ABS) plastic.

The aircraft can be autonomously controlled with waypoint navigation, including landing and takeoff, via a Pixhawk flight controller. Missions are pre-programmed and the flight status is monitored from the ground using the Mission Planner software. The UAS also includes the ability to use a first-person-view (FPV) system to monitor the flight from the ground station as seen from the UAS. To ensure safe operation, all autopilot flight plans were configured such that the UAS platform never passed directly above people or within 10 m of structures. Additionally, the system was configured to initiate an automated return-to-launch in the event that the drone lost contact to the controller or the battery reached a minimum voltage.

2.3. Infrared Camera for Thermal Imaging

An IR camera is mounted on the UAS to perform thermal imaging of the water surface as a means to visualize the overall size and contours of targeted thermal plumes. The use of thermal imaging cameras, based on uncooled (non-cryogenic) IR arrays, for quantitative temperature determination is a relatively mature technology [9] with commercial solutions readily available. The present system employs a FLIR Vue thermal imaging infrared camera specifically designed for UAS use. The camera has a 6.8 mm diameter lens, is lightweight (less than 100 g with the lens), can be powered from the UAS (5 V), and is compatible with multiple camera mounts. Data is transmitted through the First Person View (FPV) live video feed to the ground control station where the video is saved at a frame rate of 7.5 Hz. Image size and field of view are discussed in Section 3.2.

2.4. Immersible Temperature Probe

A custom made immersible probe is used to measure temperature depth profiles (relative to the water surface) at desired individual locations to map the thermal plume. The probe can be dipped into the water at multiple locations during a single UAS flight. Temperature measurements are made with a thermistor based probe (Omega HSTH-440000, Omega Engineering, Norwalk, CT, USA) that was selected due to its combination of size, mass, and ease of use in the water environment. To determine the depth of the temperature probe (relative to the water surface), pressure measurements are recorded with a transducer (Honeywell PCB 11, Honeywell, Morris Plains, NJ, USA) and hydrostatic pressure relations are used. A micro-controller (Arduino) samples data from the temperature and pressure sensors at a rate of 2 Hz and the data is stored to a memory card (microSD) for post-flight retrieval. The temperature and pressure sensors were individually tested in dedicated experiments, described in Section 3.3. The thermistor and pressure sensors are contained within a waterproof immersible housing (length ~15 cm and diameter ~3.5 cm), shown in Figure 2, produced with a 3-D printer from ABS plastic. The micro-controller is positioned remotely on the UAS platform and connected to the immersible probe via wires within an umbilical cord (length 6 m). The umbilical cord holds the probe at a fixed position below the UAS such that the probe can sample different water depths as the altitude of the UAS is changed. A safety breakaway connection allowing the lower part of the cord (and immersible probe) to detach from the UAS, in the event that the probe is snagged on the ground or underwater, was developed and flight tested as a safety feature. The mass of the entire probe (including the mounting and umbilical cord) is 900 g.

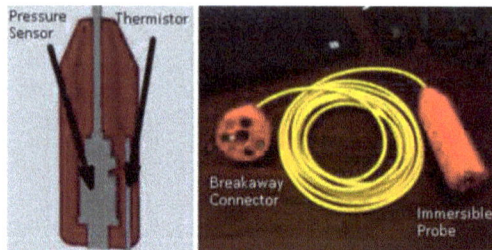

Figure 2. Underwater temperature probe. (**Left**) Probe interior design showing the temperature and pressure (depth) sensors; (**Right**) Umbilical cord with immersible sensor at one end and the breakaway connection at other.

For safety reasons, the UAS flight controller is set to limit the flight path such that the UAS has a minimum altitude of 2.7 m above the water surface. Given the uncertainty in flight position (controlled versus actual) of ~1 m, this results in limiting the actual UAS altitude to a minimum of ~1.7 m (~5 feet). The corresponding probe immersions can then be to maximum depths in the range of ~3–4 m. This depth limit is sufficient for the desired profiles in this work, but longer cord lengths could be used if needed. Note that despite the uncertainty in controlled UAS flight, the immersive probe measures its position relative to the water surface much more precisely (Section 3.3).

3. Results and Discussion

This section contains the results of UAS platform testing, sensor testing, and demonstrative results of the full proof of principle system operating at a power plant.

3.1. UAS Flights and Navigation

The majority of flight testing was done at Christman Field which is an unused air field at Colorado State University in Fort Collins, CO, USA. A series of flight tests were performed examining basic UAS operations and procedures for taking off and landing, flight maneuvers, altitude changes to dip the immersible sensor, and autonomous waypoint navigation. Figure 3 shows a photograph of the flying UAS in its final flight configuration with the sensors and umbilical cord.

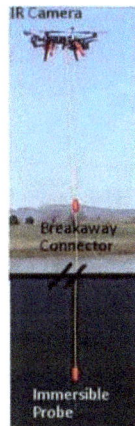

Figure 3. Photograph of the flying UAS in its final flight configuration with all sensors and the umbilical cord. (Full length of the cord is not shown).

Flight testing was used to determine the sensitivity of the UAS to changes in the center of gravity (CG) of the system. The onboard flight controller uses multiple accelerometers for precision flight adjustments. The system performs best when the overall center of gravity coincides with the location of the flight controller. Gusty winds (>~10 m/s for the present system) are an obvious risk for stable flight as they can cause the flight controller to overcompensate pitch-and-roll feedback. The system is particularly vulnerable during landing where a wind gust coinciding with touchdown can trigger an undesired response from the onboard flight controller. Other mounting schemes and/or more sophisticated control schemes can be considered if UAS control becomes problematic due to the change in CG position caused by the umbilical and immersible probe [10]. Flight tests also examined the functionality of the umbilical breakaway mechanism and verified that the mechanism did not inadvertently detach during regular flight operations, but did detach when the immersible probe could not be lifted (as tested by tying it to the ground). The accuracy of the waypoint navigation was examined in a series of flights with different atmospheric conditions. These tests were first conducted over land and then over water. The automated return to the launch system consistently returned the UAS to within 2 m of the launch point.

3.2. IR Camera Testing

Infrared camera thermal imaging systems are generally designed for the measurement of solid objects or terrestrial surfaces, not liquid surfaces (i.e., water). Initial testing of the IR camera verified that the camera could capture temperature variations of 1 K (by imaging some known scenes separately measured with thermocouples). The thermal camera was mounted and flown on the UAS to examine the field of view and the effects of the resolution and shutter speed settings. Test flights for the thermal camera were performed by mounting it to the UAS, as well as in separate flights where it was mounted to a fixed wing aircraft (Cessna, Cessna Aircraft Company, Wichita, KS, USA) flying over the discharge reservoir at the Rawhide Energy Station near Wellington, CO, USA. The field of view (on the water surface) attainable by the camera is defined by the flight altitude and angular extent of the image, where the latter is determined by the lens and camera optical design. For typical altitudes of ~120 m, the IR camera yielded image dimensions of ~100 m × 80 m based on the view angles of 44 and 36 degrees, with an image resolution of 333 × 256 pixels. This field of view is generally too small to image a full thermal plume, such that multiple images should be "stitched" (combined) together to achieve the desired larger view. In this work, we have manually combined the images using a commercial software package (Photoshop), though more sophisticate stitching combined with geo-tagged images could be used in the future. Testing confirmed that the camera can be used to provide qualitative images of the thermal plume, allowing the determination of the plume contour. Examples of IR camera images are shown in Section 3.4.

3.3. Immersible Temperature Probe Testing

Tests were performed to confirm the performance of both the temperature and depth measurements from the immersible probe. For the former, temperature readings of a water bath from the immersible probe thermistor were compared against a reference thermocouple and were found to agree to within ±1 K, which is taken as the uncertainty (error bar) for temperature readings. The approach for determining the sensor depth is to measure the local pressure and then infer depth (relative to the water surface) using hydrostatic pressure relations, i.e., pressure increases by ~96 kPa/m of depth in fresh water. The sensor depth readings were calibrated (to 5 m depth) by immersing the sensor to varying known depths within a column of water. As shown in Figure 4, the agreement (between known and measured position) was very linear with a slope close to unity and zero offset (the linear fit yields a slope of 1.002 and an offset of −3.0 cm with $R^2 = 0.99993$) and showed that depth could be measured to within ±7 cm.

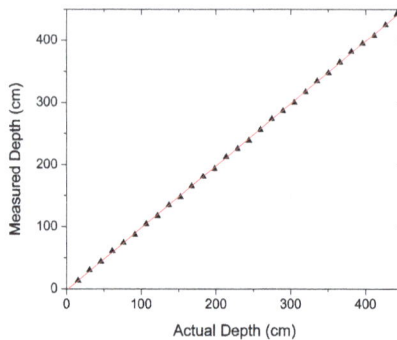

Figure 4. Test results of the depth probe showing the relationship between measured and expected depth.

3.4. Field Testing of the UAS Thermal Plume Mapping

The full UAS sensor system was tested on 21 April 2016 during three separate flights at the Rawhide Energy Station (RES) in Wellington, CO, USA. The RES has a single 280 MW coal burning unit, four 65 MW single-cycle natural gas turbines and one 128 MW single cycle gas turbine. Testing comprised of flying a survey mission at an altitude of ~120 m and obtaining both IR images as well as subsurface water temperature profiles, by dipping the temperature sensor at critical points in the plume. Winds were relatively strong at ~6 m/s to 9 m/s, and the UAS flew successfully, with the payload, under these conditions.

Imaging of the thermal plume with the IR camera was performed by ascending to a flight ceiling of ~120 m and making several passes over the water to image the extents of the plume. The left panel of Figure 5 shows a composite image of the water discharge area, revealing the thermal plume. The infrared camera used during the flight automatically scales each image based on the range of temperatures in the individual image. Note that other available UAS compatible cameras, for example ICI 9640 P-series (Infrared Cameras, INC., Beaumont, TX, USA), allow consistent color scaling over multiple image acquisitions resulting in more continuous composite images. The right panel of Figure 5 shows the results of a single infrared image.

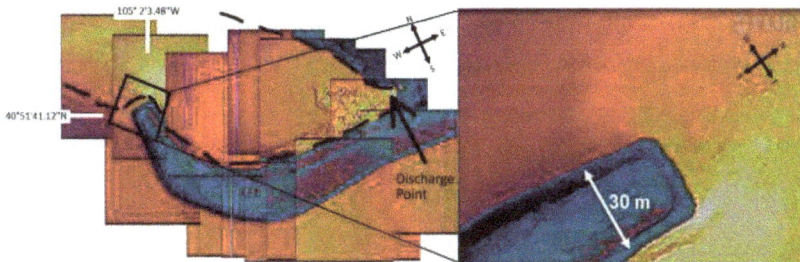

Figure 5. (Left) Composite image of the water surface showing thermal discharge from the Rawhide Energy Station obtained by the UAS IR camera. The location of the discharge point is indicated while the approximate plume contour is shown with a dashed line; **(Right)** Single image of a thermal plume at the end of the discharge guide berm.

The final aspect of UAS flight testing was to conduct three immersion dips to measure the subsurface water temperature profiles. Figure 6 shows a map view of the discharge area indicating the dip locations. The UAS took off from the nearby peninsula and flew approximately 140 m to the first dip location, 190 m to the second dip, 150 m to the third dip, and finally 80 m to return and land at the original launch location. The entire flight was autonomous and executed without input from the pilot.

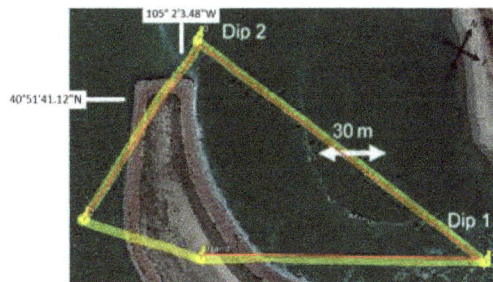

Figure 6. Flight path (actual in red and programmed in yellow) showing the three dip locations for the subsurface water temperature profiles at the Rawhide Energy Station.

Example temperature depth profiles from the immersible UAS probe are shown in Figure 7. The plotted temperature profiles are referenced to the water surface at depth zero. Dip 1 was relatively close to the outflow point while Dip 2 was further away and at the tip of a peninsula. For depths below ~1 m, the Dip 1 profile shows warmer temperatures than Dip 2, consistent with the greater proximity to the discharge source location. The profiles were obtained by lowering the probe through the water at a speed of ~50–70 cm/s (the exact value was influenced by water forces on the probe) with data gathered at a rate of 2 Hz. Each downward dip was ~2.5 m in extent, corresponding to a time extent of ~4–5 s that yielded 8–10 data points with a spacing ~25–30 cm. The top several points (recorded in the first 2 s of immersion) in each dip profile are colored grey to indicate that the thermistor readings may not have sufficiently stabilized for these specific measurements. There is a relatively large temperature jump that occurs at the instant of immersion, due to the sudden jump from cooler air (~13 °C) to warmer water, such that more time is required for stabilization to within the uncertainty (± 1 K) for points near the surface where the temperature change is largest. Overall, the system has shown the ability to dip the immersion probe and record temperature profiles. Future work should further examine the probe response time and modifications to the dip speed should be made depending on the data needs; for example, the probe can be lowered more slowly if the data at the top of the water layer is critical.

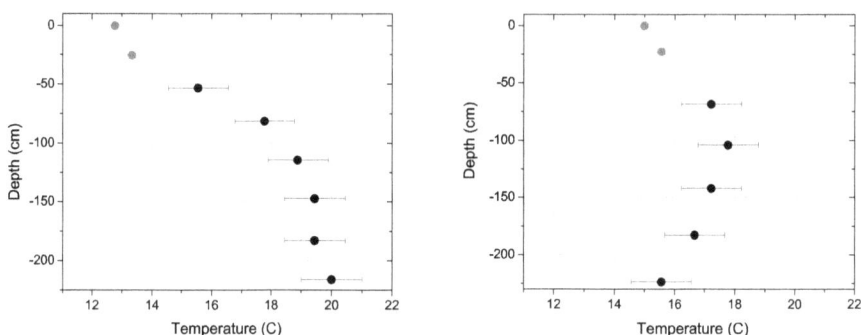

Figure 7. Temperature profiles below the water surface from two dips at the Rawhide Energy Station (**Left**: dip 1, **Right**: dip 2).

4. Conclusions

The present research has developed a UAS sensor package for the study of thermal plumes in water, particularly for the examination of thermal effluents from power plants. Such approaches can potentially displace (or complement) the boat-based operations that are currently widely used for effluent characterizations for regulatory and environmental assessments. A quad-copter UAS has been modified to carry the sensors. With the appropriate consideration of the center of gravity location, stable flights could still be achieved with the modified configuration. An attractive feature of UAS operation is the ability to navigate via automated waypoints as demonstrated in the present research. The two main required sensors for the thermal characterization are an infrared thermal imaging camera, used for qualitative imaging of thermal gradients to reveal the overall plume location, and an immersible probe to record the temperature depth profiles at different fixed point locations in the plume. Both sensor subsystems were characterized in controlled tests and were then integrated to the UAS. The immersible probe hangs below the UAS, such that it can be dipped in/out of the water by lowering/raising the overall UAS, and is connected to the UAS via an umbilical cord. For safety reasons, the umbilical cord contains a breakaway mechanism allowing the probe (and the lower part of the cord) to disconnect and fall into the water in the event that the probe is snagged or cannot be retracted.

The full system was flight tested in a proof of concept study at the Rawhide Energy Station including waypoint navigation, imaging of the thermal plume from the IR camera, and acquisition of the temperature depth profiles. The IR camera used in the present work provided a proof of concept demonstration but could ultimately be replaced with a more advanced unit allowing more capability for geotagging of the images, higher resolution, on board storage, and the use of automated image stitching software. The dip probe data demonstrates that the water temperature can be measured at the surface and below the surface using a UAS. Testing is required for each UAS system to understand the accuracy and time response of the temperature probe.

The most significant shortcomings of the system are the limited flight duration and weather and regulatory challenges. The current configuration cannot operate in precipitation and is limited to 20–35 min of flight time (depending on flight profile, wind, number of dips, etc.), which can be limiting for sampling large plumes. However, these limitations can be largely overcome by using a larger UAS and continued improvements in battery technology. The fully loaded UAS is currently limited to flying with wind speeds less than ~10 m/s. The wind speed limitation is expected to improve if a larger UAS is used and as flight controller technology improves. The most significant wind limitation is during landing, where a gust of wind can adversely affect flight performance when the UAS is at low power near the ground. More advanced flight controllers would also benefit the performance in high winds and general stability, and one could integrate the sensor control and data storage functions into the flight controller itself (obviating the need for the Arduino currently used for this purpose).

In the United States, the operation of UAS systems (without additional certification) is limited to 122 m (400 feet) above ground level by the Federal Aviation Administration. This limits the area which can be seen in a single image. A wide-angle camera lens or oblique view of the terrain can improve the coverage footprint; however, a wide-angle lens introduces distortion and oblique views are more challenging in terms of determining the position of the plume.

Author Contributions: D.G. conceived the approach and aided in experiments, analysis, and manuscript writing; A.D. was lead on the immersible probe and contributed to manuscript writing; E.P. was lead on the UAS modifications and piloting; P.L. contributed to component integration to the UAS; H.X. and E.Z. led the IR imaging; R.W. contributed to test flights and deployment on the Cessna aircraft; A.P.Y. mentored the team members on the design work and data interpretation and wrote the majority of the paper.

Conflicts of Interest: The authors declare no conflict of interest.

References

1. Memorandum on Implementation of Clean Water Act Section 316(a) Thermal Variances in National Pollutant Discharge Elimination System. Available online: https://www3.epa.gov/region1/npdes/merrimackstation/pdfs/ar/AR-338.pdf (accessed on 15 November 2016).
2. Dunbabin, M.; Marques, L. Robots for Environmental Monitoring Significant Advancements and Applications. *IEEE Robot. Autom. Mag.* **2012**, *19*, 24–39. [CrossRef]
3. DeBell, L.; Anderson, K.; Brazier, R.E.; King, N.; Jones, L. Water resource management at catchment scales using lightweight UAVs: Current capabilities and future perspectives. *J. Unmanned Veh. Syst.* **2016**, *4*, 7–30. [CrossRef]
4. News: Student Teams Develop Innovative UAV Applications. Available online: http://www.wur.nl/en/newsarticle/Student-teams-develop-innovative-UAV-applications-.htm (accessed on 12 January 2016).
5. Deitchman, R.S. Thermal Remote Sensing of Stream Temperature and Groundwater Discharge: Applications to Hydrogeology and Water Resources Policy in the State of Wisconsin. Ph.D. Thesis, University of Wisconsin, Madison, WI, USA, 2009.
6. Jensen, A.M.; Neilson, B.T.; Mckee, M.; Chen, Y.Q. Thermal Remote Sensing with an Autonomous Unmanned Aerial Remote Sensing Platform for Surface Stream Temperatures. In Proceedings of the 2012 IEEE International Geoscience and Remote Sensing Symposium (IGARSS), Munich, Germany, 22–27 July 2012.
7. Ore, J.P.; Elbaum, S.; Burgin, A.; Zhao, B.; Detweiler, C. Autonomous Aerial Water Sampling. In *Field and Service Robotics*; Springer Tracts in Advanced Robotics; Springer: Cham, Switzerland, 2015; Volume 105, pp. 137–151.

8. Chung, M.; Detweiler, C.; Hamilton, M.; Higgins, J.; Ore, J.P.; Thompson, S. Obtaining the Thermal Structure of Lakes from the Air. *Water* **2015**, *7*, 6467–6482. [CrossRef]

9. Rogalski, A. Infrared detectors: Status and trends. *Prog. Quantum Electron.* **2003**, *27*, 59–210. [CrossRef]

10. Bernard, M.; Kondak, K.; Maza, I.; Ollero, A. Autonomous Transportation and Deployment with Aerial Robots for Search and Rescue Missions. *J. Field Robot.* **2011**, *28*, 914–931. [CrossRef]

sensors

MDPI

Article

Spatial Ecology of Estuarine Crocodile (*Crocodylus porosus*) Nesting in a Fragmented Landscape

Luke J. Evans [1,2,3,*], T. Hefin Jones [1,4], Keeyen Pang [5], Silvester Saimin [3] and Benoit Goossens [1,2,3,4,*]

1 Cardiff School of Biosciences, Cardiff University, Cardiff CF10 3AX, UK; jonesth@cardiff.ac.uk
2 Danau Girang Field Centre, c/o Sabah Wildlife Department, Wisma Muis, 5th Floor, Block B, Kota Kinabalu 88100, Malaysia
3 Sabah Wildlife Department, Wisma Muis, 5th Floor, Block B, Kota Kinabalu 88100, Malaysia; silvester.saimin@sabah.gov.my
4 Sustainable Places Research Institute, Cardiff University, 33 Park Place, Cardiff CF10 3BA, UK
5 Hornbill Surveys Sdn Bhd, Lot 9, Harapan Baru Light Ind Estate, Mile 8, Jalan Labuk, Sandakan 90009, Malaysia; intrajasa@gmail.com
* Correspondence: lukeevans603@yahoo.co.uk (L.J.E.); goossensbr@cardiff.ac.uk (B.G.)

Academic Editors: Felipe Gonzalez Toro and Antonios Tsourdos
Received: 28 June 2016; Accepted: 13 September 2016; Published: 19 September 2016

Abstract: The role that oil palm plays in the Lower Kinabatangan region of Eastern Sabah is of considerable scientific and conservation interest, providing a model habitat for many tropical regions as they become increasingly fragmented. Crocodilians, as apex predators, widely distributed throughout the tropics, are ideal indicator species for ecosystem health. Drones (or unmanned aerial vehicles (UAVs)) were used to identify crocodile nests in a fragmented landscape. Flights were targeted through the use of fuzzy overlay models and nests located primarily in areas indicated as suitable habitat. Nests displayed a number of similarities in terms of habitat characteristics allowing for refined modelling of survey locations. As well as being more cost-effective compared to traditional methods of nesting survey, the use of drones also enabled a larger survey area to be completed albeit with a limited number of flights. The study provides a methodology for targeted nest surveying, as well as a low-cost repeatable flight methodology. This approach has potential for widespread applicability across a range of species and for a variety of study designs.

Keywords: drone; UAV; salt water; aerial survey; Borneo; Sabah; reptile

1. Introduction

The history of crocodilian nesting studies began in the 1960s [1,2], although detailed comparative work on the nesting behaviour of various species did not commence until a decade later. *Crocodylus porosus* nesting was first examined in detail by Webb et al. [3]. This early work focussed on the mechanics of nesting behaviour (for example, what building materials were used), and recording nest characteristics such as temperature and the number of eggs oviposited. Spatial analyses, or assessments of the distribution of crocodile nests in any given area, have not been the focus of many studies. Possibly the best example, to date, is work carried out by Harvey and Hill [4]. Using Landsat™ image analysis (see below) and aerial photography to classify areas by vegetation types, they used a Boolean overlay approach to determine potentially suitable breeding habitat [5]. During this study, aerial photography was found to be far more useful in the classification of nesting habitats [4].

There have been a couple of key factors that have limited the effectiveness of nesting studies, particularly when attempting to calculate nest density in a particular area. Firstly, it is very difficult to determine that all nests in a region have been discovered; thus, nest density calculations are always rough approximations and usually underestimates. Secondly, studies of this nature are often prohibitively expensive, as they rely on the use of helicopters and airboats, or time-costly walked surveys. The use of modern/advanced technology to examine crocodilian nesting behavior is a new development and this study builds on a drone-based methodological paper, which examined the potential use of novel drone technology in finding crocodile nests [6].

The use of drone (or unmannèd aerial vehicle (UAV)) technology in conservation and biological management programmes is a relatively recent development [7,8]. Over the past decade, drones have been employed to service a host of ecological needs. Applications such as the monitoring of Eurasian beaver (*Castor fiber*) reintroductions [9], surveying tree falls [10], identifying of forest gaps [11] and nesting of canopy birds [12] are examples of research and monitoring fields within which the technology has been applied. Drones are frequently used in active conservation practice; for example, in the monitoring of poaching activities perpetrated against both the black (*Diceros bicornis*) and white (*Ceratotherium simum*) rhinoceroses [13]. The flexibility of the technology to perform in remote and urban locations alike means that, as long as weather conditions allow, drones can be a financially accessible method of surveying most areas on Earth.

One challenge encountered when using traditional aerial photography is that some areas are obscured from view; this is particularly problematic in forested areas. The use of drones in the determination of crocodile nesting distribution could demonstrate the feasibility and applicability of the technology to other facets of crocodilian research. Martin et al. [14], for example, have already demonstrated that adult alligators could be easily identified aerially using drone technology, suggesting that count surveys and future population density estimates could, at least in part, be calculated in this way.

Selective land conversion for agricultural purposes raises a number of potential issues for crocodile nesting. Prevalent throughout tropical regions, oil palm (*Elaeis guineensis*) plantations are expanding rapidly into previously remote areas [15,16]. The implications of this for estuarine crocodile (*C. porosus*) nesting are unknown; crocodiles are frequently seen in plantation areas (*pers. obs.*). Preliminary work carried out by Evans et al. [6] indicated that crocodiles will nest in areas of medium to high anthropogenic disturbance, and within close proximity to oil palm plantation. This could be indicative of either insufficient nesting habitat being available or individuals continuing to utilise successful nest sites even after their surrounding environment has been altered. There are numerous other less-well examined effects that could influence both the likelihood of successful nesting, as well as post-hatching survival of the estuarine crocodile. Oil palm plantations require a non-natural, irrigation system to ensure sufficient water for crop development; they are, however, unable to withstand long periods of flooding [17]. These artificial hydrological systems are unlikely to benefit hatchling dispersal due to the lack of connectivity between irrigation ditches, and are also likely to bring young crocodiles into closer proximity to potential predators such as monitor lizards [18].

Utilisation of drone surveys enables high-resolution identification of crocodile nests; this allows more accurate mapping of their spatial distribution [6]. This present study aimed to identify all possible estuarine crocodile nests within a specific region of the Kinabatangan River in Sabah, Malaysia. The study site is known to harbor a resilient population of estuarine crocodiles, compared to other sections of the Kinabatangan River, as well as to neighboring rivers in Sabah [19]. It was hoped that baseline data for crocodile nesting in a tropical freshwater ecosystem could be established. Nest sites can, as previously discussed, be found in semi-predictable locations owing to their proximity to permanent water sources, as well as their prevalence in certain types of habitat [4,20,21]. Given that much of *C. porosus* habitat is comprised of closed canopy, and the aforementioned predilection for swampland, particular targeting of 'suitable' nesting habitat was possible. In an effort to validate the applicability of the predictive modeling, additional areas were surveyed within the study site.

Five hypotheses were tested. Firstly, nest site density is higher in exposed, open-canopy areas than under dense forest cover with walked "closed-canopy" recces (Hypothesis 1). Female nest site selection can be predicted in terms of habitat and can therefore be selectively surveyed through the use of predictive modelling and aerial drone technology (Hypothesis 2). Nest sites are solitary, with individuals actively choosing nest sites that are spatially independent of other nesting females (Hypothesis 3). Estuarine crocodiles select nest sites in a non-random fashion given habitat availability constraints (Hypothesis 4), thus validating the use of drones as a tool for nesting surveys. Finally, building on findings from Evans et al. [6], nesting can occur in the presence of medium to high levels of human disturbance, implying that the presence of oil palm plantations is not necessarily a barrier to successful nesting (Hypothesis 5).

2. Methods

The study was carried out over a 35 km stretch of the Kinabatangan River. The stretch was selected due to the presence of high numbers of both adult and juvenile crocodile individuals (unpublished data). The second longest river in Borneo, and, at 560 km, the longest in the eastern Malaysian state of Sabah, the Kinabatangan River has a catchment of around 16,800 km^2, an area encompassing around 23% of the total land area of Sabah [22,23]. Within this catchment area lies the Lower Kinabatangan Wildlife Sanctuary (LKWS), consisting of 10 distinct forest "lots" covering an area of 27,960 ha [15]. The landscape consists of a highly fragmented forest-oil palm matrix, with forested areas being largely degraded secondary forest. This type of forest results in patchy areas of closed canopy forest interspersed with open grassland and areas with very sparse partial tree coverage. The Kinabatangan region supports a large population of estuarine crocodiles (*C. porosus*); the population has undergone rapid recovery following state-wide protection of the species initiated in 1982 [19].

2.1. Walked Recces

Walked recces were initially carried out throughout the 35 km study region to determine whether crocodiles were likely to be utilising areas of closed canopy cover for nesting; the resultant lack of positive nest identification justified the use of aerial drones. These recces were carried out from September–November 2013. These detailed walked recces were conducted over an area totaling 120 km of riparian riverbank habitat, oxbow lake (consisting of a mixture of riparian and seasonally flooded habitats) and swampland habitats (areas where water is present year round). A total of 101 km (84.2%) covered during recces was covered under closed canopy. Each recce was conducted by two observers, each of whom walked parallel to the water source at a distance of 5 m from the water, whilst maintaining 15 m between them, creating total coverage of approximately 20 m.

Of the 120 km of walked recces conducted, 50.8% of the habitat surveyed was riparian riverbank, 27.1% oxbow lake shoreline and 22.1% swampland, across 14 separate nesting transects. These transects (4.5–10 km) were searched, in detail, for any indication of crocodile nesting or presence; for example, footprints or slide marks caused by the dragging of the body through mud, sand or vegetation. Marks found were carefully examined to distinguish them from other animals such as monitor lizards, snakes, bearded pigs or any other ground-dwelling animals found in the region. Transect locations were chosen as being potentially suitable for nesting primarily based on their proximity to permanent water sources [3,4,21]. Each transect was selected to incorporate the highest percentage of closed canopy possible; this allowed determination of whether regions with open access to direct sunlight were more likely to be selected for nesting (Hypothesis 4). The walking of large tracts of the landscape also provided an indication of the different habitats present within the region, and which areas would be most suitable for aerial analysis.

2.2. Drone Surveys

Aerial surveys were conducted on the basis that nesting under closed canopy was occurring at such low occurrence that not observing them would have no meaningful impact on estimated nest densities (based on lack of detection during walked surveys). Aerial surveys were conducted with the use of two different drone systems (during 2013 and 2014, respectively). Surveys were carried out during September and October in both survey years. These months were chosen due to observance of hatchling emergence during November and December (*pers. obs.*).

In 2013, an exploratory series of surveys were carried out using a Bormatec Maja™ drone (see Evans et al. [6] for a detailed description of it specifications). As this initial attempt proved successful, a second, more expansive, series of surveys was conducted in 2014, this time utilising the Skywalker™ drone. This equipment provides a more stable and efficient flight. Both drone systems were fixed wing aircraft, able to provide a high degree of stability, whilst ensuring desirable range capabilities. Under typical weather conditions (low wind and no rain), the Skywalker™ was able to conduct flights totaling one hour, whilst carrying a payload of up to 1 kg. In real terms, this equates to flight distances of up to 35 km and search grids of around 550 ha. Two cameras were used for surveys conducted in 2013 and 2014. However, these were of the same make and model (Model S100, Canon, Ota, Tokyo, Japan). A CHDK (Canon Hack Development Kit) script was uploaded to the camera to ensure the camera could take pictures every three seconds without manual triggering.

Other than during take-off or landing, or during emergencies, drone flights were conducted using an Auto Pilot Module (APM). Flights were conducted as close to the planned flight grid as possible (for flight grid description see Evans et al. [6]), whilst allowing for an open landing and take-off site. These "ground stations" were located in a variety of habitats, including riverbanks, grasslands and oil palm plantations. Ensuring that ground stations were located close to search grids allowed for larger areas to be covered within any particular grid, as battery power was not wasted traveling to and from the study site. Post-hoc analysis of aerial photographs taken involved stitching following the methodology explained in Evans et al. [6].

The selection of drone mapping grids was based on ensuring an array of riverine, swamp and oxbow habitat, as well as covering all the major tributaries across the study site. A predictive theoretical model of suitable nesting sites was produced to aid in the selection of these sites (Figure 1). This model was produced using the "fuzzy membership" and "fuzzy overlay" functions in ArcGIS 10; these functions allow the designation of certain required spatial prerequisites (such as distance to permanent water sources) for nesting, whilst discounting other areas based on presumed undesirable geographic traits (for example, proximity to oil palm plantations (distances tailored based on findings from Evans et al. [6]). These traits were derived from the existing literature [4]. Whilst flights were flown to include these "suitable zones", flights were also conducted outside of these areas to test the efficacy of the predictive modeling.

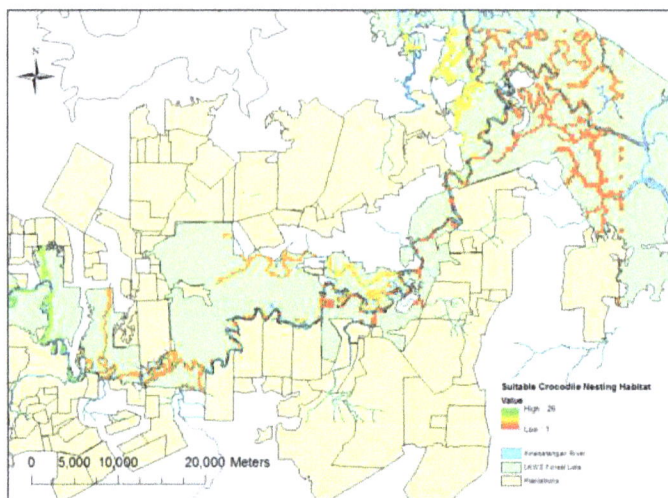

Figure 1. Nesting suitability model for the LKWS. Defined using a "fuzzy overlay" function in ArcGIS. Areas of suitability are defined by the presence of a coloured pixel with increasing suitability defined on a red (low) to green (high) scale. Suitable nesting locations are largely confined to major waterways.

Once identified, nests sites were ground-validated, where possible, and general ground habitat assessed. Both validated and non-validated nests were subsequently assessed for a series of geographic traits, such as distance to permanent water, distance to canopy and distance to plantation. These descriptive statistics were stated throughout as ±1SE. These variables were used to create a binomial generalised linear mixed model (GLMM) using R (3.1.3) (Table 1). The "lme4" package [24] was used to determine which factors were of greatest importance in determining a nesting location of estuarine crocodiles. The model was refined through the use of a "dredge" model-comparison function carried out with the use of the package "MuMIn" [25]. "Dredging" creates a series of models with subsets of variables in order to find the most reliable model. Conditional and marginal R^2 values were then used to assess the level of variance explained by both fixed and random model terms. Finally, model predictions were made to evaluate the role proximity to plantation plays in nesting location choices.

Table 1. "Fixed" and "random" model terms included in the binomial General Linear Mixed Model (GLMM) used to identify the most important factors in the presence or absence of crocodile nests. A logit link function was used for the model.

Dependent Variable	Fixed Model Terms	Random Model Terms
Presence of Nest (1/0)	Ground solidity	Year of detection
	Distance to water	
	Distance to canopy cover	
	Distance to plantation	
	Ground water presence	

3. Results

Walked recces in closed canopy failed to identify any nests, providing evidence that the majority of nesting occurs in areas of open, canopy-devoid, areas. During the 2013 field season, a total of 1550 aerial ha were surveyed using drones and three potential nests identified. Two of these were confirmed as true nesting locations. A further 5160 ha were surveyed during 2014; this resulted in a total area surveyed over two field seasons of 6710 ha.

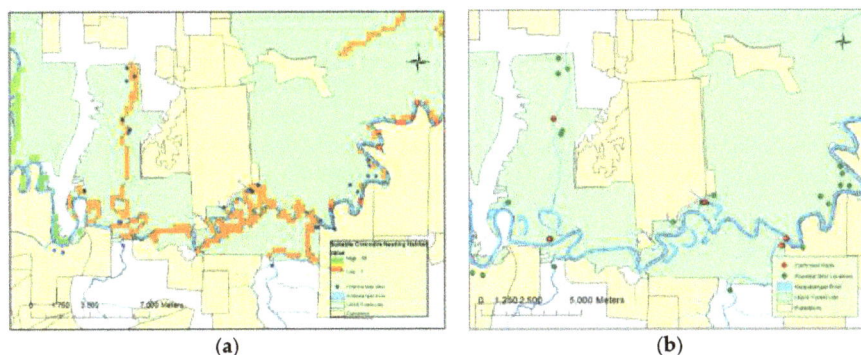

(a)   (b)

Figure 2. (a) Potential nest sites in relation to habitat suitability model; the majority of nests sites fell inside of, or close to, identified suitable areas within the study site. Suitability defined as areas of coloured pixels as in Figure 1, with potential nest sites overlaid as blue dots; (b) Locations of confirmed nest sites showing close proximity to water, as well as, on three occasions, close proximity to oil palm plantations.

Twenty six potential nests were identified during 2014; however, flooding of a large part of the study site during the field season prevented verification of 10 of these nests. Those nests that were flooded were not verified for either safety or lack of accessibility reasons. Nine of these unverified nests were excluded from analysis in order to retain the rigor of the model. One flooded nest was included despite the lack of ground verification based on similarities between its aerial image and those of previously ground-verified nests. Of the drone-facilitated nesting surveys carried out over two field seasons, 2013 and 2014, a total of 29 potential nests were identified. Of these, four were confirmed as actual crocodile nests with the addition of one unverified nest (two in 2013 and two in 2014) (Figure 2a). A total of 15 nests were highlighted as potential nests from the aerial photography that were later found to consist of accumulations of dead material unassociated with crocodilian nesting behaviour.

3.1. Habitat Suitability

Of the five nests, all were located in close proximity (mean 13.9 ± 12.9 m) to permanent water sources. They were found within small open areas and within close proximity (22.2 ± 14.3 m) of closed canopy cover (Figure 2b). While plantations were, generally, not included in the surveyed areas, one nest was found close to a plantation border; across all nests identified, the nests were a distance of 374 m (±139.7) from such boundaries. Four of the confirmed nests were located within the protected habitat lots of LKWS. One nest was located outside, in privately-owned land that could be open to conversion. Neither nest site located during 2013 was reutilised during 2014, and all nests were spatially independent (Figure 2b). Three of the five nests were located in "drying" or "old" oxbow lakes; each of these had a permanent aquatic connection to a main water body, such as a large tributary. The other two nests' locations, including the unverified nest, were directly adjacent to a major water body.

Nest sites could not be attributed to specific females and no instances were recorded of females guarding their nests. There was, however, evidence of females spending time at the nest site, and of excavation of eggs during hatching. Wallows (depressions containing mud or shallow water) were recorded around one of the nest sites. One nest located during 2013 surveys was visited the day after hatching and visual confirmation of 19 hatchlings was recorded. Egg membranes and shells were collected, and evidence of at least 24 successful hatchings was found. There was no evidence of pre- or post-hatchling mortality.

3.2. General Linear Mixed Model

The most parsimonious GLMM model structure to explain the presence/absence of crocodile nests (lowest AIC–Akaike Information Criterion–identified using the "dredge" function) included the variables "distance to water" and "ground water presence". This model also yielded the greatest model weight (W = 0.195), and was therefore considered the best model structure to explain the model variables included. Distance to water was significantly negatively correlated with the likelihood of finding a nesting site ($F_{1,29} = 5.59$, $p = 0.018$). The presence of ground water was close to significance ($F_{3,29} = 7.36$, $p = 0.061$), with the presence of less than 1 m of standing ground water resulting in a higher likelihood of nesting (Figure 3). Marginal (R^2m) and conditional (R^2c) r-squared values showed that the majority of the variance being described by the model was derived from the fixed terms (distance to water and standing ground water), with negligible variation explained by the random term, year ($R^2m = 0.571$, $R^2c = 0.571$).

Predictions based on this model suggested that whilst locations of nests in standing water could not be easily predicted, nests on solid ground were very likely to be less than 100 m away from permanent water sources (Figure 2b). In reality, these predictions are in line with those used during the original "fuzzy overlay" modelling (Figure 2a). An increase in sample size could lead to a refinement of solid-ground predictions and lead to more stringent standing water predictive sampling.

There was a marked difference in image quality between the 2013 (higher quality) and 2014 (lower quality) drone surveys, despite the use of two cameras of the same make and model. This could have been a result of different light conditions (see Evans et al. [16]), or the speed of travel of the drone. Issues with image clarity and stitching quality were largely linked to the time of day the flight was flown. Flights conducted in the hours immediately following dawn (7 a.m.–9 p.m. (GMT+8)) and preceding dusk (3 p.m.–5 p.m. (GMT+8)) appeared to produce the highest quality images.

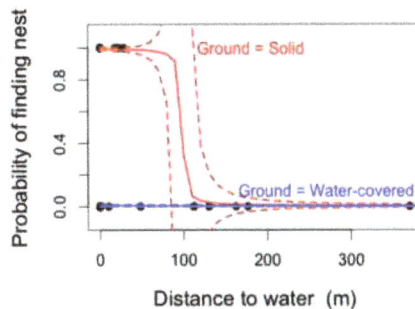

Figure 3. Plotting predictions from binomial GLMM. Model provides a binomial predictive distribution, indicating that nesting is less likely further away from permanent water sources. Solid lines denote predicted probability; with dashed lines showing the error associated with the probability levels. Data included both confirmed nest sites as well as those that were "potential" and later discounted nest sites. Despite trajectory of confidence intervals, prediction could not be less than zero.

4. Discussion

This study centered around the use of drone technology to complete an estuarine crocodile nesting survey of a section of the Kinabatangan River in Borneo. Positive identification was possible of five nests out of a total 29 potential locations; all other potential nests, not affected by flooding, were visually discounted on the ground. The approach, albeit in need of refinement, is a major improvement on the traditional techniques; for example, costly helicopter surveys entailing flying over large tracts of unsuitable habitat would still require post-event ground validation [4].

The identification of five nests within a relatively small (6710 ha) area of the study site suggests that crocodile females are actively selecting areas identified by the habitat suitability model as potential sites (Hypothesis 1). The use of such a model, as well as the selection of areas with open or semi-open canopy coverage, allowed for both highly selective and highly predictive flight mission planning (Hypothesis 3). The walked recces provided important justification of the use of drones in identifying potential nesting sites and crocodilian habitat (Hypothesis 2), and although the distance walked was relatively short (120 km), the transects' placement to encompass tributaries, ox-bow lakes and other areas of permanent water sources (essential to successful nesting), provided a clear indicator that dense forest canopy does not represent important nesting habitat (Hypothesis 4). The identification of nests in close proximity to oil palm plantations demonstrates that females can nest in medium to high levels of human disturbance (Hypothesis 5).

While the identification of five nests provided validation of the methodology, the limited 35 km river stretch probably represented too small an area to provide a clear picture of the nesting habits of the crocodiles found throughout the LKWS. Nesting appears to be occurring at low densities, and whilst it is, at least to some extent, predictable, the presence of degraded, patchy secondary forest represented a challenge to successful and encompassing nesting surveys. As a result of breaks in forest canopy coverage, secondary forest results in far larger expanses of open areas than would likely to be present were the region to have retained its original primary forest landscape. As a result, refining search grids is more challenging in secondary forest ecosystems than in pristine primary forest habitats. Despite this reduction in canopy coverage, it is unlikely that all nests will be built in open areas and some crocodilians can use alternate heat sources such as termite mounds to keep their nests at the optimal temperatures [26]; however, to date, this has not been reported in *C. porosus*. The effect of such potential behaviours on the number of nests detected should, however, be negligible.

Flooding, as also determined by Webb et al. [3], is the primary threat to *C. porosus* nests in the Kinabatangan. Ten potential nests were completely submerged during the period of ground-verification, the river having risen in excess of 1 m over one night. There is evidence that increasing global temperatures associated with global warming could lead to a doubling of the frequency of El Niño events [27]. Unstable weather during nesting periods could lead to variability in successes and failures of nesting seasons.

The application of a GLMM was intended to inform what a model habitat would be for crocodile nesting in the LKWS. The model predictions provided less stringent buffers around major water sources than used in the original habitat suitability model and, as a result, the original cut off of 150 m (based on literature) used during the original "fuzzy overlay" model remained the best predictor of nesting habitat presence or absence. That only five nests were positively confirmed provided a too limited framework to generate statistically rigorous data for generic habitat features. Similarities between nest site choices and their spatial separation determined using the GLMM, and from observations, suggest that LKWS crocodiles are choosing sites preferentially and making active selections for nest locations (Hypotheses 1 and 3). That the nests of the 2013 survey were not reused, despite being successful, suggests either that females are not nesting annually or are not nest site-fidelic. There also appeared to be a general preference for smaller open areas, rather than large expanses of open grassland or swamp.

In carrying out this study, numerous UAV-related challenges were faced. While repeatability of transect observations is one of the major benefits of the technology over traditional techniques, the huge disparity in the resolution of the images produced in the two survey years is of concern. This variation could have resulted in the omission of potential nest sites during the 2014 surveys. Conversely, low-resolution images meant that a much larger number of "potential" nest sites needed to be ground-verified, as they could not be excluded due to poor image quality. A similar image resolution to that which was achieved during the 2013 surveys (of 5–6 cm per pixel) would have allowed for the exclusion of a number of the 2014 'potential' nests. In addition, the number of nests located during this study provided limited statistical power. Expansion of the study range in order to provide a larger data set would allow for more wide-reaching conclusions to be made.

The crocodile population of the LKWS has endured fluctuations in both extent and stability; the current population size has, however, raised human-conflict concerns, with at least six anecdotal fatalities having occurred within the study area since 2010. The mapping of nesting habitats has a role in the mediation of conflict zones, especially if a further reduction in forest habitat results in a closer nesting proximity to human settlement. The identification of nests on an annual basis can also aid in the mapping of population trends. This, coupled with spotlighting surveys, could give a better indication of how the population is adapting to anthropogenic expansion. Nesting surveys of this nature could also provide estimates of the carrying capacity of both the study site and the LKWS as a whole, and how crocodile numbers could alter as forest conversion continues. An increase in sample size provided by annual nesting surveys would allow not only for a more in-depth modelling of nesting areas but also a more stringent predictive modelling. In this way, areas deemed most important to successful nesting could be protected, an action, which, in turn, would provide mediation of human-conflict issues. The status of crocodile populations throughout Sabah is also unclear so long term monitoring of nesting habits could be used as an indicator of ongoing population health.

Whilst in terms of cost, drone technology is far cheaper than many traditional survey methodologies, the main barrier to its use by small independent research projects is the cost of image stitching (a cost of around GBP1 per hectare (total area surveyed 6710 ha)). The number of images produced during flights prohibited the use of freeware image stitching software. These image data can be collected and utilised without the use of image stitching but detailed analysis of each specific image would require a far longer time-period. Additionally, the placing of any potential nest's location within the broader context of the landscape, and assessing the hydrological relations, would be far more challenging and would require a highly specified knowledge of the study region.

In summary, the nests identified were spatially exclusive, showing that *C. porosus* individuals in the LKWS are not aggregate nesters. There was, however, an element of statistical predictability to nesting site locations in terms of distance from water bodies, thus allowing search area refinement. Nests were located at least several hundred meters from each other, and from any previously-used nesting sites. Nesting sites were found at sites of medium disturbance levels, although the presence of nest sites close to oil palm plantations suggests that human disturbance is not necessarily a barrier to nesting. This does suggest that stable estuarine crocodile populations could endure even in areas of moderate to high land-use conversion. In order to monitor the stability of the population in such areas, a long term monitoring programme is required.

Acknowledgments: Thanks to Chester Zoo, Columbus Zoo and Aquarium, and the IUCN Crocodile Specialist Group for funding the project. Also to Ryan Pang for assistance during field-testing and flights, and Rob Thomas for statistical advice and Meaghan Evans for proof reading.

Author Contributions: Luke J. Evans, T. Hefin Jones, Benoit Goossens and Silvester Saimin conceived and designed the experiments; Luke Evans and Keeyen Pang performed the experiments. Luke Evans analysed the data; Keeyen Pang assisted with analysis tools. Luke Evans, Hefin Jones and Benoit Goossens wrote the paper.

Conflicts of Interest: The authors declare no conflicts of interest.

References

1. Joanen, T. Nesting ecology of alligators in Louisiana. *Proc. Southeast. Assoc. Fish Wildl. Commun.* **1964**, *23*, 141–151.
2. Pooley, A.C. Preliminary studies on the breeding of the Nile crocodile *Crocodylus niloticus* in Zululand. *Lammergeyer* **1969**, *10*, 22–44.
3. Webb, G.J.W.; Messel, H.; Magnusson, W. The nesting of *Crocodylus porosus* in Arnhem Land, Northern Australia. *Copeia* **1977**, *2*, 238–249. [CrossRef]
4. Harvey, K.R.; Hill, G.J.E. Mapping the nesting habitats of saltwater crocodiles (*Crocodylus porosus*) in Melacca Swamp and the Adelaide River wetlands, Northern Territory: an approach using remote sensing and GIS. *Wildl. Res.* **2003**, *30*, 365–375. [CrossRef]
5. Jiang, H.; Eastman, J.R. Application of fuzzy measures in multi-criteria evaluation in GIS. *Int. J. Geogr. Inf. Sci.* **2000**, *14*, 173–184. [CrossRef]
6. Evans, L.J.; Jones, T.H.; Pang, K.; Evans, M.N.; Saimin, S.; Goossens, B. Use of drone technology as a tool for behavioral research: A case study of crocodilian nesting. *Herpetol. Conserv. Biol.* **2015**, *10*, 90–98.
7. Everaerts, J. The use of unmanned aerial vehicles (UAVs) for remote sensing and mapping. *Int. Arch. Photogramm. Remote Sens. Spat. Inf. Sci.* **2008**, *37*, 1187–1192.
8. Koh, L.P.; Wich, S.A. Dawn of drone ecology: Low-cost autonomous aerial vehicles for conservation. *Trop. Conserv. Sci.* **2012**, *5*, 121–132.
9. Puttock, A.K.; Cunliffe, A.M.; Anderson, K.; Brazier, R.E. Aerial photography collected with a multirotor drone reveals impact of Eurasian beaver reintroduction on ecosystem structure. *J. Unmmaned Veh. Syst.* **2015**, *3*, 123–130. [CrossRef]
10. Inoue, T.; Nagai, S.; Yamashita, S.; Fadaei, H.; Ishii, R.; Okabe, K.; Taki, H.; Honda, Y.; Kajiwara, K.; Suzuki, R. Unmanned aerial survey of fallen trees in a deciduous broadleaved forest in eastern Japan. *PLoS ONE* **2014**, *9*, e109881. [CrossRef] [PubMed]
11. Getzin, S.; Nuske, R.S.; Wiegand, K. Using unmanned aerial vehicles (UAVs) to quantify spatial gap patterns in forests. *Remote Sens.* **2014**, *6*, 6988–7004. [CrossRef]
12. Weissensteiner, M.H.; Poelstra, J.W.; Wolf, J.B.W. Low-budget ready-to-fly unmanned aerial vehicles: An effective tool for evaluating the nesting status of canopy-breeding bird species. *J. Avian Biol.* **2015**, *46*, 1–6. [CrossRef]
13. Mulero-Pázmány, M.; Stolper, R.; van Essen, L.D.; Negro, J.J.; Sassen, T. Remotely piloted aircraft systems as a rhinoceros anti-poaching tool in Africa. *PLoS ONE* **2014**, *9*, e83873. [CrossRef] [PubMed]
14. Martin, J.; Edwards, H.H.; Burgess, M.A.; Percival, H.F.; Fagan, D.E.; Gardner, B.E.; Ortega-Ortiz, J.G.; Ifju, P.G.; Evers, B.S.; Rambo, T.J. Estimating distribution of hidden objects with drones: from tennis balls to manatees. *PLoS ONE* **2012**, *7*, e38882. [CrossRef] [PubMed]

15. Fitzherbert, E.B.; Struebig, M.J.; Morel, A.; Danielsen, F.; Brühl, C.A.; Donald, P.F.; Phalan, B. How will oil palm expansion affect biodiversity? *Trends Ecol. Evol.* **2008**, *23*, 538–545. [CrossRef] [PubMed]

16. Wilcove, D.S.; Koh, L.P. Addressing the threats to biodiversity from oil-palm agriculture. *Biodivers. Conserv.* **2010**, *19*, 999–1007. [CrossRef]

17. Abram, N.K.; Xofis, P.; Tzanopoulos, J.; MacMillan, D.C.; Ancrenaz, M.; Chung, R.; Knight, A.T. Synergies for improving oil palm production and forest conservation in floodplain landscapes. *PLoS ONE* **2014**, *9*, e95388. [CrossRef] [PubMed]

18. Somaweera, R.; Shine, R. Nest-site selection by crocodiles at a rocky site in the Australian tropics: Making the best of a bad lot. *Austral Ecol.* **2012**, *38*, 313–325. [CrossRef]

19. Sabah Wildlife Department. *Crocodile Management Plan*; Sabah State Government: Kota Kinabalu, Malaysia, 2010; pp. 1–67.

20. Magnusson, W.; Grigg, G.; Taylor, J. An aerial survey of potential nesting areas of the saltwater crocodile, *Crocodylus porosus* Schneider, on the North Coast of Arnehem Land, Northern Australia. *Wildl. Res.* **1978**, *5*, 401–415. [CrossRef]

21. Webb, G.J.W.; Manolis, S.C.; Buckworth, R.; Sack, G.C. An examination of *Crocodylus porosus* nests in two northern Australian freshwater swamps, with an analysis of embryo mortality. *Wildl. Res.* **1983**, *10*, 571–605. [CrossRef]

22. Scott, D.A. *A Directory of Asian Wetlands*; IUCN: Gland, Switzerland, 1989.

23. WWF. The Kinabatangan: Summary of Basin Characteristics. *Managing Rivers Wisely*. Available online: http://www.gwp.org/Global/ToolBox/Case%20Studies/Asia%20and%20Caucasus/Malaysia-kinabatangancasestudy_256.pdf (accessed on 19 September 2016).

24. Kuznetsova, A.; Brockhoff, P.B.; Christensen, R.H.B. lmerTest: Tests for Random and Fixed Effects for Linear Mixed Effect Models (Lmer Objects of lme4 Package). Available online: https://cran.r-project.org/web/packages/lmerTest/index.html (accessed on 19 September 2016).

25. Barton, K. MuMIn: Multi-model Inference. R Package Version. Available online: https://rdrr.io/cran/MuMIn/man/MuMIn-package.html (accessed on 19 September 2016).

26. Magnusson, W.E.; Lima, A.P.; Sampaio, R.M. Sources of heat for nests of *Paleosuchus trigonatus* and a review of Crocodilian nest temperatures. *J. Herpetol.* **1985**, *19*, 199–207. [CrossRef]

27. Cai, W.; Borlace, S.; Lengaigne, M.; van Rensch, P.; Collins, M.; Vecchi, G.; Jin, F. Increasing frequency of extreme El Niño events due to greenhouse warming. *Nat. Clim. Chang.* **2014**, *4*, 111–116. [CrossRef]

sensors

MDPI

Article

Onboard Robust Visual Tracking for UAVs Using a Reliable Global-Local Object Model

Changhong Fu [1,2], **Ran Duan** [1,2], **Dogan Kircali** [1,2] **and Erdal Kayacan** [1,*]

1 School of Mechanical and Aerospace Engineering, Nanyang Technological University (NTU),
   50 Nanyang Avenue, Singapore 639798, Singapore; changhongfu@ntu.edu.sg (C.F.);
   duanran@ntu.edu.sg (R.D.); dkircali@ntu.edu.sg (D.K.)
2 ST Engineering-NTU Corporate Laboratory, Nanyang Technological University, 50 Nanyang Avenue,
   Singapore 639798, Singapore
* Correspondence: erdal@ntu.edu.sg; Tel.: +65-9728-8774

Academic Editor: Felipe Gonzalez Toro
Received: 13 July 2016; Accepted: 25 August 2016; Published: 31 August 2016

Abstract: In this paper, we present a novel onboard robust visual algorithm for long-term arbitrary 2D and 3D object tracking using a reliable global-local object model for unmanned aerial vehicle (UAV) applications, e.g., autonomous tracking and chasing a moving target. The first main approach in this novel algorithm is the use of a global matching and local tracking approach. In other words, the algorithm initially finds feature correspondences in a way that an improved binary descriptor is developed for global feature matching and an iterative Lucas–Kanade optical flow algorithm is employed for local feature tracking. The second main module is the use of an efficient local geometric filter (LGF), which handles outlier feature correspondences based on a new forward-backward pairwise dissimilarity measure, thereby maintaining pairwise geometric consistency. In the proposed LGF module, a hierarchical agglomerative clustering, i.e., bottom-up aggregation, is applied using an effective single-link method. The third proposed module is a heuristic local outlier factor (to the best of our knowledge, it is utilized for the first time to deal with outlier features in a visual tracking application), which further maximizes the representation of the target object in which we formulate outlier feature detection as a binary classification problem with the output features of the LGF module. Extensive UAV flight experiments show that the proposed visual tracker achieves real-time frame rates of more than thirty-five frames per second on an i7 processor with 640×512 image resolution and outperforms the most popular state-of-the-art trackers favorably in terms of robustness, efficiency and accuracy.

Keywords: unmanned aerial vehicle; visual object tracking; reliable global-local model; local geometric filter; local outlier factor; robust real-time performance

---

1. Introduction

Visual tracking, as one of the most active vision-based research topics, can assist unmanned aerial vehicles (UAVs) to achieve autonomous flights in different types of civilian applications, e.g., infrastructure inspection [1], person following [2] and aircraft avoidance [3]. Although numerous visual tracking algorithms have recently been proposed in the computer vision community [4–9], onboard visual tracking of freewill arbitrary 2D or 3D objects for UAVs remains as a challenging task due to object appearance changes caused by a number of situations, inter alia, shape deformation, occlusion, various surrounding illumination, in-plane or out-of-plane rotation, large pose variation, onboard mechanical vibration, wind disturbance and aggressive UAV flight.

To track an arbitrary object with a UAV, the following four basic requirements should be considered to implement an onboard visual tracking algorithm: (1) real-time: the tracking algorithm must process onboard captured live image frames at high speed; (2) accuracy: the tracking algorithm must track the object accurately even with the existence of the aforementioned challenging factors; (3) adaptation: the tracking algorithm must adapt real object appearance online; (4) recovery: the tracking algorithm must be capable of re-detecting the object when the target object becomes visible in the field of view (FOV) of the camera again after the object is lost.

In this paper, the following three main modules are proposed to have a reliable visual object model under the aforementioned challenging situations and to achieve those basic requirements in different UAV tracking applications:

- A global matching and local tracking (GMLT) approach has been developed to initially find the FAST [10] feature correspondences, i.e., an improved version of the BRIEF descriptor [11] is developed for global feature matching, and an iterative Lucas–Kanade optical flow algorithm [12] is employed for local feature tracking between two onboard captured consecutive image frames based on a forward-backward consistency evaluation method [13].
- An efficient local geometric filter (LGF) module has been designed for the proposed visual feature-based tracker to detect outliers from global and local feature correspondences, i.e., a novel forward-backward pairwise dissimilarity measure has been developed and utilized in a hierarchical agglomerative clustering (HAC) approach [14] to exclude outliers using an effective single-link approach.
- A heuristic local outlier factor (LOF) [15] module has been implemented for the first time to further remove outliers, thereby representing the target object in vision-based UAV tracking applications reliably. The LOF module can efficiently solve the chaining phenomenon generated from the LGF module, i.e., a chain of features is stretched out with long distances regardless of the overall shape of the object, and the matching confusion problem caused by the multiple moving parts of objects.

Extensive UAV flight experiments show that the proposed visual tracker achieves real-time frame rates of more than thirty-five frames per second on an i7 processor with 640×512 image resolution and outperforms the most popular state-of-the-art trackers favorably in terms of robustness, efficiency and accuracy.

The outline of the paper is organized as follows: Section 2 presents the recent works related to the visual object tracking for UAVs. Section 3 introduces the proposed novel visual object tracking algorithm. The performance evaluations in various UAV flight tests and its comparisons with the most popular state-of-the-art visual trackers are discussed in Section 4. Finally, the concluding remarks are given in Section 5.

2. Related Works

2.1. Color Information-Based Method

Color information on the image frame has played an important role in visual tracking applications. A color-based visual tracker is proposed in [16] for UAVs to autonomously chase a moving red car. A visual tracking approach based on the color information is developed in [17] for UAVs to follow a red 3D flying object. Similarly, a color-based detection approach is employed in [18] for UAVs to track a red hemispherical airbag and to achieve autonomous landing. Although all of these color-based visual tracking approaches are very efficient and various kinds of color spaces can be adopted, this type of visual tracker is very sensitive to illumination changes and noise on the image, and it is preferably applicable for target tracking with a monotone and distinctive color.

2.2. Direct or Feature-Based Approach

A static or moving object tracked by a UAV has often been represented by a rectangle bounding box. Template matching (TM) has usually been applied to visual object tracking tasks for UAVs. It searches a region of interest (ROI) in the current image frame that is similar to the template defined in the first image frame. The TM approach for UAVs can be categorized into two groups:

the direct method [19] and featured-based approach [20–23]. The direct method uses the intensity information of pixels directly to represent the tracked object and to track the target object, whereas the feature-based approach adopts a visual feature, e.g., Harris corner [24], SIFT [25], SURF [26] or ORB [27] feature, to track the target object. However, these existing visual trackers are not robust to the aforementioned challenging situations since the object appearance is only defined or fixed in the first image frame, i.e., they cannot update the object appearance during the UAV operation. What is more, these trackers are not suitable for tracking 3D or deformable objects.

2.3. Machine Learning-Based Method

Machine learning methods have also been applied to vision-based UAV tracking applications. In general, these approaches can be divided into two categories based on the learning methods: offline and online learning approaches.

2.3.1. Offline Machine Learning-Based Approach

An offline-trained face detector is utilized in [28] to detect the face of a poacher from a UAV in order to protect the wildlife in Africa. An offline-learned visual algorithm is applied in [29] to detect a 2D planar target object for a UAV to realize autonomous landing. A supervised learning approach is presented in [30] to detect and classify different types of electric towers for UAVs. However, a large number of image training datasets, i.e., positive and negative image patches with all aforementioned challenging conditions, should be cropped and collected to train these trackers, thereby guaranteeing their high detection accuracies. Moreover, labeling the image training datasets requires much experience, and it is a tedious, costly and time-consuming task. In addition, an offline-trained visual tracker is only capable of tracking specifically-trained target objects instead of freewill arbitrary objects.

2.3.2. Online Machine Learning-Based Method

Recently, online learning visual trackers have been developed as the most promising tracking approaches to track an arbitrary 2D or 3D object. Online learning visual tracking algorithms are generally divided into two categories: generative and discriminative methods.

Generative approaches learn only a 2D or 3D tracking object online without considering the background information around the tracking object and then apply the online-learned model for searching the ROI on the current image frame with minimum reconstruction error. Visual tracking algorithms based on incremental subspace learning [31] and hierarchical methods are developed in [28,32] to track a 2D or 3D object for UAVs. Although these works obtained promising tracking results, a number of object samples from consecutive image frames should be collected, and they have assumed that the appearance of the target object does not change significantly during the image collection period.

Discriminative methods treat the tracking problem as a binary classification task to separate the object from its background using an online updating classifier with positive and negative (i.e., background) information. A tracking-learning-detection (TLD) [8] approach is utilized in [33] for a UAV to track different objects. A real-time adaptive visual tracking algorithm, which is based on a vision-based compressive tracking approach [6], is developed in [1] for UAVs to track arbitrary 2D or 3D objects. A structured output tracking with kernels (STRUCK) algorithm is adopted in [2] for a UAV to follow a walking person. However, updating on consecutive image frames is prone to include noises, and the drift problem is likely to occur, thereby resulting in tracking failure. Although an online multiple-instance learning (MIL) [4] approach is developed in [3] to improve tracking performance for UAVs, the update step may still not effectively eliminate noises. Additionally, most of the discriminative methods cannot estimate the scale changes of the target object.

3. Proposed Method

The proposed method in this paper mainly includes three modules: (1) global matching and local tracking (GMLT); (2) local geometric filter (LGF); and (3) local outlier factor (LOF).

3.1. Global Matching and Local Tracking Module

Let b_1 be a bounding box around an online selected 2D or 3D target object, e.g., a moving person on the first image frame I_1 shown in Figure 1. The FAST features on each RGB image frame are detected using a bucketing approach [34], i.e., the captured image frame is separated into pre-defined non-overlapped rectangle regions, i.e., buckets. The bucketing approach keeps the FAST features evenly distributed in the image frame and guarantees the real-time visual tracking performance. A group of the FAST features detected in b_1 is denoted as $\{x_1^1, x_1^2, ..., x_1^n\}$, where $x_1^i \in \mathbb{R}^2$, $i = 1, 2, ..., n$; they compose the global model $\mathcal{M}_g$ of the target object. The model $\mathcal{M}_g$ is utilized to globally match the candidate FAST features detected on each image frame with the improved version of the BRIEF descriptor. It is to be noted that the global matching is able to achieve the re-detection of the object when the target object becomes visible in the camera FOV again after the object is lost. For the i-th FAST feature on the k-th image frame x_k^i, its improved BRIEF descriptor, i.e., $\mathcal{B}(x_k^i) = \{B_1(x_k^i), B_2(x_k^i), ..., B_{N_b}(x_k^i)\}$, is defined as follows:

$$B_j(x_k^i) = \begin{cases} 1 & \text{if } I_k(x_k^i + p_j) < I_k(x_k^i + q_j) \\ 0 & \text{otherwise} \end{cases}, \forall j \in [1, ..., N_b] \tag{1}$$

where $B_j(x_k^i)$ is the j-th bit of the binary vector in $\mathcal{B}(x_k^i)$, $I_k(*)$ is the intensity of the pixel on the k-th image frame and (p_j, q_j) is sampled in a local neighbor region $S_r \times S_r$ based on the location of the i-th FAST feature. $p_j = \mathcal{N}(0, (\frac{1}{5}S_r)^2)$ and $q_j = \mathcal{N}(p_j, (\frac{2}{25}S_r)^2)$. If the intensity of the pixel on the location $x_k^i + p_j$ is smaller than the one on the location $x_k^i + q_j$, then the $B_j(x_k^i)$ is one, otherwise, it is zero. The parameter N_b is the length of the binary vector $\mathcal{B}(x_k^i)$, i.e., the number of comparisons to perform. It is to be noted that the distance d of two binary vectors is computed by counting the number of different bits between these two vectors, i.e., the Hamming distance, which has less computation cost compared to the Euclidean distance. For searching the FAST feature correspondences on the k-th ($k \geq 2$) image frame using the model $\mathcal{M}_g$, the FAST feature that has the lowest Hamming distance d_1 is the best feature candidate; the FAST feature that has the second-lowest Hamming distance d_2 is the second-best feature candidate; when the ratio between the best and second-best match d_1/d_2 is less than a threshold ρ, the best FAST feature candidate is accepted as a matched FAST feature.

An iterative Lucas–Kanade optical flow algorithm with a three-level pyramid has been utilized to track each FAST feature between the $(k$-1)-th image frame and the k-th image frame based on the forward-backward consistency evaluation approach within a $S_h \times S_h$ local search window. These tracked FAST features constitute the local model $\mathcal{M}_l$ of the target object. It is to be noted that in the local tracking stage, the model $\mathcal{M}_l$ is updated frame-by-frame, i.e., the model $\mathcal{M}_l$ is adaptive.

Let $\mathcal{F}_1$ be the matched and tracked FAST features on the k-th image frame, denoted as $\mathcal{F}_1 = \{x_k^1, x_k^2, ..., x_k^m\}$, where $x_k^i \in \mathbb{R}^2$, $i = 1, 2, ..., m$. In this work, the scale s_k of the target object is estimated based on [13], i.e., for each pair of FAST features, a ratio between the FAST feature distance on the current image frame and the corresponding FAST feature distance on the first image frame is calculated:

$$s_k^{ij} = \frac{||x_k^i - x_k^j||}{||x_1^i - x_1^j||}, i \neq j \tag{2}$$

and then, the median of $\{s_k^{ij}\}$ is the estimated scale s_k of the target object, since it is more robust with respect to outliers.

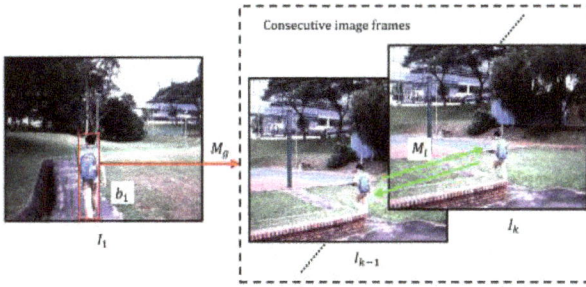

Figure 1. Illustration of the global matching and local tracking (GMLT) module. The bounding box b_1 is shown with the red rectangle. The FAST features detected in the b_1 compose the global object model $\mathcal{M}_g$, which is employed to globally find the feature correspondences on onboard captured consecutive image frames with the improved BRIEF descriptor. The green arrow is the Lucas–Kanade optical flow algorithm-based tracking between the (k-1)-th frame and the k-th frame. $\mathcal{M}_l$ is the local object model, which is updated frame-by-frame.

In our extensive visual tracking tests, we find that the $\mathcal{F}_1$ includes certain outlier FAST features, as one example is shown in the left image in Figure 2, i.e., I_k^{GMLT}. Let a matching with the model $\mathcal{M}_g$ be a global FAST feature correspondence, a tracking with model $\mathcal{M}_l$ be a local FAST feature correspondence and the combination of the global matching and local tracking correspondences be $\mathcal{C}_k^\cup$; mt_k^i is the i-th FAST feature correspondence in $\mathcal{C}_k^\cup$. The next subsection introduces the second main module to detect these outliers.

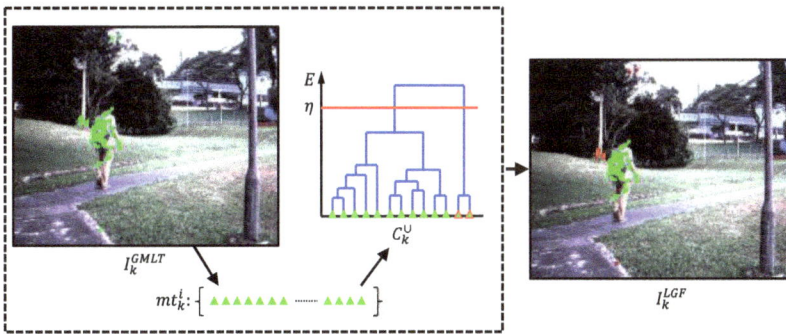

Figure 2. Illustration of the local geometric filter (LGF) module. The green points on the I_k^{GMLT} are the matched and tracked FAST features from the GMLT module, i.e., $\mathcal{F}_1$. The green triangles are the FAST feature correspondences, i.e., $\mathcal{C}_k^\cup$. The LGF module is utilized to filter outlier correspondences, as the green triangles with red edges shown in the dendrogram. The red points on the I_k^{LGF} are the outliers filtered by the LGF module.

3.2. Local Geometric Filter Module

The second main module in the proposed method is a novel efficient local geometric filter (LGF), which utilizes a new forward-backward pairwise dissimilarity measure E_{LGF} between correspondences mt_k^i and mt_k^j based on pairwise geometric consistency, as illustrated in Figure 3.

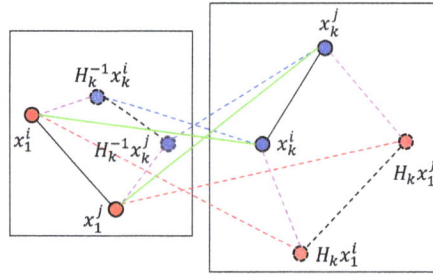

Figure 3. Forward-backward pairwise dissimilarity measure E_{LGF} between correspondence mt_k^i and mt_k^j (green solid lines). The dashed red and blue points are transformed by the homography H_k and the inversion of homography H_k^{-1}.

The E_{LGF} for every pair of correspondences is defined as below:

$$E_{\text{LGF}}(mt_k^i, mt_k^j) = \frac{1}{2}\left[E_{\text{LGF}}(mt_k^i, mt_k^j | H_k) + E_{\text{LGF}}(mt_k^i, mt_k^j | H_k^{-1})\right], i \neq j \tag{3}$$

where:

$$E_{\text{LGF}}(mt_k^i, mt_k^j | H_k) = \left\| (x_k^i - x_k^j) - H_k(x_1^i - x_1^j) \right\|, i \neq j$$

$$E_{\text{LGF}}(mt_k^i, mt_k^j | H_k^{-1}) = \left\| (x_1^i - x_1^j) - H_k^{-1}(x_k^i - x_k^j) \right\|, i \neq j$$

$\|*\|$ is the Euclidean distance; H_k is a homography transformation [35] estimated by the $\mathcal{C}_k^\cup$; and H_k^{-1} is the inversion of this homography transformation.

To reduce the ambiguity correspondences and filter the erroneous correspondences, a hierarchical agglomerative clustering (HAC) approach [14] is utilized to separate outlier correspondences from inliers based on an effective single-link approach with the forward-backward pairwise dissimilarity measure E_{LGF}. Let $S(G, G')$ be a cluster dissimilarity for all pairs of clusters, $G, G' \subset \mathcal{C}_k^\cup$; the single-link HAC algorithm is defined:

$$S(G, G') = \min_{mt_k^i \in G, mt_k^j \in G'} E_{\text{LGF}}(mt_k^i, mt_k^j) \tag{4}$$

It defines the cluster dissimilarity S as the minimum among all of the forward-backward pairwise dissimilarities between the two correspondences of the two clusters. A dendrogram generated from the single-link HAC approach is shown in the middle of Figure 2; the $\mathcal{C}_k^\cup$ is divided into some subgroups based on a cut-off threshold η, and the biggest subgroup is considered as the correspondences for the target object. The green points shown on the right side of Figure 2, i.e., I_k^{LGF}, are the FAST features output from the LGF module, denoted as $\mathcal{F}_2 = \{\tilde{x}_k^1, \tilde{x}_k^2, ..., \tilde{x}_k^w\}$, where $\tilde{x}_k^i \in \mathbb{R}^2$, $i = 1, 2, ..., w$, while the red points are the outliers.

The bottom-up aggregation in the single link-based clustering method is strictly local. The single-link HAC approach is easy to generate the chaining phenomenon, i.e., a chain of correspondences is stretched out for long distances without considering the real shape of the target object, especially in cluttered environments, leading to inefficient exclusion of the outliers. Additionally, multiple parts of objects, e.g., moving hands, are prone to confuse the FAST feature matching in the next new image frame. In this work, the local outlier factor (LOF) is developed to handle the chaining and confusion problems efficiently. The following subsection introduces the third main module.

3.3. Local Outlier Factor Module

The third main module of this work is a heuristic local outlier factor (LOF) [15], which is developed for the first time in a visual tracking application to further remove outliers, thereby maximizing target object representation and solving the matching confusion problem. The LOF is based on local density, i.e., the outlier is considered when its surrounding space contains relatively few FAST features.

As shown in Figure 4, the local density of the FAST feature $\tilde{x}_k^i$ is compared to the densities of its neighborhood FAST features. In this case, if the FAST feature $\tilde{x}_k^i$ has much lower density than its neighbors, then it is an outlier.

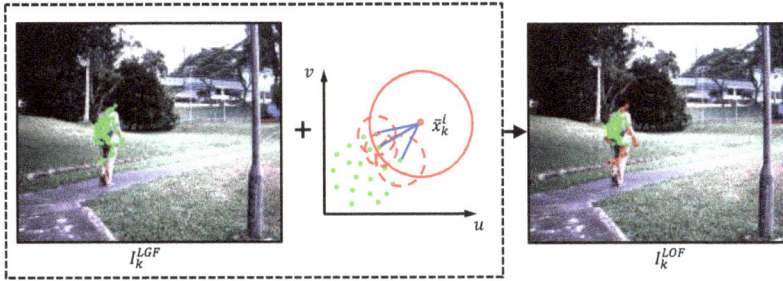

Figure 4. Illustration of local outlier factor (LOF) module. The green points on the I_k^{LGF} are the FAST features from the LGF module, i.e., $\mathcal{F}_2$. The LOF module is developed to further remove the outliers, as the red points shown on the I_k^{LOF}. The green points on the I_k^{LOF} are final reliable output FAST features in our visual tracking application.

In this work, we formulate outlier feature detection as a binary classification problem. The binary classifier is defined as follows:

$$f(\tilde{x}_k^i) = \begin{cases} \text{target object,} & E_{\text{LOF}}(\tilde{x}_k^i) \le \mu \\ \text{outlier,} & E_{\text{LOF}}(\tilde{x}_k^i) > \mu \end{cases} \tag{5}$$

where $E_{\text{LOF}}(\tilde{x}_k^i)$ is a density dissimilarity measure of FAST feature $\tilde{x}_k^i$, $\tilde{x}_k^i \in \mathcal{F}_2$, and μ is a cut-off threshold to classify that a FAST feature belongs to the target object or an outlier. If the value of the $E_{\text{LOF}}(\tilde{x}_k^i)$ is larger than μ, then $\tilde{x}_k^i$ is the outlier, otherwise, $\tilde{x}_k^i$ belongs to the target object.

The LOF module includes three steps to calculate the $E_{\text{LOF}}(\tilde{x}_k^i)$:

- Construction of the nearest neighbors: the nearest neighbors of the FAST feature $\tilde{x}_k^i$ are defined as follows:

$$\mathcal{NN}(\tilde{x}_k^i) = \{\tilde{x}_k^j \in \mathcal{F}_2 \setminus \{\tilde{x}_k^i\} | D(\tilde{x}_k^i, \tilde{x}_k^j) \le R_t(\tilde{x}_k^i)\} \tag{6}$$

where $D(\tilde{x}_k^i, \tilde{x}_k^j)$ is the Euclidean distance between the FAST features $\tilde{x}_k^i$ and $\tilde{x}_k^j$. $R_t(\tilde{x}_k^i)$ is the Euclidean distance from $\tilde{x}_k^i$ to the t-th nearest FAST feature neighbor.

- Estimation of neighborhood density: the neighborhood density δ of the FAST feature $\tilde{x}_k^i$ is defined as:

$$\delta(\tilde{x}_k^i) = \frac{|\mathcal{NN}(\tilde{x}_k^i)|}{\sum_{\tilde{x}_k^j \in \mathcal{NN}(\tilde{x}_k^i)} \max\{R_t(\tilde{x}_k^j), D(\tilde{x}_k^i, \tilde{x}_k^j)\}} \tag{7}$$

where $|\mathcal{NN}(\tilde{x}_k^i)|$ is the nearest neighbor number of $\tilde{x}_k^i$.

- Comparison of neighborhood densities: the comparison of neighborhood densities results in the density dissimilarity measure $E_{LOF}(\hat{x}_k^i)$, which is defined below:

$$E_{LOF}(\hat{x}_k^i) = \frac{\sum_{\hat{x}_k^j \in \mathcal{NN}(\hat{x}_k^i)} \frac{\delta(\hat{x}_k^i)}{\delta(\hat{x}_k^j)}}{|\mathcal{NN}(\hat{x}_k^i)|} \tag{8}$$

Figure 4 shows the illustration of the LOF module; the green points on the I_k^{LOF} are final reliable output FAST features for our visual tracking application, denoted as $\mathcal{F}_3 = \{\hat{x}_k^1, \hat{x}_k^2, ..., \hat{x}_k^o\}$, where $\hat{x}_k^i \in \mathbb{R}^2, i = 1, 2, ..., o$. The FAST features in $\mathcal{F}_3$ and their corresponding features in b_1 compose final FAST feature correspondences $\hat{\mathcal{C}}_k^{\cup}$. Then, the center c_k of the target object is calculated as follows:

$$c_k = \frac{\sum_{mt_k^i \in \hat{\mathcal{C}}_k^{\cup}} (\hat{x}_k^i - Hx_1^i)}{|\hat{\mathcal{C}}_k^{\cup}|} \tag{9}$$

4. Real Flight Tests and Comparisons

In the UAV flight experiments, a Y6 coaxial tricopter UAV equipped with a Pixhawk autopilot from 3D Robotics is employed; the onboard computer is an Intel NUC Kit NUC5i7RYH Mini PC, which has a Core i7-5557U processor with dual-core, 16 GB RAM and a 250-GB SATA SSD drive. Both forward- and downward-looking cameras are USB 3.0 RGB cameras from Point Grey, i.e., Flea3 FL3-U3-13E4C-C, which capture the image frames with a resolution of 640 × 512 at 30 Hz. The whole UAV system is shown in Figure 5.

Figure 5. Robust real-time accurate long-term visual object tracking onboard our Y6 coaxial tricopter UAV. No. 1 and 2 in (**a**) show downward- and forward-looking monocular RGB cameras. Some 2D or 3D objects with their tracking results are shown in (**b–g**).

To practically test and evaluate the robustness, efficiency and accuracy of the proposed onboard visual tracker, we have developed our visual tracker in C++ and conducted more than fifty UAV flights in various types of environments of Nanyang Technological University, including challenging situations. As shown in Figure 5, target objects include a moving car (**b**), walking people (**c** and **d**), a container (**e**), a gas tank (**f**) and a moving unmanned ground vehicle (UGV) with a landing pad (**g**). In this paper, six recorded image sequences are randomly selected which contain 11,646 image frames, and manually labeled for the ground truth. The challenging factors of the each image sequence are listed in Table 1.

Table 1. Challenging factors of each image sequence. MV: mechanical vibration; AF: aggressive flight; IV: illumination variation; OC: partial or full occlusion; SV: scale variation; DE: deformation, i.e., non-rigid object deformation; IR: in-plane rotation; OR: out-of-plane rotation; OV: out-of-view; CB: cluttered background. The total number of evaluated image frames in this paper is 11,646.

Sequence	Number	MV	AF	IV	OC	SV	DE	IR	OR	OV	CB
Container	2874	√				√			√	√	√
Gas tank	3869	√	√	√		√		√	√	√	√
Moving car	582	√					√			√	
UGV_{lp}	1325	√			√		√		√		
$People_{bw}$	934	√				√	√	√			
$People_{fw}$	2062	√		√	√	√	√		√		√

To compare our proposed visual tracker, we have employed the most popular state-of-the-art visual trackers, e.g., MIL [4], STRUCK [5], CT [6], Frag [7], TLD [8] and KCF [9], which have adaptive capabilities for appearance changes of the target objects and have been utilized to achieve the real UAV tracking applications. For all of these state-of-the-art trackers, we have utilized the source or binary programs provided by the authors with default parameters. In our proposed visual tracker, the main parameters are defined in Table 2 below. In addition, all visual trackers are initialized with the same parameters, e.g., initial object location.

Table 2. Main parameters in our presented visual tracker.

Parameter Name	Value	Parameter Name	Value
Bucketing configuration	10×8	FAST threshold	20
Sampling patch size (S_r)	48	BRIEF descriptor length (N_b)	256
Ratio threshold (ρ)	0.85	Local search window (S_h)	30
LGF cut-off threshold (η)	18	LOF cut-off threshold (μ)	1.5

In this work, the center location error (CLE) of the tracked target object has been utilized to evaluate all visual trackers. It has been defined as the Euclidean distance between the estimated target object center and the manually-labeled ground truth center on each image frame, i.e.:

$$\text{CLE} = \left\| O_k^E - O_k^{GT} \right\| \tag{10}$$

where O_k^E and O_k^{GT} are the estimated and ground truth centers of the target object. Figures 6–11 show the CLE evolutions of all visual trackers in different image sequences. Specifically, we note that the TLD tracker easily loses the target completely for certain image frames when the target object is still in the FOV of the onboard camera; since it is able to re-detect the target object, we show the CLE error for the image sequence that the TLD tracker can track more than 96% of frames as a reference. Table 3 shows the CLE errors of all visual trackers. To visualize the tracking precisions of all visual trackers, Figures 12a, 13a, 14a, 15a, 16a and 17a show the precision plots of all image sequences.

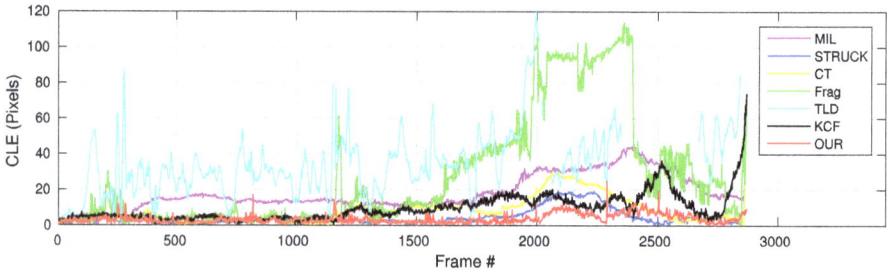

Figure 6. Center location error (CLE) error evolution plot of all visual trackers with the container image sequence.

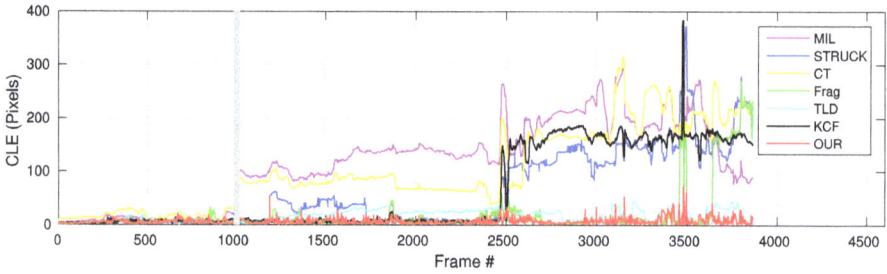

Figure 7. CLE evolution plot of all visual trackers with gas tank image sequence. The grey area represents that the whole target object is out of view.

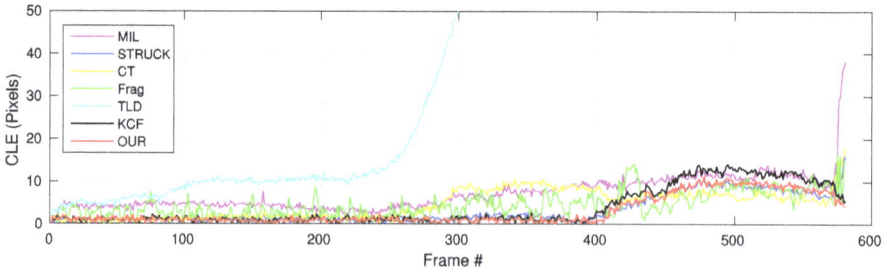

Figure 8. CLE error evolution plot of all trackers with the moving car image sequence.

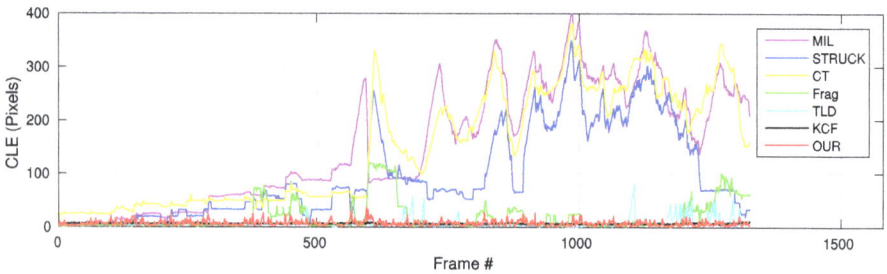

Figure 9. CLE error evolution plot of all trackers with the UGV$_{lp}$ image sequence.

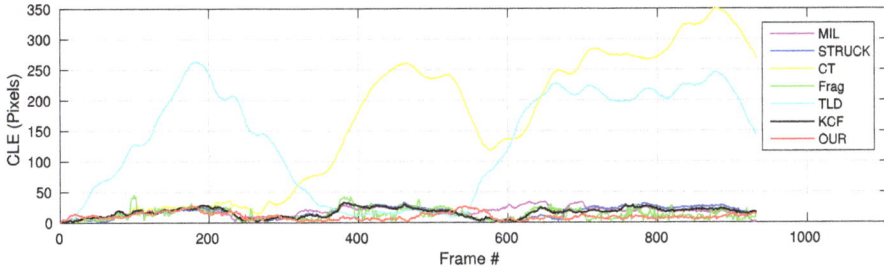

Figure 10. CLE error evolution plot of all trackers with the People$_{bw}$ image sequence.

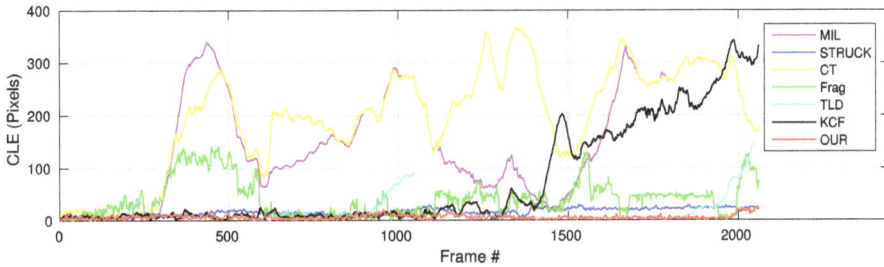

Figure 11. CLE error evolution plot of all trackers with the People$_{fw}$ image sequence.

Table 3. Center location error (CLE) (in pixels) and frames per second (FPS). Red, blue and green fonts indicate the best, second best and third best performances in all visual trackers. The total number of evaluated image frames in this paper is 11,646.

Sequence	MIL	STRUCK	CT	Frag	TLD	KCF	Our
Container	17.7	4.5	8.2	27.4	-	9.3	3.6
Gas tank	118.1	62.4	103.4	22.7	16.6	63.1	6.8
Moving Car	7.1	3.1	4.5	4.5	105.1	3.7	3.3
UGV$_{lp}$	152.2	97.1	150.1	20.6	7.6	4.8	7.3
People$_{bw}$	17.8	16.7	157.9	13.8	130.8	16.6	10.2
People$_{fw}$	153.3	15.7	197.1	41.7	-	73.0	6.1
CLE$_{Ave}$	96.4	41.6	107.1	28.4	-	18.9	6.5
FPS$_{Ave}$	24.8	16.2	28.7	13.1	23.9	149.8	38.9

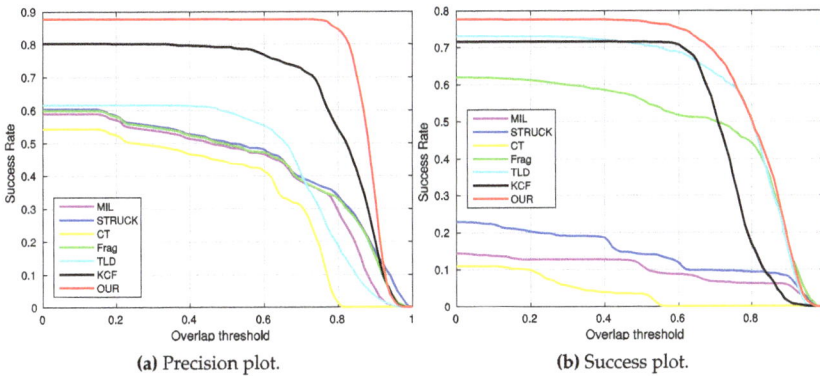

(a) Precision plot.

(b) Success plot.

Figure 12. Precision and success plots of all visual trackers with the container image sequence.

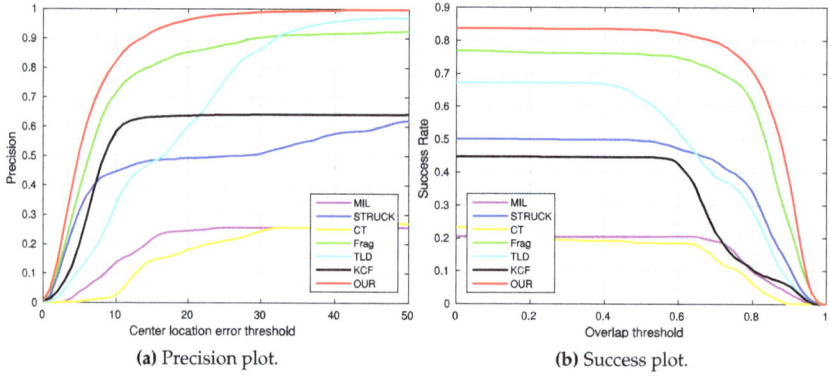

(a) Precision plot.

(b) Success plot.

Figure 13. Precision and success plots of all visual trackers with the gas tank image sequence.

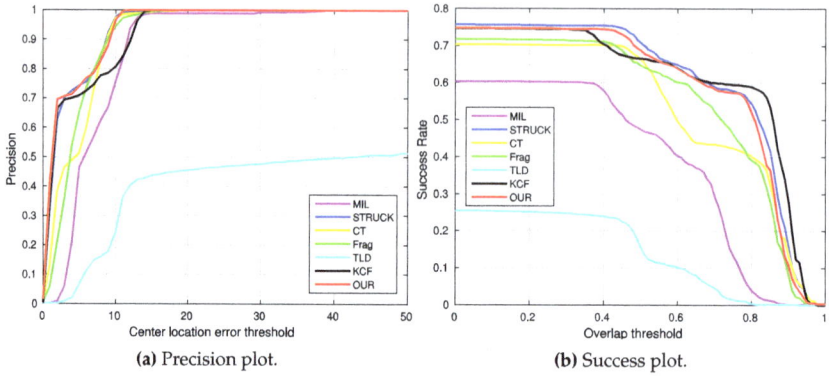

(a) Precision plot.

(b) Success plot.

Figure 14. Precision and success plots of all trackers with the moving car image sequence.

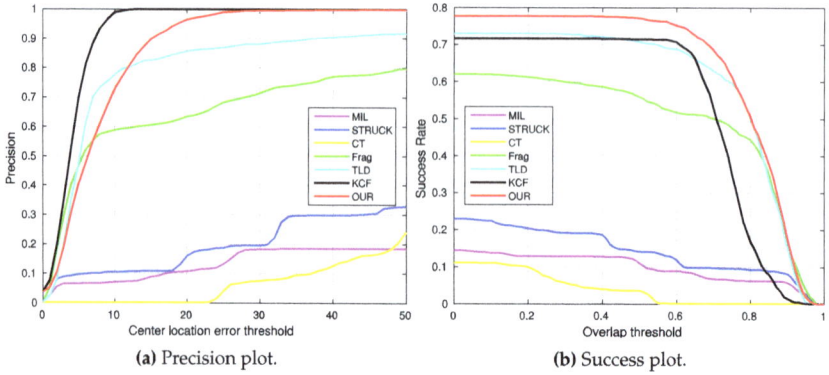

(a) Precision plot.

(b) Success plot.

Figure 15. Precision and success plots of all trackers with the UGV_{lp} image sequence.

Sensors **2016**, *16*, 1406

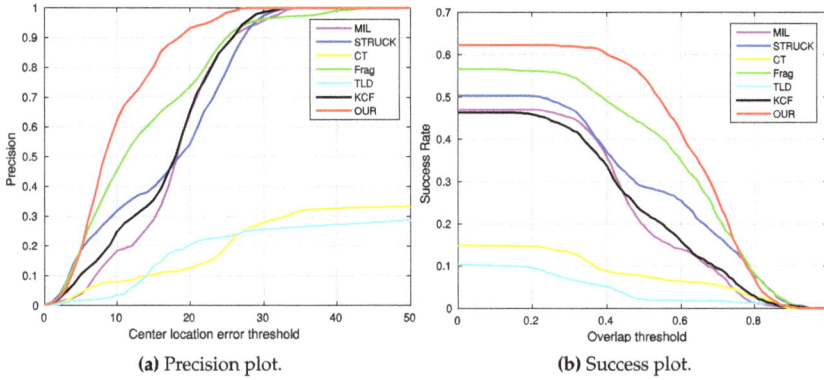

(a) Precision plot.

(b) Success plot.

Figure 16. Precision and success plots of all trackers with the People$_{bw}$ image sequence.

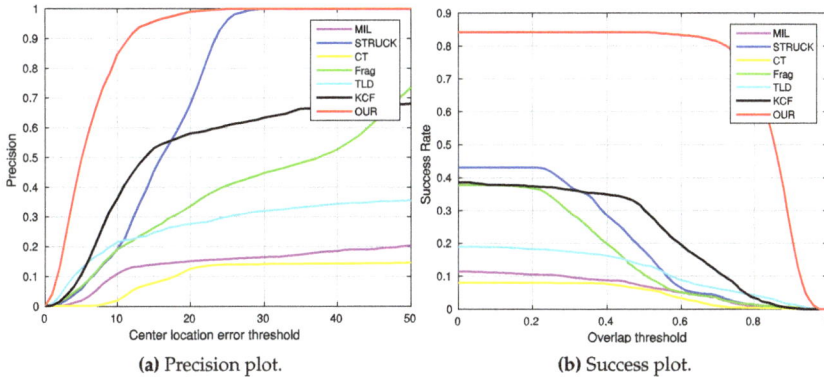

(a) Precision plot.

(b) Success plot.

Figure 17. Precision and success plots of all trackers with the People$_{fw}$ image sequence.

In addition, the success score (SS) has also been employed to evaluate the performances of all visual trackers in this paper, as it can evaluate the scales of the target object. The SS has been defined as below:

$$SS = \frac{|ROI_E \cap ROI_{GT}|}{|ROI_E \cup ROI_{GT}|} \tag{11}$$

where ROI_E and ROI_{GT} are the estimated and ground truth sizes of the target object. $\cap$ and $\cup$ are the intersection and union operators. $|*|$ represents the number of pixels in a region. If the SS is larger than ζ in an image frame, the tracking result is considered as a success. Table 4 shows the tracking results ($\zeta = 0.5$) of all visual trackers in terms of success rate, which is defined as the ratio between the number of success frames and the total number of image frames. Moreover, we have shown the area under area (AUC) of each success plot, which is defined as the average of the success rates based on the overlap thresholds, as the results shown in Figures 12b, 13b, 14b, 15b, 16b and 17b.

Table 4. Success rate (SR) (%) (ξ = 0.5). Red, blue and green fonts indicate the best, second best and third best performances in all visual trackers. The total number of evaluated image frames in this paper is 11,646.

Sequence	MIL	STRUCK	CT	Frag	TLD	KCF	Our
Container	62.9	62.7	62.7	62.5	81.2	96.6	99.8
Gas tank	25.7	61.8	24.5	90.8	85.6	62.7	97.3
Moving car	68.0	88.1	89.9	82.5	24.7	78.4	85.7
UGV_{lp}	15.1	18.7	6.8	70.3	89.9	99.6	98.6
$People_{bw}$	30.0	40.5	10.5	63.2	2.67	34.7	81.7
$People_{fw}$	10.6	29.5	9.9	16.9	19.6	46.8	99.4
SR_{Ave}	32.9	50.1	30.8	65.3	51.8	71.0	95.9

4.1. Test 1: Visual Tracking of The Container

In this test, a static container is selected as the target object for our Y6 coaxial tricopter UAV to carry out the visual tracking application, and the onboard forward-looking camera is employed to track the locations of the target object. As the challenging factors concluded in Table 1, the container image sequence includes mechanical vibration (all image frames), scale variation (e.g., Figure 18, Frames 1827 and 2000), out-of-plane rotation (e.g., Figure 18, Frames 175 and 2418), out-of-view (e.g., Figure 18, Frame 2874) and cluttered background (all image frames).

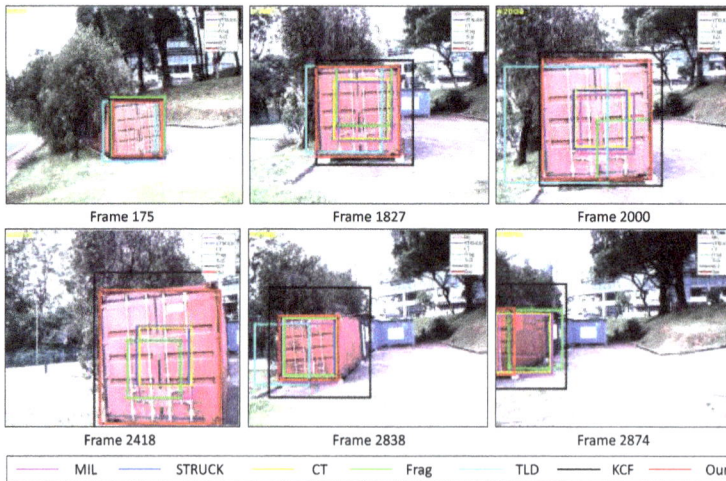

Figure 18. Some tracking results in the container image sequence. The total number of frames: 2874.

As can be seen in Figure 6, we can find that the CLE errors of our presented visual tracker (red line) and STRUCK (blue line) are always less than 20 pixels, i.e., the precisions of these two visual trackers can achieve almost one when the CLE threshold is 20 pixels, as shown in Figure 12a. Moreover, the tracking performance of the CT tracker (yellow line) is ranking as No. 3 in this image sequence; its CLEs are changing extensively when the target object is out-of-plane; the maximum CLE error of the CT tracker is 29.8 pixels. The KCF tracker is ranking No. 4; its performance is decreasing when the flying UAV is approaching the target object. Additionally, the MIL tracker (magenta line) outperforms the TLD (cyan line) and Frag (green line) trackers, and the tracking performance of Frag is better than that of the TLD tracker; it is noticed that the TLD tracker completely loses track of the target object when some portion of the target object is out of view, i.e., some parts of the target object are not shown in the FOV of the onboard forward-looking camera, as Frame 2874 shown in Figure 18. In addition, the

2874th frame also shows that our presented visual tracker is able to locate the target object accurately even under the out-of-view situation.

Figure 12b shows the average success rates of all visual trackers. Since the MIL, STRUCK, CT and Frag trackers cannot estimate the scales of the target object, their average success rates are relatively low. Conversely, the TLD, KCF and our presented visual tracker can estimate the target object scales. However, the accuracies of the TLD and KCF trackers for estimating the center locations are lower. Therefore, the average success rates of the TLD and KCF trackers are also lower than ours.

4.2. Test 2: Visual Tracking of the Gas Tank

A static gas tank is employed as the tracking object of our UAV in this test. As shown in Figure 19, this target object does not contain much texture information. Additionally, the aggressive flight (e.g., Figure 19, Frames 3343 and 3490), out-of-view (e.g., Figure 19, Frames 999, 1012 and 3490), scale variation (e.g., Figure 19, Frames 500 and 2587) and cluttered background (all of the frames) are the main challenging factors.

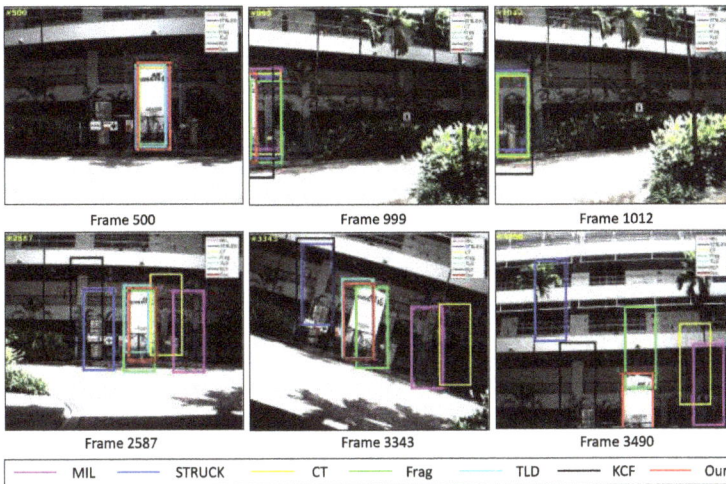

Frame 500	Frame 999	Frame 1012
Frame 2587	Frame 3343	Frame 3490

| —— MIL | —— STRUCK | —— CT | —— Frag | —— TLD | —— KCF | —— Our |

Figure 19. Some tracking results in the gas tank image sequence. The total number of frames: 3869.

As shown in Figure 7, the CLE errors of the MIL and CT trackers are relatively high after the whole target object has been out-of-view (the period is shown as the gray area). In this case, the MIL and CT trackers have completely learned new appearances for their target objects, resulting in losing the target object, which they should track. Moreover, the STRUCK tracker has some drifts from Frame 1191 because of the sudden large displacement.

From Frame 2452, our UAV has carried out the first aggressive flight. The CLE errors of the STRUCK and KCF trackers started to increase due to larger displacement, since the STRUCK and KCF trackers have also adapted to the new appearances of the target object during the first aggressive flight, leading to losing their target objects until the end of the UAV tracking task.

The second aggressive flight has started from Frame 2903. Although the movements are even larger than the ones in the first aggressive flight, the Frag and our presented trackers can locate the target object well; their CLE errors are 7.8 and 7.5 pixels, respectively. The strongest aggressive flight in this test is from Frame 3318, as Frames 3343 and 3490 shown in Figure 19; its maximum flight speed has reached 3.8 m/s. Figure 13a shows the precision plot of all visual trackers. It can be seen that our presented visual tracker has achieved 90% precision when the CLE threshold is 14 pixels. In addition, Figure 13b also shows that our presented visual tracker is ranked as No.1.

4.3. Test 3: Visual Tracking of the Moving Car

In Tests 3–6, moving target objects are selected for our UAV to conduct the vision-based tracking applications. In this test, our UAV is utilized to track one moving car from an 80 m height over a traffic intersection. The main challenging factors are mechanical vibration (all image frames), in-plane rotation (e.g., Figure 20, Frame 410), out-of-view (e.g., Figure 20, Frame 582) and similar appearances of other moving cars (all image frames).

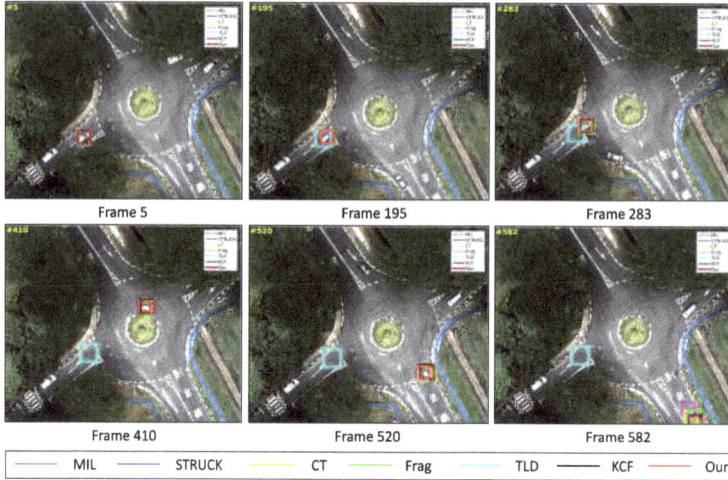

| Frame 5 | Frame 195 | Frame 283 |
| Frame 410 | Frame 520 | Frame 582 |

| —— MIL | —— STRUCK | —— CT | —— Frag | —— TLD | —— KCF | —— Our |

Figure 20. Some tracking results in the moving car image sequence. The total number of frames: 582.

Figure 8 shows the CLE error evolutions of all visual trackers. We can find that all of the trackers can track the target object well in all image frames, except for the TLD tracker, as also shown in Figure 14a; the CT, STRUCK, Frag and our trackers have achieved 95% precision when the CLE threshold is 10.2 pixels.

From Frame 301, the TLD tracker has lost its target object completely because of its adaptation to a new target appearance, which is similar to the background information around the moving car, as shown in Figure 20, Frames 5 and 410. Additionally, the MIL and Frag trackers have generated slightly higher drifts compared to the STRUCK, CT, KCF and our tracker at the beginning of the UAV tracking application. When the moving car is conducting the in-plane rotation movement, these six trackers have started to generate larger drifts, as shown in Figure 20, Frame 520; they are not able to locate the head of the moving car. Before the moving car is out-of-view, the MIL also lost track of the moving car; it located the "HUMP" logo on the road, as shown in Figure 20, Frame 582. When some portion of the moving car is out-of-view, only KCF and our tracker can continue to locate the moving car well, achieving better tracking performances. As can be seen from Figure 14b, we can also find that the MIL, STRUCK, CT, Frag, KCF and our presented tracker have outperformed the TLD tracker.

4.4. Test 4: Visual Tracking of the UGV with the Landing Pad (UGV$_{lp}$)

In this test, a moving UGV with a landing pad is chosen. During the UAV tracking process, the direction of the UGV is manually controlled by an operator, as Frames 543, 770 and 1090 shown in Figure 21. Thus, some portion of target object is occluded by the operator's hand.

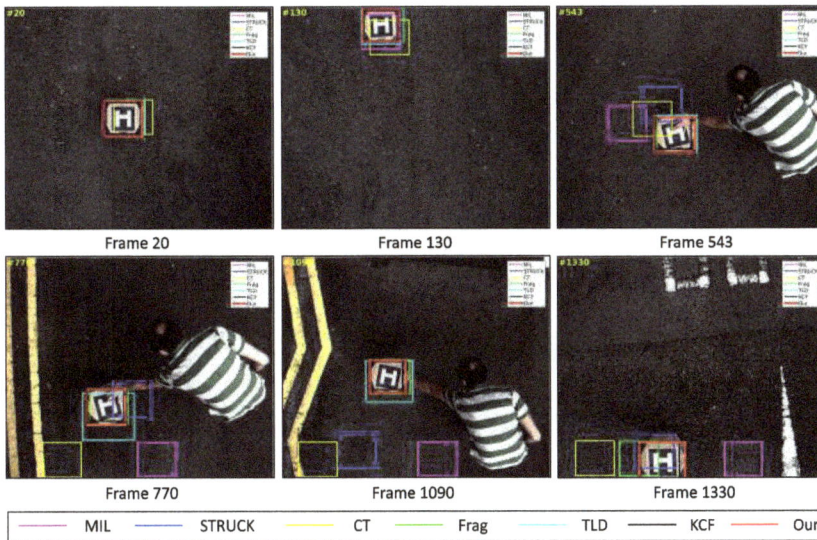

Figure 21. Some tracking results in the UGV$_{lp}$ image sequence. The total number of frames: 1325.

As can be seen from Figures 9 and 15, the KCF, Frag, TLD and our presented tracker have outperformed the CT, MIL and STRUCK trackers. Especially, the CT and our presented tracker have generated the drifts on Frame 3 because of the sudden roll rotation of our UAV. However, our presented tracker has tracked the target object back on Frame 4, while the CT tracker has learned a new appearance on Frame 3, i.e., the background information around the target object has been included as positive samples to train and update its model. Therefore, the CT tracker cannot locate the target object well from Frame 4. For the MIL and STRUCK trackers, their performances are similar to that of the CT tracker in a way that both of them have also adapted to the new appearances of target objects when our UAV is quickly moving forward. Although the CLE error of the KCF tracker is less than ours, the average success rate of our presented tracker is better than the one of the KCF tracker.

4.5. Test 5: Visual Tracking of Walking People Below (People$_{bw}$)

Recently, different commercial UAVs have been developed to follow a person. However, most of these UAVs still mainly depend on the GPS and IMU sensors to achieve the person following application. Therefore, a moving person is selected as the target object for our vision-based UAV tracking task in this paper. The task of the fifth test is that our UAV is utilized to locate a moving person from a high altitude using its onboard downward-looking camera. The main challenging factors include deformation and in-plane rotation, as shown in Figure 22.

As shown in Figure 10, we can find that all visual trackers except the TLD and CT trackers, can locate the moving person in all image frames. Their CLE errors are less than 46 pixels. As can be seen in Figure 16a, their precisions have achieved more than 90% when the CLE threshold is 27 pixels. For the TLD tracker, it has lost the target object from Frame 72, since it has adapted to a new appearance of the target object when the deformable target object is conducting in-plane rotation. After the target object moved back to the previous positions, the TLD tracker has learned the appearance of the target object back. Therefore, the TLD tracker can locate the target object well. From Frame 566, the TLD tracker has lost its deformable target object again until the end of the visual tracking application because of in-plane rotation. For the CT tracker, its tracking performance has also been influenced by Figure 16b; although the MIL, STRUCK, KCF and our tracker can track the target object well, their average success rates are relatively low because of in-plane rotation and deformation.

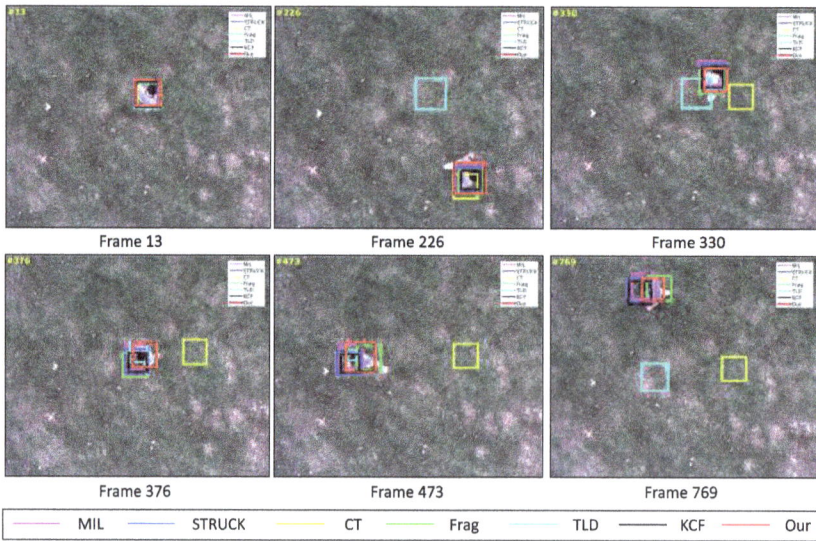

Figure 22. Some tracking results in the People$_{bw}$ image sequence. The total number of frames: 934.

4.6. Test 6: Visual Tracking of Walking People in Front (People$_{fw}$)

In this test, our UAV is employed to follow a moving person using its onboard forward-looking camera. The main challenging factors include deformation, scale variation, cluttered background and out-of-plane rotation (e.g., Figure 23, Frames 312, 1390 and 1970).

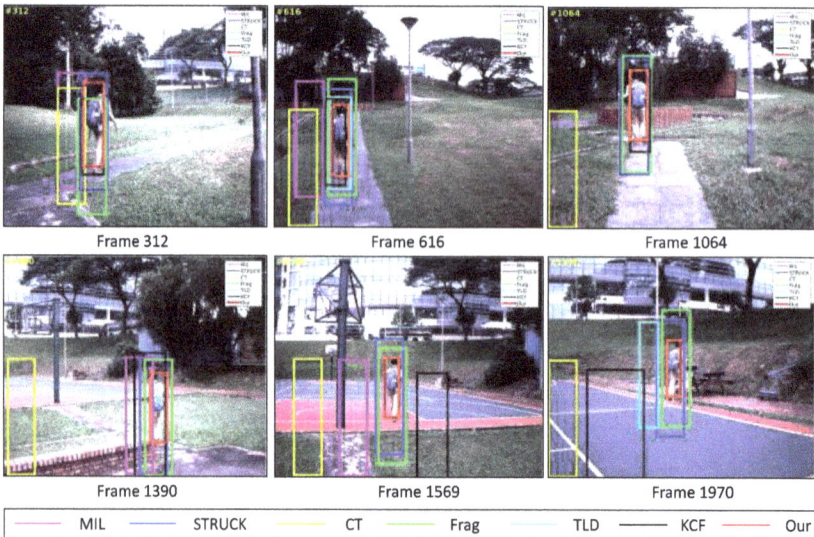

Figure 23. Some tracking results in the People$_{fw}$ image sequence. The total number of frame: 2062.

As can be seen in Figure 11 and Frame 312 shown in Figure 23, the MIL and CT trackers have lost their target object because of the similar appearance in the background, e.g., a tree. Although the

MIL tracker has relocated its target object from Frame 1151 and 1375, it has continued to lose the target object from Frame 1290 and 1544 since the appearances of the brick fence and road are similar to the ones of the target object. For the KCF tracker, it can locate the target object well at the beginning of the UAV tracking application. However, it also lost its target from Frame 1432 due to the similar appearance of the background, e.g., brick fence. For the TLD tracker, it is prone to lose the target object when the target object is conducting out-of-plane rotation. On the other hand, the other visual trackers can always locate the target object, especially the STRUCK and our presented tracker, which are able to track their target object within 30-pixel CLE errors. However, our presented tracker has outperformed the STRUCK tracker in the average success rate, as shown in Figure 17b, since the STRUCK tracker cannot estimate the scale of the target object.

4.7. Discussion

4.7.1. Overall Performances

Tables 3 and 4 show the overall performances of all visual trackers.

For the average CLE error (i.e., CLE_{Ave}) of all image sequences, our presented tracker, the KCF and Frag trackers are ranked as No. 1, No. 2 and No. 3 in all visual trackers. For the average FPS (i.e., FPS_{Ave}), the KCF, our presented tracker and CT tracker have achieved 149.8, 38.9 and 28.7 frames per second, resulting in rankings of No. 1, No. 2 and No. 3 among all visual trackers. For the average success rate (SR) (i.e., SR_{Ave}) when the ζ is set to 0.5, our presented tracker, the KCF and Frag trackers are ranked as No.1, No. 2 and No. 3 again, especially for our presented visual tracker, which has achieved a 95.9% success rate in all image sequences.

Figure 24 shows the overall performances of all visual trackers in 11,646 image frames with precision and success plots. It can be seen that our presented tracker has obtained the best performance. In addition, the KCF and Frag trackers are ranked No. 2 and No. 3.

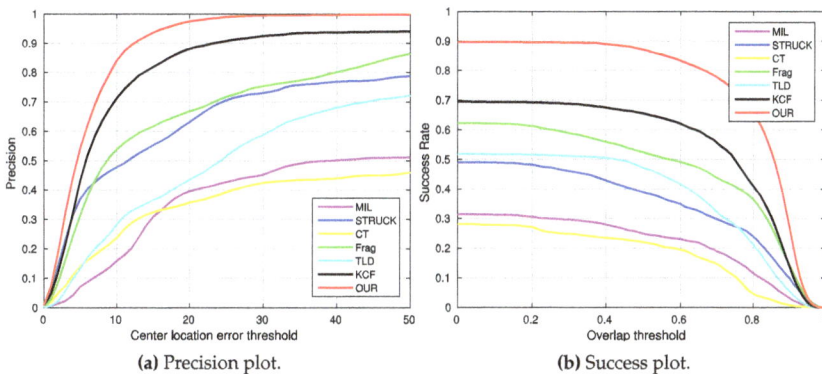

(a) Precision plot.

(b) Success plot.

Figure 24. Overall performances of all visual trackers in 11,646 image frames.

The video related to the tracking results of all visual trackers can be checked at the following YouTube link: https://youtu.be/cu9cUYqJ1P8.

4.7.2. Failure Case

Our presented visual tracker cannot handle the below situations properly as it has employed the features to represent the target object: (1) strong motion blur; and (2) large out-of-plane rotation. These cases can result in cases in which the feature detector cannot detect many features, leading to imprecise estimation for the bounding box of the target object.

Sensors **2016**, *16*, 1406

5. Conclusions

In this paper, a novel robust onboard visual tracker has been presented for long-term arbitrary 2D or 3D object tracking for a UAV. Specifically, three main modules, i.e., the GMLT, LGF and LOF modules, have been developed to obtain a reliable global-local feature-based visual model efficiently and effectively for our visual tracking algorithm, which has achieved the real-time frame rates of more than thirty-five FPS. The extensive UAV flight tests show that our presented visual tracker has outperformed the most promising state-of-the-art visual trackers in terms of robustness, efficiency and accuracy and overcome the object appearance change caused by the challenging situations. It is to be noted that the KCF tracker has achieved an average of 149.8 FPS, but its tracking precision and average success rate are less than ours in all image sequences. Additionally, the UAV can achieve good control performance with more than 20 FPS in real flights. In addition, our visual tracker does not require software or hardware stabilization systems, e.g., a gimbal platform, for stabilizing the onboard captured consecutive images throughout the UAV flights.

Acknowledgments: The research was partially supported by the ST Engineering-NTU Corporate Lab through the NRF corporate lab@university scheme.

Author Contributions: Changhong Fu has developed the presented visual tracking algorithm and carried out the UAV flight experiments in collaboration with Ran Duan and Dogan Kircali. Erdal Kayacan has supervised the work.

Conflicts of Interest: The authors declare no conflict of interest.

References

1. Fu, C.; Suarez-Fernandez, R.; Olivares-Mendez, M.; Campoy, P. Real-time adaptive multi-classifier multi-resolution visual tracking framework for unmanned aerial vehicles. In Proceedings of the 2nd Workshop on Research, Development and Education on Unmanned Aerial Systems (RED-UAS), Compiegne, France, 20–22 November 2013; pp. 99–106.
2. Lim, H.; Sinha, S.N. Monocular localization of a moving person onboard a Quadrotor MAV. In Proceedings of the IEEE International Conference on Robotics and Automation (ICRA), Seattle, WA, USA, 26–30 May 2015; pp. 2182–2189.
3. Fu, C.; Carrio, A.; Olivares-Mendez, M.; Suarez-Fernandez, R.; Campoy, P. Robust real-time vision-based aircraft tracking from Unmanned Aerial Vehicles. In Proceedings of the IEEE International Conference on Robotics and Automation (ICRA), Hong Kong, China, 31 May–7 June 2014; pp. 5441–5446.
4. Babenko, B.; Yang, M.H.; Belongie, S. Visual tracking with online Multiple Instance Learning. In Proceedings of the IEEE Conference on Computer Vision and Pattern Recognition (CVPR), Miami, FL, USA, 20–25 June 2009; pp. 983–990.
5. Hare, S.; Saffari, A.; Torr, P. Struck: Structured output tracking with kernels. In Proceedings of the IEEE International Conference on Computer Vision (ICCV), Barcelona, Spain, 6–13 November 2011; pp. 263–270.
6. Zhang, K.; Zhang, L.; Yang, M.H. Real-time compressive tracking. In Proceedings of the European Conference on Computer Vision (ECCV), Florence, Italy, 7–13 October 2012; pp. 864–877.
7. Adam, A.; Rivlin, E.; Shimshoni, I. Robust fragments-based tracking using the integral histogram. In Proceedings of the IEEE Conference on Computer Vision and Pattern Recognition (CVPR), New York, NY, USA, 17–22 June 2006; pp. 798–805.
8. Kalal, Z.; Mikolajczyk, K.; Matas, J. Tracking-learning-detection. *IEEE Trans. Pattern Anal. Mach. Intell.* **2012**, *34*, 1409–1422.
9. Henriques, J.F.; Caseiro, R.; Martins, P.; Batista, J. High-speed tracking with kernelized correlation filters. *IEEE Trans. Pattern Anal. Mach. Intell.* **2015**, *37*, 583–596.
10. Rosten, E.; Drummond, T. Machine learning for high-speed corner detection. In Proceedings of the 9th European Conference on Computer Vision (ECCV), Graz, Austria, 7–13 May 2006; pp. 430–443.
11. Calonder, M.; Lepetit, V.; Strecha, C.; Fua, P. BRIEF: Binary robust independent elementary features. In Proceedings of the 11th European Conference on Computer Vision (ECCV), Heraklion, Greece, 5–11 September 2010; pp. 778–792.

12. Lucas, B.D.; Kanade, T. An iterative image registration technique with an application to stereo vision. In Proceedings of the 7th International Joint Conference on Artificial Intelligence (IJCAI), Vancouver, BC, Canada, 24–28 August 1981; pp. 674–679.

13. Kalal, Z.; Mikolajczyk, K.; Matas, J. Forward-backward error: automatic detection of tracking failures. In Proceedings of the 20th International Conference on Pattern Recognition (ICPR), Istanbul, Turkey, 23–26 August 2010; pp. 2756–2759.

14. Mullner, D. Fastcluster: Fast hierarchical, agglomerative clustering routines for R and Python. *J. Stat. Softw.* **2013**, *53*, 1–18.

15. Breunig, M.M.; Kriegel, H.P.; Ng, R.T.; Sander, J. LOF: Identifying density-based local outliers. *ACM SIGMOD Rec.* **2000**, *29*, 93–104.

16. Teuliere, C.; Eck, L.; Marchand, E. Chasing a moving target from a flying UAV. In Proceedings of the IEEE/RSJ International Conference on Intelligent Robots and Systems (IROS), San Francisco, CA, USA, 25–30 September 2011; pp. 4929–4934.

17. Olivares-Mendez, M.A.; Mondragon, I.; Cervera, P.C.; Mejias, L.; Martinez, C. Aerial object following using visual fuzzy servoing. In Proceedings of the 1st Workshop on Research, Development and Education on Unmanned Aerial Systems (RED-UAS), Sevilla, Spain, 30 November–1 December 2011.

18. Huh, S.; Shim, D. A vision-based automatic landing method for fixed-wing UAVs. *J. Intell. Robotic Syst.* **2010**, *57*, 217–231.

19. Martínez, C.; Campoy, P.; Mondragón, I.F.; Sánchez-Lopez, J.L.; Olivares-Méndez, M.A. HMPMR strategy for real-time tracking in aerial images, using direct methods. *Mach. Vis. Appl.* **2014**, *25*, 1283–1308.

20. Mejias, L.; Campoy, P.; Saripalli, S.; Sukhatme, G. A visual servoing approach for tracking features in urban areas using an autonomous helicopter. In Proceedings of the IEEE International Conference on Robotics and Automation (ICRA), Orlando, FL, USA, 15–19 May 2006; pp. 2503–2508.

21. Campoy, P.; Correa, J.; Mondragon, I.; Martinez, C.; Olivares, M.; Mejias, L.; Artieda, J. Computer vision onboard UAVs for civilian tasks. *J. Intell. Robotic Syst.* **2009**, *54*, 105–135.

22. Mondragón, I.; Campoy, P.; Martinez, C.; Olivares-Mendez, M. 3D pose estimation based on planar object tracking for UAVs control. In Proceedings of the IEEE International Conference on Robotics and Automation (ICRA), Anchorage, AK, USA, 3–7 May 2010; pp. 35–41.

23. Yang, S.; Scherer, S.; Schauwecker, K.; Zell, A. Autonomous landing of MAVs on an arbitrarily textured landing site using onboard monocular vision. *J. Intell. Robotic Syst.* **2014**, *74*, 27–43.

24. Harris, C.; Stephens, M. A combined corner and edge detector. In Proceedings of the Fourth Alvey Vision Conference (AVC), Manchester, UK, 31 August–2 September 1988; pp. 147–151.

25. Lowe, D. Distinctive image features from Scale-Invariant keypoints. *Int. J. Comput. Vis.* **2004**, *60*, 91–110.

26. Bay, H.; Ess, A.; Tuytelaars, T.; Van Gool, L. Speeded-up robust features (SURF). *Comput. Vis. Image Underst.* **2008**, *110*, 346–359.

27. Rublee, E.; Rabaud, V.; Konolige, K.; Bradski, G. ORB: An efficient alternative to SIFT or SURF. In Proceedings of the International Conference on Computer Vision (ICCV), Barcelona, Spain, 6–13 November 2011; pp. 2564–2571.

28. Olivares-Mendez, M.A.; Fu, C.; Ludivig, P.; Bissyandé, T.F.; Kannan, S.; Zurad, M.; Annaiyan, A.; Voos, H.; Campoy, P. Towards an autonomous vision-based unmanned aerial system against wildlife poachers. *Sensors* **2015**, *15*, 31362–31391.

29. Sanchez-Lopez, J.; Saripalli, S.; Campoy, P.; Pestana, J.; Fu, C. Toward visual autonomous ship board landing of a VTOL UAV. In Proceedings of the International Conference on Unmanned Aircraft Systems (ICUAS), Atlanta, GA, USA, 28–31 May 2013; pp. 779–788.

30. Sampedro, C.; Martinez, C.; Chauhan, A.; Campoy, P. A supervised approach to electric tower detection and classification for power line inspection. In Proceedings of the International Joint Conference on Neural Networks (IJCNN), Beijing, China, 6–11 July 2014; pp. 1970–1977.

31. Ross, D.; Lim, J.; Lin, R.S.; Yang, M.H. Incremental learning for robust visual tracking. *Int. J. Comput. Vis.* **2008**, *77*, 125–141.

32. Fu, C.; Carrio, A.; Olivares-Mendez, M.A.; Campoy, P. Online learning-based robust visual tracking for autonomous landing of Unmanned Aerial Vehicles. In Proceedings of the International Conference on Unmanned Aircraft Systems (ICUAS), Orlando, FL, USA, 27–30 May 2014; pp. 649–655.

33. Pestana, J.; Sanchez-Lopez, J.L.; Saripalli, S.; Campoy, P. Computer vision based general object following for GPS-denied multirotor unmanned vehicles. In Proceedings of the American Control Conference (ACC), Portland, OR, USA, 4–6 June 2014; pp. 1886–1891.

34. Kitt, B.; Geiger, A.; Lategahn, H. Visual odometry based on stereo image sequences with RANSAC-based outlier rejection scheme. In Proceedings of the IEEE Intelligent Vehicles Symposium (IV), San Diego, CA, USA, 21–24 June 2010; pp. 486–492.

35. Hartley, R.I.; Zisserman, A. *Multiple View Geometry in Computer Vision*, 2nd ed.; Cambridge University Press: Cambridge, England, 2004.

sensors

MDPI

Article

Gimbal Influence on the Stability of Exterior Orientation Parameters of UAV Acquired Images

Mateo Gašparović * and Luka Jurjević

Chair of Photogrammetry and Remote Sensing, Faculty of Geodesy, University of Zagreb, Zagreb 10000, Croatia; ljurjevic@geof.hr
* Correspondence: mgasparovic@geof.hr; Tel.: +385-1-4639-223

Academic Editor: Ayman F. Habib
Received: 21 December 2016; Accepted: 16 February 2017; Published: 18 February 2017

Abstract: In this paper, results from the analysis of the gimbal impact on the determination of the camera exterior orientation parameters of an Unmanned Aerial Vehicle (UAV) are presented and interpreted. Additionally, a new approach and methodology for testing the influence of gimbals on the exterior orientation parameters of UAV acquired images is presented. The main motive of this study is to examine the possibility of obtaining better geometry and favorable spatial bundles of rays of images in UAV photogrammetric surveying. The subject is a 3-axis brushless gimbal based on a controller board (Storm32). Only two gimbal axes are taken into consideration: roll and pitch axes. Testing was done in a flight simulation, and in indoor and outdoor flight mode, to analyze the Inertial Measurement Unit (IMU) and photogrammetric data. Within these tests the change of the exterior orientation parameters without the use of a gimbal is determined, as well as the potential accuracy of the stabilization with the use of a gimbal. The results show that using a gimbal has huge potential. Significantly, smaller discrepancies between data are noticed when a gimbal is used in flight simulation mode, even four times smaller than in other test modes. In this test the potential accuracy of a low budget gimbal for application in real conditions is determined.

Keywords: gimbal; exterior orientation parameters; photogrammetry; Inertial Measurement Unit (IMU); Unmanned Aerial Vehicle (UAV)

1. Introduction

Unmanned Aerial Vehicle (UAV) photogrammetry has been advancing at a fast pace in recent years. This is primarily the result of the development of Micro Electro Mechanical Systems and Nano Electro Mechanical Systems sensors [1], whose performances have improved several dozen times over the last two decades, while computers, batteries, and cameras are the main limiting factors of UAV photogrammetry. Based on the previous statement, a gimbal soon became a mandatory part of UAV equipment. A gimbal can be used both with fixed wing [2] and multirotor UAVs [3], even though the vibration influence is not significant in fixed wing application and is not usually necessary to use. A gimbal smooths the angular movements of a camera and provides advantages for acquiring better images; this subject and the importance of using and testing gimbals on UAVs is discussed by various authors [4,5]. Additionally, a gimbal dampens vibrations, which is significantly beneficial for real time image stabilization applications [6]. Apart from smoothing angular movements and dampening vibrations, a gimbal maintains a camera in a predefined position. This is mostly the position in which the camera's axis is horizontal or vertical, but all other positions according to the gimbal's technical capabilities are possible. This very fact has encouraged interested into the research of gimbal capabilities in maintaining angular parameters of exterior orientation. The main interest is focused on the current capabilities of low budget gimbals, and therefore both potential and possible technological applications, not only in geodesy, but also in other professions, can be foreseen [3,7–9].

The primary potential of this technology in geodesy is seen in photogrammetry application without the use of Ground Control Points (GCP) [10,11] and in aerial photogrammetry simulation, for maintaining a parallel camera axis (i.e., normal case). Photogrammetry without the use of GCP requires known exterior orientation parameters. Furthermore, there are different engineering applications, for example high accuracy stakeouts without the use of a total station, and so forth.

Considering that gimbals are nothing new in photogrammetry, one would expect this topic to be well researched. However, most authors are not researching UAV gimbals in this way. Most of the papers are based on the use of gimbal or exterior orientation parameters determination through Global Navigation Satellite System (GNSS) and Inertial Navigation System (INS) integration [12,13]. Also, most of the papers are based on determining only the spatial parameters of the exterior orientation [14]. In these papers, authors commonly use expensive gimbals with sophisticated IMUs. However, in this paper the impact of a low budget gimbal on the determination of only part of the rotational parameters of the exterior orientation will be researched using photogrammetric and IMU acquired data. Only roll and pitch (ω, φ) will be researched; yaw (κ) will not be the subject of this research because in photogrammetry yaw is mostly referred to the UAV, and not absolutely from the reference coordinate system.

2. Technology

Regarding the UAV photogrammetry, two types of UAVs are used: fixed wing and multirotor. Multirotor UAVs are characterized by better maneuverability and generally have better accuracy of object reconstruction because they are able to approach closer to the object. On the other hand, fixed wing UAVs are characterized by larger area coverage and longer flight time. The flight of fixed wing UAVs is generally more stable, so gimbals are rarely used on fixed wing UAVs. However, a camera system on a multirotor UAV is exposed to motor vibrations and sudden altitude changes, so it is recommended to use a gimbal in order to assure quality cadre, endlap, sidelap, and the acquisition of sharp images. In this paper the exterior orientation stability is analyzed, which is directly transferred to an image perspective, respectively, quality cadre, endlap, and sidelap. An upgraded multirotor UAV Cheerson CX-20 Open Source was used in this paper. The upgrades done on the UAV include mounting a gimbal and connecting it to a flight controller, extending the landing gears, using advanced GNSS antenna and a bigger capacity battery, and mounting new and better propellers.

One of the limiting factors of UAVs is the weight of the payload, and the main payload of the UAV is the camera. The maximal payload depends on the UAV's parameters, motor power, and battery capacity. In the past, it was mandatory to use metric cameras, but nowadays the situation is very different because the manufacturing quality of digital cameras has improved. Considering this, it is advised to assume the need to use a metric camera, because an amateur camera would not be sufficient. The greatest impact on the accuracy of a photogrammetric survey based on amateur cameras is lens distortion [15]. Cameras are differentiated by various parameters, but for photogrammetric applications the most important parameters are optics quality, sensor quality, and lens compatibility [16].

In order to use a camera in photogrammetric applications it is necessary to know its interior orientation parameters. Interior orientation is determined by camera calibration. For the purposes of this paper, the calibration of an improved camera, Xiaomi Yi, with declared low distortion lens and focal length of approximately 4.35 mm, was done in Orpheus (version 3.2.1). A Xiaomi Yi is a low budget action camera with a fixed lens. The specifications of the improved Xiaomi Yi camera are specified in Table 1, and both uses of the camera and the UAV are shown in Figure 1a,b.

Table 1. Improved Xiaomi Yi camera.

Processor	Ambarella A7LS
Focal length	4.35 mm (distortion < 1%)
Aperture	F2.8
FOV [1] (diagonal)	86°
Sensor	Sony Exmor R BSI CMOS 16 MP
Size	$6 \times 2.1 \times 4.2$ cm/$2.36 \times 0.83 \times 1.65$ inches
Battery	1010 mAh
Weight	72 g
Video	Up to 1080 p 60 fps
Memory	Up to 64 GB SD card
Connectivity	Wi-Fi, Bluetooth 4.0 v, USB, micro HDMI
Raw data	Yes

[1] FOV—Field of View.

(a)   (b)

Figure 1. A Xiaomi Yi action camera; (**a**) in improving phase; (**b**) on a 3-axes gimbal on an Unmanned Aerial Vehicle (UAV).

The Xiaomi Yi camera was exposed to software enhancement by scripts that were developed for storing the images taken in raw + JPG (100% quality) format. The Xiaomi Yi camera by default only has the capability to store filtrated JPG images of limited quality. The post-processed raw format potentially allows access to better quality images than the original JPG images. Furthermore, a rolling shutter effect was encountered and eliminated with a combination of appropriate illumination, reduced exposure time to 1/1002 s (fixed value), and lower flight velocity. The exposure time was manipulated via scripts development.

In the UAV domain, a gimbal is a device that maintains orientation and dampens vibrations. The task of a gimbal is to calculate a correction for every detected movement in the unit of time *i*, and compensate for it while meeting the following requirement:

$$\omega_T \approx \omega_i, \varphi_T \approx \varphi_i, \kappa_T \approx \kappa_i \tag{1}$$

$\omega_T, \varphi_T, \kappa_T$ are predefined rotation angles for each gimbal axis. Unlike angles ω_T, φ_T, angle κ_T is not absolutely defined by the direction of gravity, but relatively to the UAV or absolutely in reference to the direction north. The frequency of the gimbal is 700 Hz, which means that the correction rate is calculated using the following expression:

$$700 Hz \Rightarrow i = 1/700s \tag{2}$$

At the same time, the gimbal forms the connection between the UAV and the camera system. Depending on the number of gimbal axes, it stabilizes the camera along two or three axes. These are pitch and roll (φ, ω), or pitch, roll, and yaw $(\varphi, \omega, \kappa)$ axes. The gimbal functionality is controlled by a gimbal controller which calculates corrections for the camera position based on the horizon deflection of the IMU mounted on the camera holder.

For the purposes of flight simulation and indoor flight tests, a chessboard test field of the dimensions 81 × 57 cm, with a square size of 3 cm, has been created. During the flight simulation and indoor flight test the chessboard test field was continuously shot. Exterior orientation parameters were calculated in Orpheus software based on the images taken. A detailed description of the test procedure is described in Section 4. Part of the Matlab toolbox implemented in C and included in OpenCV was used for the chessboard's GCP detection [17]. Automatic GCP detection was the subject of research of numerous authors apart from Jean-Yves Bouguet [18,19]. Automatic detection software was used in order to save time and to eliminate the influence of operator errors. In order to import data into the Orpheus software, a Matlab (version 9.0) script for data preparation has been created. Figure 2 represents a visualization of the chessboard test field GCP detection. The exterior orientation parameters for outdoor flight were calculated in Agisoft PhotoScan software (version 1.2.6, 64 bit).

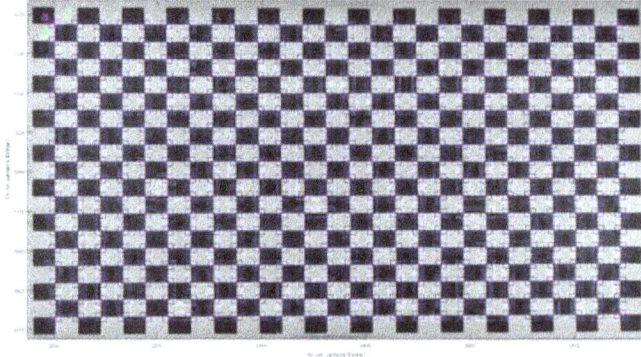

Figure 2. Chessboard test field Ground Control Point (GCP) detection in Matlab.

3. Methodology

The main goal of this study is to examine the possibility of obtaining better geometry and favorable spatial rays of bundles of images, and respectively better image cadre, endlap, and sidelap in the UAV photogrammetric survey. The research methodology and a detailed description, with an explanation of the mathematical models used in order to carry out the tests and design a new method, is given in this section. In order to explain how the analyzed data are collected, it is necessary to discuss the photogrammetric methods that were used and the IMU operating principles, which are presented in Sections 3.1 and 3.2.

3.1. Photogrammetric Mathematical Models

The pinhole camera model is the most frequently used mathematical model of a camera in photogrammetry (Figure 3). It is a model of projection where a light ray passes from every point on an object through the center of a projection and ends in the image plane. According to this mathematical model, the point on the object, the center of the projection, and the point in the image plane are collinear.

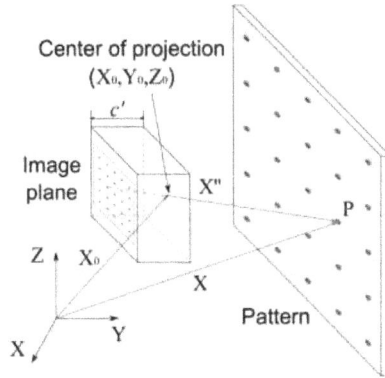

Figure 3. Pinhole camera model.

The center of the projection is defined by vector X_0 in reference to the coordinate system. Point P (vector X), located on object, can be derived based on vector X_0 and vector X'', according to Equation (3). Vector X'' is defined with the point on the object and the center of projection.

$$X = X_0 + X'' \tag{3}$$

If only one image of the given object is available, only the direction of point P can be defined. In order to determine the coordinates of point P, at least one more image of the given object from a different position is required. This kind of projection is described with collinearity equations [20].

$$
\begin{aligned}
x' &= x'_0 - c' \frac{r_{11} \cdot (X - X_0) + r_{21} \cdot (Y - Y_0) + r_{31} \cdot (Z - Z_0)}{r_{13} \cdot (X - X_0) + r_{23} \cdot (Y - Y_0) + r_{33} \cdot (Z - Z_0)} + \Delta x', \\
y' &= y'_0 - c' \cdot \frac{r_{12} \cdot (X - X_0) + r_{22} \cdot (Y - Y_0) + r_{32} \cdot (Z - Z_0)}{r_{13} \cdot (X - X_0) + r_{23} \cdot (Y - Y_0) + r_{33} \cdot (Z - Z_0)} + \Delta y',
\end{aligned}
\tag{4}
$$

where x', y' are the point P image coordinates, x_0', y_0' are the Principal Point of Autocollimation (PPA) image coordinates, c' is the camera constant, r_{ij} is the spatial rotational matrix parameter, X, Y, Z are point P coordinates in reference to the coordinate system, X_0, Y_0, Z_0 are the projection center coordinates in reference to the coordinate system, and $\Delta x'$, $\Delta y'$ are the distortion parameters.

Collinearity equations express image coordinates (x', y') as a function of the interior orientation $(x_0', y_0', c', \Delta x', \Delta y')$ and the exterior orientation $(X_0, Y_0, Z_0, \varphi, \omega, \kappa)$ parameters of a single image. Bundle Block Adjustment (BBA) is applied when adjusting interior and exterior orientation parameters for an arbitrary number of images, which are connected in a single 3D model. Within BBA, observations (image coordinates), classic survey measurements (GCP), and the referent coordinates of the object are adjusted simultaneously. Equation (5) represents the indirect measurements function model.

$$
\begin{aligned}
x\prime_i + vx\prime_i &= F\left(X_{oj}, Y_{oj}, Z_{oj}, \omega_j, \varphi_j, \kappa_j, x\prime_{0k}, c_k, \Delta x\prime_k, X_i, Y_i, Z_i\right), \\
y\prime_i + vy\prime_i &= F\left(X_{oj}, Y_{oj}, Z_{oj}, \omega_j, \varphi_j, \kappa_j, y\prime_{0k}, c_k, \Delta y\prime_k, X_i, Y_i, Z_i\right),
\end{aligned}
\tag{5}
$$

where i is the point of index, j is the image index, and k is the camera index.

The approximate unknowns of the exterior orientation are determined analytically out of at least three known non-collinear points on the object. Images are tied up together with different image matching techniques [21] (the automatic detection of tie points) or by the observation of GCP. A combination of image matching and GCP observation is possible if the GCP are not visible on every image. Images are oriented mutually by intersecting all corresponding rays (i.e., homologous rays). This way, a strong geometry is created. Interior orientation $(x_0', y_0', c', \Delta x', \Delta y')$, exterior orientation $(X_{0j}, Y_{0j}, Z_{0j}, \varphi_j, \omega_j, \kappa_j)$, and object points (X_i, Y_i, Z_i) in reference to the coordinate system are determined in a single adjustment. The statements above indicate that BBA is mathematically the most acceptable method of image orientation in the domain of photogrammetry.

The approach described above is used in an outdoor flight test. In flight simulation and indoor flight exterior orientation determination a similar approach was used, only the interior orientation parameters were not adjusted. Therefore, the interior orientation parameters were predetermined with the camera calibration and entered the exterior orientation adjustment (space resection) as fixed values. The Xiaomi Yi camera was calibrated with a test field calibration method in Orpheus software on a test field with 468 points. The calibration algorithm is explained by Gašparović and Gajski [22]. Agisoft PhotoScan software was used in order to determine the exterior orientation parameters during the outdoor flight due to its ability to automatically determine a large number of tie points that enter adjustment and consequently improve the quality of the external orientation parameters. Brown [23] and TU Wien [24] distortion models were used in this paper.

3.2. IMU Data Integration

The primary gimbal IMU consists of a 3-axis accelerometer and a 3-axis gyroscope, and it is mounted onto the gimbal's camera holder. The gimbal's control loop is shown in Figure 4. The raw measurements are processed with filters, calibration, and orientation corrections. Usually a Kalman filter is used [25]. The Attitude Heading and Reference System (AHRS) calculates the orientation angles based on the corrected IMU measurements. Based on the AHRS data, the PID (Proportional Integral Derivative) angles are calculated by a PID controller and sent via Pulse-Width Modulation (PWM) to a motor driver, which is a moving camera, to correct the position.

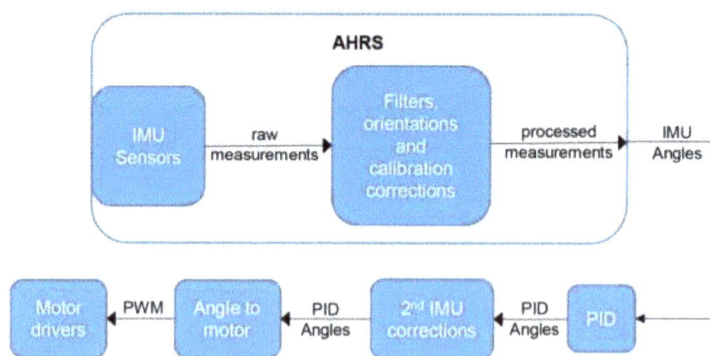

Figure 4. Control loop of the gimbal controller.

Besides the primary IMU, the gimbal can use another, secondary IMU that is located on the controller board. The secondary IMU has to be mounted independently of any motor. The use of a secondary IMU is remarkably beneficial because it contributes to a significant accuracy and increase in operational range [26], as well as maintaining a more stable yaw axis.

Various IMUs based on cheap Micro Electro Mechanical System sensors are available on the market and almost every smartphone is equipped with a cheap IMU. They are subject to systematic influence errors due to imprecise scaling factors [27] and non-perpendicular axes, which results in a reduced accuracy of its position and direction [28].

In order to reduce the aforementioned errors, the IMU should be calibrated. IMU calibration is a crucial step when ensuring the gimbal's optimal performance. If both the primary and the secondary IMU are used, they should be calibrated. For the purposes of this paper, a 1-point calibration method was used. A 1-point calibration method was conducted by stabilizing the IMU on the horizontal plane and measuring the IMU accelerometer data while it was completely still. During calibration, it is important to take into account exterior influences that are causing vibrations. The ambient temperature at which the calibration is done should be as close as possible to the ambient temperature in which the gimbal will be used.

The IMU was stabilized on an autograph WILD A7 image insertion plane, whose horizontality was tested with a calibration level of 3″ sensitivity. The primary and secondary IMUs were mounted to the horizontal plane and were left for some time while their measurements were stabilized. The calibration was performed based on the measurements after they became stabilized, while the IMU measurements were logged and analyzed in order to control the IMU. The IMU measurements can be logged as long as the gimbal controller is connected via a USB to the computer. In this paper, a gimbal controller Storm32 (Olliw.eu, Denzlingen, Germany) with a primary and integrated secondary IMU MPU6050 (InvenSense, San Jose, USA) was used. The integration between these two IMUs are based on the I2C communication protocol. Figure 5a represents the gimbal used and Figure 5b represents the gimbal controller and the primary IMU that were used.

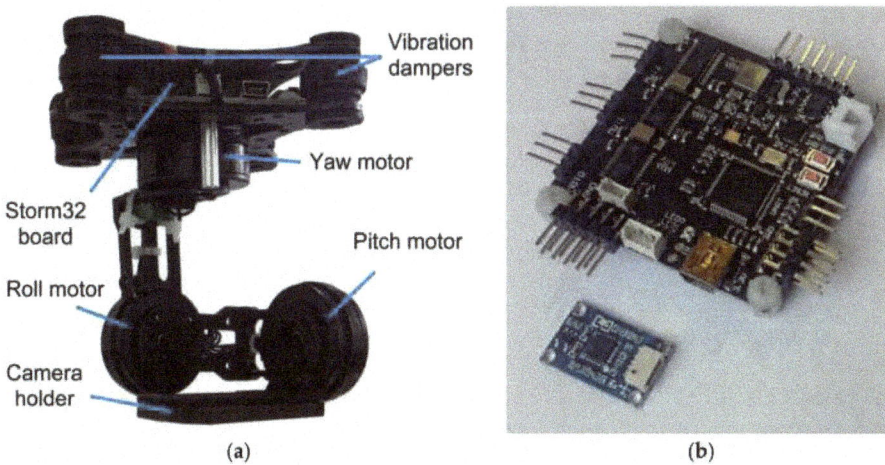

Figure 5. (**a**) Used gimbal; (**b**) gimbal controller (Storm32) board and primary Inertial Measurement Unit (IMU) (MPU6050).

The used gimbal is based on two IMUs for the calculation of horizon deflection and correction. During the research a drift along the roll axis was noticed and it was approximately 1° per minute. The noticed drift was almost totally eliminated by adjusting the IMU AHRS parameter, which controls the coefficient of the accelerometer and the gyroscope data when entering the AHRS algorithm. Significant improvement was achieved by setting the IMU AHRS parameter so that the algorithm uses only accelerometer data. This is probably a consequence of the inability of the software used to calibrate the IMU's gyroscopes.

Other than that which has been mentioned, the measurements and usage of the IMU data were controlled by Gyro Low Pass Filter and IMU 2 Feed Forward Low Pass Filter parameters. The Low Pass Filter passes data with a frequency lower than the one that is set, and attenuates data with higher frequency. The Gyro Low Pass Filter parameter is used to set the aforementioned frequency for the gyroscope data. This process is used in order to filter the vibration from the real data. The IMU 2 Feed Forward Low Pass Filter is also used to set the Low Pass Filter frequency for the attitude data before entering the Feed Forward channel.

4. Experimental Approach

The exterior orientation stability was tested in three different environments:

- Flight simulation,
- Indoor flight,
- Outdoor flight.

The main goal of this experiment is to determine the improvement of exterior orientation stability using a gimbal, and to define a new method for testing gimbal stability. In all three tests the Xiaomi Yi camera with identical predetermined internal orientation parameters was used. In order to ensure the quality of the research, all tests were shot in more than one session. While the camera was mounted onto the UAV without a gimbal, its attitude was directly correlated to the rotational parameters of the camera's exterior orientation. Considering this, the UAV body movement is logged as secondary IMU data for the flight simulation test due to having an achievable cable connection with a computer, while for indoor as well as outdoor test flights the UAV body movement was logged as flight controller IMU data. This data also represents the camera's exterior orientation parameters without the use of the gimbal. The UAV's flight controller is APM 2.52 (ArduPilot, Indianapolis, USA), which uses onboard MPU6000 IMU (InvenSense, San Jose, USA). The used Storm32 gimbal controller has the capability to log both of the IMUs' data and these data is used to analyze the exterior orientation parameters. In the case of an ideal gimbal, the exterior orientation parameters would not change during the time. The analyzed data is acquired in three different tests, with flight simulation, indoor flight, and outdoor flight. During the tests, it was noticed that the gimbal's initial position changes by a small angular value every time it is powered, but this fact is noted and ignored because the subject of this paper is stability analysis, which this change does not influence. Therefore, the photogrammetrical data presented is normalized in order to minimize the influence of difference in the initializations for each session.

4.1. Flight Simulation

Flight simulation was taken in front of the chessboard test field that contains 468 GCP in total. During the flight simulation, the UAV was turned on in order to power the gimbal, but the UAV motors were disabled. The movement of the UAV is simulated manually by rotating it up to a maximum of $\pm 45°$ across the roll and pitch axes. Analysis was carried out on the secondary IMU data during the 15 min of flight simulation. In addition to the IMU data, the exterior orientation stability was analyzed via data acquired with the photogrammetric method. During the 15 min of flight simulation, 36 images were taken and the secondary IMU data was logged. The camera and IMU are synchronized manually with sufficient precision for this type of test by starting to record the IMU data when the camera signals the start of the time lapse data acquisition. The time lapse interval was set to 25 s. It is worth noting that data synchronization is not subject of this paper and that the analysis was done with reference to a range of data. The flight simulation was done in order to log the gimbal IMU data, which was not possible during the real flight due to the requirement of a cable connection between the gimbal controller and the computer. The exterior orientation parameters were calculated for every image with Orpheus software on the test field with 468 GCP, as it is described in Section 3.1.

4.2. Indoor Flight

After analyzing the exterior orientation stability during the simulated indoor flight, real flight testing was conducted. The flight was performed in front of the already mentioned test field with 468 GCP, in an indoor environment. Three sessions were carried out and the duration of them was 15 min, in which 204 test field images were taken, approximately one image for every four seconds. Images and IMU data synchronization is done by coincidencing the system time of the camera and flight controller. For data analysis, only images taken inside an imaginary sphere with a radius of 0.5 m were chosen. The center of the imaginary sphere is the average value of the exterior orientation spatial parameter out of all 204 images taken. It was done in order to exclude all images that were taken too close or too far from the test field due to the demands of controlling the UAV indoors.

For the purpose of smoother indoor flight, the UAV maximal pitch and roll tilt was limited to ±15° via the flight controller angle maximal value. Figure 6 shows the UAV during the indoor flight test.

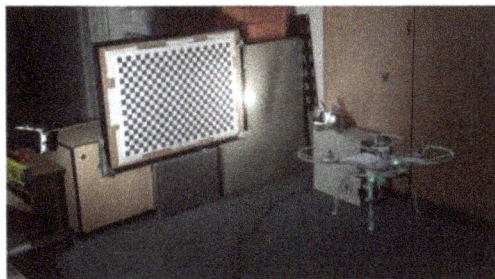

Figure 6. Indoor flight test.

In total, 137 images were used, which makes 67% out of all 204 images. As described in the above sections, the exterior orientation parameters were calculated with Orpheus software on the test field with 468 GCP for every image. The flight controller IMU data was used to test the exterior orientation accuracy improvement.

4.3. Outdoor Flight

Outdoor flight was carried out in two sessions. The test area was a rocky terrain shot with a camera axis directed vertically to the ground. The UAV was controlled by an autonomous mission through a predefined trajectory in both sessions. This makes outdoor flight much smoother and more stable than indoor flight. Because the UAV is influenced by weather conditions (e.g., wind), the maximal pitch and roll tilt value is higher than it is in indoor flight, and in this outdoor flight test it was limited to ±25°. The first session was carried out on the relative height of 40 m above the area of 4.5 ha, with 80% endlap and 60% sidelap, while the second session was carried out on the relative height in the range from 50 m to 70 m above of 6 ha, with 85% endlap and 50% sidelap. In total, 5 GCP were regularly distributed over the subject area, and 13211 tie points were used to align the first session of data, and to align the second session data 7 GCP regularly distributed over the subject area and 18,324 tie points were used. The flight velocity was 4 m/s in both sessions. During the first session 191 images were shot, and during the second session 141 images were shot, approximately one image for every two seconds for both sessions. The images and IMU data synchronization is done by coinciding the system time of the camera and the flight controller. Images on the ends of the strips were not taken into consideration because the exterior orientation data of those images would corrupt the statistical data, and they do not represent real values because on the ends of the strip the terrain is afforested and lacks the GCP and tie points necessary for quality orientation. Additionally, the turns from one strip to another are the most challenging maneuvers for a gimbal to compensate. The analysis was carried out on 47 images from the first session and 51 images from the second session, as in the flight controller IMU data. The exterior orientation parameters were calculated with Agisoft PhotoScan software because it can automatically determine a large number of tie points that enter adjustment.

5. Results

The results of the tests taken in order to analyze the impact of the gimbal on the stability of exterior orientation, and therefore its ability to achieve a better geometry of bundles of rays in UAV photogrammetry, which are explained in the above section, will be represented here.

5.1. Flight Simulation

Presented in Table 2 are the statistical data acquired photogrammetrically and by the secondary IMU, which represents the movement of the camera in use without the gimbal.

Table 2. Statistics of the Inertial Measurement Unit (IMU) and photogrammetry data for the flight simulation.

	Pitch (°)	Roll (°)	IMU Pitch (°)	IMU Roll (°)
Average	91.13797	2.71445	2.72	−0.71
St. dev.	0.46591	0.36282	13.36	10.94
Min	90.03401	1.75086	−32.17	−41.34
Max	92.59373	3.71907	37.73	51.84

The photogrammetrically and secondary IMU acquired data ratio for roll and pitch parameters are represented in Figures 7 and 8. Photogrammetrically acquired data are represented in reference to the arithmetic mean (normalized data) in order to eliminate the influence of the gimbal initialization and the non-verticality of the test field.

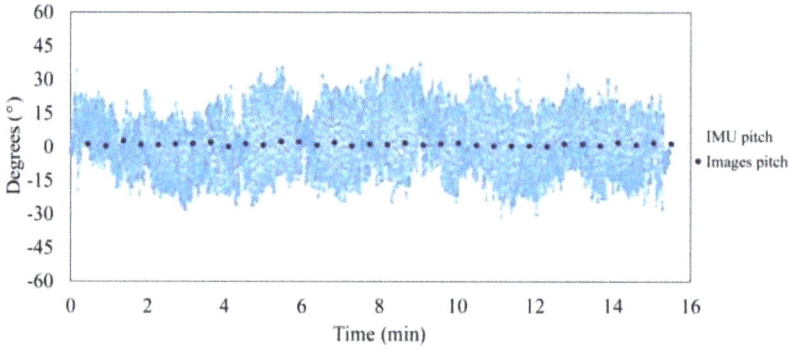

Figure 7. Exterior orientation parameters acquired photogrammetrically and by the Inertial Measurement Unit (IMU) for pitch parameter.

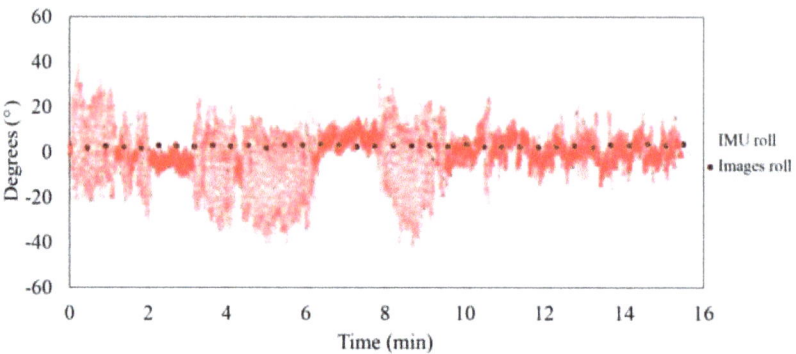

Figure 8. Exterior orientation parameters acquired photogrammetrically and by the Inertial Measurement Unit (IMU) for roll parameter.

From the presented data, improvement of the exterior orientation stability as a consequence of gimbal use can be seen. The roll and pitch parameters determined independently with a photogrammetric method (the camera is stabilized with a gimbal) are between 2.56° and 1.97°, with a standard deviation from 0.46° and 0.36°, while the secondary IMU (platform instability) data range from 69.90° and 93.18°. From the previous data value, the advantage of using a gimbal is evident. In addition, it is shown that the pitch parameter statistic is slightly worse than the roll parameter statistic, which can be explained as a consequence of the planar test field usage, which benefits roll parameter determination in opposition to the 3D test field, which benefits pitch parameter determination.

5.2. Indoor Flight

Statistical data acquired photogrammetrically using a gimbal is presented in Table 3.

Table 3. Statistics for all three sessions of indoor flight.

	1st Session		2nd Session		3rd Session	
	Pitch (°)	Roll (°)	Pitch (°)	Roll (°)	Pitch (°)	Roll (°)
Average	91.67314	−0.28787	91.31183	−0.24313	90.71399	0.39360
St. dev.	1.99739	0.80951	2.14277	0.91155	2.04433	0.82016
Min	87.40141	−1.82238	84.62633	−2.67715	85.32468	−1.64857
Max	96.35258	1.44858	95.02984	1.41679	95.50489	2.89549

The exterior orientation parameters calculated for three sessions of indoor flight, all in reference to the arithmetic mean for every session (normalized data) and trend line for the roll and pitch parameters, are represented in Figure 9. In this visualization, the gimbal initialization is taken into account. After every start-up, the gimbal is initialized and in this visualization the difference in initializations is disregarded. In Figure 9 the repeatability of the data from the three different sessions is shown. In all three sessions, the range value of the roll and pitch measurements and the standard deviation is very similar. The arithmetic mean of the third session differs from the arithmetic mean of the first and second sessions, which can be interpreted as a significantly different initialization of the gimbal. Due to specific flight conditions (indoor flight), the flight controller was limited to a ±15° tilt, which resulted in three times improved stability of the exterior orientation with the use of a gimbal compared to without it.

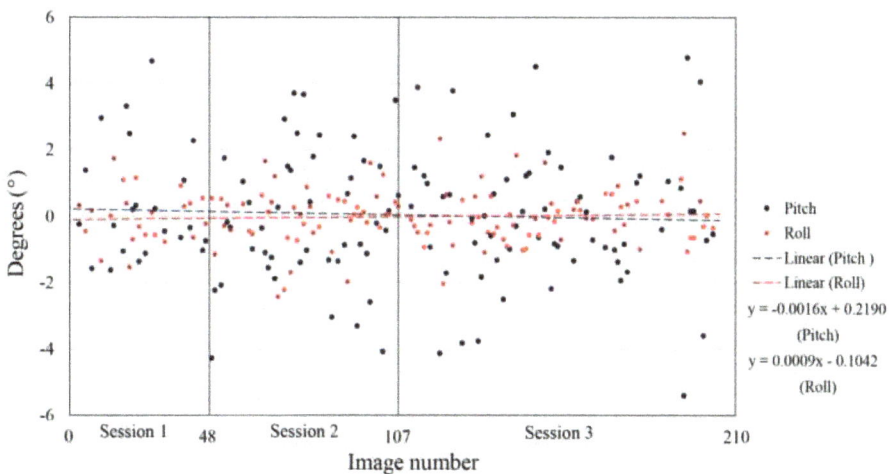

Figure 9. Roll and pitch parameter values and trend line for three sessions of indoor flight.

5.3. Outdoor Flight

Data acquired during the two sessions of outdoor flight, both in reference to the arithmetic mean (normalized data) and trend lines of every session, are represented in Figure 10.

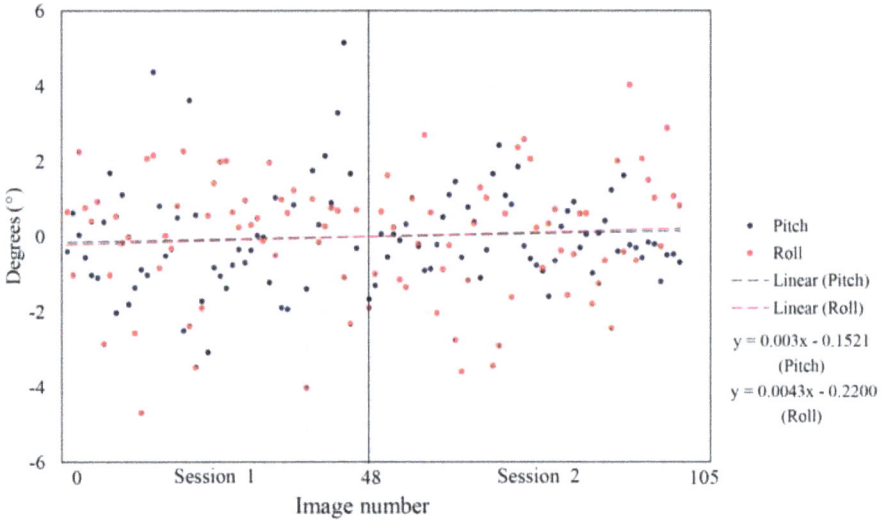

Figure 10. Roll and pitch parameter values and trend line for both sessions of outdoor flight.

The statistical data of the measurements visualized in Figure 10 are represented in Table 4, in which the repeatability of the data from the two different sessions is shown. The only significant difference in the data is a better standard deviation of pitch parameter for the second session compared to the first session. This is very likely to be a consequence of lower flight and the determination of better automatic tie points, in opposition to the higher flight of the first session. For this flight, the flight controller was limited to a ±25° tilt as a standard value for outdoor flights due to external influences, which resulted in six times improved stability of the exterior orientation with the use of a gimbal compared to without it.

Table 4. Statistics for both sessions of outdoor flight.

	1st Session		2nd Session	
	Pitch (°)	Roll (°)	Pitch (°)	Roll (°)
Average	−0.38750	−1.32690	−0.60653	−1.61740
St. dev.	1.72962	1.66566	0.92112	1.69534
Min	−3.47135	−6.02212	−2.28265	−5.20398
Max	4.75898	0.93433	1.81105	2.41166

6. Discussion

The analyzed data were acquired during three different tests, by flight simulation, and indoor and outdoor flight. The range of the minimal and maximal values of roll and pitch parameters calculated photogrammetrically for flight simulation is 2.56° and 1.97°, with a standard deviation of 0.46° and 0.36°. On the other hand, the data acquired during the indoor flight ranges between 8.95°, 10.40° and 10.18°, with a standard deviation of 1.99°, 2.14° and 2.04° for pitch parameter trough sessions, and for the roll parameter data range between 3.27°, 4.09° and 4.54° with a standard deviation of 0.81°, 0.91°

and 0.82°. The data acquired during the two sessions of outdoor flight range between 8.23° and 4.09° with a standard deviation of 1.73° and 0.92° for pitch parameter, and the roll parameter data range between 6.65° to 7.61° with a standard deviation of 1.66° and 1.69°. The improvement achieved due to the use of a gimbal depends on the flight conditions, which follows that the level of improvement achieved during the flight simulation is over 20 times, during indoor flight is three times and during outdoor flight is six times compared to the exterior orientation stability without the use of a gimbal. Note that during the indoor and outdoor test flights, the controller was limited to a ±15° and a ±25° tilt, and that during the flight simulation there were no active restrictions. From previously presented results, it is clear that a gimbal improves the exterior orientation stability of the camera. Therefore, using a gimbal on UAVs contributes to a better geometry of bundles of rays, better sidelap and endlap, and better image cadre.

Data from Tables 3 and 4 are represented in Figures 11 and 12. The horizontal line defines the arithmetic mean of the session data, the vertical line defines the minimal and maximal measured value of the session, and the top and bottom of the rectangle define 1σ area for every single session.

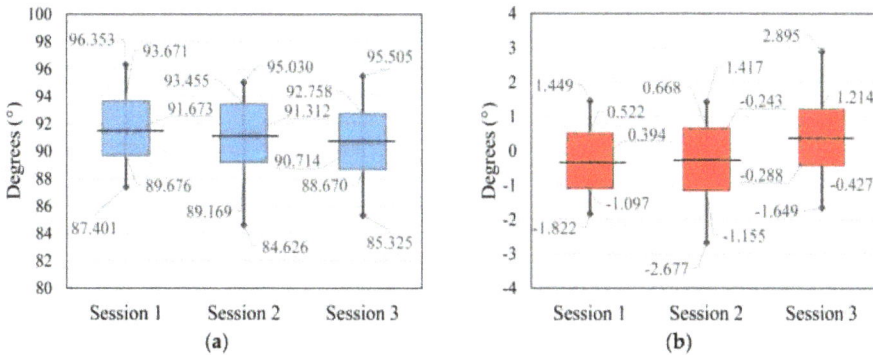

Figure 11. Display of statistical data of indoor flight for: (**a**) pitch parameter and (**b**) roll parameter.

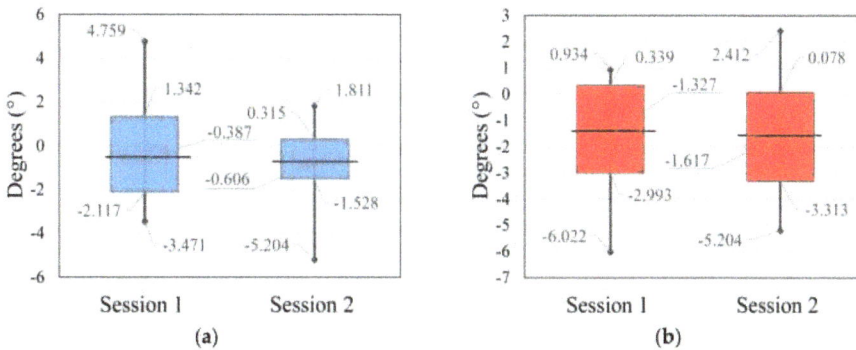

Figure 12. Display of statistical data of outdoor flight for: (**a**) pitch parameter and (**b**) roll parameter.

The data acquired during the flight simulation proved to be the best. The range of the data is smaller and the standard deviation is better than in other tests. This is a consequence of unreal conditions, and almost no shift in the spatial parameters of the exterior orientation and lower vibrations. Regardless, a huge improvement in the exterior orientation stability occurs when using a gimbal. The gimbal's secondary IMU and the flight controller IMU give us an insight. In contrast, data acquired during the indoor flight were slightly worse due to the nature of flying indoors. The indoor flight was totally manual, a GNSS lock and position hold were not available, and the spatial parameters of the exterior orientation were rapidly and almost randomly changing, which influenced the angular exterior orientation parameters due to correlation. Perhaps the greatest influence on the exterior orientation determination was caused by the test field, which did not fill the whole image cadre, and the consequence is four times worse results for the pitch parameter and two times worse for the roll parameter of the indoor flight in comparison to the flight simulation. In order to confirm the obtained results, an outdoor flight test was taken. The statistics of the pitch parameter of the second session were better than the statistics of the first session. On the other hand, the roll parameter statistics of the second session are slightly worse than the statistics of the first session. The better pitch statistics for the second session is a consequence of a lower flight height in comparison to the first session. Due to the lower flight height, which caused larger perspective differences in the adjacent images, and consequently worse tie point determination, the second session's roll statistics are slightly worse than first session's roll statistics. It is important to be aware of small error influences due to the impossibility of matching the projection center and the gimbal axis intersections [29].

7. Conclusions

A new approach to exterior orientation stability research using a gimbal is presented in this paper. The new method consists of photogrammetric methods and therefore gives us an external evaluation of stability, in contrast to former research papers on the external orientation of cameras mounted on a gimbal which are based on the IMU data, and only gives us an insight into the internal evaluation of the gimbal's stability. The expected improvement of the exterior orientation directly depends on the flight mode and gimbal quality, but the expected stability ranges $10°$ for flight. With the flight simulation test, the goal of stabilization in a real environment is to set up the reachable stabilization accuracy. With the aforementioned test, the data range from $2.56°$ and $1.97°$, and a standard deviation from $0.46°$ and $0.36°$ for pitch and roll parameters is acquired. The exterior orientation stability is enhanced 3 or 6 times in comparison to the flight tests, depending on the conditions of the flight. Considering the fact that a gimbal is low budget equipment and that used technology is quickly developing, this goal will be reached soon. According to this new method for gimbal testing, it is clear that the gimbal contributes to better stability of the predefined position of the camera (the exterior orientation) and because of this it can be concluded that the gimbal contributes to better geometry of the bundles of rays, better sidelap and endlap, and better image cadre. The gimbal testing method presented here is a key contribution, and is knowledge that is of significant importance for the UAV photogrammetric survey.

The full potential of used technology and its applications is much bigger than just the stabilization of exterior orientation parameters. Technology development can lead to various applications in geodesy and other professions and the most obvious example of this application is the potential stakeout on challenging terrain using a combination of GNSS measurements and a gimbal to define direction.

Because this paper did not research the influence of gimbals on the linear exterior orientation parameters, it is recommended to determine the linear camera shift constants in reference to the GNSS antenna phase center in further research. Furthermore, the impact of gimbals on radiometric characteristics of images is not researched enough, and future research could focus on this problem due to the high gimbal frequency that influences camera radiometric performance.

Acknowledgments: The authors would like to thank the anonymous reviewers for their constructive feedback that has helped to improve the work and its presentation.

Author Contributions: M.G. conceived and designed the experiments; M.G. and L.J. performed the experiments; M.G. and L.J. analyzed the data; M.G. and L.J. wrote the paper.

Conflicts of Interest: The authors declare no conflict of interest.

References

1. Brake, N.J. Control System Development for Small UAV Gimbal. Ph.D. Thesis, California Polytechnic State University, San Luis Obispo, CA, USA, August 2012.
2. Leira, F.S.; Trnka, K.; Fossen, T.I.; Johansen, T.A. A Ligth-Weight Thermal Camera Payload with Georeferencing Capabilities for Small Fixed-Wing UAVs. In Proceedings of the International Conference on Unmanned Aircraft Systems, Denver, CO, USA, 9–12 June 2015.
3. Gonzalez, L.F.; Montes, G.A.; Puig, E.; Johnson, S.; Mengersen, K.; Gaston, K.J. Unmanned Aerial Vehicles (UAVs) and artificial intelligence revolutionizing wildlife monitoring and conservation. *Sensors* **2016**, *16*. [CrossRef] [PubMed]
4. Detert, M.; Weitbrecht, V. A low-cost airborne velocimetry system: Proof of concept. *J. Hydraul. Res.* **2015**, *53*, 532–539. [CrossRef]
5. Flavia, T.; Pagano, C.; Phamduy, P.; Grimaldi, S.; Porfiri, M. Large-scale particle image velocimetry from an unmanned aerial vehicle. *IEEE ASME Trans. Mechatron.* **2015**, *20*, 3269–3275.
6. Windau, J.; Itti, L. Multilayer Real-Time Video Image Stabilization. In Proceedings of the International Conference on Intelligent Robots and Systems, San Francisco, CA, USA, 25–30 September 2011.
7. Jędrasiak, K.; Bereska, D.; Nawrat, A. The Prototype of Gyro-Stabilized UAV Gimbal for Day-Night Surveillance. In *Advanced Technologies for Intelligent Systems of National Border Security*; Nawrat, A., Simek, K., Swierniak, A., Eds.; Springer: Berlin/Heidelberg, Germany, 2013; pp. 107–115.
8. Shahbazi, M.; Sohn, G.; Théau, J.; Menard, P. Development and evaluation of a UAV-photogrammetry system for precise 3D environmental modeling. *Sensors* **2015**, *15*, 27493–27524. [CrossRef] [PubMed]
9. Blois, G.; Best, J.L.; Christensen, K.T.; Cichella, V.; Donahue, A.; Hovakimyan, N.; Kennedy, A.; Pakrasi, I. UAV-Based PIV for Quantifying Water-Flow Processes in Large-Scale Natural Environments. In Proceedings of the 18th International Symposium on the Application of Laser and Imaging Techniques to Fluid Mechanics, Lisabon, Portugal, 4–7 July 2016.
10. Barazzetti, L.; Remondino, F.; Scaioni, M.; Brumana, R. Fully Automatic UAV Image-Based Sensor Orientation. In Proceedings of the International Society for Photogrammetry and Remote Sensing Commission I Mid-Term Symposium, Image Data Acquisition – Sensors and Platforms, Calgary, AB, Canada, 15–18 June 2010.
11. Turner, D.; Lucieer, A.; Watson, C. An automated technique for generating georectified mosaics from ultra-high resolution unmanned aerial vehicle (UAV) imagery, based on structure from motion (SfM) point clouds. *Remote Sens.* **2012**, *4*, 1392–1410. [CrossRef]
12. Fuse, T.; Matsumoto, K. Self-localization Method by Integrating Sensors. *Int. Arch. Photogramm. Remote Sens. Spat. Inf. Sci.* **2015**, *XL-4/W5*, 87–92. [CrossRef]
13. Kraft, T.; Geßner, M.; Meißner, H.; Cramerb, M.; Gerke, M.; Przybilla, H. Evaluation of a Metric Camera System Tailored for High Precision UAV Applications. *Int. Arch. Photogramm. Remote Sens. Spat. Inf. Sci.* **2016**, *XLI-B1*, 901–907. [CrossRef]
14. Kawasakia, H.; Anzaib, S.; Koizumic, T. Study on Improvement of Accuracy in Inertial Photogrammetry by Combining Images with Inertial Measurement Unit. *Int. Arch. Photogramm. Remote Sens. Spat. Inf. Sci.* **2016**, *XLI-B5*, 501–505. [CrossRef]
15. Gajski, D.; Gašparović, M. Examination of the Influence of Lens Distortion of Non-Metric Digital Cameras on the Accuracy of Photogrammetric Survey. *Geod. List* **2015**, *69*, 27–40.
16. McGlone, C. *Manual of Photogrammetry*, 6th ed.; ASPRS: Bethesda, MD, USA, 2013.
17. Bouguet, J.Y. Camera Calibration Toolbox for Matlab. Available online: https://www.vision.caltech.edu/bouguetj/calib_doc/ (accessed on 19 December 2016).
18. De la Escalera, A.; Armingol, J.M. Automatic Chessboard Detection for Intrinsic and Extrinsic Camera Parameter Calibration. *Sensors* **2010**, *10*, 2027–2044. [CrossRef] [PubMed]
19. Fiala, M.; Shu, C. Self-identifying patterns for plane-based camera calibration. *Mach. Vis. Appl.* **2008**, *19*, 209–216. [CrossRef]
20. Luhmann, T.; Robson, S.; Kyle, S.; Boehm, J. *Close Range Photogrammetry and 3D Imaging*, 2nd ed.; Walter De Gruyter: Boston, MA, USA, 2013; p. 327.
21. Gruen, A. Development and Status of Image Matching in Photogrammetry. *Photogramm. Rec.* **2012**, *27*, 36–57. [CrossRef]

22. Gašparović, M.; Dubravko, G. The Algorithm for the Precise Elimination of Lens Distortion Influence on Digital Cameras. *Geod. List* **2016**, *70*, 25–38.

23. Fryer, J.G.; Brown, D.C. Lens distortion for close-range photogrammetry. *Photogramm. Eng. Remote Sens.* **1986**, *52*, 51–58.

24. Kager, H.; Rottensteiner, F.; Kerschner, M.; Stadler, P. *Orpheus 3.2.1 User Manual*; Institute of Photogrammetry and Remote Sensing, Vienna University of Technology: Vienna, Austria, 2002.

25. Li, W.; Wang, J. Effective Adaptive Kalman Filter for MEMS-IMU/Magnetometers Integrated Attitude and Heading Reference Systems. *J. Navig.* **2013**, *66*, 99–113. [CrossRef]

26. Gašparović, M.; Gajski, D. Analysis of the Gimbal Impact on the Determination of Camera External Orientation Elements on the UAV. *Geod. List* **2016**, *70*, 161–172.

27. Cai, Q.; Song, N.; Yang, G.; Liu, Y. Accelerometer calibration with nonlinear scale factor based on multi-position observation. *Meas. Sci. Technol.* **2013**, *24*, 105002:1–105002:9. [CrossRef]

28. Syed, Z.F.; Aggarwal, P.; Goodall, C.; Niu, X.; El-Sheimy, N. A new multi-position calibration method for MEMS inertial navigation systems. *Meas. Sci. Technol.* **2007**, *18*, 1897–1907. [CrossRef]

29. Greve, C.W. *Digital Photogrammetry: An Addendum to the Manual of Photogrammetry*; ASPRS: Bethesda, MD, USA, 1996; pp. 52–57.

sensors

MDPI

Article

Atmospheric Sampling on Ascension Island Using Multirotor UAVs

Colin Greatwood [1], Thomas S. Richardson [1,*], Jim Freer [2,3], Rick M. Thomas [4], A. Rob MacKenzie [4], Rebecca Brownlow [5], David Lowry [5], Rebecca E. Fisher [5] and Euan G. Nisbet [5]

[1]   Department of Aerospace Engineering, University of Bristol, Bristol BS8 1TR, UK; colin.greatwood@bristol.ac.uk
[2]   School of Geographical Sciences, University of Bristol, Bristol BS8 1SS, UK; jim.freer@bristol.ac.uk
[3]   Cabot Institute, University of Bristol, Bristol BS8 1SS, UK
[4]   School of Geography, Earth & Environmental Sciences, University of Birmingham, Birmingham B15 2TT, UK; r.thomas@bham.ac.uk (R.M.T.); A.R.Mackenzie@bham.ac.uk (A.R.M.)
[5]   Royal Holloway, University of London, Egham TW20 0EX, UK; Rebecca.Brownlow.2009@live.rhul.ac.uk (R.B); d.lowry@es.rhul.ac.uk (D.L.); r.fisher@es.rhul.ac.uk (R.E.F.); e.nisbet@es.rhul.ac.uk (E.G.N.)
*   Correspondence: thomas.richardson@bristol.ac.uk; Tel.: +44-117-33-15532

Academic Editors: Felipe Gonzalez Toro and Antonios Tsourdos
Received: 15 November 2016; Accepted: 15 May 2017; Published: 23 May 2017

Abstract: As part of an NERC-funded project investigating the southern methane anomaly, a team drawn from the Universities of Bristol, Birmingham and Royal Holloway flew small unmanned multirotors from Ascension Island for the purposes of atmospheric sampling. The objective of these flights was to collect air samples from below, within and above a persistent atmospheric feature, the Trade Wind Inversion, in order to characterise methane concentrations and their isotopic composition. These parameters allow the methane in the different air masses to be tied to different source locations, which can be further analysed using back trajectory atmospheric computer modelling. This paper describes the campaigns as a whole including the design of the bespoke eight rotor aircraft and the operational requirements that were needed in order to collect targeted multiple air samples up to 2.5 km above the ground level in under 20 min of flight time. Key features of the system described include real-time feedback of temperature and humidity, as well as system health data. This enabled detailed targeting of the air sampling design to be realised and planned during the flight mission on the downward leg, a capability that is invaluable in the presence of uncertainty in the pre-flight meteorological data. Environmental considerations are also outlined together with the flight plans that were created in order to rapidly fly vertical transects of the atmosphere whilst encountering changing wind conditions. Two sampling campaigns were carried out in September 2014 and July 2015 with over one hundred high altitude sampling missions. Lessons learned are given throughout, including those associated with operating in the testing environment encountered on Ascension Island.

Keywords: Ascension Island; atmospheric sampling; methane; UAV; SUAS; multirotor; BVLOS

1. Introduction

This study reports the successful development of an important new sampling technique for atmospheric methane in the mid-troposphere. Methane is a major greenhouse gas, which is rising rapidly, particularly in the Tropics [1,2]. The reasons for the rise remain unclear, but tropical wetlands may be a major contributor [3]. These wetlands, many of which are in the Congo and Amazon basins, are relatively inaccessible to integrating studies of emissions: field access is difficult, and in some regions of Africa, aircraft surveys are both challenging and not favourably viewed by local security forces. Furthermore, such integration of the more global atmospheric signal is problematic where sampling might occur near a range of source areas and convection processes.

Therefore, to obtain a more representative global-scale signal, sampling is preferred in zones that are downwind of mixed regional emissions that can then be benchmarked against local background conditions not attributed to the sources being characterised. One such unique place for such a sampling laboratory is Ascension Island in the South Atlantic. In order to sample the atmosphere downwind, a very promising research sampling platform based on a Small Unmanned Air System (SUAS) has a number of benefits. The challenge is to realise safe, repeatable and reliable operations to a significant altitude that is Beyond Visual Line Of Sight (BVLOS). This paper describes the successful demonstration of the use of SUAS to sample equatorial air up to almost 3000 m Above Sea Level (ASL) on Ascension Island, detailing the SUAS used, key operational requirements and lessons learned throughout the build-up and during the field campaigns. Using SUAS, greenhouse gas measurement on Ascension can in principle access both air at ground level from a very wide swathe of the southern oceans and also sample air from above the Trade Wind Inversion, thereby addressing emissions from a significant part of the global tropical land masses.

SUAS and UAVs are increasingly being developed and deployed for a range of environmental applications [4–9]. In particular, significant traction is being realised in the areas of remote sensing [10,11], mapping 2D/3D structures [12–14] and atmospheric sampling [15–25] using a range of emerging sensor technologies [26–29]. However, most applications to date that have been used for atmospheric sampling have been at lower altitudes in the 500 m–1000 m range [15,19] or involve longer range fixed wing platforms that often require considerable resources to deploy [30–32]. This paper critically evaluates the development and deployment of a multirotor-based system to investigate the feasibility of collecting mid-tropospheric air samples from above the Trade Wind Inversion (TWI) layer on Ascension Island for the purposes of identifying methane mole fractions and isotopic composition. The objectives were to capture air samples from below, within and above the inversion layer above Ascension Island in the mid Atlantic. The minimum altitude requirement was to be sufficiently above the TWI layer - which changes altitude seasonally and has some daily fluctuations - to ensure that the air captured was free of boundary layer air from below the TWI that may have mixed upward. Therefore, the higher the sampling altitude achieved, the more confidence that we can determine the mixed air back-trajectory [33,34] "reach" of the method, potentially sampling wide source regions in Africa in the right synoptic conditions.

2. Campaign Field Site

Ascension Island, situated in the mid-Atlantic just south of the Equator (8° South) (Figure 1), approximately 1500 km from Africa, is ideally located. At sea level, the SE Trade Winds are in the South Atlantic marine boundary layer. The Trade Winds themselves are almost invariant, derived from the deep South Atlantic and with little contact with Africa. Above the TWI at about 1200 m–2000 m above sea level (depending on seasonal meteorology and diurnal cycle), the air masses are very different, of equatorial origin; see Figure 2. Dominantly, they have been last in contact with the ground in tropical Africa, but at times from South America. In detail, depending on season, air above the TWI is sourced mainly from tropical and southern Africa with some inputs of air also from southern tropical South America. African and South American methane sources are major contributors to the global methane budget [3,35], but although local campaign studies have been made, these emissions are not well known in bulk. Understanding the changing greenhouse gas burden of the atmosphere demands sustained long-term measurement. Ascension is ideal for this, both in location and in security of access.

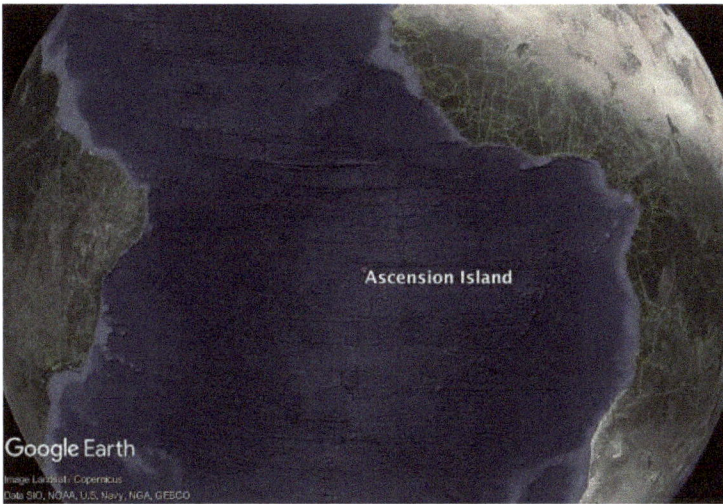

Figure 1. Ascension Island location in the mid-Atlantic.

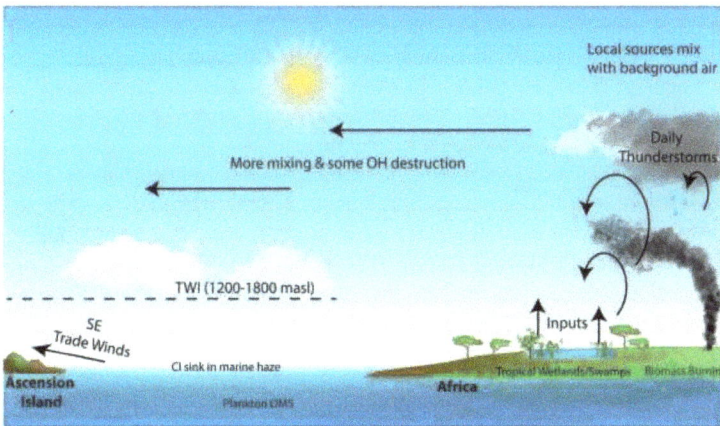

Figure 2. Ascension atmospheric sources.

The island hosts one of the very few equatorial high precision measurement facilities for CO_2 and CH_4, worldwide; see: [3]. The cavity ring-down greenhouse gas analyser and calibration suite are installed at the UK Met Office base at the Airhead on Ascension and, in normal operation, continuously measure CO_2 and CH_4 in the ambient marine boundary air of the Trade Wind. However, the highest point on Ascension is the top of Green Mountain, which is 859 m above sea level and is therefore not high enough to sample air above the TWI. Thus, the purpose of this SUAS project, by demonstrating the usefulness of the instrument in measuring air samples from above the TWI, was to show that Ascension potentially becomes a virtual mountain with access to air from sea level to nearly 3000 m: it can become the UK equivalent of the USA Hawaiian observatory on Mauna Loa, at 3397 m above the Pacific, if an effective SUAS sampling platform is fully demonstrated. In addition to being an ideal location for the sample flights from a science perspective; the remote location, military air base and size of the island mean it is possible to arrange for segregated airspace. This is a key requirement for allowing current SUAS to operate Beyond Line of Sight (BLOS).

3. Rationale for the SUAS Platform

Key targets of the field campaign and the proof of concept system can be identified as:

1. To be able to operate on Ascension Island, with the required associated logistics and support.
2. To be able to operate in a tropical equatorial environment, in close proximity to the sea, at high altitude (for SUAS) and with wind speeds averaging 8 ms^{-1} at ground level.
3. To be able to sample repeatedly at a minimum altitude 100 m or more above the inversion layer, identified on Ascension as varying seasonally between 1200 m and 2000 m ASL through the year.
4. To be able to identify the lower and upper boundaries of the inversion during flight in order to be able to target samples within, above and below.
5. To be able to remotely trigger the air sample collection.
6. To be able to fly multiple times per day, nominally six samples per day at specified altitudes in a safe and reliable manner.

The SUAS approach presented here was chosen because of its inexpensive flexibility. In previous work by the Royal Holloway, University of London (RHUL) group, air sampling has been carried out at altitude by using the UK Facility for Airborne Atmospheric Measurements (FAAM) aircraft facility to fly air sampling equipment at the required altitudes. Sampling with full size aircraft enables greater flexibility than land-based measurements offer due to the ability to climb to required altitudes, but incurs very high costs, especially in remote locations. Moreover, the FAAM aircraft barely has the range to reach Ascension with a full load of instruments. The frequency at which samples may be collected would also be very low. Whilst Ascension Island hosts the 3000 m runway of Wideawake Airfield, most flights are large aircraft en route to the Falklands, and it is not suitable for sustained sampling. No commercial light aircraft or helicopters are based on the island that could be used for the air sampling campaign. In contrast, SUAS have the potential to offer fast turn-around times, remote deployment with small teams and inexpensive sampling [36].

A number of different aerial solutions to the atmospheric sampling problem were considered, including kites, helikites and fixed-wing SUAS; however, a combination of flexibility, low cost and potential ease of operation led to the choice of a small electric unmanned multirotors. Although other options could have been made to work, the electric multirotor could fly directly to the altitude required, sample and return to base, pausing only for the sample collection at altitude. Key advantages of the multirotor were identified as:

*   Potentially low cost.
*   Flights could be carried out in a matter of minutes, thereby accounting for rapid changes in conditions and allowing for multiple samples at specific times throughout the day.
*   Flight profiles can be near vertical, allowing for easy airspace integration and de-confliction.
*   Transport and ground support for the vehicles are relatively easy to deploy.
*   Maintenance is relatively easy due to a modular design.
*   The design allows for flexibility in the payload integration.

There are, however, key challenges to operating a small electric SUAS in this way. These include the requirement to fly what is defined as BVLOS (Beyond Visual Line of Sight); the requirement to fly through saturated air to allow sampling above the lower cloud layers; the requirement to climb and descend at relatively rapid rates of 5 ms^{-1} continuously; the requirement to continuously monitor temperature and humidity to allow clear identification of the temperature inversion and the requirement to operate in relatively high wind speeds. Throughout this paper however, it is shown that these challenges can be overcome, and in the right situations and conditions, a SUAS multirotor is an excellent vehicle for sensing applications to over 3000 m. Two sampling campaigns were carried out in September 2014 and July 2015 with over one hundred high altitude sampling missions. Lessons learned are given throughout, including those associated with operating in the harsh environment encountered on Ascension Island.

4. System Description

The aircraft used for the sampling campaign was an eight motor multirotor (or octocopter) in an X-8 configuration, as shown in Figure 3a. The airframe is a custom design from the University of Bristol that provides enough space for a large battery capacity (typically 533 Wh, but tested with up to 710 Wh) and air sampling equipment. Situated above the main aircraft, the temperature and humidity sensors are located towards the centre and away from the body in order to minimise the effect of the local flow on the sensor readings. The Tedlar bags are held underneath the vehicle to allow for inflation at the selected altitudes. A summary of the vehicle specifications is given in Table 1, and additional key features are described as in the following section. The ground-based element of the system is shown in Figure 3b and is indicative of the operating environment encountered. In close proximity to the sea and with near constant wind speed, this required significant weatherising of the onboard electronics.

(a)　　　　　　　　　　　　　　　　(b)

Figure 3. Flight operations on Ascension Island (a) University of Bristol X-8 Multirotor; (b) Field site.

Table 1. Key UAV specifications.

Maximum Take Off Weight (MTOW) (inc. batteries)	10 kg
Diagonal rotor-rotor distance	1.07 m
Maximum battery capacity	32,000 mAh 6-cell Lithium Polymer
Motors	T-Motor MN3515 400 KV
Propellers	T-Motor 16x5.4"
Electronic Speed Controllers (ESC)	RCTimer NFS ESC 45 A (OPTO)
Autopilot	Pixhawk by 3DRobotics
Autopilot software	ArduCopter v3.1.5
Safety pilot control link	FrSky L9R 2.4 GHz
Ground Control Station (GCS) link	Ubiquiti 5-GHz directional
Onboard computing	BeagleBone Black
Sampling pump	KNF Diaphragm pump (NMP 850 KNDC)

4.1. Airborne Vehicle

Vehicle configuration: An eight-rotor vehicle was chosen to achieve reasonable redundancy against loss of a motor or speed controller during flight. The vehicle could theoretically sustain a loss of four motors, provided that none were on the same arm, and tests on disconnecting three motors demonstrated that the vehicle retained good control in flight provided there was sufficient overall thrust. Users of octocopter platforms have also anecdotally suggested that the four-arm coaxial configuration was likely to provide better gust tolerance than a flat eight-arm configuration. It was found that the vehicle did perform well in the wind, but further research is being conducted at the University of Bristol in order to quantify the differences.

Vehicle size: The size of the vehicle was driven by three factors: the mass of the payload; the endurance required to reach high altitudes; and overall vehicle stability in high winds. Tests showed that the final flight vehicle could operate in wind speeds of up to 20 ms^{-1}. The payload required to conduct the experiments was designed to be under 0.5 kg, and the maximum take off weight given in the risk assessment and the application for BVLOS operations was 10 kg. This was never exceeded on Ascension, and the typical take-off weight was in the region of 8.5–9.5 kg depending on the number of batteries that were used.

Battery capacity: The vehicle was configured to be able to operate with either two, three or four 8 Ah 22.2 V Lithium Polymer (LiPo) batteries in parallel. Fewer installed battery packs would lead to a lighter more agile vehicle, but with reduced endurance. Installing more batteries increases the flight time, but due to the additional mass, the additional endurance reduces with each battery. Four batteries was deemed as an acceptable upper limit in terms of endurance achieved and stability of the aircraft. During the campaign, the vehicle flew with three batteries, as this provided enough endurance to reach beyond the maximum expected TWI altitude of 2.0 km ASL.

Autopilot: The ArduCopter autopilot software was selected due to its reliable performance and flexible operation. The telemetry protocol is well documented and enabled integration with a long-range telemetry link. Additionally, a comprehensive flight dataset is logged for each flight, enabling the analysis presented with this paper. The hardware selected was the Pixhawk by 3D Robotics.

Onboard computing: An onboard computer was required to store and forward the analogue and digital data from the onboard sensing, as well as control the sample collection pumps. This enables the ground operators to see the sensor data in real time for decision making in flight. MAVProxy software was installed for forwarding the telemetry data, and a custom module was written to allow integrated monitoring and control of the payload. All collected data were stored on-board to allow for dropped packets. The architecture of the on-board system components and how they communicate is shown in Figure 4.

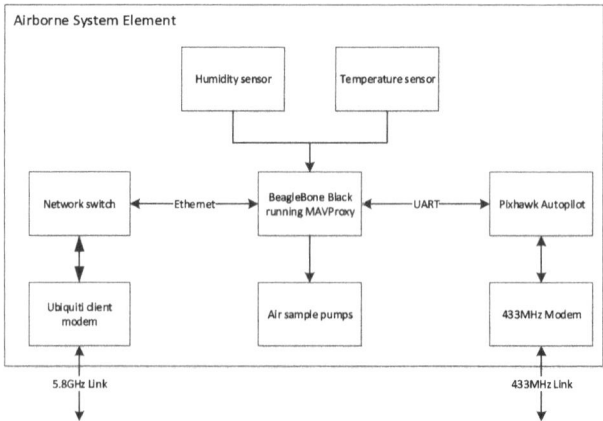

Figure 4. Airborne system diagram.

Sensors: The payload had three core functions: measure temperature; measure relative humidity; and pump air into sample bags from an external demand signal. Temperature and humidity measurements were used to characterise the boundary layer profile on the ascent and to indicate the location and characteristics of the inversion layer. The ascent rates were reasonably fast (5 ms^{-1}), and so it was necessary to select sensors that responded fast enough in order to measure the boundary layer height. To target a minimum of 100 m above the trade wind inversion when ascending at this rate, at least one of the sensors needs to be fast enough to respond to a step change in temperature within 20 s for this to be the case. Ideally, this is within 10 s to account for uncertainty in the down leg measurement when assessing the sample height post-flight (we do not profile up and down and then choose a sampling height), and we test this for the temperature sensor using derived boundary layer heights in Section 8.

Sensors used in radiosonde measurements were sourced due to having very similar design requirements. The temperature sensor was easy to interface through the onboard computer analogue inputs and proved to be reliable; however, the relative humidity sensor required more attention to integrate due to the limited I2C address range available, and custom addresses had to be programmed ahead of time. The temperature sensor itself was a GE fast-tip FP07 glass bead thermistor (analogue) <0.2-s response time with a spectral response close to 5 Hz (Figure 15 in [37]). Figure 5 shows a three-point calibration for this sensor in a standard Weiss WKL 34/40 calibration oven. The calibration was performed at three set points (0 °C, 10 °C, 20 °C), and the reference was a NIST traceable temperature logger combined with a standard thermistor probe (accurate to ±0.2 °C). Based on this data and in the configuration used, this was found to be accurate to ±1 °C and therefore suitable for the purposes of this campaign.

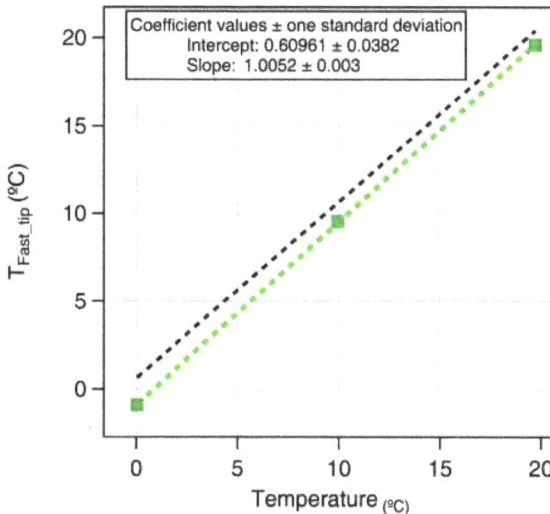

Figure 5. Temperature sensor calibration.

The relative humidity sensor was significantly more sensitive to general handling and dust ingress than the temperature sensor. The reliability of the sensor was poor, and so, it was replaced daily and protected between flights. The humidity sensor used was an IST P-14 Rapid capacitance humidity sensor (I2C) <1.5 s response time, which has been extensively tested for fast-response humidity measurements on UAVs by Wildmann et al. [38]. As with the Fasttip temperature sensor, we expect the humidity sensor characteristics to satisfy our minimum 100 m vertical resolution requirements, as mentioned above. The manufacturer states that these are calibration free following their factory calibration to ±3% relative humidity, and no changes were made to the sensors themselves prior to flight.

As mentioned previously, the temperature and humidity sensors were placed toward the centre of the vehicle and above the main chassis. This was done to minimise the effect of the rotor wake on the sensor readings, specifically in the climb to the sample altitudes. Whilst ascending, the air is essentially accelerated vertically downwards, and the sensors themselves are in the relatively smooth air being induced from above (Figure 3 in [39]). It was also found that inspection of the data collected during ascent and descent showed no systematic bias with regards to temperature and pressure. We conclude, therefore, that rotor-induced local air flows have no significant effect on the performance of these sensors when placed in this position. In addition, with the response time of the temperature sensor used at <0.2 s, the resolution of the data collected at (5 ms^{-1}) ascent rate was higher than required in order to pinpoint the temperature inversion and the sample heights to target during the flight. It was also found that when a second flight was carried out in quick succession to a first flight, the temperature and humidity data followed the first very closely.

Air collection: The air samples were collected into five-litre Tedlar bags by directly pumping in air through NMP 850 KNDC diaphragm pumps, in accordance with internationally-agreed best practice [3]. Two pumps were installed, each plumbed directly to a sampling bag. The pumps were controlled via a P-Channel MOSFET load switch circuit triggered from the onboard computer. Whilst the pumps could have been triggered automatically upon the vehicle reaching specific waypoints, it was decided that the triggering should happen remotely, allowing the payload operator the chance to decide during flight on the best sampling locations. A custom box was laser cut out of corrugated correx to house the two bags and was simply attached to the underside of the vehicle. Methane mixing ratios in the Tedlar bag samples were measured within 1–2 days of collection using an in-house Picarro 1301 CRDS (cavity ring-down spectrometer) with an NOAA traceable six-gas calibration suite (on Ascension Island) giving a precision of ± 0.5 ppb [40]. Samples are measured for 240 s with the last 120 s being used to determine the mixing ratio. For bags containing less air, the bags were run for 120 s, with 60 s being used to determine the mixing ratio.

With regards to mixing of the air sample due to the vehicle rotors, the induced velocities at a distance of greater than three rotor diameters are very low (note that this does vary depending on disk loading [39]), and so, although the air is mixed locally due to the air vehicle whilst in the hover and when collecting an air sample, the volume over which that sample is collected is actually relatively small, with an outer sample collection diameter (in the absence of wind) of no more than a few metres. With the vertical distances of up to 3 km involved in this campaign, the sample volume itself is relatively very small. With the vehicle itself drawing in air for the sample from only a few metres, the atmospheric conditions themselves will typically have a much greater effect on the mixing of the sample than the vehicle itself.

The air masses under study have been transported several thousand km from interior Africa and have mixed en route. Thus, though on 3000 m scale, there are strong vertical changes depending on source inputs; each air mass is generally homogeneous on the 10–50 m scale. The UAV causes local mixing on a metre scale, but this scale of mixing is unlikely to be significant in sampling separate air masses, unless there is a sharp laminar boundary present that has survived the transport from Africa and consequent boundary mixing between air masses.

Tedlar bag samples were also taken approximately 1 m above ground level each day from the UAS site and were analysed together with the samples from the UAS. These samples were then compared with the continuous ground measurements made by RHUL at the Met Office on Ascension Island. The measurements by RHUL are long-standing, both with the in situ continuous system installed on the island and by regular flask sampling analysed in London. RHUL measurements are subject to ongoing intercomparison with the parallel co-located flask collection by U.S. NOAA, measured in Boulder Colorado. Please see [3] for additional information.

Safety pilot link: The safety pilot link is used for manual control of the aircraft, as well as selecting flight modes, such as waypoint following or Return To Home (RTH). The link is required to maintain communication with the vehicle at all times so that the safety pilot could always command an RTH. Two off-the-shelf systems were identified as suitable for the safety pilot link: Immersion RC EzUHF and FrSky L12R systems. The Immersion RC system transmits on 459 MHz (in the UK firmware), whilst the FrSky system uses 2.4 GHz. During testing, it was found that the Immersion RC system was sensitive to interference from the onboard systems. The FrSky system proved extremely robust, both during testing, as well as the campaign. Ground-based tests were conducted before the campaign with a line of sight horizontal separation of 3 km during which the signal strength was consistently strong.

Telemetry: Ubiquiti radio modules were selected due to the long range and high bandwidth offered. The primary function of the telemetry link was to enable monitoring of system health data on the ground, as well as interaction with the payload. The directional antennas used both on the ground, as well as the aircraft were selected to provide a stronger link. A second telemetry link was used for redundancy, transmitting on a separate frequency using a simple omnidirectional antenna. Radio modems by 3D Robotics were used to communicate with the vehicle on 433 MHz, with duplicate vehicle health information being transmitted.

Environmental protection: The motors are able to operate in water and did not require special consideration other than inspection of bearing smoothness prior to flight. Electronics however, required protection, and the onboard computer was sprayed with PCB lacquer; both the autopilot and onboard computer were encased within plastic containers with vents on the underside. The vents would enable air to come in, which would contain moisture from the clouds, but signs of any water deposited within the containers was closely monitored and found to be minimal. Connecting wires were sealed, and loops to below the entry points were included to minimise the run down of collected water droplets.

4.2. Ground Station

A ground station network was set up as shown in Figure 6. This was designed in order to provide the mission and payload operators with the information required to operate the vehicles safely and reliably. Key elements of this system are as follows.

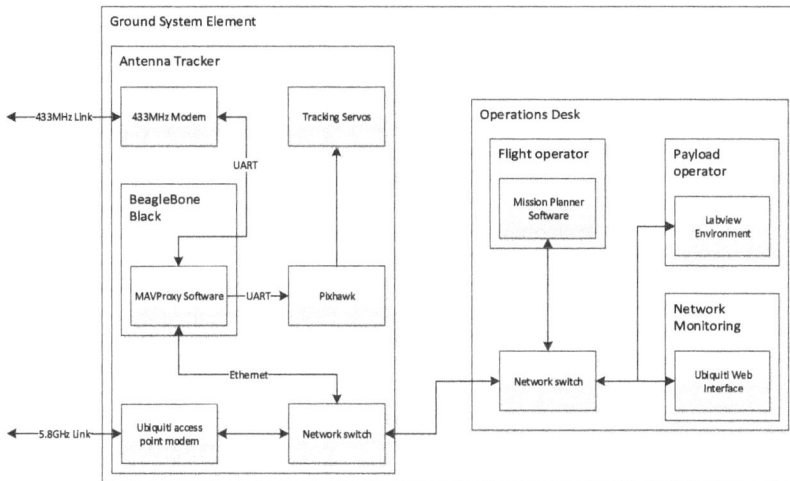

Figure 6. Ground system diagram.

Antenna tracker: A servo-driven pan and tilt system was constructed with both the 433 MHz and 5.8 GHz antennas attached. A Pixhawk running the ArduTracker software was installed to control the antenna tracking, which compares the UAV position with the position and attitude of the antennas and sends corrective signals to the pan and tilt servos. The Pixhawk obtains information about the position of the UAV from the MAVProxy software, which combines information about the UAV from both telemetry links. The combined information is also shared with the flight and payload operators through a two-way data link over the local area network.

Visitor and road management: Road blocks were setup each day before flight operations commenced as agreed with the Ascension Island police force. During operations, there was always at least one person ready to handle any external interruptions, allowing the rest of the team to continue focussing on the mission. Typically, this would involve answering the radio when road access was requested, but also included talking to visitors that had arranged to watch the flights.

Weather monitoring: A portable weather station was used during the first campaign, primarily to measure wind speed and direction. During the second campaign, a Gill Instruments R3 Sonic Anemometer was used, which enabled remote monitoring of the same information.

5. Operational Considerations

5.1. Field Site Operations on Ascension Island

In September 2013, a three-person team from Birmingham and Bristol visited Ascension Island for a reconnaissance trip to identify possible operational sites. This proved to be invaluable for the subsequent field campaign, and the authors highly recommend this approach for any significant UAS operations. Three possible sites were initially identified, of which one was selected as the most suitable to operate from. The area found was located by the road leading out to the old NASA site and provides excellent access, isolation and the ability to block the road and control access for third parties. It is approximately 350 m ASL, which reduces the height required to climb compared to a sea level launch and is on the windward side of the island, providing clean, unobstructed airflow from the prevailing wind direction: south, southwest.

Ideally, all of the equipment including both the vehicles and the ground support equipment would have been shipped to Ascension prior to the field campaign. Due to extended development and early shipping dates, however, only the ground equipment and maintenance equipment were sent out ahead of time. This included all of the lithium polymer batteries (necessary due to airfreight restrictions) and all the heavy items, such as portable shelters. Flights to Ascension Island depart from Royal Air Force Brize Norton, with the Air Bridge to Ascension Island and the Falkland Islands.

With the permission of the military personnel and police on Ascension, a base was set up, as shown in Figure 7, just off the road to the old NASA site. Given the strong continuous wind speeds on Ascension, two 3.66 m (12 ft) by 3.66 m (12 ft) shelters were shipped from the UK and assembled on site. During the course of the campaign, basic equipment was stored on site throughout, with the aircraft being bought up from Georgetown on a daily basis. Battery charging was carried out both on site and in Georgetown depending on daily usage. Only minor maintenance was required throughout the campaign, and this was carried out in Georgetown.

Safety was the focus of the build-up and operations through the project. Although there is an element of redundancy through the air vehicle configuration chosen, there are still multiple single points of failure on the airframe. Because of this, all operations were carried out with the worst safety case based on a total failure of the onboard power systems. Given the maximum permitted BVLOS altitude of 3048 m (10,000 ft), the worse case scenario considered given the conditions encountered throughout the campaign resulted in a safety radius of at least 1.2 km from the flight path. The road was blocked at 1.4 km (straight line) from the point of operation, and the nearest inhabited site was 3.0 km from the flight path.

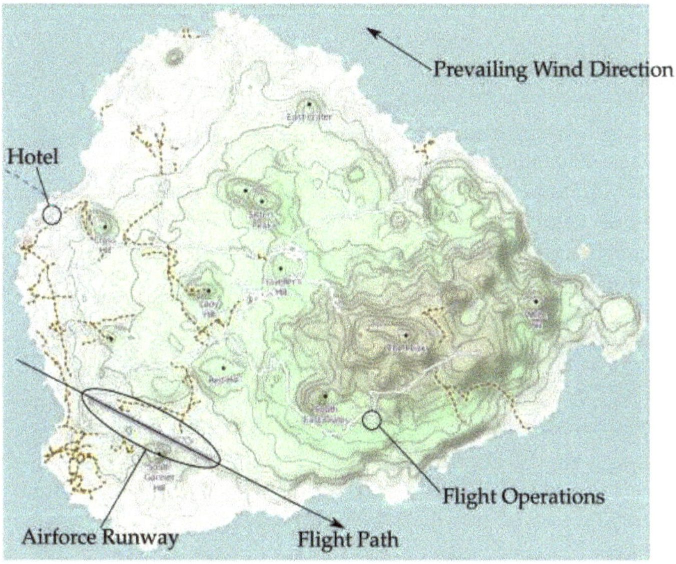

Figure 7. Map of Ascension Island.

5.2. Typical Flight Operations

The objectives required the multirotor to climb to a commanded altitude and loiter in position whilst an air sample was taken, pausing for ten seconds prior to the sample collection. This air sample was collected using a diaphragm pump that inflated a bag in just under one minute. The vehicle would then return to base, and the air sample would be removed for analysis. Throughout the flights, the temperature and Relative Humidity (RH) were measured and transmitted to the ground station at a rate of 10 Hz, and it was possible to ascertain the location of the trade wind inversion and the different air masses from the temperature and humidity readings during the ascent. The real-time measurement of the atmospheric profile is a significant capability of the system and enabled the atmospheric scientists to accurately target key parts of the profile to collect the samples.

For both campaigns, the flights on Ascension Island were carried out under Beyond Visual Line Of Sight (BVLOS) conditions. This required an exemption to the Air Navigation Order, which was granted by Air Safety Support International (ASSI) for the period of both campaigns. The exemption itself was based on a safety case, which the University of Bristol put together, and was granted subject to a number of conditions including: contact with Wideawake Airport would be maintained at all times; all flights would be operated in accordance with the permission/procedures agreed with the Royal Air Force and the United States Air Force. These measures put in, both prior to the campaigns and during flight operations, were designed to ensure complete separation from other air traffic. For example, no flights were carried out during a given window encompassing an aircraft arriving or departing from the island. Whilst this approach to BVLOS operations requires airspace separation and close communication with all other airspace operators and users, it allows small SUAS vehicles to be operated safely in challenging environments. Ongoing improvements with communications, sensors and computing will allow more closely-integrated BVLOS operations in the future; however, for the present, the use of segregated airspace for these types of operational flights is likely to be required.

The flight operations themselves followed a pattern, established over the course of the first week. This was based on the following tasks:

1. Preparation of the flight vehicle including physical and system checks and preparation of the air sample bags including air evacuation.
2. Preparation of the flight plan, including alternative routes for descent depending on the upper wind conditions. See Figure 8 for a visual depiction of a typical mission.
3. Pre-flight checks, arming the vehicle and manual take-off to 20 m AGL undergoing manual flight checks by the safety pilot.
4. Switching the vehicle into automatic mode and carrying out the ascent to the pre-determined altitude at 5 ms^{-1} ascent speed.
5. Monitor temperature and relative humidity profiles throughout the climb, confirming the location and thickness of the trade wind inversion. The sensor operator at this time would confirm the predetermined sample heights or adjust depending on the altitude at which the trade wind inversion was encountered.
6. On agreement with the ground station and sensor operators, the air samples were triggered at the required altitudes. Pump times varied between 40 s and 60 s depending on the target altitudes for the samples.
7. The descent was carried out automatically at -5 ms^{-1}, reverting to manual at 20 m Above Ground Level (AGL).
8. Post flight checks were then carried out, battery voltages recorded, flight data stored and the air sample(s) retrieved.

For all flights on Ascension Island, the aircraft was taken off and landed manually, under the direct control of the safety pilot. Although the system is fully capable of an automatic take-off and landing, a manual approach allowed for the safety pilot to carry out flight checks prior to initializing the mission. The take-off point itself was situated across and downwind from the operations tent, allowing the ground crew to remain upwind and yet control full access to the site. One of the benefits of operating on Ascension Island is that the prevailing wind direction and strength at ground level are consistent and predictable, allowing the flight operations to be consistent and refined over time.

Figure 8. Sample flight path over Ascension Island.

6. Lessons Learned

All anomalies with any part of the system were fully investigated prior to any operations. The following is a list of the key problems encountered during the field campaign and the actions taken as a result.

1. Ground station power was lost during one of the initial flights. The power sources taken to support operations malfunctioned, and a return to launch was triggered by the safety pilot. All subsequent operations in the first campaign were powered from two inverters. For the second field campaign, a generator was sourced to provide power on site.

2. One of the ESCs (Electronic Speed Controller) failed as the aircraft was prepared for flight. This was identified by the safety pilot and replaced before further operations continued.

3. Difficulties were identified at one point with the telemetry downlink; this was traced to a file size limit being reached on the onboard computer. Once identified, this was corrected, and flights continued.

4. High wind speeds above the inversion were identified in the second half of the first field camp again. These are outlined in detail together with mitigation strategies in the subsequent section.

5. One of the GPS batteries became loose during flight and was identified in the pre-flight checks prior to the following operation. This was triggered as a magnetometer error, and the unit was replaced and checked before the subsequent flight.

6. One of the commercially-purchased battery connections failed during flight, which was identified from the flight director observing an unusually high voltage drop on the initial climb out; therefore, a return to home was triggered by the safety pilot. All additional battery connections were checked and modified prior to resumption of flight operations.

7. Flight Envelope

7.1. Achieving Maximum Altitude and on-Board Power Management

At the start of the campaign, the maximum commanded altitude of the UAV was increased gradually as knowledge and confidence in the system grew. The maximum altitude achieved was 2500 m AGL, which surpassed the required altitude (typical of the local TWI) by approximately 1000 m. Figure 9 shows some of the recorded data for a 2500 m AGL flight. The current drawn from the flight battery illustrates that, as one might expect, the majority of the energy is expended during the ascent. The UAV drew 45 A in the hover at 20 m AGL (370 m ASL) and peaked at 90 A at the top of the ascent, before dropping to just under 70 A whilst hovering at 2500 m AGL. The increase in motor speeds required to hover at 2500 m AGL is also apparent from the bottom of the plots, where the speed to hover at 20 m AGL is 50% compared to 66% at 2500 m. On this particular flight, the UAV landed with 23% of the battery capacity remaining.

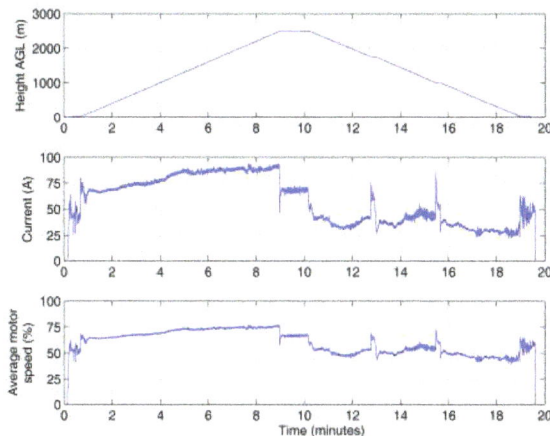

Figure 9. Altitude, battery voltage and motor speeds for a sample flight to 2500 m AGL.

7.2. Performance in Wind

The trade wind conditions were found to be very consistent from day to day at ground level. The winds above the TWI, however, were seen to change over the course of the campaign. The different wind conditions therefore enabled the team to test the UAV performance in conditions ranging from light wind to prohibitively strong winds. It is possible to estimate the speed and direction of the wind from the aircraft orientation after estimating some key parameters found by performing a slow orbit manoeuvre, such as that shown in Figure 10b. Making some assumptions about steady wind conditions and resolving free body diagram forces, a simple mapping function can be created between attitude angles reported by the autopilot and the estimated wind conditions. Full details on the procedure are outlined in [41].

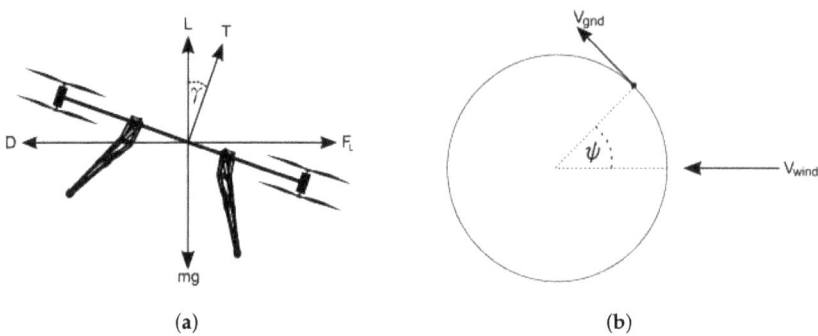

Figure 10. Forces, angles and velocities used for wind estimation. (**a**) Free body diagram of forces on UAV; (**b**) angle around track during orbit manoeuvre.

7.3. Flight through Still Air

The ideal flight path for the UAV from a mission scripting perspective would be to ascend vertically to the required altitude and then descend back along the same path to the take-off site. Rapid descent through still air for rotary wing aircraft however can lead to instability. The wind speeds on Ascension Island were typically around 8 ms^{-1} at ground level, which is equivalent to the UAV travelling at a reasonable forward velocity in still air conditions. As previously mentioned, the wind conditions varied above the TWI, and over the course of a few days of the first campaign, the wind speeds were found to be extremely low. Conditions varied significantly reaching less than 2.5 ms^{-1} at times, such as in the flight shown in Figure 11 where aircraft bank angles and estimated wind speeds have been plotted against height above ground.

Figure 12a shows the roll and pitch angles during the same flight, overlaid on the altitude. The wind speed on the ground was approximately 8 ms^{-1}, meaning that the vehicle had to roll and pitch in order to maintain position during the loiter and vertical climb-out. An interesting result can be seen around the four- to seven-minute mark (i.e., between 600 m and 1500 m), where the roll and pitch angles change significantly during the ascent. During this stage of the ascent, the angle of the vehicle has rotated from around −10° roll and 4° pitch to approximately 0° roll and −2° pitch. The change in attitude is attributed to a dramatic reduction in wind speed as the vehicle enters a different air mass.

During the ascent, the reduction in wind speed poses no problems. Descending through still air, however, requires the vehicle to constantly make corrective actions, and it has to work harder to maintain a level attitude. The attitude is shown in Figure 12b for a 15 s period half way through the still-air descent phase (at around 1600 m). Although the magnitude of the attitude variations is manageable, the persistent rate of change in attitude is undesirable. The constant sharp changes in attitude observed in the still-air descent will lower the vehicle's endurance. Figure 12a also shows the current being drawn by the UAV, which during the initial descent through still air, is roughly the same as that required to hover at 1800 m. Upon descending into the different airmass with higher wind speeds, the current drops to nearly half as the motors do not have to work as hard to maintain the stability of the aircraft. The current only increases again at the end of the flight when the UAV hovers briefly prior to touch down.

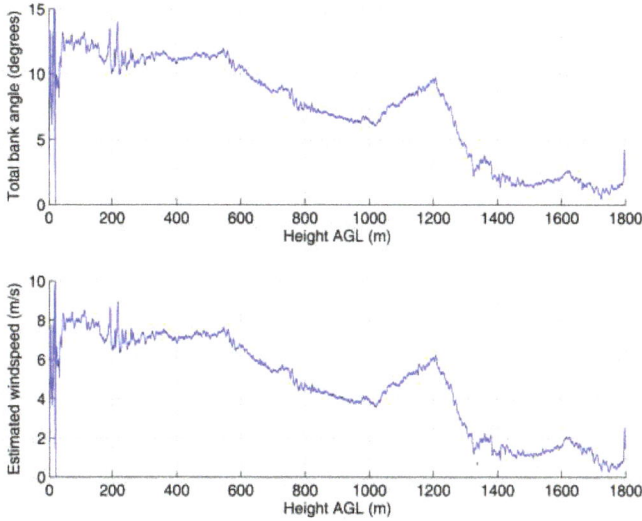

Figure 11. Estimated wind speeds from the aircraft attitude during a flight up to 1800 m AGL. (**a**) measured bank angle; (**b**) estimated wind speed.

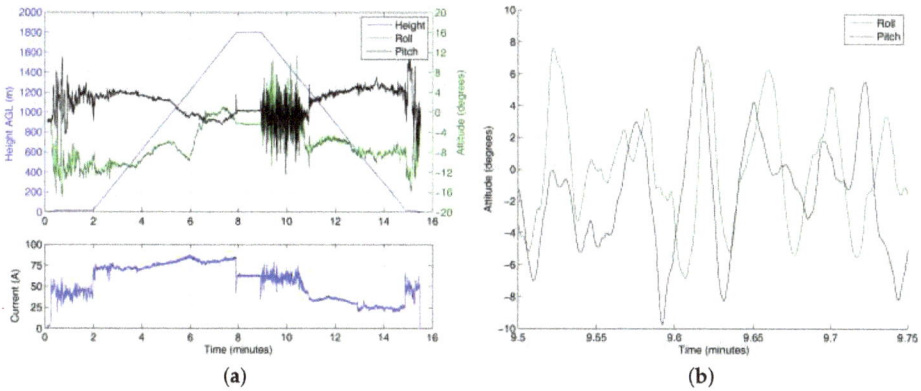

Figure 12. Sample at 1800 m AGL, with low wind speeds above the TWI. (**a**) Attitude and voltage profiles; (**b**) attitude during initial descent.

7.4. Trajectory Design for Descent in Still Air Conditions

It is well known by helicopter pilots that one should not descend vertically, as this would entail entering the helicopter's own wake and lead to instabilities as described in [42]. The descent through still air described in Section 7.3 is undesirable as it reduces the endurance of the vehicle, which in turn reduces the altitudes attainable. To provide the rotors with clean airflow, the mission script was modified to include a lateral manoeuvre during the descent. This lateral manoeuvre, referred to here as a dog-leg, typically consisted of translating away from the launch site whilst descending, pausing and then translating back towards the launch location. This can be seen clearly in Figure 8 with a climb to the sample altitude, followed by a descent to the waypoint identified, then a return to the original trajectory. The profile shown is to the same scale as Ascension Island.

The current consumption shown in Figure 13a suggests that the inclusion of a dog-leg reduced the power consumed, as the current drops significantly once the descent has initiated and, apart from two brief sections of hovering, stays low. The pauses at the additional waypoints required the UAV to arrest its movement and draw additional power, but this is believed to have significantly less impact on the endurance than omitting the dog-leg and descending vertically in still air. It should be noted that the dog-leg does not increase the time taken to descend as the vertical velocity is unchanged. The key reason for including this manoeuvre however is the reduction in the roll rates that are experienced during the descent in the low speed air mass. This reduction in rolling and pitching during the initial descent can by observed by comparing those angles previously shown in Figure 12 and those found by including a dog-leg in Figure 13b.

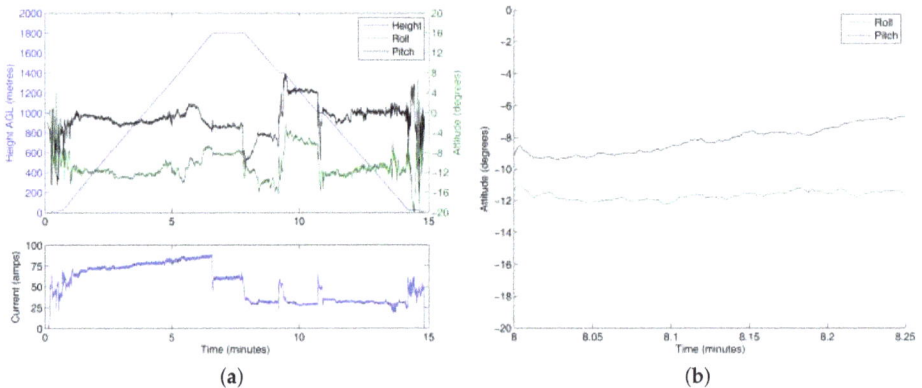

Figure 13. Sample flight to 1800 m AGL, with a translation manoeuvre through low wind speeds above the trade wind inversion. (**a**) Attitude and voltage profiles; (**b**) attitude during initial descent.

The improvements observed by updating the flight plan with knowledge of the wind conditions suggests that some form of automatic trajectory generation would be highly beneficial. The wind direction and magnitude can be estimated using the attitude of the vehicle during the ascent, and as previously observed, the still air masses can clearly be spotted by the dramatic reduction in roll and pitch angles. An automatic trajectory planner could therefore introduce flight paths to ensure a stable descent. Furthermore, the planner should be constrained to guarantee that the trajectory does not pass into prohibited airspace and that the trajectory brings the vehicle back to the launch site.

7.5. Upper Wind Speed Limits

The wind speeds above the TWI were, on some of the days during the campaign, observed to increase rather than decrease, as discussed in Section 7.3. Figure 14 shows flight data collected on such a day, where the wind speed increases significantly at around 1350 m AGL.

Figure 14 shows how the climb rate remained constant at 4.5 ms^{-1} until around 1350 m AGL when it was seen to decrease to just 1 ms^{-1}. The decrease in climb rate was a direct result of the increase in wind speed, which caused the vehicle to pitch and roll to maintain the commanded ground course. The figure also shows the total bank angle of the vehicle during the ascent by taking the magnitude of the roll and pitch angles. The low level strong winds required the aircraft to bank by an average of 20° during the main portion of the climb. At around 1350 m AGL, however, the attitude required to maintain ground course increased to nearer 30° from level, dramatically reducing the available thrust to maintain the desired climb rate. Upon inspection of the motor speeds in the bottom of the figure, it could be reasonably assumed that the flight controller has saturated the motor outputs within the requirements for stability, i.e., the motors are spinning at their maximum speeds. Throughout these high winds, the UAV was still able to maintain an ascent, albeit at a reduced 1 ms^{-1}.

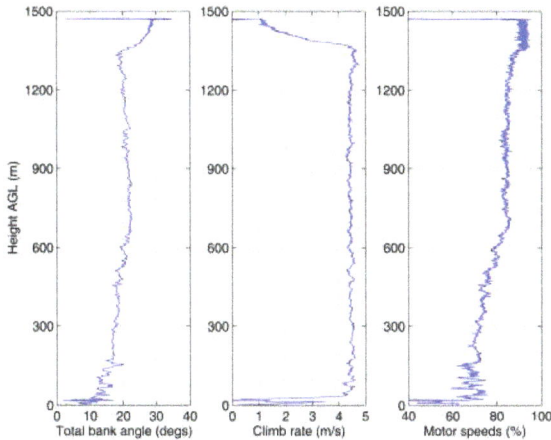

Figure 14. Sample flight during high wind speeds above the TWI.

8. Payload

8.1. Meteorological Sensor Assessment

The fast-response temperature and humidity sensors generally performed well during the field campaigns, and we assess their performance in terms of Targets 3 and 4. However, the normally robust capacitance RH sensors were adversely affected by volcanic wind-blown dust during the first campaign. For the second campaign, an improved shielding cap (with sufficient holes for rapid air ingress, but not allowing large dust particles to enter) was designed. The shield was constructed from white correx with an extended circular plate over a perforated tube. There is no evidence, e.g., from a discrepancy between the upward and downward legs, to suggest the shield had a significant effect on the time response of the temperature and humidity sensor system. In addition, the high wind speeds encountered during most flights aspirate the sensor to make solar heating effects unlikely [43].

Figure 15 shows a typical flight profile of temperature (red) and humidity (light blue) and also the pitch (green) and roll (dark blue) angles during a flight made on 14 September 2014. A kestrel 4500 weather station under manufacturer's calibration was operated at the measurement site during each flight day, and these baseline measurements are shown for reference and comparison with the onboard sensors. Pitch and roll angles are included in 15 to enable the reader to identify where the samples are being taken; which in this case corresponds to the two altitudes where there are rapid variations in roll and pitch angles. In this example, the base of the TWI, where temperature begins to rise with altitude, is apparent at 1670 m ASL and marks the top of the turbulently-mixed atmospheric boundary layer below. The inversion zone acts as a cap on the upward motions of the boundary layer. Figure 15 shows the temperature increasing from 12 °C–17 °C over this transition to the unmixed free-troposphere above, and correspondingly, humidity drops from near saturation in the boundary layer to 30 percent at at 1880 m ASL, the top of the TWI. This thickness of 210 m for the inversion (entrainment) zone is at the larger end of inversion thickness observations during both campaigns, with 50 m–70 m being more typical.

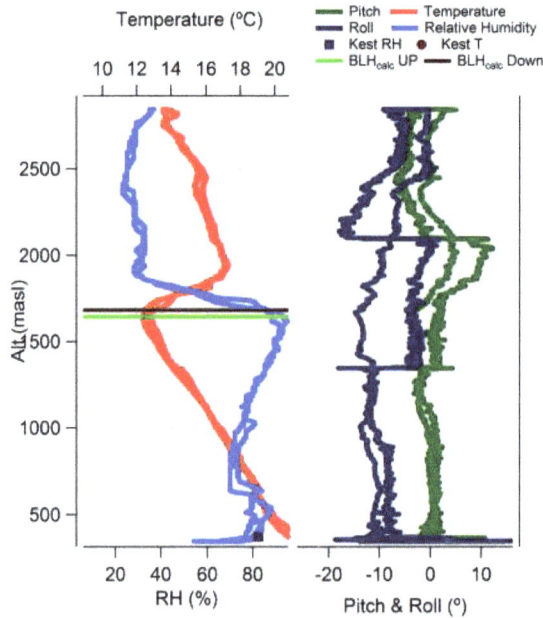

Figure 15. Meteorological data: temperature (red), relative humidity (light blue), pitch (green) and roll (dark blue) angles for a sample flight made at 16:54 p.m. on 14 September 2014.

The temperature and humidity traces show both the upward and downward data and clearly demonstrate the suitability of the sensors for this task with a temperature difference of less than a degree and a minimal relative humidity difference for most, some of the larger differences most likely represent real variability, e.g., the proximity to clouds/proto-clouds in the lower boundary layer at about 600 m ASL. Given the narrow inversion layer thickness, this campaign clearly demonstrates the ability of small UAVs to measure small-scale atmospheric features (<200 m) with minimal disturbance. With such a system, there is potential for new insights into such features, which may have implications for numerical weather prediction and global pollutant transport models.

With the data shown in Figure 15 relayed to the ground station in real time, it was possible during the ascent to identify the exact location of the inversion in terms of the lower and upper limits. Based on these, the sample positions were then chosen, allowing the air samples to be taken in the upper air mass and clear of the inversion, within the inversion itself or below it. This capability, in the presence of uncertainty in the meteorological data prior to the flight, provided the required confidence for the selection of the air sample collection points relative to the TWI.

To demonstrate the performance of this system, the inversion height for the up and down legs of flights from September 2014 have been calculated and compared. Sensor lag time will manifest as a systematic difference in calculated boundary layer height when two profiles are compared; up-leg Boundary Layer Height (BLH) will be higher than down legs. This will give an indication of the Boundary Layer Height accuracy based on sensor lag time.

A potential temperature method was used to identify the start of the potential temperature gradient maximum, similar to Hennemuth and Lammert [44]. Methods to calculate the inversion height are varied [45] and are beyond the scope of this paper to resolve; however, using the same algorithm and thresholds to calculate the up- and down-leg boundary layer heights offers directly comparable results in this test. The result of this analysis for 25 flights that crossed the TWI in the September 2014 campaign demonstrates a positive bias; two out of 25 flights analysed demonstrating a negative difference between BLH-up minus BLH-down. The mean vertical difference was 30.7 m with a standard deviation of 21.1 m (maximum 76.5 m, minimum 2.3 m), which is well within the stated aim of targeting 100 m above the TWI.

To assess the influence of propeller interference in the sensor response, we contrast power spectra for two periods, one when the Octocopter was stationary on the surface and one when in a hover at 300 m above ground level. Figure 16 shows the raw and smoothed power spectra for each period a $-5/3$ line is shown for comparison. The propeller rotation rate for the motors and propellers used, Table 1, was between 4050 (50% throttle) and 6250 (100% throttle) rpm, or 67.5–104 Hz. These fundamental frequencies and associated harmonics are not seen in the temperature sensor, which from the spectra in Figure 16 responds up to 2 Hz in our system. There is also no indication that the propellers or the aircraft's movement during hover influence the temperature signal.

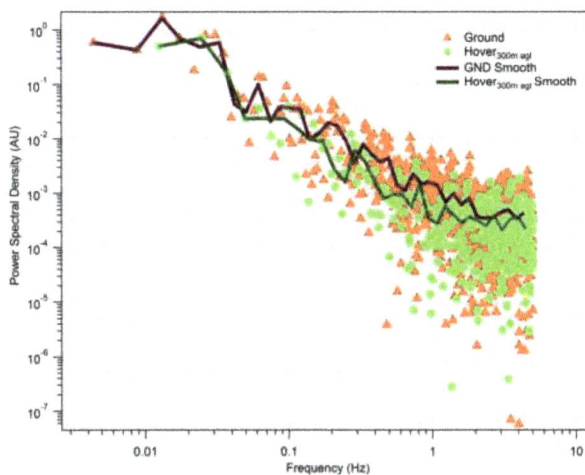

Figure 16. Power spectral density (arbitrary units) for the fast tip temperature sensor when stationary on the ground (motors off) and at a level hover at 300 m. Averaging periods were 226 and 80 s, respectively. Propellers rotate at 67.5–104 Hz, and this fundamental frequency is greater than the approximately 2-Hz response time seen here. Agreement between the spectra indicate no influence of either the propellers or the aircraft movement on the measured temperature.

8.2. Methane Sample Results

Figure 17 demonstrates the good agreement, within expected variability and vertical profile, between the ground bag samples and the continuous values measured by the permanent cavity-ringdown system installed on the ground at the Airhead. It also shows that there is good consistency between the ground bag samples and the bag samples taken below the TWI with the r^2 values for the two plots greater than the critical value showing significant correlation. Details for each campaign are:

Figure 17a, September 2014 campaign: correlation between the ground bag samples and CRDS samples: $r^2 = 0.558$ where critical $r^2 = 0.247$ at 0.05 significance. Correlation between ground samples and samples taken below the TWI: $r^2 = 0.329$ where critical $r^2 = 0.283$ at 0.05 significance.

Figure 17b, July 2015 campaign: correlation between the ground bag samples and CRDS samples: $r^2 = 0.77$ where critical $r^2 = 0.171$ at a 0.05 significance. This correlation is better than for September 2014 because ground samples were collected at both take-off site and co-located with the CRDS. Correlation between ground samples and samples taken below the TWI: $r^2 = 0.493$ where critical $r^2 = 0.305$ at 0.05 significance.

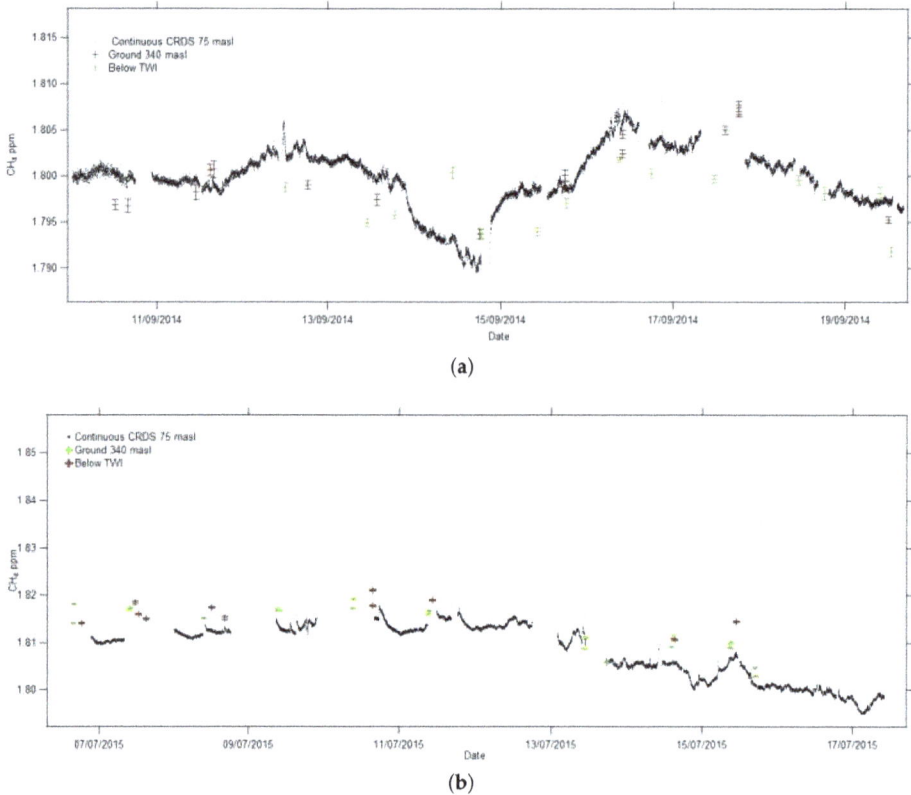

(a)

(b)

Figure 17. Continuous cavity ring-down spectroscopy (CRDS) CH_4 ground level (75 m ASL) compared with bag samples taken from the ground (340 m ASL) and below the TWI. Date markers are 00:00 UTC. (**a**) 2014 campaign; (**b**) 2015 campaign.

Figure 18 shows the CH_4 mole fraction (ppm) variation with both altitude ASL and relative to the boundary layer for both campaigns. Both the September 2014 and July 2015 campaigns show consistently higher CH_4 mole fractions above the TWI, Figure 18, with increments up to 31 ppb. Mixed source emissions (e.g., wetlands, agriculture and biomass burning) from north of the intertropical convergence zone and Africa may be influencing the air masses. Samples from July 2015 have higher CH_4 mole fractions and ranges compared to September 2014, which is likely to reflect the year on year growth and seasonality [46]. Different mixing ratios across the TWI may also be inferred when comparing the continuous ground level monitoring and samples with CH_4 mole fractions above the TWI [46].

Green mountain on Ascension Island is not high enough to enable samples to be taken above the TWI without the use of an air vehicle. With the increased levels of CH_4 shown in Figure 18, there is strong evidence a system is required in order to collect additional high altitude air samples. SUAS offers the potential for a low cost, repeatable and flexible sample system, which in the longer term could be used routinely by a non-specialist for air sample collections of this type.

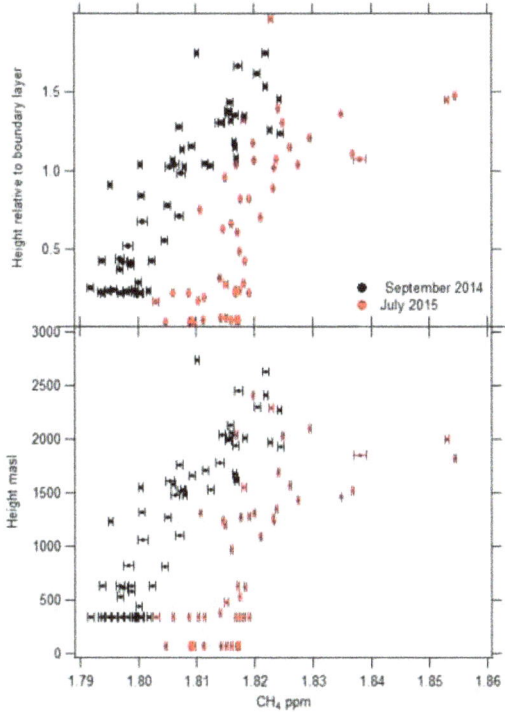

Figure 18. CH_4 mole fraction (ppm) with altitude during the two campaigns (bottom). CH_4 mole fraction normalised to the boundary layer height for the two campaigns [46].

9. Conclusions and Future Work

A multirotor has been successfully flown on Ascension Island up to 2500 m above ground level collecting air samples for analysis over the course of two campaigns in September 2014 and July 2015. The system was shown to operate reliably, amassing over one hundred flights without incidence. This paper has highlighted key aspects of the system from the vehicle specification through to flight performance data and lessons learned. Recommendations have been made for using the system to identify wind conditions at the edge of the vehicle's flight envelope, as well as a flight profile that can alleviate issues encountered with still air.

Throughout the flights, the temperature and relative humidity were measured and transmitted to the ground station making it possible to identify the location of the trade wind inversion and the different air masses during the ascent. This real-time measurement of the atmospheric profile is a significant capability of the system and enabled the atmospheric scientists to accurately target key parts of the profile during the flights in order to collect the samples required.

Using SUAS, it has been shown that in principle greenhouse gas measurement on Ascension can access both air at ground level from a very wide swathe of the southern oceans and also sample air from above the trade wind inversion, thereby addressing emissions from a significant part of the global tropical land masses. These air samples can be used for the purposes of identifying methane mole fractions and isotopic composition.

Whilst the campaigns in September 2014 and July 2015 required a highly specialised team of engineers and scientists for successful operations, it is expected that a fully-automatic atmospheric sampling SUAS could be developed for long-term sampling requirements. This could also include the development of more capable vehicles for sampling in higher wind conditions and in more challenging environments.

One of the key requirements for the campaigns to be successful was to be able to operate the platforms Beyond Line of Sight (BLOS). The approach taken in this work was to manage the question of BLOS operations through the operational strategy, procedures and flight within restricted airspace. BLOS operations in unrestricted airspace would significantly expand the potential applications of SUAS, including air sampling; however, this integration will take time and will require significant research and development effort into platform reliability, autonomous operations, sense and avoid, as well as the legal framework. As each of these is addressed, the range of applications for which SUAS are applicable will increase, including air sampling, together with the range of airspace that is open for operations.

Acknowledgments: This work is supported by the Natural Environment Research Council Grant NE/K006045/1. We would also like to say thank you for the huge amount of support we gained from the RAF (Royal Air Force) and USAF (United States Air Force) base commanders, as well as the Ascension Police, Government, CAA (Civil Aviation Authority) , ASSI (Air Safety Support International) , Met Office, Ascension Conservation and the people of Ascension Island.

Author Contributions: Colin Greatwood, Thomas S. Richardson, Jim Freer, Rick M. Thomas and A. Rob MacKenzie designed, built and operated the SUAS platforms and sensors in Ascension. Rebecca Brownlow, David Lowry and Rebecca E. Fisher devised the sampling strategy and analysed the air samples. Euan G. Nisbet created and led the Investigation of the Southern Methane Anomaly project.

Conflicts of Interest: The authors declare no conflict of interest.

Abbreviations

The following abbreviations are used in this manuscript:

AGL	Above Ground Level
ASL	Above Sea Level
BBB	Beagle Bone Black
BVLOS	Beyond Visual Line of Sight
ESC	Electronic Speed Controller
RH	Relative Humidity
RTH	Return to Home
SUAS	Small Unmanned Air System
TWI	Trade Wind Inversion
UAV	Unmanned Air Vehicle

References

1. Nisbet, E.G.; Dlugokencky, E.J.; Bousquet, P. Methane on the Rise-Again. *Science* **2014**, *343*, 493–495.
2. Dlugokencky, E.J.; Nisbet, E.G.; Fisher, R.; Lowry, D. Global atmospheric methane: Budget, changes and dangers. *Philos. Trans. R. Soc. A Math. Phys. Eng. Sci.* **2011**, *369*, 2058–2072.
3. Nisbet, E.G.; Dlugokencky, E.J.; Manning, M.R.; Lowry, D.; Fisher, R.E.; France, J.L.; Michel, S.E.; Miller, J.B.; White, J.W.C.; Vaughn, B.; et al. Rising Atmospheric Methane: 2007–2014 Growth and Isotopic Shift. *Glob. Biogeochem. Cycles* **2016**, *30*, 1356–1370.
4. Bhardwaj, A.; Sam, L.; Akanksha; Martin-Torres, F.J.; Kumar, R. UAVs as remote sensing platform in glaciology: Present applications and future prospects. *Remote Sens. Environ.* **2016**, *175*, 196–204.
5. Pajares, G. Overview and Current Status of Remote Sensing Applications Based on Unmanned Aerial Vehicles (UAVs). *Photogramm. Eng. Remote Sens.* **2015**, *81*, 281–329.
6. Klemas, V.V. Coastal and Environmental Remote Sensing from Unmanned Aerial Vehicles: An Overview. *J. Coast. Res.* **2015**, *31*, 1260–1267.
7. Detweiler, C.; Ore, J.P.; Anthony, D.; Elbaum, S.; Burgin, A.; Lorenz, A. Bringing Unmanned Aerial Systems Closer to the Environment. *Environ. Pract.* **2015**, *17*, 188–200.

8. Vivoni, E.R.; Rango, A.; Anderson, C.A.; Pierini, N.A.; Schreiner-McGraw, A.P.; Saripalli, S.; Laliberte, A.S. Ecohydrology with unmanned aerial vehicles. *Ecosphere* **2014**, *5*, 1–14.

9. Fladeland, M.; Sumich, M.; Lobitz, B.; Kolyer, R.; Herlth, D.; Berthold, R.; McKinnon, D.; Monforton, L.; Brass, J.; Bland, G. The NASA SIERRA science demonstration programme and the role of small-medium unmanned aircraft for earth science investigations. *Geocarto Int.* **2011**, *26*, 157–163.

10. Tamminga, A.D.; Eaton, B.C.; Hugenholtz, C.H. UAS-based remote sensing of fluvial change following an extreme flood event. *Earth Surf. Process. Landf.* **2015**, *40*, 1464–1476.

11. Immerzeel, W.W.; Kraaijenbrink, P.D.A.; Shea, J.M.; Shrestha, A.B.; Pellicciotti, F.; Bierkens, M.F.P.; de Jong, S.M. High-resolution monitoring of Himalayan glacier dynamics using unmanned aerial vehicles. *Remote Sens. Environ.* **2014**, *150*, 93–103.

12. Zweig, C.L.; Burgess, M.A.; Percival, H.F.; Kitchens, W.M. Use of Unmanned Aircraft Systems to Delineate Fine-Scale Wetland Vegetation Communities. *Wetlands* **2015**, *35*, 303–309.

13. Stocker, C.; Eltner, A.; Karrasch, P. Measuring gullies by synergetic application of UAV and close range photogrammetry—A case study from Andalusia, Spain. *Catena* **2015**, *132*, 1–11.

14. Nagai, M.; Chen, T.; Shibasaki, R.; Kumagai, H.; Ahmed, A. UAV-Borne 3-D Mapping System by Multisensor Integration. *IEEE Trans. Geosci. Remote Sens.* **2009**, *47*, 701–708.

15. Peng, Z.R.; Wang, D.S.; Wang, Z.Y.; Gao, Y.; Lu, S.J. A study of vertical distribution patterns of PM2.5 concentrations based on ambient monitoring with unmanned aerial vehicles: A case in Hangzhou, China. *Atmos. Environ.* **2015**, *123*, 357–369.

16. Diaz, J.A.; Pieri, D.; Wright, K.; Sorensen, P.; Kline-Shoder, R.; Arkin, C.R.; Fladeland, M.; Bland, G.; Buongiorno, M.F.; Ramirez, C.; et al. Unmanned Aerial Mass Spectrometer Systems for in-Situ Volcanic Plume Analysis. *J. Am. Soc. Mass Spectrom.* **2015**, *26*, 292–304.

17. Alvarado, M.; Gonzalez, F.; Fletcher, A.; Doshi, A. Towards the Development of a Low Cost Airborne Sensing System to Monitor Dust Particles after Blasting at Open-Pit Mine Sites. *Sensors* **2015**, *15*, 19667–19687.

18. Martin, S.; Beyrich, F.; Bange, J. Observing Entrainment Processes Using a Small Unmanned Aerial Vehicle: A Feasibility Study. *Bound.-Layer Meteorol.* **2014**, *150*, 449–467.

19. Cassano, J.J. Observations of atmospheric boundary layer temperature profiles with a small unmanned aerial vehicle. *Antarct. Sci.* **2014**, *26*, 205–213.

20. Bates, T.S.; Quinn, P.K.; Johnson, J.E.; Corless, A.; Brechtel, F.J.; Stalin, S.E.; Meinig, C.; Burkhart, J.F. Measurements of atmospheric aerosol vertical distributions above Svalbard, Norway, using unmanned aerial systems (UAS). *Atmos. Meas. Tech.* **2013**, *6*, 2115–2120.

21. Reuder, J.; Jonassen, M.O.; Olafsson, H. The Small Unmanned Meteorological Observer SUMO: Recent Developments and Applications of a Micro-UAS for Atmospheric Boundary Layer Research. *Acta Geophys.* **2012**, *60*, 1454–1473.

22. Karion, A.; Sweeney, C.; Tans, P.; Newberger, T. AirCore: An Innovative Atmospheric Sampling System. *J. Atmos. Ocean. Technol.* **2010**, *27*, 1839–1853.

23. Villa, T.; Gonzalez, F.; Miljievic, B.; Ristovski, Z.; Morawska, L. An Overview of Small Unmanned Aerial Vehicles for Air Quality Measurements: Present Applications and Future Prospectives. *Sensors* **2016**, *16*, 1072.

24. Brady, J.; Stokes, M.; Bonnardel, J.; Bertram, T. Characterization of a Quadrotor Unmanned Aircraft System for Aerosol-Particle-Concentration Measurements. *Environ. Sci. Technol.* **2016**, *50*, 1376–1383.

25. Roldan, J.; Joossen, G.; Sanz, D.; Cerro, J.; Barrientos, A. Mini-UAV Based Sensory System for Measuring Environmental Variables in Greenhouses. *Sensors* **2015**, *15*, 3334–3350.

26. Detert, M.; Weitbrecht, V. A low-cost airborne velocimetry system: Proof of concept. *J. Hydraul. Res.* **2015**, *53*, 532–539.

27. Hill, S.L.; Clemens, P. Miniaturization of High-Spectral-Spatial Resolution Hyperspectral Imagers on Unmanned Aerial Systems. *Proc. SPIE* **2015**, *9482*, doi:10.1117/12.2193706.

28. Wildmann, N.; Mauz, M.; Bange, J. Two fast temperature sensors for probing of the atmospheric boundary layer using small remotely piloted aircraft (RPA). *Atmos. Meas. Tech.* **2013**, *6*, 2101–2113.

29. Thornberry, T.D.; Rollins, A.W.; Gao, R.S.; Watts, L.A.; Ciciora, S.J.; McLaughlin, R.J.; Fahey, D.W. A two-channel, tunable diode laser-based hygrometer for measurement of water vapor and cirrus cloud ice water content in the upper troposphere and lower stratosphere. *Atmos. Meas. Tech.* **2015**, *8*, 211–224.

30. Corrigan, C.E.; Roberts, G.C.; Ramana, M.V.; Kim, D.; Ramanathan, V. Capturing vertical profiles of aerosols and black carbon over the Indian Ocean using autonomous unmanned aerial vehicles. *Atmos. Chem. Phys.* **2008**, *8*, 737–747.

31. Ramana, M.V.; Ramanathan, V.; Kim, D.; Roberts, G.C.; Corrigan, C.E. Albedo, atmospheric solar absorption and heating rate measurements with stacked UAVs. *Q. J. R. Meteorol. Soc.* **2007**, *133*, 1913–1931.

32. De Boer, G.; Palo, S.; Argrow, B.; LoDolce, G.; Mack, J.; Gao, R.S.; Telg, H.; Trussel, C.; Fromm, J.; Long, C.N.; et al. The Pilatus unmanned aircraft system for lower atmospheric research. *Atmos. Meas. Tech.* **2016**, *9*, 1845–1857.

33. Freitag, S.; Clarke, A.D.; Howell, S.G.; Kapustin, V.N.; Campos, T.; Brekhovskikh, V.L.; Zhou, J. Combining airborne gas and aerosol measurements with HYSPLIT: A visualization tool for simultaneous evaluation of air mass history and back trajectory consistency. *Atmos. Meas. Tech.* **2014**, *7*, 107–128.

34. Chambers, S.D.; Zahorowski, W.; Williams, A.G.; Crawford, J.; Griffiths, A.D. Identifying tropospheric baseline air masses at Mauna Loa Observatory between 2004 and 2010 using Radon-222 and back trajectories. *J. Geophy. Res. Atmos.* **2013**, *118*, 992–1004.

35. Saunois, M.; Bousquet, P.; Poulter, B.; Peregon, A.; Ciais, P.; Canadell, J.G.; Dlugokencky, E.J.; Etiope, G.; Bastviken, D.; Houweling, S.; et al. The Global Methane Budget: 2000–2012. *Earth Syst. Sci. Data Discuss.* **2016**, *2016*, 1–79.

36. Watai, T.; Machida, T.; Ishizaki, N.; Inoue, G. A Lightweight Observation System for Atmospheric Carbon Dioxide Concentration Using a Small Unmanned Aerial Vehicle. *J. Atmos. Ocean. Technol.* **2006**, *23*, 700–710.

37. Katsaros, K.B.; Decosmo, J.; Lind, R.J.; Anderson, R.J.; Smith, S.D.; Kraan, R.; Oost, W.; Uhlig, K.; Mestayer, P.G.; Larsen, S.E.; et al. Measurements of Humidity and Temperature in the Marine Environment during the HEXOS Main Experiment. *Atmos. Ocean. Technol.* **1994**, *11*, 964.

38. Wildmann, N.; Kaufmann, F.; Bange, J. An inverse-modelling approach for frequency response correction of capacitive humidity sensors in ABL research with small remotely piloted aircraft (RPA). *Atmos. Meas. Tech.* **2014**, *7*, 3059–3069.

39. Chen, R. *A Survey of Nonuniform Inflow Models for Rotorcraft Flight Dynamics and Control Applications*; National Aeronautics and Space Administration: Washington, DC, USA, 1989.

40. Lowry, D.; Fisher, R.; France, J.; Lanoiselle, M.; Nisbet, E.; Brunke, E.; Dlugokencky, E.; Brough, N.; Jones, A. Continuous monitoring of greenhouse gases in the South Atlantic And Southern Ocean: Contributions from the equianos network. In Proceedings of the 17th WMO/IAEA Meeting of Experts on Carbon Dioxide, Other Greenhouse Gases and related Tracers Measurement Techniques, Beijing, China, 10–13 June 2013; Volume 213, pp. 109–112.

41. Greatwood, C.; Richardson, T.; Freer, J.; Thomas, R.; Brownlow, R.; Lowry, D.; Fisher, R.E.; Nisbet, E. Automatic Path Generation for Multirotor Descents Through Varying Air Masses above Ascension Island. In Proceedings of the AIAA SciTech Forum, San Diego, CA, USA, 4–8 January 2016.

42. Padfield, G.D. *Helicopter Flight Dynamics*; John Wiley & Sons: Chichester, UK, 2008.

43. Anderson, S.; Baumgartner, M. Radiative Heating Errors in Naturally Ventilated Air Temperature Measurements Made from Buoys. *J. Atmos. Ocean. Technol.* **1998**, *15*, 157–173.

44. Hennemuth, B.; Lammert, A. Determination of the Atmospheric Boundary Layer Height from Radiosonde and Lidar Backscatter. *Bound.-Layer Meteorol.* **2006**, *120*, 181–200.

45. Dai, C.; Wang, Q.; Kalogiros, J.A.; Lenschow, D.H.; Gao, Z.; Zhou, M. Determining Boundary-Layer Height from Aircraft Measurements. *Bound.-Layer Meteorol.* **2014**, *152*, 277–302.

46. Brownlow, R.; Lowry, D.; Thomas, R.M.; Fisher, R.E.; France, J.L.; Cain, M.; Richardson, T.S.; Greatwood, C.; Freer, J.; Pyle, J.A.; et al. Methane mole fraction and $\delta^{13}C$ above and below the Trade Wind Inversion at Ascension Island in air sampled by aerial robotics. *Geophys. Res. Lett.* **2016**, *43*, 11893–11902.

sensors

MDPI

Article

Development of Cloud-Based UAV Monitoring and Management System

Mason Itkin [1], Mihui Kim [2] and Younghee Park [1],*

[1]   Computer Engineering Department, San Jose State University, One Washington Square,
      San Jose, CA 95192, USA; mason.itkin@sjsu.edu
[2]   Department of Computer Science & Engineering, Computer System Institute,
      Hankyong National University, 327 Jungang-ro, Anseong-si, Gyeonggi-do 456-749, Korea;
      mhkim@hknu.ac.kr
*   Correspondence: younghee.park@sjsu.edu; Tel.: +1-408-924-7854

Academic Editors: Felipe Gonzalez Toro and Antonios Tsourdos
Received: 4 September 2016; Accepted: 8 November 2016; Published: 15 November 2016

Abstract: Unmanned aerial vehicles (UAVs) are an emerging technology with the potential to revolutionize commercial industries and the public domain outside of the military. UAVs would be able to speed up rescue and recovery operations from natural disasters and can be used for autonomous delivery systems (e.g., Amazon Prime Air). An increase in the number of active UAV systems in dense urban areas is attributed to an influx of UAV hobbyists and commercial multi-UAV systems. As airspace for UAV flight becomes more limited, it is important to monitor and manage many UAV systems using modern collision avoidance techniques. In this paper, we propose a cloud-based web application that provides real-time flight monitoring and management for UAVs. For each connected UAV, detailed UAV sensor readings from the accelerometer, GPS sensor, ultrasonic sensor and visual position cameras are provided along with status reports from the smaller internal components of UAVs (i.e., motor and battery). The dynamic map overlay visualizes active flight paths and current UAV locations, allowing the user to monitor all aircrafts easily. Our system detects and prevents potential collisions by automatically adjusting UAV flight paths and then alerting users to the change. We develop our proposed system and demonstrate its feasibility and performances through simulation.

Keywords: unmanned aerial vehicles (UAVs); monitoring and management; collision avoidance; cloud-based application

1. Introduction

Unmanned aerial vehicles (UAVs) are an emerging technology with strong implications for improving many common public and private processes. Public departments (i.e., police, public safety and transportation management) are beginning to use UAVs to deliver timely disaster warnings and improve the efficiency of rescue and recovery operations when a telecommunication infrastructure in a region is damaged or otherwise unavailable [1]. Additionally, UAVs can be used as a tool for convenience, allowing autonomous delivery systems (e.g., Amazon Prime Air [2]) to provide goods quickly to people in geographically-isolated areas. UAVs, often referred to as drones, are a type of aircraft that can fly without the need for a physical pilot onboard. UAVs are typically controlled in one of two ways: autonomously or remotely by a trained operator. The UAVs that we will be referring to are of the autonomous variety. To fly autonomously without the need for expensive visual sensors, UAVs must be constantly connected to a controlling entity, either directly or through a representative UAV, to monitor current flight status and to set an appropriate flight path.

As the density of UAVs in large urban centers increases, it becomes increasingly necessary to have a control and management system to provide an intelligent collision avoidance system [1,3]. Eventually, as UAVs are more frequently used in modern airspace, flight monitoring and collision avoidance systems will face issues of scale. Large commercial and government entities that control thousands of UAVs would want systems that can monitor all of the air traffic in real time with as little infrastructure as possible. To enable safe autonomous flight, collision avoidance must be included at the very core of monitoring solutions. These monitoring systems must be enabled to act without user approval to ensure the best possible response time and, thus, a higher probability of uninterrupted flight.

Mobility management, control and monitoring methodologies for multiple UAVs have been well studied and documented [1,3], but the implementation of such systems coupled with a comprehensive monitoring solution has not. Some UAV monitoring and control systems focus on wireless radio as the most crucial aspect of the control unit [3,4]. Other research groups have concentrated on UAV collision avoidance methods, developing complex algorithms for UAVs to avoid each other, obstacles or intruders [5–11]. Moreover, remote measurement and control technologies for UAVs present certain problems [12,13]. Existing solutions for UAV monitoring are difficult to deploy and are often extremely resource intensive, limiting their scalability. We aim to provide a lightweight UAV monitoring solution that can leverage the power of highly distributed cloud computing platforms to provide a UAV monitoring and collision avoidance system suitable for thousands of concurrently connected UAVs.

To pursue our goal in this paper, we develop a scalable cloud-based control and management system for UAVs, called UAV Flight Tracker. UAV Flight Tracker has a client and server model. The client allows the user to control added UAVs, receive real-time sensor updates, monitor the visualized UAVs and receive priority alerts in collision detection. The server provides sensor and collision information to the client, implements collision detection algorithms, manages user profiles and updates UAV control information. The contributions of this work are as follows:

- Design of real-time flight monitoring and visualization of multiple UAVs
- Collision detection and avoidance with automatic adjustment of UAV flight paths
- Scalability support via cloud systems
- Implementation and performance simulation

The remainder of this paper is organized as follows: Section 2 is the problem statement. Section 3 introduces existing work related to UAV monitoring and management. Section 4 proposes our UAV monitoring and management system. Section 5 explains our system in detail. Section 6 demonstrates simulation results for the operation and performance of the system. Section 7 provides concluding remarks and future work.

2. Problem Statement

Issues regarding various protocols for long- and short-range wireless communication between UAV systems and UAV control entities have been well researched, but solutions for a scalable monitoring system are lacking. Modern systems to support UAV communication over existing 3G and 4G cellular networks pave the way for commercial entities to deploy large fleets of autonomous UAVs in dense urban centers [12]. However, not many research projects offer a solution for a scalable monitoring and collision avoidance system that can handle a large number of concurrently connected UAVs. In addition to scalability, the ease of deployment is an important factor for modern UAV monitoring systems. As more UAVs are utilized, the number of monitoring entities will also increase, making quick, lightweight and flexible deployment an essential feature of monitoring systems.

The purpose of this paper is to address the need for a scalable UAV monitoring and collision avoidance system by providing a web application that can leverage the power of highly distributed, cloud-based systems. To properly address the problem statement, the UAV control system must provide: (1) a detailed, visually-driven graphic user interface that provides an interactive map interface and a detailed view of individual sensor data; (2) a scalable framework that can accommodate multiple concurrently connected UAV systems and with frequent sensor data updates; (3) a collision avoidance system that can send UAV flight commands and react quickly to potential collision situations; and (4) a well-designed framework that can be deployed, cleared and redeployed easily on cloud-based computing services.

3. Related Work

Though the concept of UAVs is promising as an emerging technology, the field is relatively new to industry and is much less developed than other military applications. There are various issues to address in order to realize the effective, stable and reliable use of UAVs, i.e., network topology, routing, seamless handover, energy efficiency and management [1]. As the number of potential users of UAVs increases, it is especially important to provide mobility management, control and monitoring including collision avoidance of intruders or obstacles [8,10].

Some UAV monitoring and control systems focus on wireless radio as the most crucial aspect of the control unit. In a centrally-designed system, the controller radio facilitates communication between every UAV in the system; without the controller radio, the system could not function. Lam, T.M.; et al. [3] consider a radio delay to control UAVs. Tele-operation of a UAV may involve time delay due to signal transmission. This would result in poor operator performance and control difficulties. Thus, the authors describe a theoretical analysis using wave variables with a collision avoidance system for UAV tele-operation with time delay. Kim, H.N.; et al. unburden the controller radio by allowing every UAV in a particular swarm to connect to and manage one another [4]. To manage this type of peer-to-peer control over several UAVs, a comprehensive methodology is needed to replace nodes seamlessly in the swarm. This is accomplished by creating a second revised routing table sent to the swarm and saving it in the background until the new UAV is connected. Once the new UAV is connected, the new routing table is used by the swarm, and the process to optimize this table begins. The obvious advantage to the system described in [4] is that there is no central communication point; thus, an outage of the ground control system or controller radios would not cause a system-wide failure. However, restricting UAV control to the swarm means that all UAVs must start within a specific range of one another, making routing complicated. Additionally, distance constraints on UAVs limit the full utilization of the system, as it cannot track UAVs that are flying long distances in opposite directions. Further, scaling of this system can be expensive and ultimately impractical with the routing process becoming more and more complex as the number of concurrently connected UAVs grows. Eventually, the routing process could exhaust the computational power of the UAVs, and all UAVs in the system would need to be upgraded. It is also important to note that in an ideal system, only small computations are performed on the UAV. This is preferred as onboard computations quickly increase UAV hardware costs and flight time by draining battery power.

Many research groups attempting to develop innovative ways to control UAVs have concentrated on collision avoidance, anticipating a large number of UAVs, obstacles or intruders. The paper by Kuriki, Y.; et al. [5] presents a cooperative control strategy for collision avoidance in a multi-UAV system by using decentralized predictive and consensus-based control. Liu, W.; et al. develop a novel path planner to control multiple UAVs synchronously based on distributed path generation and a bi-level programming technique. Patel, A.; et al. propose an autonomous collision avoidance system using an acceleration-matching algorithm to increase autonomy in connected UAVs [7]. The paper by Ling, L.; et al. utilizes a particle filter to estimate the target state accurately for collision avoidance [8]. The paper by Alejo, D.; et al. [9] presents a collision avoidance method for multiple UAVs and other non-cooperative aircraft based on velocity planning and trajectory prediction of a particle filter. Yoo, C.S.; et al. implement a collision detection and avoidance algorithm on a digital flight control computer [10]. Sharon Lozano at NASA has implemented a cloud-based Unmanned Aircraft Traffic Management (UTM) system that provides a way for civilian pilots to reserve airspace [14]. This system maintains a database of reserved and active flights, providing information to pilots about adverse weather conditions and restricted airspace. The UTM project consists of four technical capability phases, ultimately enabling the management of UAVs in high-density urban areas with large-scale contingency mitigation. The paper by Zhu, L.; et al. [12] assumes UAVs may be operating in joint missions with manned aircraft, e.g., helicopters, and use the concept of automatic dependent surveillance broadcasting to collect aircraft data used in collision avoidance systems. Based on flight maneuvering, the proposed detection algorithm creates a sector range to cover possible flight direction changes of UAVs and helicopters.

Moreover, remote measurement and control technology for UAVs are vital, leading Ling, L.; et al. [8] to build an onboard embedded computing platform that realizes the remote measurement and control of the UAV. Their solution utilizes a UAV position compensation method based on network time delay prediction. This remote measurement platform increases the accuracy of control and management, but fails to consider collision avoidance and provides no mention of performance when connecting a large number of UAVs. The paper by Roldán, J.J.; et al. [13] focuses on using a central control system to move several mini-UAVs around a greenhouse, gathering sensor information. The employed system has only a central service that performs all of the sensor computations, manages flight paths and controls external actuators for the greenhouse operation. The advantage of this system is that a centralized control point controls all UAVs and performs all resource intensive computations. This results in each UAV being inexpensive, but at greater risk, as a failure of the central control point would cause the entire system to fail. Additionally, relying on a central control radio either limits the UAVs to a relatively short distance from the controller or drives up the cost of the radio.

While those solutions explained earlier [3–13] may satisfy some of the use cases presented, they do not fully address the need for a truly scalable UAV control system. The aforementioned solutions present no way to deploy the monitoring system easily, nor do they mention the use of a highly distributed system, a key component of UAV Flight Tracker. Most importantly, the related papers did not thoroughly test their UAV control systems to prove that their solutions are reliable on a large scale.

4. Architecture of the UAV Monitoring and Management System (UAV Flight Tracker)

This section presents our cloud-based UAV monitoring and management system, called UAV Flight Tracker. The novelty of our system is to allow monitoring of real-time UAV sensor data, a visual display for current flight paths for multiple concurrent UAV flights and support for automatic collision avoidance. The UAV Flight Tracker architecture consists of three entities: the client, server and UAV. Normal operation of our system utilizes a single cloud-hosted server with a varying number of clients and UAVs.

The client includes UAV monitoring and UAV management modules that allow the user to control added UAVs and receive real-time sensor updates, as well as user and server interfaces with which to communicate, as shown in Figure 1.

The UAV monitoring module allows users to receive real-time sensor information and visually monitor their UAVs using a 2D mapping interface. Data from the server communication interface of the client system is passed to the UAV monitoring module for processing before being sent to the map and displayed on the user interface. The map uses GPS calculations, map provisioning and other visualization services to display the current UAV location, UAV destination and a direct flight path.

The UAV management module allows the user to add any number of UAVs to the server and send live path data, utilizing three sub-modules: UAV management, UAV control and collision management. The sub-module for UAV management allows the application to associate UAVs with users and grant access to monitoring and control features. The UAV control sub-module is responsible for sending the live path data to the server, where it can be deployed to UAVs. The collision management sub-module provides a way for the user to set preferences for automatic path redirection and receive priority alerts from the collision detection algorithm.

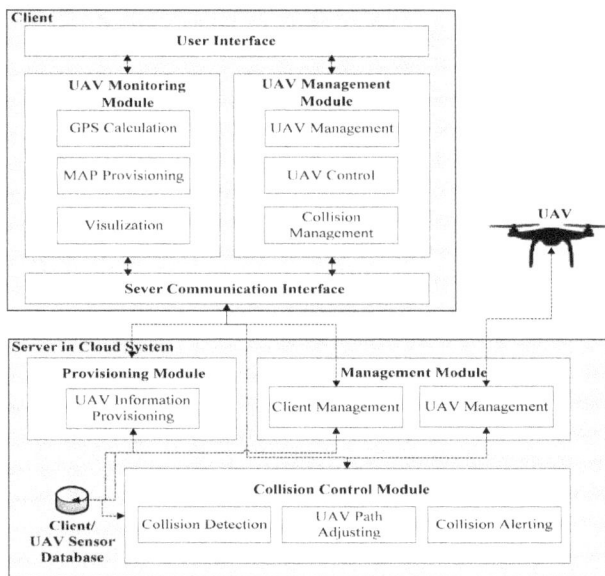

Figure 1. System Architecture of UAV Flight Tracker.

The server of UAV Flight Tracker provides a provisioning module, a management module, a collision control module and a database. Together, these modules serve sensor and collision information to the client system, implement the collision detection algorithm, manage user profiles and update UAV control information.

The server provisioning module facilitates communication of sensor data between the databases of server and client. The UAV information provisioning sub-module defines application program interface (API) routes to the database. This allows simple requests to perform create, read, update and delete (CRUD) operations on sensor datasets in the database.

The management module keeps track of the clients and associated UAVs that are currently being utilized in the application. The client management sub-module controls the association between a specific client instance and flight control permissions for the connected UAV. The UAV management sub-module tracks the connected UAVs and sends authorized commands in real-time to deployed UAVs.

The collision control module implements the collision control algorithm, performs necessary UAV path adjustments and alerts the client to potential collisions and UAV path changes. The collision detection sub-module implements the collision detection algorithm by reading real-time sensor information from the database, checking for collisions and then preparing data to make the needed UAV path adjustments. The UAV path-adjusting sub-module receives collision information from the collision detection sub-module, performs the path adjustments and sends the updated path to the management module where it can be deployed to the UAV. The collision alerting sub-module reports detected collisions and adjusted path data to the collision management module of the client, where it can then be displayed to the user.

To design the collision detection algorithm on the server, we assume that each UAV has a safe zone, a redirection zone and an emergency zone, as shown in Figures 2 and 3. Of these three zones, the safe zone has the biggest radius around each UAV, e.g., 1000 m, and is used to designate a completely safe flying distance for adjacent UAVs. All of the connected UAVs that are flying within the safe zone of another UAV are flagged as being adjacent and are checked at frequent intervals. The redirection zone has a middle radius around each UAV, e.g., 500 m, meant to indicate the distance at which adjacent flying UAVs should be redirected. UAV redirection is an adjustment of the connected UAV heading as needed to redirect the UAV away from the UAV or UAVs whose redirection or emergency zone it has entered. This new temporary heading will be held until the offending UAV moves out of the redirection zone. UAV redirection occurs with the UAV moving at the slowest speed. If all UAVs are traveling at the same speed, then one of the violating UAVs can be selected with a specific criterion (e.g., randomly, lower ID or more energy) to have its course redirected. The emergency zone has the smallest radius around each UAV, e.g., 300 m, where it is determined that no UAV should be flying. When any UAV is found to be within the emergency zone of another UAV, both aircrafts will be immediately stopped mid-flight before being redirected, thus preventing a potential collision. The radii of the three zones can be altered according to the average speed, wingspan and movement accuracy of the monitored UAVs. Algorithm 1 shows the pseudo code for the collision avoidance.

Algorithm 1: Collision Avoidance

1.   void CollisionAvoidance(UAV[i], UAV[j]) {
2.   // **Input:** UAV[i], UAV[j]
3.   // **Case 1:** Monitor closely for changes in flight path
4.   if UAV[i] is within UAV[j].safeZone{
5.   UAV[i].setAdjacent(UAV[j]);
6.
7.   // **Case 2:** UAV in emergency zone
8.   if UAV[i] is within UAV[j].emergencyZone {
9.   Stop UAV[i] for 10 s;
10.   RedirectHeading(UAV[i],UAV[j]);
11.   }
12.   // **Case 3:** UAV in redirection zone
13.   else if UAV[i] is within UAV[j].redirectionZone{
14.   RedirectHeading(UAV[i],UAV[j]);
15.   }
16.   }
17.   }

In Case 1 of Figure 2, UAV B is flying within the safe zone of UAV A, causing UAV B to be monitored at a more frequent interval by the collision avoidance module. These more frequent updates ensure that redirection can take place quickly if UAV B moves into the redirection zone of UAV A.

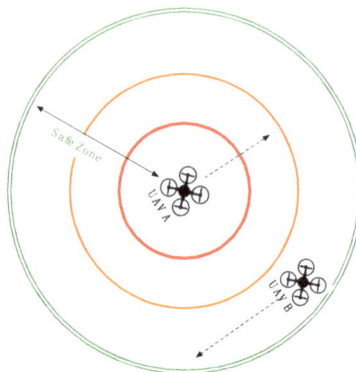

Figure 2. Case 1: closely monitor UAV B in the safe zone.

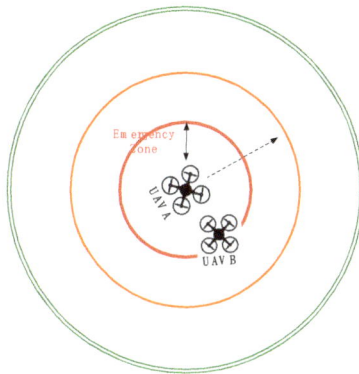

Figure 3. Case 2: stop movement for UAV B in the emergency zone.

In Case 2 of Figure 3, UAV B has entered the emergency zone of UAV A and is prevented from moving for 10 s in order to prevent immediate collision. The stop time can be raised according to the average speed of the UAVs. The collision avoidance module sends a signal to the management module of the server on UAV Flight Tracker in order to stop the movement of UAV B and keep the aircraft hovering in place.

In Case 3 of Figure 4, UAV B has entered the redirection zone of UAV A and has its flight path redirected to avoid any potential collision with UAV A. The collision avoidance module informs the management module to redirect UAV B so that UAV B does not enter the emergency zone of UAV A.

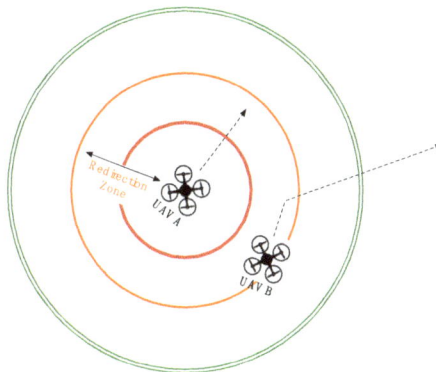

Figure 4. Case 3: change flight path of UAV in the redirection zone.

5. Development of the UAV Flight Tracker

This section details the development of our proposed UAV Flight Tracker by discussing the general web application packages utilized by the system.

5.1. Sketch for Development of UAV Flight Tracker

UAV Flight Tracker is built on the MEAN stack [15] using MongoDB [16], ExpressJS [17], AngularJS [18] and NodeJS [19], as shown in Figure 5. The provisioning module of the server provides HTTP connection listening, routes for the API endpoints and a constant connection with the MongoDB database to read and write JSON documents [20]. The management module of the server manages active client sessions and connected UAVs, controlling communication privileges between the two.

The collision control module implements the collision avoidance algorithm by reading sensor data from the database to determine UAV locations and sending the appropriate commands for UAV redirection (i.e., direction and speed) to the management module. Additionally, collision avoidance reports and UAV redirection information are sent to the provisioning module where it can be presented to the user. NodeJS and the Express framework are especially well suited to a real-time UAV tracking application because they function using non-blocking I/O calls, providing low response times for multiple concurrent connections; this is a key factor that allows UAV Flight Tracker to easily scale up.

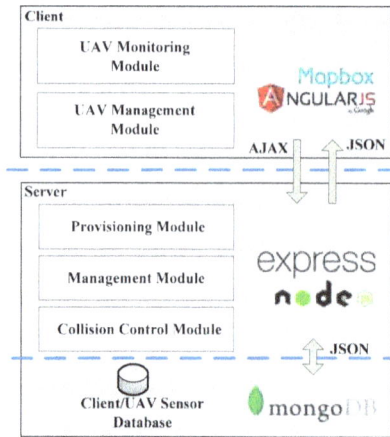

Figure 5. Development design of UAV Flight Tracker.

The client system is implemented as an AngularJS web application that provides real-time UAV sensor monitoring and interactive location mapping. The client's UAV monitoring module uses Angular directives to manipulate the webpage document object model (DOM) [21] to display the sensor data retrieved from the server. Data in the DOM are displayed and formatted on the HTML page or used to draw objects on the map. The client uses Leaflet [22], an interactive JavaScript map library, to plot UAV coordinates, paths and destination markers. A community library for Leaflet Directives [23] is used, as Leaflet does not natively support Angular Directives. The client uses asynchronous JavaScript and XML (AJAX) for HTTP requests. AJAX is utilized as part of the AngularJS client application by providing a way for the client to make asynchronous HTTP requests to the server.

5.2. Implementation of UAV Flight Tracker

The implementation of the client and server of UAV Flight Tracker can be split into three main files. The server is composed of a single JavaScript file (server.js) running on NodeJS, while the client is composed of two files, the main application JavaScript file (app.js), written with AngularJS, and an HTML file (map-view.html).

The server.js file hosts the frontend application, manages the MongoDB database connection and provides API routes for the sensor data. UAV Flight Tracker uses the Express framework to aid in the construction of the NodeJS server by drastically reducing the code complexity of the server. The first task for the server as it first runs is to connect to the MongoDB database. Connection to the database is done with a call to the MongoDB uniform resource identifier (URI), found in the environment file, which will return a database object. After securing communication with the database, the server is initialized and is ready to identify incoming connections over the port specified in the environment file. HTTP GET requests that the '/map' URI return the frontend code (i.e., HTML, CSS and JavaScript), while all requests to the '/fc/sensors' URI are routed to the API endpoint where the specific request (i.e., GET, PUT and POST) can be handled. Any errors the server encounters when looking up a specific URI are caught by a generic error handler and the appropriate error code or a generic HTTP 500 error is returned. All of the routes to the '/fc/sensors' endpoint, including GET, PUT and POST, are all defined in the 'server.js' file as separate functions. In each of these functions, the specific type of HTTP request along with the request data are passed to the function body, where it is used to perform the specific request. In the example of an HTTP GET request, the requested sensor dataset is searched for by unique id in the database, returned to the server and then sent back to the requestor. Similarly, in the example of an HTTP POST request, the data to be written are parsed by the server from the POST request body to the database, and then, a successful status code is sent to the requestor. The Figure 6 shows the flowchart for the operation of the server and client.

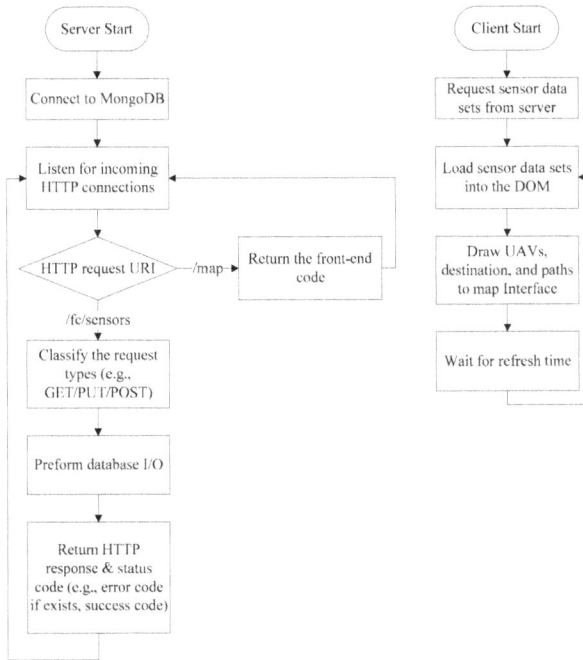

Figure 6. Flowchart for server and client.

On the client-side, the app.js file dictates the behavior of the front-end of the application by using a route provider, a set of services for API requests and a JavaScript controller definition for the main mapping interface. The route provider defines the content template URL, service resolutions and the JavaScript controller for a specific URI of the web application. The content template is the HTML content page that is loaded inside of the main HTML template when a page is selected. By modeling the front end in this way, multiple monitoring pages can be displayed without having to duplicate any HTML code. Service resolutions, in this application, are the sensor datasets pulled from the API that the JavaScript controller needs in order to function properly; the resolutions occur before the JavaScript controller runs. The JavaScript controller, in this case "MapBoxController", is selected by the route provider to be paired with the html template.

All of the map rendering, GPS coordinate calculations and sensor formatting are done in the "MapBoxController". When run, the JavaScript controller initializes all of the variables used by the Leaflet mapping interface, including the map view location, all displayed markers and all drawn paths.

Next, a call to the "/fc/sensors" endpoint is made using a service in the "Sensors" services set to pull and update the map display information that has already been initialized. It is important to note that communication between the JavaScript controller and the API is defined according to the functions of the service set. These functions formulate GET and PUT requests to the API endpoints that are provided by the NodeJS server. Inside the callback from the sensors' GET request, data for the latitude and longitude of all UAVs and paths pass to the map directive to be drawn.

Figure 7 shows the main screen used to visualize UAVs from sensor data. This client graphic user interface (GUI) consists of a top bar filled with three buttons, a main map interface on the left-hand side and a detailed data view on the right-hand side. The top bar of the client GUI provides three buttons for the user to start a simulation based on the UAV sensor data that exist in the database. During the simulation, UAVs move along their simulation path, and all aspects of UAV Flight Tracker operate normally. Options are available for the simulation to update local sensor values from the database or update all sensor datasets to the database on an interval of 100 ms. The main map interface shows connected UAV locations, set destinations and paths. The user can click on any of the map items to see the id (identity) of the UAV to which a particular piece of data refers. On the right-hand side of the GUI, a listing for all information from the sensor data variables being tracked can be viewed in real time. An input box for the currently selected UAV at the top of the pane allows the user to change the UAV sensor dataset being viewed.

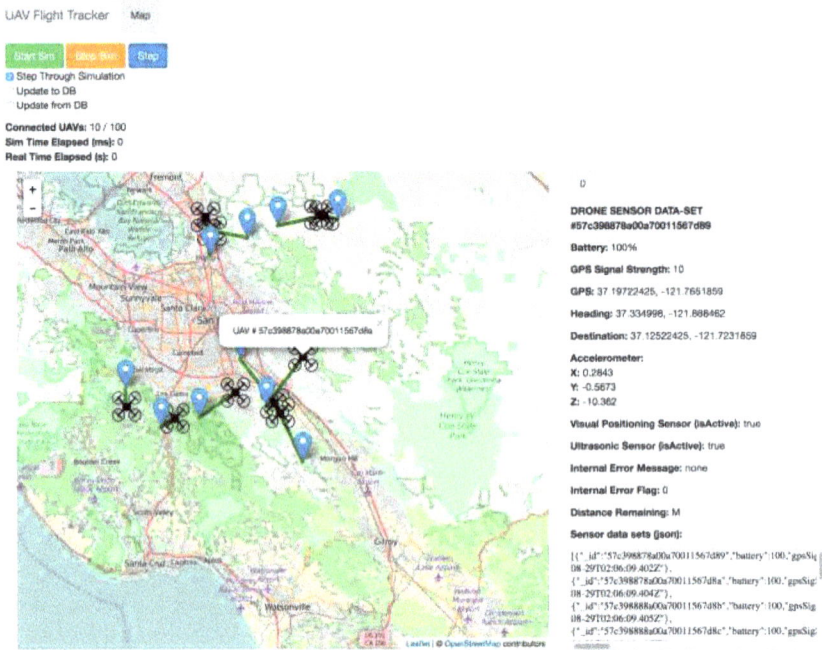

Figure 7. A screen on the client system.

Figure 8 shows the performance results provided by Chrome DevTools [24] while we were simulating the flight of UAVs. We can see the mobility of UAVs together with the variance in performance as shown in Figure 8a. The performance timeline in Figure 8b details nearly all of the technical information gathered from the server and client performance testing. From this performance-monitoring tool, the maximum and minimum values for client memory utilization and a timeline with the timing of all round trips for all requests sent from the client to the server could be easily monitored.

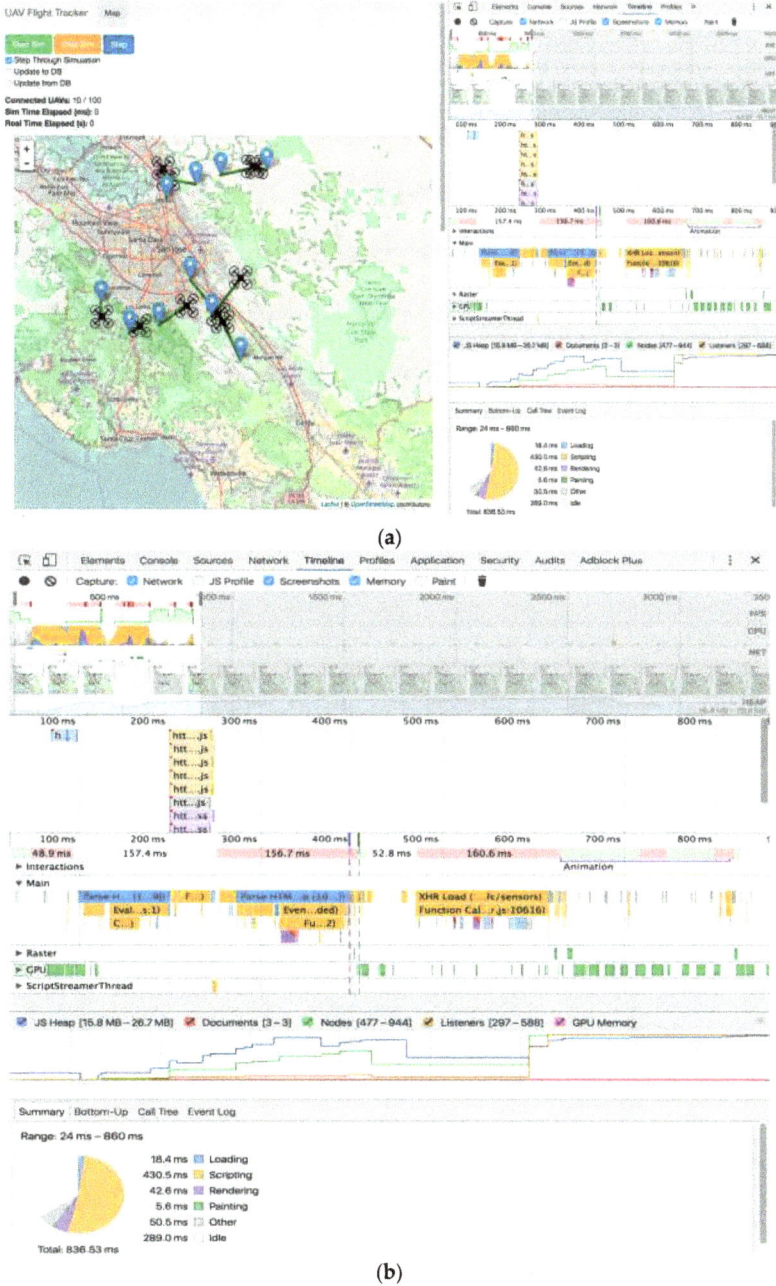

(a)

(b)

Figure 8. Client performance testing results by Chrome DevTools: (**a**) screen of client system with performance results and (**b**) performance results.

6. Performance Evaluation

We implement UAV Flight Tracker and evaluate the performance in terms of response time, storage/memory overhead and collision avoidance.

6.1. Setup and Testing Methodology

UAV Flight Tracker runs on Heroku [25], a cloud application platform that provides a NodeJS runtime and MongoDB service. A JSON package file lists the versions of NodeJS (Version 4.4.7), Express (Version 4.13.3) and MongoDB (Version 2.1.6) that will run on the cloud service once deployed. Heroku automatically manages all server provisioning and redundancies. On the client, minified versions of Jquery (Version 2.1.4), Bootstrap (Version 3.3.4) and AngularJS (Version 1.4.6) are used along with Leaflet (Version 0.7.7) and Angular Leaflet Directives. The client machine used for development and testing has an Intel Core i7 @ 2.3 GHz with 16 GB of DDR3 memory clocked at 1600 MHz.

Testing of UAV Flight Tracker is conducted with the server and database hosted on Heroku and a laptop (with specifications in the setup description) to run the client. All tests in the performance evaluation are conducted with the same test set of 100 UAV sensor datasets. Additionally, all subsets of the initial 100 datasets are also identical, meaning that a subset of 20 datasets used in one test is identical to the subset of 20 datasets used in another test. Each sensor dataset used for testing contained unique, randomized GPS coordinates for the UAV start location and the destination. A random point generator provided by the 'Geographic Midpoint Calculator' website [26] is used to generate the initial GPS coordinates for UAVs. Coordinates are generated around a center point in San Jose, CA, USA, and a maximum distance of 25 km. For each individual set of coordinates, a unique destination 5 to 14 km from the UAV GPS location was randomly generated using the equations set forth as Equations (1) and (2). The variable *gpsCoordinate* is the UAV starting location, and the *destinationCoordinate* is the resulting randomized destination through a random function with a range between *a* and *b*, RANDBETWEEN (*a*, *b*). Equation (1) uses the GPS latitude to calculate a randomized destination latitude, and Equation (2) uses GPS longitude to calculate a randomized destination longitude.

$$destinationCoordinateN = gpsCoordinateN + \text{RANDBETWEEN}\,(-0.009, 0.009) \tag{1}$$

$$destinationCoordinateW = gpsCoordinateW + \text{RANDBETWEEN}\,(-0.009, 0.009) \tag{2}$$

To express the performance and scalability of UAV Flight Tracker adequately, the latency of data sent between the client, server and UAV must be recorded along with the size of data being held on the server. This means that the roundtrip request latency must be measured for communication between the client and server, as well as between the UAV and server. Additionally, the memory utilization of the client is monitored to gauge front-end performance and ensure that the UAV Flight Tracker client runs on portable machines with fewer available resources. The number of concurrently active UAVs is the most significant factor in determining server response time and client performance. Therefore, the number of connected UAVs is used as the domain over which we gauge HTTP request time, required client memory size and database size on the server. We increase the number of UAVs from one UAV to 100 UAVs.

We then measure the effects for collision avoidance as follows:

- Server performance: response times in PUT, GET and POST operations
- Overhead: required database size on the server and required memory size on the client
- Collision avoidance: total UAV collisions and UAV mission completion time

All HTTP request times and required client memory information are provided using Google Chrome DevTools. The MongoDB size is provided by the web interface for mLab Database Service [27]. For testing of collision avoidance, we set up 300 m, 500 m and 1000 m as the radii of the safe, redirection and emergency zones, respectively.

6.2. Results Analysis

6.2.1. Results on Server Performance

(1) Sensor data update time (PUT) vs. connected UAVs

Sensor data update time is the total time that the client needs to update every sensor dataset in a series. For every connected UAV, a single HTTP PUT request is sent to the server. Each request is sent in a series, meaning that the total roundtrip time of all PUT requests should scale linearly with the number of connected UAVs. The total roundtrip time is the total time needed to send all of the requests, update them in the database and receive back a response from the server. The total sensor data update time starts with the exact time that the first PUT request is sent and ends when the server response from the last PUT request is received. Total time is a good indicator of sensor update performance and system performance in general, because the client sends all PUT requests in a series, one after the other.

As shown in Figure 9 the time taken to update all of the connected UAV sensor datasets is linearly proportional to the number of connected UAVs. The fitted linear approximation of the recorded data has an R-squared value of $R^2 = 0.97417$; in statistics, R-squared is the coefficient of determination, i.e., the number that indicates the proportion of the variance in the dependent variable that is predictable from the independent variable [28]. This is expected as each additional connected UAV provides another dataset to be updated via an HTTP PUT request to the server.

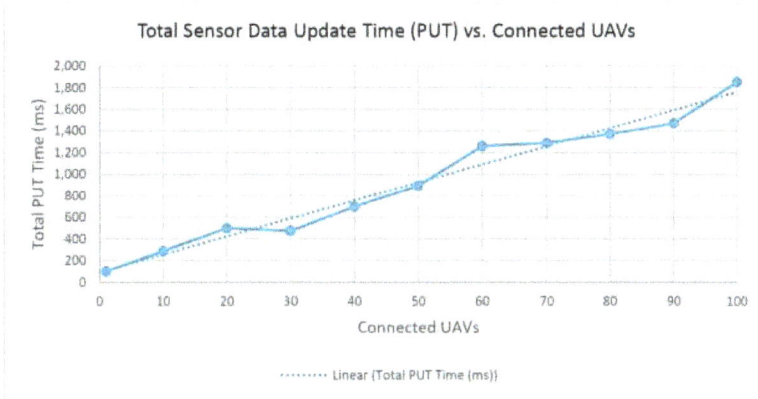

Figure 9. Total time to complete sensor data update.

(2) Total sensor retrieval time (GET) vs. connected UAVs

The sensor retrieval time is the roundtrip time for a single HTTP GET request from the client to reach the server, query the database and return with the requested sensor datasets. This total is used to describe the sensor retrieval time because all of the UAV sensor datasets are being retrieved, though only one HTTP GET request is sent from the client. Unlike the previous test for total sensor update time, where a series of HTTP PUT requests are being sent, this test requires only a single HTTP GET request to retrieve all sensor datasets.

As shown in Figure 10, the fitted linear approximation of this graph has an R-squared value of $R^2 = 0.93346$. This linearly increasing trend is expected, because the database query time and the time needed to send the server response increase as the dataset becomes larger with more connected UAVs.

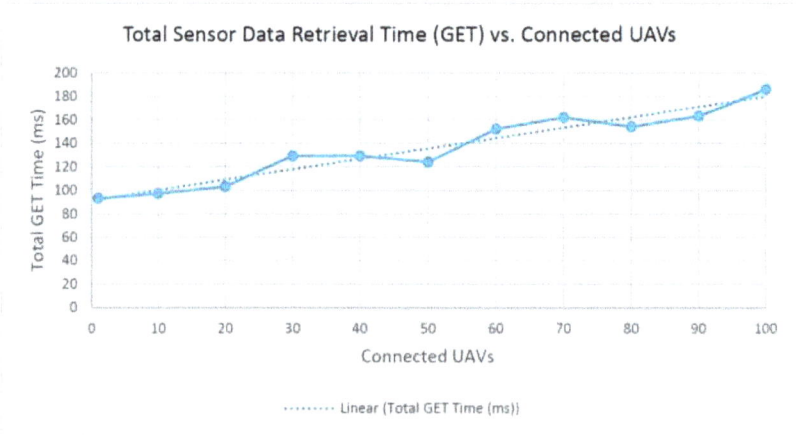

Figure 10. Total time to complete sensor data retrieval.

(3) Average sensor data insertion time (POST) vs. connected UAVs

The sensor data insertion time is the roundtrip time for an HTTP POST request issued by a newly connected UAV to add its sensor data to the database for the first time and return an HTTP code to the requesting entity. After receiving the request, the server copies the new JSON data to the database, and a unique ID and timestamp are then associated with the document. The time taken to perform an HTTP POST request to the server should not depend on the current number of connected UAVs, as the database is able to add JSON documents without modifying existing data. In a production environment, sensor data insertion requests are likely to be spread throughout a large interval of time with the occasional cluster of new UAVs connected to the system. As a result, the test for sensor data insertion time is calculated as the average of requested roundtrip times to display performance in a more realistic scenario. As expected, Figure 11 shows similar values even as the number of UAVs increases. Though the results vary a bit, it is important to note that all of the measured insertion times are within 5% of 426.83 ms.

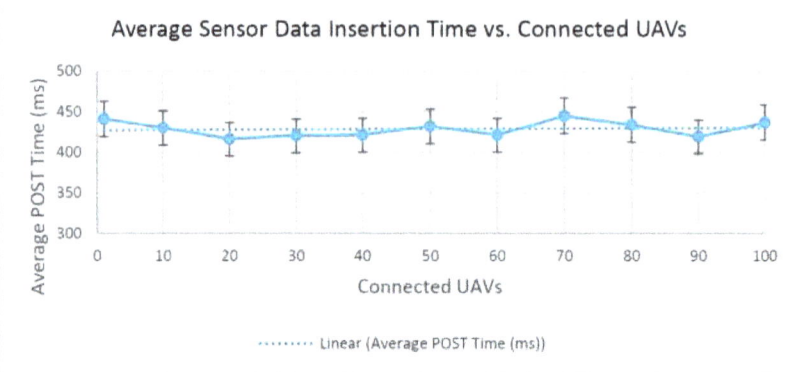

Figure 11. Average time for sensor data insertion.

Server performance testing highlights a few perceived inconsistencies in the roundtrip times for HTTP requests. When scaling the number of connected UAVs, HTTP PUT requests for updating all of the sensor datasets prove to be the costliest operation, taking 1850 ms to update 100 connected UAVs. Compared to the HTTP GET request's high of only 186 ms at 100 connected UAVs, processing PUT requests to update sensor data seems abnormally high. However, this abnormality is explained by considering the number of HTTP requests needed to insert, update or retrieve data from the server. Sensor data retrieval (HTTP GET) requests use an array of JSON sensor data sent within the body of a single request, while individual sensor update (HTTP PUT) requests are sent for each sensor set that needs to be updated. Therefore, the added latency of PUT requests is not unexpected, but could be avoided in the future if PUT requests were sent in parallel.

6.2.2. Results on Overhead

(1) Required server database size vs. connected UAVs

The database size is the total size of the MongoDB that holds all sensor datasets in individual JSON documents. As shown in Figure 12, the size of the DB collection in kilobytes is linearly proportional to the number of connected UAVs. This result shows that each UAV has the same number of key and value pairs held in its respective sensor dataset. The fitted linear approximation of this graph has a perfect R-squared value of $R^2 = 1$.

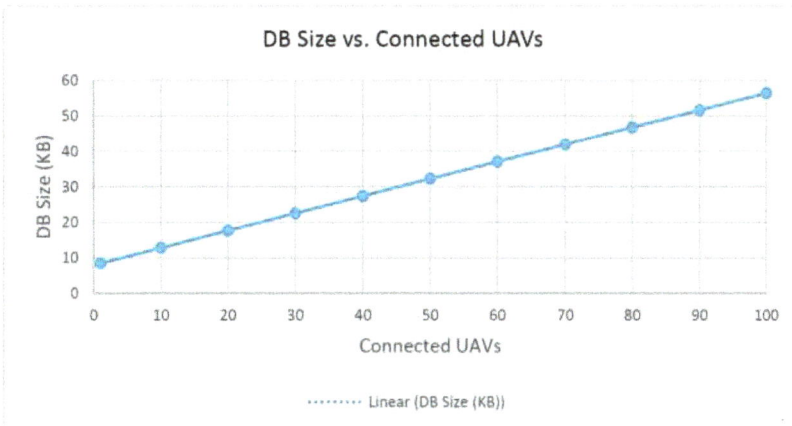

Figure 12. Required DB size on server vs. connected UAVs.

(2) Required client memory size vs. connected UAVs

The client memory size is the minimum and maximum client system memory used by the UAV Flight Tracker during normal operation. The maximum consumed memory trends higher as the number of connected UAVs increases, while the minimum or baseline memory stays relatively static, as shown in Figure 13. The fitted linear approximation of the maximum client memory trend line has an R-squared value of $R^2 = 0.85785$. Most of the client memory is used to draw the map interface and maintain a list of sensor data objects. Thus, as more UAVs and destinations are added to the map, the client uses more memory. When the client interface initially loads, there are no objects drawn to the map until the server fulfills the HTTP GET request for sensor data. Therefore, the baseline memory usage does not depend on the number of connected UAVS and instead stays consistent.

Client Memory vs. Connected UAVs

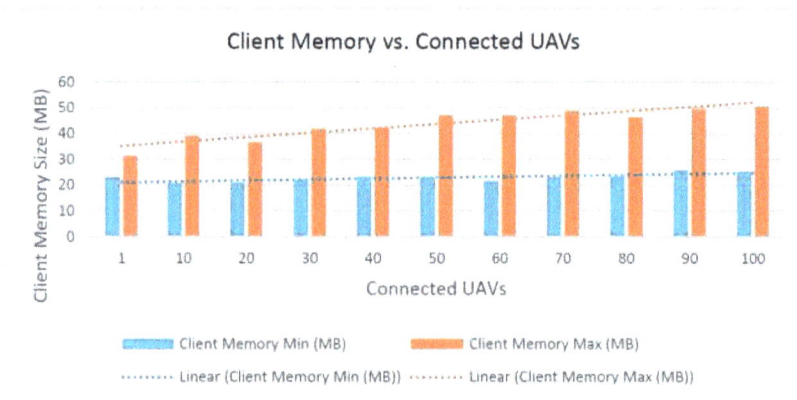

Figure 13. Required memory size on client vs. connected UAVs.

6.2.3. Results on Collision Avoidance

(1)   Total UAV collisions (without collision avoidance)

The total number of UAV collisions that occur over time is a measurement of how many active UAVs are within another UAV emergency zone at any given time during their respective flight paths. Any given UAV can only crash once, so any subsequent crashes reported for that UAV will be ignored as erroneous in this dataset. Every UAV path is linear, spanning from a randomly-generated origin to a randomly-generated destination according to Equations (1) and (2). As shown in Figure 14, the number of actively connected UAVs increases the likelihood of collisions occurring.

Total UAV Collisions vs. Time (Without Collision Avoidance)

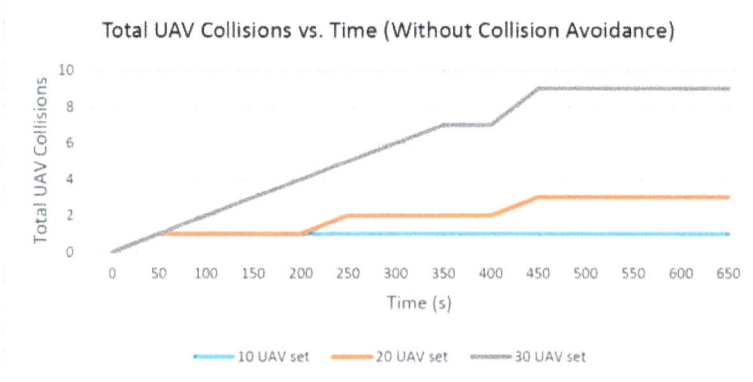

Figure 14. Total UAV collisions.

(2)   UAV mission completion time (with/without collision avoidance)

UAV mission completion is the total number of connected UAVs that have completed their missions by reaching their respective destinations. In the experiments, all UAVs move at the same speed and their path distances change depending on the randomly-generated origin and destination points. This test is repeated twice with or without the collision avoidance. In other words, we enable the collision avoidance to redirect active UAVs while disabling the collision avoidance system, rendering it unable to change UAV flight paths. As a result, Figure 15 shows that the finishing time that the set of UAVs without collision avoidance reach a target or a destination is quicker than one for the set of UAVs with collision avoidance. This is the reason that some UAVs would collide and never reach their destination. Redirection in the collision avoidance test occurs as the collision avoidance algorithm description detailed in the Section 4. It normally adds anywhere from two seconds to one minute of extra flight time per incident. In Figure 15, with a set of 20 connected UAVs, the normal, uninfluenced time of completion is 900 s, while the time of completion for redirected flight is 1075 s. This test is based off of the same sample flight data used in the previous test for total UAV collisions. This result shows that the redirection algorithm only adds a total of 175 s to the overall completion time.

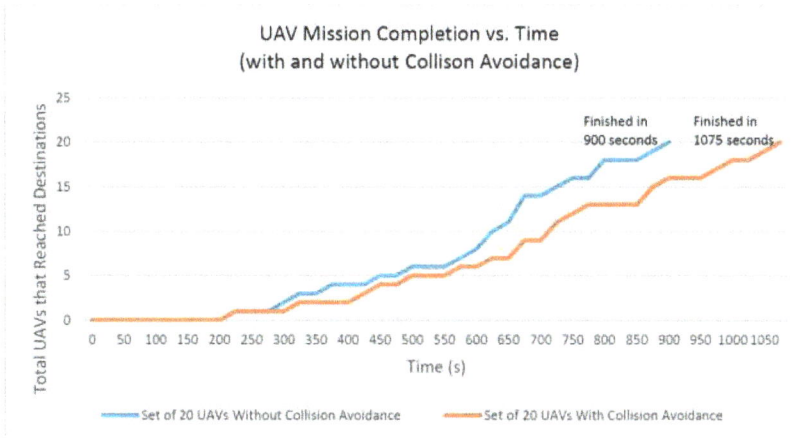

Figure 15. UAV mission completion time.

7. Conclusions

Developing a UAV monitoring and collision avoidance system that is both scalable and simple to deploy is a challenging task. This paper has proposed a UAV flight tracking solution that allocates critical and non-critical computing intensive processes between the server and the client, allowing for optimal performance on a large scale. The server handles critical processes for collision avoidance, UAV control and sensor updates. Following this distribution, more trivial tasks (i.e., map rendering and sensor display) are left for the client to compute. This allows the server to run on a distributed cloud-hosting platform, ensuring that the most important aspects of UAV tracking run with a scalable resource pool.

As proposed in [8], the future of commercial and other non-military UAV flights will use 3G and 4G cellular networks to facilitate UAV communication with the controlling entities. Our work aligns well with this proposal by providing a cloud-based web application that monitors and controls UAVs over the HTTP protocol. As UAVs are easily connected to cellular networks, providing them with nearly limitless flight possibilities, our flight monitoring solution matches that flexibility by allowing for easy deployment of the entire system on cloud-based computing services. This flexibility continues from the server deployment to the client, as the user-facing application can run in any modern web browser and requires a small amount of system memory.

Future work on this paper will continue to provide collision detection in three dimensions, improve the performance of the client application and conduct tests with non-simulated UAVs. We will optimize the radii of the three zones (i.e., safe zone, redirection zone and emergency zone) to enhance the performance of UAVs, i.e., in terms of mission completion time. Moreover, collision detection accuracy depends on the specifications of the UAV system utilized. The accuracy of the GPS signal and average radio delay affect the radii of collision avoidance zones needed to ensure safe flight.

Sensors **2016**, *16*, 1913

Continued research on the specific attributes of UAVs that determine safe flight distances would complement our UAV management system design.

Acknowledgments: This work is supported by grants from National Science Foundation (NSF-CNS-1637371) and the San Jose State University Research Foundation.

Author Contributions: This work is done by Mason Itkin, Mihui Kim and Younghee Park. Mason and Younghee Park developed their idea throughout the project year. Mason implemented and tested the proposed system based on their idea while writing his paper. Mihui Kim helped to write his paper and designed the paper structure while guiding him in the paper writing, simulation and its experimental result analysis. As a corresponding author, Younghee Park managed this project during the project year and advised the direction of this work while reviewing the paper.

Conflicts of Interest: The authors declare no conflict of interest.

References

1. Gupta, L.; Jain, R.; Vaszkun, G. Survey of Important Issues in UAV Communication Networks. *IEEE Commun. Surv. Tutor.* **2016**, *18*, 1123–1152. [CrossRef]
2. Amazon Prime Air. Available online: https://www.amazon.com/b?node=8037720011&ref=producthunt (accessed on 1 August 2016).
3. Lam, T.M.; Mulder, M.; Paassen, M.M.V. Collision avoidance in UAV tele-operation with time delay. In Proceedings of the IEEE International Conference on Systems, Man and Cybernetics, Montreal, QC, Canada, 7–10 October 2007; pp. 997–1002.
4. Kim, H.N.; Yoo, S.H.; Kim, K.H.; Chung, A.Y.J.; Lee, J.Y.; Lee, S.K.; Jung, J.T. Method for Controlling Hand-over in Dron Network. U.S. 14/709226, 11 December 2015. Available online: http://www.freepatentsonline.com/y2015/0327136.html (accessed on 1 August 2016).
5. Kuriki, Y.; Namerikawa, T. Formation control with collision avoidance for a multi-UAV system using decentralized MPC and consensus-based control. In Proceedings of the European Control Conference (ECC), Linz, Austria, 15–17 July 2015; pp. 3079–3084.
6. Liu, W.; Zheng, Z.; Cai, K.Y. Distributed on-line path planner for multi-UAV coordination using bi-level programming. In Proceedings of the Chinese Control and Decision Conference (CCDC), Guiyang, China, 25–27 May 2013; pp. 5128–5133.
7. Patel, A.; Winberg, S. UAV Collision Avoidance: A Specific Acceleration Matching control approach. In Proceedings of the AFRICON, Livingstone, Zambia, 13–15 September 2011; pp. 1–6.
8. Ling, L.; Niu, Y. A particle filter based intruder state estimation method for UAV collision avoidance. In Proceedings of the Chinese Automation Congress (CAC), Wuhan, China, 27–29 November 2015; pp. 1110–1115.
9. Alejo, D.; Conde, R.; Cobano, J.A.; Ollero, A. Multi-UAV collision avoidance with separation assurance under uncertainties. In Proceedings of the IEEE International Conference on Mechatronics (ICM), Malaga, Spain, 14–17 April 2009; pp. 1–6.
10. Yoo, C.S.; Cho, A.; Park, B.J.; Kang, Y.S.; Shim, S.W.; Lee, I.H. Collision avoidance of Smart UAV in multiple intruders. Control. In Proceedings of the International Conference on Automation and Systems (ICCAS), JeJu Island, Korea, 17–21 October 2012; pp. 443–447.
11. Lin, C.E.; Lai, Y.-H.; Lee, F.-J. UAV collision avoidance using sector recognition in cooperative mission to helicopters. In Proceedings of the Integrated Communications, Navigation and Surveillance Conference (ICNS), Herndon, VA, USA, 8–10 April 2014; pp. 1–9.
12. Zhu, L.; Yin, D.; Yang, J.; Shen, L. Research of remote measurement and control technology of UAV based on mobile communication networks. In Proceedings of the IEEE International Conference on Information and Automation (ICIA), Lijiang, China, 8–10 August 2015; pp. 2517–2522.
13. Roldán, J.J.; Joossen, G.; Sanz, D.; del Cerro, J.; Barrientos, A. Mini-UAV Based Sensory System for Measuring Environmental Variables in Greenhouses. *Sensors* **2015**, *15*, 3334–3350. [CrossRef] [PubMed]
14. Sharon Lozano. First Steps Toward Drone Traffic Management. 19 November 2015. Available online: http://www.nasa.gov/feature/ames/first-steps-toward-drone-traffic-management (accessed on 1 August 2016).
15. MeanIO. Available online: http://mean.io/#!/ (accessed on 1 August 2016).
16. MongoDB. Available online: https://www.mongodb.com/ (accessed on 1 August 2016).

17. ExpressJS. Available online: http://expressjs.com/ (accessed on 1 August 2016).

18. AngularJS. Available online: https://angularjs.org/ (accessed on 1 August 2016).

19. NodeJS. Available online: https://nodejs.org/en/ (accessed on 1 August 2016).

20. JavaScript Object Notation (JSON). Available online: http://www.json.org/ (accessed on 1 August 2016).

21. Document Object Model (DOM). Available online: https://www.w3.org/DOM/ (accessed on 1 August 2016).

22. Leaflet. Available online: http://leafletjs.com/ (accessed on 1 August 2016).

23. Angular-Leaflet-Directive. Available online: https://github.com/tombatossals/angular-leaflet-directive (accessed on 1 August 2016).

24. Chrome DevTools. Available online: https://developers.google.com/web/tools/chrome-devtools/ (accessed on 1 August 2016).

25. Heroku. Available online: https://www.heroku.com/ (accessed on 1 August 2016).

26. Geographic Midpoint Calculator. Available online: http://geomidpoint.com/ (accessed on 1 August 2016).

27. mLAB Database Service. Available online: https://mlab.com/ (accessed on 1 August 2016).

28. Coefficient of Determination. Available online: https://en.wikipedia.org/wiki/Coefficient_of_determination (accessed on 1 August 2016).

sensors

MDPI

Article

Fast Orientation of Video Images of Buildings Acquired from a UAV without Stabilization

Michal Kedzierski and Paulina Delis *

Department of Remote Sensing and Photogrammetry, Faculty of Civil Engineering and Geodesy, Military University of Technology, Warsaw 00908, Poland; michal.kedzierski@wat.edu.pl
* Correspondence: paulina.delis@wat.edu.pl; Tel.: +48-261-837-148

Academic Editors: Felipe Gonzalez Toro and Antonios Tsourdos
Received: 16 March 2016; Accepted: 9 June 2016; Published: 23 June 2016

Abstract: The aim of this research was to assess the possibility of conducting an absolute orientation procedure for video imagery, in which the external orientation for the first image was typical for aerial photogrammetry whereas the external orientation of the second was typical for terrestrial photogrammetry. Starting from the collinearity equations, assuming that the camera tilt angle is equal to 90°, a simplified mathematical model is proposed. The proposed method can be used to determine the X, Y, Z coordinates of points based on a set of collinearity equations of a pair of images. The use of simplified collinearity equations can considerably shorten the processing tine of image data from Unmanned Aerial Vehicles (UAVs), especially in low cost systems. The conducted experiments have shown that it is possible to carry out a complete photogrammetric project of an architectural structure using a camera tilted 85°–90° (φ or ω) and simplified collinearity equations. It is also concluded that there is a correlation between the speed of the UAV and the discrepancy between the established and actual camera tilt angles.

Keywords: close range photogrammetry; collinearity equations; exterior orientation; image sequence; UAV

---

1. Introduction

The assurance of providing adequate protection of cultural heritage sites, as well as preserving their authenticity, can only be obtained by creating a comprehensive inventory of a given site. This includes the definition of the type, shape, dimensions and geospatial location of the given structure. One of the ways to create such an inventory for a cultural heritage building is based on its photogrammetric documentation in the form of a three-dimensional model.

Generating a 3D model of a historical structure using photogrammetric methods can be troublesome due to the height of the structure. Whether a camera is used or a terrestrial laser scanner, too great a height of the structure will greatly limit the possibility of using terrestrial photogrammetry techniques. In such instances, better results can be obtained by integrating terrestrial imagery data with data acquired from a low altitude flights. Using an Unmanned Aerial Vehicle (UAV) as a platform for the sensor ensures that relatively large scale imagery can be acquired, which warrants high-quality end products.

In recent years UAVs are being increasingly used in architectural photogrammetry, being a fundamental module of management and conservation of national cultural heritage sites [1]. Together with terrestrial laser scanning [2], imagery is the main source of information when producing cultural heritage inventory. The purpose of photogrammetric systems based on UAVs, in which the data acquisition module is a video camera, is to acquire data to develop orthophotomaps [3,4], digital terrain models [5–7], 3D city models [8], 3D models of buildings [9] and sculptures [10]. UAV systems prove particularly useful where the access to an architectural object is difficult, which may be due to the topography or close proximity to other architectural objects. The complexity of the shapes of buildings is also a frequent obstacle during the implementation of photogrammetric techniques. It is necessary to use a relative image orientation obtained from multiple camera positions or a terrestrial laser scanner. In the case of an insufficient number of camera positions, detection of tie points in the images may not be feasible. Then, it is very helpful to use a sequence of video images. Finding similarities in adjacent images is much easier than in images taken from distant camera positions [11]. In addition, in the case of video data, the redundant number of video images makes it possible to eliminate blurred images [12,13], images with a low radiometric quality index [14] or those, for which the values of the exterior orientation parameters significantly differ within an image sequence which may be a result of flight instability.

In the literature, one can often find descriptions of systems involving the integration of data from the two altitudes: terrestrial and aerial. These methods often relate to the integration of point clouds from terrestrial and aerial laser scanning [15], point clouds and imagery data [16] or image data from terrestrial and aerial levels. An example includes the studies conducted by Bolognesi et al. [17]. The authors have developed a 3D model of a historical architectural structure based on image data acquired using a Canon EOS M high resolution digital camera mounted on UAV platforms and from ground level. In studies conducted by Püschel et al. [18] a method was developed of documenting an architectural monument, consisting of the integration of terrestrial images with images captured using a UAV. The image data were recorded by a non-metric digital camera in video mode.

The large majority of video image sequence orientation methods are based on Structure from Motion (SFM) algorithms [19]. SFM relies on determining the 3D point coordinates and the camera projection matrix simultaneously, based on homologous points measured on a large number of images [20]. It should be mentioned that the Rodriquez matrix method can be used to solve the problem of nonlinearity models of absolute and relative orientation. One of the advantages this method is the lack of gimble effect [21].

To ensure high accuracy of orientation of the video images, methods of image orientation based on the fundamental equation of photogrammetry, i.e., the collinearity equation should be used [22].

This paper presents the issues which occur when performing an orientation of two images, of which one was acquired with a tilt angle close to $90°$. Such a situation takes place quite often when conducting photogrammetric measurements of buildings based on video imagery. During such measurements, first the side walls of the building are filmed and then the camera is put into aerial orientation in order to acquire imagery from the roof. The aim of this research was to assess the possibility of conducting an orientation procedure of video imagery, in which the external orientation for the first image was typical for aerial photogrammetry whereas the external orientation of the second was typical for terrestrial photogrammetry. This type of orientation (Figure 1) of video images is a special case of processing two images for photogrammetric documentation of architectural structures. It is closely associated with the problem of integrating aerial and terrestrial imagery data. This issue is especially important, due to the fact that it occurs very often when conducting photogrammetric measurements of buildings. The problem of integrating aerial and terrestrial imagery data stems from using two different coordinate systems (and two rotation matrices): typical for aerial photogrammetry and for terrestrial photogrammetry in order to photogrametrically process both the walls and the roof of a given architectural structure in one unified coordinate system.

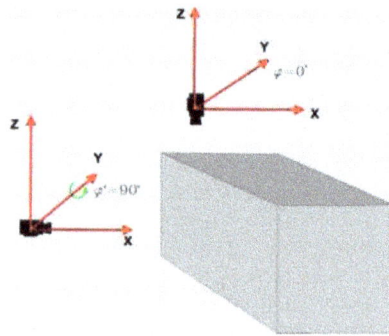

Figure 1. Transition of the video camera mounted on the UAV from aerial to terrestrial orientation.

2. Proposed Method of Orientation for Terrestrial and Aerial Video Images

Assuming, that the axes of the terrestrial coordinate system are parallel to the building's walls, a rotation occurs about one of the axes of the terrestrial coordinate system X or Y (when the video camera makes the transition from aerial orientation to terrestrial orientation). In this case, the tilt angle (ω or φ) for the terrestrial image is close to 90°.

By adopting the rotation matrix and maintaining the order of the rotation ω φ κ [22]:

$$A_{\omega\varphi\kappa} = \begin{bmatrix} a_{11} & a_{12} & a_{13} \\ a_{21} & a_{22} & a_{23} \\ a_{31} & a_{32} & a_{33} \end{bmatrix} \tag{1}$$

The rotation matrix coefficient equations can be greatly simplified and take on the form shown in (Table 1):

Table 1. Simplified rotation matrix coefficient equations when $\phi = 90°$ or $\omega = 90°$.

Rotation Matrix Coefficient	Equation of Rotation Matrix Coefficient in the Classic Form (Rotation Sequence: $\omega'\kappa$)	Simplified Equation of Rotation Matrix Coefficient for $\lim\limits_{\varphi \to 90} f(\varphi)$	Simplified Equation of Rotation Matrix Coefficient for $\lim\limits_{\omega \to 90} f(\omega)$
a_{11}	$\cos\varphi\cos\kappa$	0	$\cos\varphi\cos\kappa$
a_{12}	$\cos\omega \sin\kappa + \sin\omega\sin\varphi\cos\kappa$	$\cos\omega\sin\kappa + \sin\omega\cos\kappa$	$\sin\varphi\cos\kappa$
a_{13}	$\sin\omega\sin\kappa - \cos\omega \sin\varphi \cos\kappa$	$\sin\omega\sin\kappa - \cos\omega\cos\kappa$	$\sin\kappa$
a_{21}	$-\cos\varphi\sin\kappa$	0	$-\cos\varphi\sin\kappa$
a_{22}	$\cos\omega\cos\kappa - \sin\omega\sin\varphi\sin\kappa$	$\cos\omega\cos\kappa - \sin\omega\sin\kappa$	$-\sin\varphi\sin\kappa$
a_{23}	$\sin\omega\cos\kappa + \cos\omega\sin\varphi\sin\kappa$	$\cos\omega\cos\kappa - \sin\omega\sin\kappa$	$\cos\kappa$
a_{31}	$\sin\varphi$	1	$\sin\varphi$
a_{32}	$-\sin\omega\cos\varphi$	0	$-\cos\varphi$
a_{33}	$\cos\omega\cos\varphi$	0	0

By implementing the inverse of the collinearity equations for a stereo made up of video frames with two different orientations (aerial and terrestrial), the following set of equations was created:

$$\begin{bmatrix} X = X'_0 + (Z - Z'_0) \frac{(x'-x_0)a'_{11}+(y'-y_0)a'_{21}-a'_{31}\ ck}{(x'-x_0)a'_{13}+(y'-y_0)a'_{23}-a'_{33}\ ck} \\ Y = Y'_0 + (Z - Z'_0) \frac{(x'-x_0)a'_{12}+(y'-y_0)a'_{22}-a'_{32}\ ck}{(x'-x_0)a'_{13}+(y'-y_0)a'_{23}-a'_{33}\ ck} \\ X = X''_0 + (Z - Z''_0) \frac{(x''-x_0)a''_{11}+(y''-y_0)a''_{21}-a''_{31}\ ck}{(x''-x_0)a''_{13}+(y''-y_0)a''_{23}-a''_{33}\ ck} \\ Y = Y''_0 + (Z - Z''_0) \frac{(x''-x_0)a''_{12}+(y''-y_0)a''_{22}-a''_{32}\ ck}{(x''-x_0)a''_{13}+(y''-y_0)a''_{23}-a''_{33}\ ck} \end{bmatrix} \tag{2}$$

After substituting the third and fourth Equations from Equation (2) with the simplified equations (Table 1), the set of collinearity equations for the aerial (first image) and terrestrial (second image) imagery will look as follows, if $\varphi'' = 90°$:

$$
\begin{bmatrix}
X = X'_0 + (Z - Z'_0) \frac{(x'-x_0)a'_{11}+(y'-y_0)a'_{21}-a'_{31}\ ck}{(x'-x_0)a'_{13}+(y'-y_0)a'_{23}-a'_{33}\ ck} \\
Y = Y'_0 + (Z - Z'_0) \frac{(x'-x_0)a'_{12}+(y'-y_0)a'_{22}-a'_{32}\ ck}{(x'-x_0)a'_{13}+(y'-y_0)a'_{23}-a'_{33}\ ck} \\
X = X''_0 + (Z - Z''_0) \frac{-ck}{(x''-x_0)b''_{13}+(y''-y_0)b''_{23}} \\
Y = Y''_0 + (Z - Z''_0) \frac{(x''-x_0)b''_{12}+(y''-y_0)b''_{22}}{(x''-x_0)b''_{13}+(y''-y_0)b''_{23}}
\end{bmatrix}
\tag{3}
$$

For $\omega'' = 90°$ the set of equations will look as follows:

$$
\begin{aligned}
X &= X'_0 + (Z - Z'_0) \frac{(x'-x_0)a'_{11}+(y'-y_0)a'_{21}-a'_{31}\ ck}{(x'-x_0)a'_{13}+(y'-y_0)a'_{23}-a'_{33}\ ck} \\
Y &= Y'_0 + (Z - Z'_0) \frac{(x'-x_0)a'_{12}+(y'-y_0)a'_{22}-a'_{32}\ ck}{(x'-x_0)a'_{13}+(y'-y_0)a'_{23}-a'_{33}\ ck} \\
X &= X''_0 + (Z - Z''_0) \frac{(x''-x_0)c''_{11}+(y''-y_0)c''_{21}-a''_{31}\ ck}{(x''-x_0)c''_{13}+(y''-y_0)c''_{23}} \\
Y &= Y''_0 + (Z - Z''_0) \frac{(x''-x_0)c''_{12}+(y''-y_0)c''_{22}-c''_{32}\ ck}{(x''-x_0)c''_{13}+(y''-y_0)c''_{23}}
\end{aligned}
\tag{4}
$$

where:

X, Y, Z—ground control point coordinates

$X_0', Y_0', Z_0', \omega'\ \varphi'\ \kappa'$—exterior orientation parameters of aerial image

$X_0'', Y_0'', Z_0'', \omega'', \varphi'', \kappa''$—exterior orientation parameters of terrestrial image

x', y'—image coordinates of a point on the aerial image

x'', y''—image coordinates of a point on the terrestrial image

c_k, x_0, y_0—camera's interior orientation parameters

$a'_{11}, \ldots, a'_{33}$—rotation matrix coefficients for the aerial image

$a''_{11}, \ldots, a''_{33}$—rotation matrix coefficients for the terrestrial image

$b''_{11}, \ldots, b''_{33}$—coefficients of the simplified rotation matrix for the terrestrial image when $\varphi'' = 90°$

$c''_{11}, \ldots, c''_{33}$—coefficients of the simplified rotation matrix for the terrestrial image when $\omega'' = 90°$.

The above Equations (3) and (4) can be rewritten in a simplified form:

$$
\begin{bmatrix}
X = X'_0 + (Z - Z'_0)\ F \\
Y = Y'_0 + (Z - Z'_0)\ G \\
X = X''_0 + (Z - Z''_0)\ H \\
Y = Y''_0 + (Z - Z''_0)\ I
\end{bmatrix}
\tag{5}
$$

After transforming Equation (5), new equations are obtained for calculating the ground coordinates X, Y, Z of a point with known image-space coordinates:

$$
X = \frac{X''_0 F - Z''_0 FH - X'_0 H + Z'_0 FH}{F - H} \quad Y = \frac{Z'_0 GI - Y'_0 I + Y''_0 G - Z''_0 GI}{G - I} \quad Z = \frac{X - X'_0}{F} + Z'_0 \tag{6}
$$

In order to determine the ground coordinates XYZ, those equations for which XYZ can be calculated independently from other ground coordinates were selected. When the following are known: the interior orientation parameters of the camera: c_k, x_0, y_0, the exterior orientation parameters of the video frame with the aerial orientation: $X_0', Y_0', Z_0', \omega'\ \phi'\ \kappa'$, and with the terrestrial orientation: $X_0'', Y_0'', Z_0'', \omega'', \phi'', \kappa''$, it is possible to calculate the ground coordinates of points, whose image coordinates had been measured on the video frames (x', y', x'', y'') based on the simplified collinearity equations.

Such a simplification of the collinearity equations is done to limit the amount of intermediate calculations when determining the coordinates of architectural structures in a ground coordinate system. It had therefore been decided, that the video camera's lens errors, such as radial and tangential distortion, which could increase the number of calculations, would not be taken into account in this prototype version of system.

Of course, the situation in which the difference in tilt angles of the camera during video registration is equal to 90° still remains only theoretical. The possibility of processing such a stereoimage was verified in a laboratory experiment. It was feared that because of the low coverage between the images and a large tilt angle, the absolute orientation of video images could be impossible.

3. Test Data Used in the Study

In order to verify these assumptions and check the possibility of conducting an absolute orientation of a pair of images for which the difference between tilt angle is near 90°, research was carried out on the test object in the shape of a 1 m × 2 m × 1 m cuboid. The test object was filmed with a Sony NEX-VG10 E video camera with Sony E 16 mm F2.8 fixed focal length lens [23] (Table 2).

Table 2. The main parameters of the Sony Handycam NEX-VG10E video camera.

Camera	Sony Handycam NEX-VG10E
Sensor size	CMOS 23.4 mm × 15.6 mm
Camera resolution	1920 × 1080
Pixel size	10.8 μm
Number of frames per second	25 fps
Video format	AVCHD (MPEG-4 AVC (H.264))

A simulated UAV flight was performed to acquire the test object video data from the terrestrial and aerial level. Six video frames were selected from the videos: three with a terrestrial orientation and three with an aerial orientation (Figure 2).

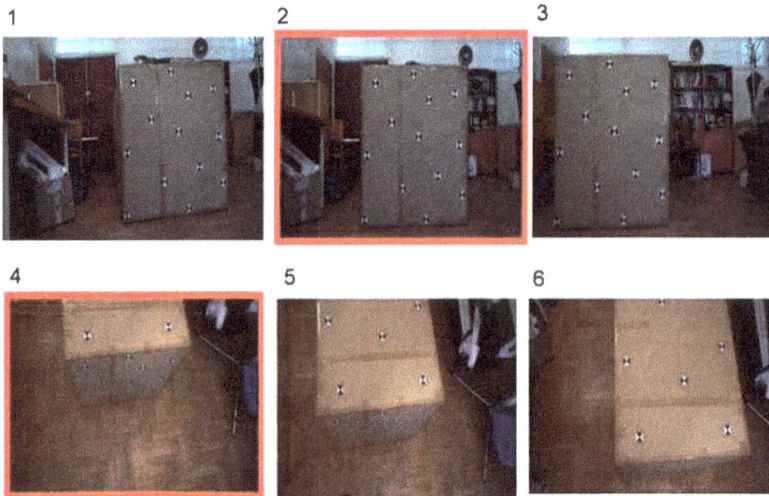

Figure 2. Test object video frames acquired from the simulated terrestrial level (frames **1–3**) and aerial level (frames **4–6**) Images number **2** and **4** are framed in red because for this pair of images, it is possible to conduct an absolute orientation when difference of tilt angle is near 90°.

Based on the acquired image data, the following processes were performed: (1) orientation of video frames from the terrestrial level, creating stereo: 1–2 and 2–3; (2) aerotriangulation of video frames from the UAV level, creating stereo: 10–11 and 11–12; (3) orientation of video frames acquired from terrestrial and UAV levels creating stereo 2–10. For the stereo 2–10 the difference in camera rotation angle about the Y axis was 74°. Table 3 presents values of angles of video frames' exterior orientation parameters. Values of angles have been calculated based on standard collinearity equations.

Table 3. Angular exterior orientation parameters of the video frames of the test object.

Frame No.	φ (°)	ω (°)	κ (°)
1	−4.9	−90.3	95.0
2	−1.9	−92.2	90.7
3	−2.5	−92.6	90.3
10	6.1	−17.7	97.0
11	8.8	−20.1	96.6
12	5.5	−19.1	92.3

Table 4 below shows the base-distance ratio of each of the stereos as well as the spatial resolution in the XY plane (ΔXY) of the geometric model and along the Z axis (ΔZ) calculated from the following equations:

$$\Delta XY = \frac{H}{ck}\delta\rho \tag{7}$$

$$\Delta Z = \frac{H}{B}\Delta XY \tag{8}$$

where:

H—distance from the camera to the object
ck—camera/s focal length
δp—pixel size
B—base

Table 4. Base-distance ratios and resolution of the stereo geometric model of test object.

Stereo	B (m)	H (m)	B/H	ΔXY (m)	ΔZ (m)
1–2	0.411	1.912	0.220	0.001	0.006
2–3	0.532	1.789	0.300	0.001	0.004
2–10	2.206	1.789	1.230	0.001	0.001
11–10	0.244	1.604	0.150	0.001	0.007
11–12	0.364	1.587	0.230	0.001	0.005

For all the stereos, a planar resolution (ΔXY) of 1 mm was obtained. Moreover, the resolution (ΔZ) of the 2–10 stereo proved to be the lowest compared to other stereos because of the favourable base-to height ratio. Research conducted on the test object confirmed the validity of the assumption that it is possible to perform the orientation of video images acquired using a video camera mounted on a UAV.

4. Experiment on an Architectural Structure

Positive orientation results of video images obtained for the test object prompted the authors to investigate further. To this end, a photogrammetric flight of an unmanned mini-copter was performed over a building with a sloping roof and rectangular base with the following dimensions: 13 m × 20 m × 6 m (Figure 3).

(a)   (b)

Figure 3. (a) Measured architectural structure; (b) UAV—octocopter used in studies.

4.1. Photogrammetric Flight Planning

In order obtain the video data, a Sony NEX-5N non-metric digital camera and Sony E 16 mm F2.8 lens were used. The camera weight with the lens was 336 g. The following factors were decisive in the choice of altitude: building dimensions, GSD, and terrestrial frame dimensions. The key equipment and flight plan parameters are listed in Table 5.

Table 5. Flight plan parameters.

Parameter	Value
Camera	Sony NEX-5N
Lens	Sony E 16 mm F2.8.
Height above terrain W	16 m
Scale denominator of a video frame M_z	1000
Scale denominator of a video frame M_z for the roof	625
GSD roof/wall/terrain (mm)	3/5/7
Image swath	5.29 m × 9.41 m
Across base (q = 60%)	3.8 m

4.2. Choosing Targets for the Photogrammetric Network

Due to the low accuracy of the GPU-IMU systems in low-budget UAVs, it is recommended to use control points. In order to select the appropriate targets to act as control points, four groups of targets were designed: crosses, circles, squares and checkerboards. It was assumed that the targets for both aerial and terrestrial level applications should allow for clear identification of their centres, regardless of the distance and angle of imaging. At the testing stage, several possibilities for visualizing the GCP signals were examined. After considering both criteria, the checkerboard targets proved to be the best. A design was created for the placement of photogrammetric terrestrial network points on building walls and around the measured structure (Figure 4). The signals were made from a white board painted with black paint. The targets placed on the walls of the building were printed on A4 sheets, then laminated with non-reflective foil (Figure 5). The signals were made in two sizes: 30 cm × 30 cm (horizontal network) and 20 cm × 20 cm (points on the wall).

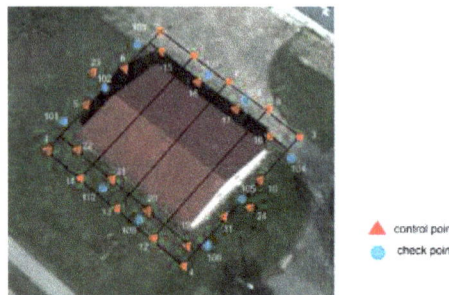

▲ control point
● check point

Figure 4. Photogrammetric terrestrial network around the building.

Figure 5. A part of a video image with targets placed on two surfaces: the building wall and on the ground.

The terrestrial photogrammetric network consisted of 24 control points and 10 check points. Targets were also placed on the walls of the building—a total of 47 signals on the walls of the building and 12 signals on the roof (Figure 6). The coordinates of the points were measured using the Topcon Total Station with an error of ±6 mm.

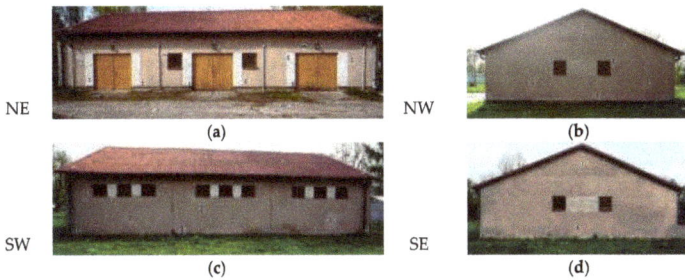

Figure 6. Photogrammetric network on the building walls and around the building. (**a**) Northeastern wall (NE); (**b**) Northwestern wall (NW); (**c**) Southwestern wall (SW); (**d**) Southestern wall (SE).

Two different types of UAV flights were conducted. The first flight over the building, with the aerial orientation, was performed along 27 pre-designed POIs (Figure 7a). The second flight, with the terrestrial orientation, was performed along the building's four walls (Figure 7b). In this case the average flight altitude was 3 m above the ground level. For each wall, a set of two flights was performed at two different camera tilt angles.

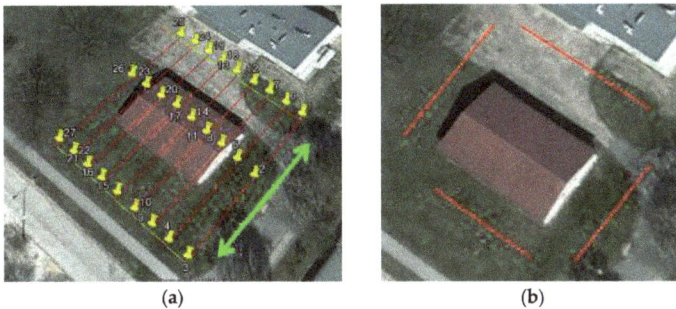

Figure 7. Flight plan for the aerial orientation (**a**) and terrestrial orientation (**b**). The numbers 1–27 represent the POIs. The red lines represent the UAV flight path, with the green line showing the direction of flight of the UAV.

Due to insufficient data (the targets were covered by vegetation), the south east (SE) wall of the building was not included further in this research. Pairs of images had been chosen from the acquired image sequences from both the aerial and terrestrial orientations to create stereopairs (Figure 8). The overlap area between images of two different orientations (aerial and terrestrial) includes the following features of the building: the basement, fragments of the side walls and the roof's edge.

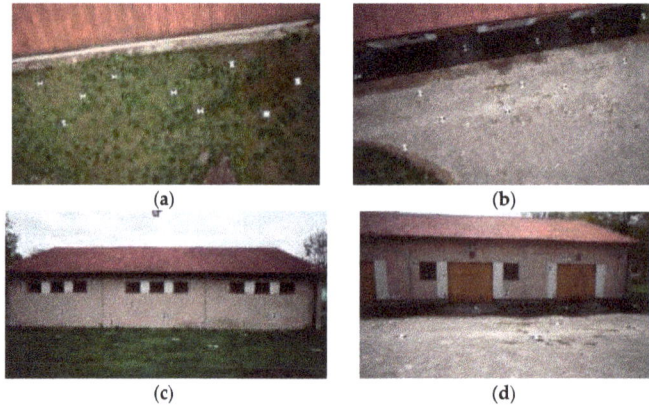

Figure 8. Example pairs of images from the terrestrial and aerial orientations: NE wall (a,b); SW wall (c,d).

4.3. Control Point and Tie Point Configurations

Modern UAV systems are all equipped with a GPS/INS receiver, which records information about the location of the camera during acquisition. However, the precision of these systems is sometimes so low, that it is better to determine the exterior orientation parameters of an image using space resection, i.e., based on control point measurements. When performing measurements in field conditions, it is sometimes impossible to place measurement targets where they are most needed as the surface may be inaccessible (with very high structures) or it may simply be forbidden to do so (with cultural heritage structures). In the research work, three configurations of control points and tie points were considered: (a) control points on the facade (b) control points on the facade and around the object (c) control points around the object (Figure 9).

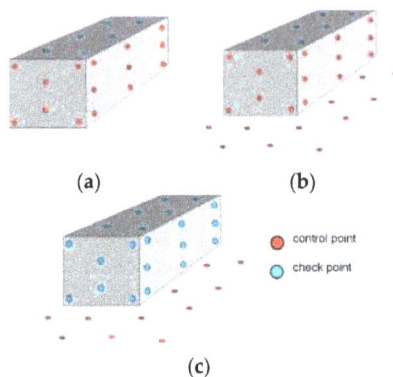

Figure 9. Three configurations of control points and tie points: (a) I—control points on the facade; (b) II—control points on the facade and around the object; (c) III—control points around the object.

The exterior orientation parameters were determined from control point measurements, using three different configurations (Figure 9). When transitioning from the aerial orientation to the terrestrial orientation, the angle which changes by close to 90° will differ between the different walls. Namely, for the north eastern (NE) and south western (SW) walls it will be the φ angle, whereas for the north western (NW) wall it will be the ω angle.

Tables 6 and 7 present values of angular exterior orientation and differences between the linear exterior orientation parameters. These values have been calculated based on standard collinearity equations.

Table 6. Angular exterior orientation parameters calculated based on measuring control points distributed in different configurations.

Wall	Configuration	ω (°)		φ (°)		κ (°)	
		N	L	N	L	N	L
NE	I	−49.54	−3.59	−79.72	−3.34	−136.78	84.34
	II	−43.72	−2.92	−78.90	−2.58	−130.75	84.64
	III	-	-	-	-	-	-
NE	I	−46.23	−3.78	85.47	−4.67	−134.78	84.34
	II	−41.24	−2.71	−85.14	−2.24	−132.75	84.64
	III	-	-	-	-	-	-
NW	I	80.80	−11.72	3.26	−9.05	−3.52	77.20
	II	80.00	−11.55	3.29	−9.00	−3.51	77.27
	III	-	-	-	-	-	-
NW	I	86.04	−10.63	2.43	−8.12	−1.52	74.20
	II	86.34	−11.61	5.28	−9.42	−5.51	76.27
	III	-	-	-	-	-	-
SW	I	−70.04	−4.62	79.10	1.23	161.65	88.14
	II	−56.20	−5.95	79.49	−1.57	147.87	87.92
	III	-	-	-	-	-	-
SW	I	−71.67	−5.66	84.80	0.77	162.24	87.51
	II	−57.08	−5.58	83.69	−1.03	147.66	87.27
	III	-	-	-	-	-	-

Table 7. Differences between the linear exterior orientation parameters determined using different control point distribution configurations: I and II.

Configuration	Control Points			Check Points		
	dX (m)	dY (m)	dZ (m)	dX (m)	dY (m)	dZ (m)
I	0.005	0.008	0.003	0.014	0.014	0.005
II	0.010	0.014	0.018	0.037	0.046	0.016
III	-	-	-	-	-	-
I	0.007	0.007	0.007	0.016	0.004	0.009
II	0.014	0.008	0.018	0.017	0.003	0.028
III	-	-	-	-	-	-
I	0.009	0.007	0.004	0.019	0.027	0.096
II	0.015	0.016	0.015	0.003	0.012	0.001
III	-	-	-	-	-	-

The stereo orientation of video frames from two different levels was successful. It turned out to be impossible to perform an absolute orientation of aerial and terrestrial imagery with the control point distributed in accordance with configuration III—i.e., when control points are located only on the ground around the building.

This research has shown that for a pair of images with an aerial and terrestrial exterior orientation and having an unfavourable configuration control points, it is impossible to obtain reliable values for the exterior orientation parameters calculated using only control points located around the building (configuration III).

When using the other two control point configurations (I—control points only on the walls and II—control points on the walls and around the building) similar values for the exterior orientation parameters were obtained (differences for the linear parameters of between 0–0.1 m and 0.5°–15° for the angular parameters).

5. Results of the Experiment

A verification was performed, whether it is possible to apply the simplified collinearity equations to determine the X, Y, Z coordinates of the building. Matlab software was used to run an algorithm for calculating the ground coordinates of points (X, Y, Z) based on classic equations sets Equation (2) and the simplified collinearity Equations (3) and (4). The process of determining coordinates used the following input data: known interior and exterior orientation parameters of the camera and the measured image coordinates of the points on the building's walls. The calculated coordinates were compared to theoretical values, which is shown in Table 8 below.

Table 8. RMSE values of the point coordinates calculated using a simplified set of collinearity equations and RMSE values of the point coordinates calculated using the standard collinearity equations for two variants of distribution of the control and tie points.

Tilt Angle	Configuration	mX (cm)	mY (cm)	mZ (cm)	msX (cm)	msY (cm)	msZ (cm)
NE ′ = 80°	I	1	1	3	60	35	260
	II	2	2	2	70	42	260
NE ′ = 85°	I	2	2	3	6	8	17
	II	2	4	4	9	11	15
NW ω = 80°	I	1	1	2	40	20	140
	II	1	1	3	40	20	150
NW ω = 86°	I	2	4	4	7	11	12
	II	3	3	4	8	10	14
SW ′ = 79°	I	2	2	2	39	33	180
	II	2	3	2	32	40	210
SW ′ = 84°	I	2	3	2	6	11	15
	II	2	4	5	9	12	17

mX—RMSE of the X coordinate calculated using the classic set of collinearity equations; mY—RMSE of the Y coordinate calculated using the classic set of collinearity equations; mZ—RMSE of the Z coordinate calculated using the classic set of collinearity equations; msX—RMSE of the X coordinate calculated using the simplified set of collinearity equations; msY—RMSE of the Y coordinate calculated using the simplified set of collinearity equations; msZ—RMSE of the Z coordinate calculated using the simplified set of collinearity equations.

Figures 10 and 11 below illustrate the ratios of the RMSE of the point coordinates (X, Y, Z) calculated using a classic set of collinearity equations to the RMSE of the point coordinates (X, Y, Z) calculated using the simplified collinearity equations.

Similar error values were obtained for all measured walls for tilt angles φ, $\omega \approx 85°$. This means that for tilt angles close to 90°, the mean errors of determining coordinates based on simplified collinearity equations are close to 10 cm. The closer the tilt angle is to 90°, the RMSE value increases for coordinates calculated using the classic set of equations.

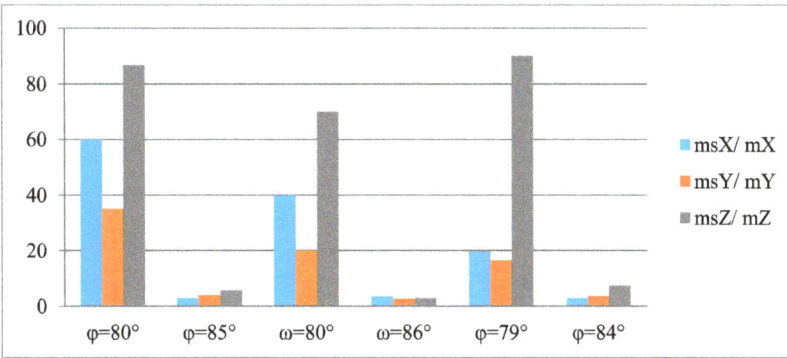

Figure 10. Ratios of the RMSE of the point coordinates calculated using simplified set of collinearity equations to the RMSE of the point coordinates calculated using classic collinearity equations for configuration I.

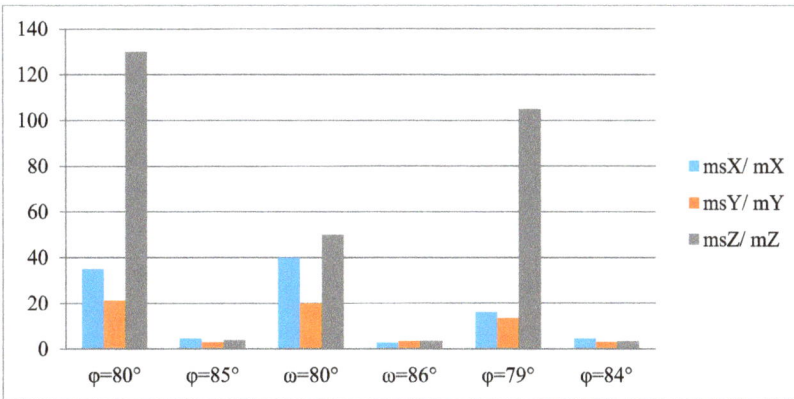

Figure 11. Ratios of the RMSE of the point coordinates calculated using simplified set of collinearity equations to the RMSE of the point coordinates calculated using classic collinearity equations for configuration II.

An analysis of the results obtained using the simplified collinearity equations shows, that the worst results were achieved for a $\varphi = 80°$ tilt angle. Slightly better results were obtained for the ω angle of the same value (80°). This means that when simplifying a collinearity equation using $\omega = 90°$, the new collinearity equation is simplified by a smaller number of rotation matrix coefficients, than when the same is done with angle φ. It should be noticed, that when the tilt angle $\varphi = 90°$, 4 out of 5 rotation matrix coefficients are equal to zero (0). However, when the tilt angle $\omega = 90°$, only one of these coefficients is equal to zero (0).

The greatest RMSE m_{0Z}, both for the classic equation set and for the simplified collinearity equations, had been obtained for the Z coordinate. This is due to the fact that the Z coordinate is calculated based on the values of the other two coordinates (X and Y). Slightly higher errors had been observed for configuration II of the distribution of control and tie points.

6. Determining the Relation between the Established and Actual Tilt Angle of the Video Camera

The research described above had shown that for tilt angles φ and ω close to 90°, the calculated ground coordinates of points on the building using the simplified set of equations are very close to the

Sensors **2016**, *16*, 951

results obtained using the classic set of equations. This means that when performing a flight over a building using a copter, (in case of using the simplified equations) the tilt angle of the camera φ (ω) for all video frames should be within the 85°–90° range.

Therefore, for low-cost unmanned aerial copter systems, which do not have a flight stabilization system, it is essential to determine the difference between the established camera tilt angle ϕ (ω) defined during flight planning and the actual tilt angle value defined during the image orientation process. The difference between their values can differ for different copter systems. This difference depends on many factors, including the speed with which the platform is moving, it's mass, payload capacity, number of engines, etc. Most importantly, this difference in angles will also be affected by the speed and direction of winds. This difference can be determined empirically by conducting a series of test flights at different speeds.

In the article, an experiment had been conducted to show the change in the difference between the established and actual tilt angles of a camera mounted on a UAV computer system in relation to the speed of the platform. A series of 6 flights were performed at chosen speeds: 1, 3, 5, 7, 9, 11 m/s. The experiment was performed during a windless day.

For the UAV system flying at a speed of 3 m/s, the difference between the established camera tilt angle and the same angle derived from the exterior orientation parameters was equal to 3°. The experiment shows that with an increase in the UAV's speed, the difference between the established and actual tilt angles lessens, therefore making the platform's flight more stable (Figure 12). During video acquisition of an architectural structure at high flying speeds, the possibility of image blur must be taken into account. The assumption that the actual camera tilt angle ϕ (ω) is equal to the established angle is only possible with high-end copters, equipped with modern stabilization systems.

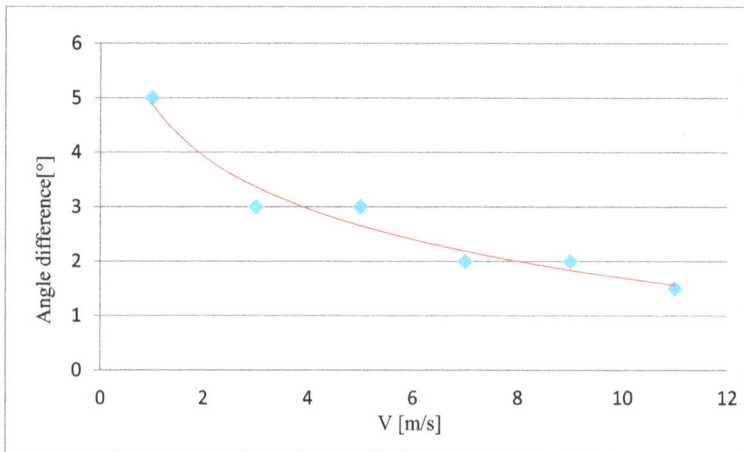

Figure 12. Chart of the relation between the difference in the established and actual camera tilt angles and the UAV speed.

7. Discussion

Starting from the collinearity equations, which are the basis for photogrammetry, simplified mathematical model was proposed. The use of a simplified collinearity equations can significantly shorten the processing of image data from UAV. It was found that it is possible to perform an absolute orientation of video images acquired using a video camera mounted on a UAV, with a smooth transition from a terrestrial to aerial exterior orientation.

Experiments were conducted on a geometrically simple architectural structure. The video camera was mounted on a UAV copter system not equipped with a high precision GPS/INS measurement system. Therefore the exterior orientation parameters were determined using space resection—by measuring the location of control and tie points in two different distribution configurations. The target used to signalise these points made it possible to identify control points in two planes—on the wall and on the ground. The experiments have proven, that for two out of the three proposed control point distribution configurations—I and II (I—control points on the facade, II—control points on the facade and around the object, III—control points around the object) it is possible to obtain good results when processing video imagery. Slightly better results were obtained for configuration I.

The errors in point ground coordinates calculated using simplified collinearity equations oscillate about a few to a few hundred centimetres. When using the classic set of collinearity equations, the errors in coordinates are greater when the tilt angle ϕ (ω) $\approx 85°$, i.e., when the value of this angle is close to the angle (90°). The opposite is true when the coordinates are calculate using the simplified collinearity equations, where the errors for all coordinates (m_{0X} m_{0Y}, m_Z) become smaller.

The proposed method deals with the problem of integrating 3D models of the roof and walls of an architectural structure. The presented algorithm for determining the X, Y, Z coordinates of a structure had been established with the thought of designing an unmanned system, the aim of which would be to determine the coordinates on the surface of a structure in near-real-time. The designed algorithm could be implemented in a device onboard the UAV. Given known exterior orientation parameters, its purpose would be to automatically determine the X, Y, Z coordinates of points on the surface of a building. In the situation that one of the camera tilt angles ϕ or ω would be between 85°–90°, the set of simplified collinearity equations would be used to calculated the X, Y, Z coordinates of the points (Figure 13).

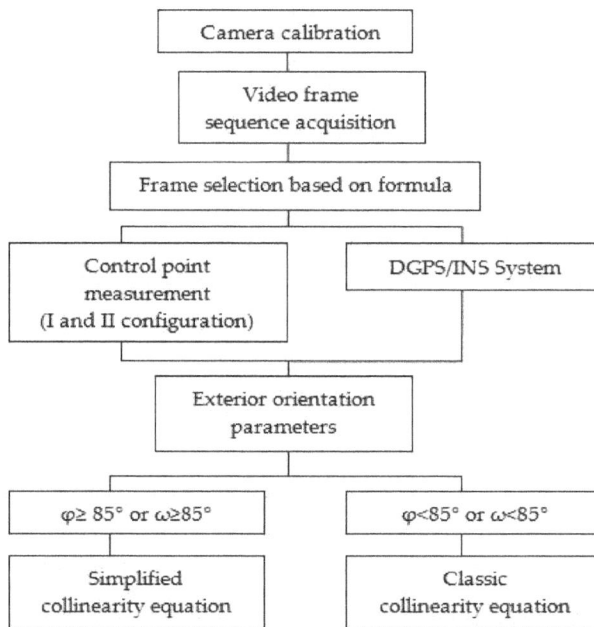

Figure 13. Proposed methodology for integrating video images acquired from terrestrial and aerial orientations of the camera mounted on the UAV.

The proposed methods will shorten the calculation time by over 50%. In the case of a low cost system, the number of equations has a great influence on the computational speed.

8. Conclusions

This paper describes the issue of performing absolute orientation of video images which have two different exterior orientations of the camera mounted on the UAV: terrestrial and aerial. The proposed method can be used to determine the X, Y, Z coordinates of points based on a set of collinearity equations of a pair of images, with the equations for the terrestrial exterior orientation image being greatly simplified, assuming that the camera tilt angle is equal to $90°$. The aim of this simplification is to limit the number of intermediate computations when calculating the coordinates of points on the surface of architectural structures in a ground coordinate system in the special case of image orientation mentioned.

Acknowledgments: This paper has been supported by the Military University of Technology, the Faculty of Civil Engineering and Geodesy, Department of Remote Sensing and Photogrammetry as a research project 933/2016.

Author Contributions: The work presented in this paper was carried out in collaboration between both authors. M.K. and P.D. designed the method, acquired data and wrote this paper. P.D. carried the laboratory experiments, analyzed data and interpreted the results. Both authors have contributed to seen and approved the manuscript.

Conflicts of Interest: The authors declare no conflict of interest.

References

1. Brutto, M.; Garraffa, A.; Meli, P. UAV Platforms for Cultural Heritage Survey: First Results. *ISPRS Ann. Photogramm. Remote Sens. Spat. Inf. Sci.* **2014**, *II-5*, 227–234. [CrossRef]
2. Wilińska, M.; Kędzierski, M.; Fryśkowska, A.; Deliś, P. Noninvasive Methods of Determining Historical Objects Deformation Using TLS. In *Structural Analysis of Historical Construction*; Jasieńko, J., Ed.; Dolnośląskie Wydawnictwo Edukacyjne: Wrocław, Poland, 2012; Volume 3, pp. 2582–2588.
3. Kędzierski, M.; Fryśkowska, A.; Wierzbicki, D. *Opracowania Fotogrametryczne z Niskiego Pułapu*; Redakcja Wydawnictw Wojskowej Akademii Technicznej: Warszawa, Poland, 2014.
4. Markiewicz, J.S.; Podlasiak, P.; Zawieska, D. A New Approach to the Generation of Orthoimages of Cultural Heritage Objects—Integrating TLS and Image Data. *Remote Sens.* **2015**, *7*, 16963–16985. [CrossRef]
5. Haarbrink, R.B.; Eisenbeiss, H. Accurate DSM Production from Unmanned Helicopter Systems. *Int. Arch. Photogramm. Remote Sens. Spat. Inf. Sci.* **2008**, *B1*, 1259–1264.
6. Eltner, A.; Schneider, D. Analysis of different methods for 3D reconstruction of natural surfaces from parallel-axes UAV images. *Photogramm. Rec.* **2015**, *30*, 279–299. [CrossRef]
7. Shahnazi, M.; Sohn, G.; Theau, J.; Menard, P. Development and Evaluation of a UAV—Photogrammetry System for Precise 3D Environmental Modeling. *Sensors* **2015**, *15*, 27493–27524. [CrossRef] [PubMed]
8. Gruen, A.; Huang, X.; Qin, R.; Du, T.; Fang, W.; Boavida, J.; Oliveira, A. Joint Processing of UAV Imagery and Terrestrial Mobile Mapping System Data for very High Resolution City Modeling. *Int. Arch. Photogramm. Remote Sens. Spat. Inf. Sci.* **2013**, *XL-1/W2*, 175–182. [CrossRef]
9. Wefelscheid, C.; Hansch, R.; Hellwich, O. Three-Dimensional Building Reconstruction Using Images Obtained by Unmanned Aerial Vehicles. *Int. Arch. Photogramm. Remote Sens. Spat. Inf. Sci.* **2011**, *38–1/C22*. [CrossRef]
10. Jang, H.S.; Lee, J.C.; Kim, M.S.; Kang, I.J.; Kim, C.K. Construction of National Cultural Heritage Management System Using RC Helicopter Photographic Surveying System. In Proceedings of the XXth International Society for Photogrammetry and Remote Sensing Congress, Istanbul, Turkey, 12–23 July 2004.
11. Hartley, R.I.; Zisserman, A. *Multiple View Geometry in Computer Vision*, 2nd ed.; Cambridge University Press: Cambridge, UK, 2004.
12. Dardi, F.; Abate, L.; Ramponi, G. No-Reference Measurement of Perceptually Significant Blurriness in Video Frames. *Signal. Imag. Video Process.* **2011**, *5*, 271–282. [CrossRef]
13. Quian, Q.; Gunturk, B. Blind super-resolution restoration with frame by frame nonparametric blur estimation. *Multidimens. Syst. Signal Process.* **2013**, *27*, 255–273. [CrossRef]
14. Kędzierski, M.; Wierzbicki, D. Radiometric quality assessment of images acquired by UAV's in various lighting and weather conditions. *Measurement* **2015**, *76*, 156–169.
15. Fryśkowska, A.; Kędzierski, M. Methods of laser scanning point clouds integration in precise 3D building modeling. *Measurement* **2015**, *74*, 221–232.

16. Tong, X.; Liu, X.; Peng, C.; Liu, S.; Luan, K.; Li, L.; Liu, S.; Liu, X.; Xie, H.; Jin, Y.; Hong, Z. Integration of UAV-Based Photogrammetry and Terrestrial Laser Canning for the Three-Dimensional Mapping and Monitoring of Open-Pit Mine Areas. *Remote Sens.* **2015**, *7*, 6635–6662. [CrossRef]

17. Bolognesi, M.; Furini, A.; Russo, V.; Pellegrinelli, A.; Russo, P. Accuracy of Cultural Heritage 3D Models by RPAs and Terrestrial Photogrammetry. In Proceedings of the International Archives of the Photogrammetry, Remote Sensing and Spatial Information Sciences 2014, ISPRS Technical Commission V Symposium, Riva del Garda, Italy, 23–25 June 2014.

18. Püschel, H.; Sauerbier, M.; Eisenbeiss, H. A 3D model of Castle Landenberg (CH) from Combined Photogrammetric Processing of Terrestrial and UAV-Based Images. In Proceedings of the International Archives of Photogrammetry, Remote Sensing and Spatial Information Sciences, Riva del Garda, Italy, 23–25 June 2014; pp. 96–98.

19. Robertson, D.P.; Varga, R.M. Structure from Motion. In *Practical Image Processing and Computer Vision*; Halsted Press: New York, NY, USA, 2009; Volume 13, pp. 1–49.

20. Barazzetti, L.; Scaioni, M. Automatic Orientation of Image Sequences for 3D Object Reconstruction: First Results of a Method Integrating Photogrammetric and Computer Vision Algorithms. In Proceedings of 3D-ARCH 2009, Trento, Italy, 25–28 February 2009.

21. Pozzoli, A.; Mussio, L. Quick solutions particularly in close range photogrammetry. *Int. Arch. Photogramm. Remote Sens. Spat. Inf. Sci.* **2003**, *XXXIV*, 273–278.

22. McGlone, J.C.; Mikhail, E.M.; Bethel, J.S.; Mullen, R. *Manual of Photogrammetry*; American Society for Photogrammetry and Remote Sensing: Bethesda, MD, USA, 2004.

23. Dąbrowski, R.; Deliś, P.; Wyszyński, M. Analysis of the possibility of using a video camera as a UAV sensor. In Proceedings of the 9th International Conference Environmental Engineering, Vilnius, Lithuania, 22–23 May 2014.

sensors

MDPI

Article

Towards Autonomous Modular UAV Missions: The Detection, Geo-Location and Landing Paradigm

Sarantis Kyristsis [1], Angelos Antonopoulos [2], Theofilos Chanialakis [2], Emmanouel Stefanakis [2], Christos Linardos [1], Achilles Tripolitsiotis [1,3] and Panagiotis Partsinevelos [1,4,*

1   School of Mineral Resources Engineering, Technical University of Crete, Chania 73100, Greece; sarkyritsis@gmail.com (S.K.); linardos_thegreat@hotmail.com (C.L.); atripol@mred.tuc.gr (A.T.)
2   School of Electrical and Computer Engineering, Technical University of Crete, Chania 73100, Greece; a_antonopoulos@outlook.com (A.A.); theofilos13@hotmail.com (T.C.); emmanuelstef95@gmail.com (E.S.)
3   Space Geomatica Ltd., Chania 73133, Greece
4   School of Mineral Resources Engineering, Laboratory of Geodesy and Geomatics Engineering, SenseLAB Research Group, Chania 73100, Greece
*   Correspondence: ppartsi@mred.tuc.gr; Tel.: +30-697-447-8823

Academic Editors: Felipe Gonzalez Toro and Antonios Tsourdos
Received: 27 August 2016; Accepted: 25 October 2016; Published: 3 November 2016

Abstract: Nowadays, various unmanned aerial vehicle (UAV) applications become increasingly demanding since they require real-time, autonomous and intelligent functions. Towards this end, in the present study, a fully autonomous UAV scenario is implemented, including the tasks of area scanning, target recognition, geo-location, monitoring, following and finally landing on a high speed moving platform. The underlying methodology includes AprilTag target identification through Graphics Processing Unit (GPU) parallelized processing, image processing and several optimized locations and approach algorithms employing gimbal movement, Global Navigation Satellite System (GNSS) readings and UAV navigation. For the experimentation, a commercial and a custom made quad-copter prototype were used, portraying a high and a low-computational embedded platform alternative. Among the successful targeting and follow procedures, it is shown that the landing approach can be successfully performed even under high platform speeds.

Keywords: UAV; search and rescue; autonomous landing; smart-phone drone

1. Introduction

The global commercial unmanned aerial vehicle (UAV) market has shown considerable growth [1], gradually embracing a wide range of applications. Irrespective of the underlying application (media and entertainment, energy, government, agriculture, search and rescue (SAR), environment, etc.), the majority of common UAV operations involve image and video acquisition, maneuvering, navigational attitude, routing to predefined waypoints, landing and failsafe features, which are processed and fulfilled through the standard flight controller systems. Nevertheless, in order to address a broader range of applications, one has to migrate to integrated processing add-ons that would carry out on demand, on-board, collaborative or autonomous functions, towards an intelligent UAV functionality. Increasing the computational potential of the UAV does not solely address computationally demanding tasks, but transforms the UAV into a potential decision operating system that can take control of its own flight and perform optimized missions.

The significance of UAV autonomy and processing power is demonstrated in several applications including technical infrastructure inspection [2–6], indoor navigation using simultaneous localization and mapping [7–10], obstacle avoidance [11–14], terrain reconstruction [15–17], and real-time environmental monitoring [18,19]. The flexibility, human safety, cost effectiveness and ease of use facilitates the usage of UAVs to support humanitarian actions in disaster situations [20,21]. The identification and rescue of potential victims in the shortest possible time constitutes the main objective of search and rescue operations [22,23].

In the current study, we address an autonomous operating scenario of a UAV that takes off, performs an area scan, and finally lands onto a moving platform. This concept was partially addressed in an ongoing challenge by DJI (2016 DJI Developer Challenge). More specifically, the overall scenario includes a UAV that takes off from a moving platform to extract, recognize, track and follow possible targets where markers in the form of AprilTags are scattered to represent survivors (true IDs) and non-survivors (false IDs). After real-time geo-location of the survivors through geographic coordinates, the UAV approaches, follows and autonomously lands on the moving platform. Both the UAV and platform are equipped with a Global Navigation Satellite System (GNSS). Landing on a high-speed moving platform constitutes a highly demanding task and is addressed in several studies under varying environments. In [24], optical flow is used for indoor landing of a vertical takeoff and landing (VTOL) UAV on a moving platform. Apparently, the platform was moving manually at low speeds only in the vertical direction. No details are given about the respective speed, but this optical flow approach was simulated with a speed most certainly under 6 m/s. An algorithm based on robust Kalman filters for vision-based landing of a helicopter on a moving target is given in [25]. Extended Kalman and Extended H_∞ filters were used in [26] to fuse visual measurements with inertial data in order to maintain a high sampling rate and cover short-period occlusions of the target. This algorithm assumes that when in hover, the aerial vehicle presents zero roll, pitch and yaw values, along with zero movement in the northing and easting. However, as the authors also recognize, this is almost impossible to achieve in real life. IR light pattern detection is used in [27], but the landing platform was moving with a speed as low as 40 cm/s. In [28], a rope is used to link the UAV with a moving platform and sensor fusion is applied to provide the relative position and velocity state estimation. A similar tethering approach is presented in [29]. More recently, linear programming-based path planning is proposed in [30], whereas the usage of an omnidirectional camera for outdoor autonomous landing on a moving platform is investigated in [31].

The velocity and size of the moving platform are determinative parameters on the success of the previous approaches, especially when different types (i.e., car, ship, etc.) of platforms are examined. The size of a UAV landing onboard a ship (i.e., [32–35]) is several levels of magnitude smaller than the available landing space. At the same time, to the best of our knowledge, there are only a few studies that propose techniques for automated UAV landing on fast moving (i.e., >20 km/h) cars. For example, in [36], an unmanned aircraft travelling at 75 km/h lands on the roof of a moving car with the assistance of a relatively large landing pad that seems impractical to be attached in common cars, while the car speed provides a moving air runway that actually facilitates landing for these types of UAVs.

The present work differs from previous studies since the vehicle size is comparable to the landing platform dimensions and the landing platform moves at more than 20 km/h. Furthermore, the main contributions of this work include: (1) improvement of visual marker detection rate to 30 frames per second; (2) accurate target geolocation; (3) implementation of a novel "aggressive" landing approach technique inspired from standard practices of airplane pilots; and (4) implementation of a low-computational landing approach with our "smart-drone" prototype setting, which constitutes a real revolution towards the delivery of a low-cost, small size personalized UAV with enhanced capabilities, not even tackled from large scale commercial UAVs.

The rest of the paper is structured as follows: Section 2 presents the hardware and software components used in the underlying approaches, followed by Section 3 where the identification and geo-location methodology is detailed. In Section 4, the implementation of the follow-up and landing on a high speed moving platform is presented along with the processing capabilities of a UAV smart-phone device system to provide a low-cost, modular alternative to commercial UAV products. Finally, in Section 5, the main findings of the present study and future plans are discussed.

2. Materials and Methods

In the subsequently described approaches, we use two different hardware and software settings representing a high and low computational potential of the underlying processing power. Both scenarios are based on the same architectural design (Figure 1). The flight controller (FC) of the UAV is responsible for the flying attitude and stability of the vehicle. An embedded device interacts with the FC and the peripheral devices (i.e., camera, sonars). The embedded device collects the data from the peripherals, and after processing, it decides about the movement of the UAV (i.e., obstacle avoidance, landing pad detection). The decisions are forwarded to the FC as flight commands (i.e., moving to x, y, z or starting to descend at x m/s).

In the first scenario, the following hardware and software components were used: a DJI Matrice 100 UAV platform (Shenzhen, China), X3 Zenmuse camera (Shenzhen, China) with a gimbal and 4 k @ 30 fps and 1080 p @ 60 fps, a NVIDIA Tegra K1 SOC embedded processor (Santa Clara, CA, USA) (DJI Manifold component), and the DJI Guidance Sensor (Shenzhen, China). In the second scenario, a Quadrotor UAV (Senselab, Chania, Greece), an ATMEGA 2560 flight controller (San Jose, CA, USA) and the processing power of a mobile device are used. The latter constitutes our "smart-drone" implementation [37], which integrates a standard smart-phone into a custom made UAV carrier. The DJI Matrice 100 comes with the A2 GPS PRO Plus GPS receiver (Shenzhen, China), whereas the custom built "smart-drone" setting is powered by the smart-phone's GPS receiver. Apparently, custom settings may incorporate more advanced multi-frequency, multi-constellation GNSS boards if sub-centimeter accuracy in AprilTag geolocation is required, yet this implementation is far beyond the scope of the current study.

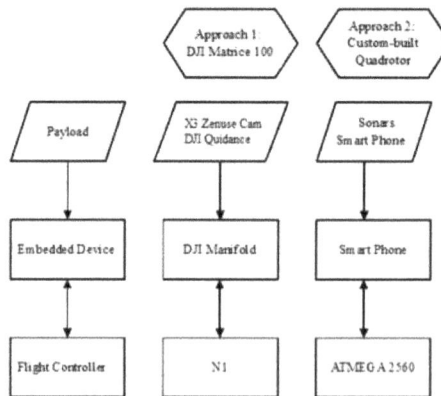

Figure 1. The general architectural design for the two alternative hardware/software approaches.

Various types and sizes of AprilTags are used to denote the survivors and landing targets. The AprilTag markers [38] are high-contrast, two-dimensional tags designed to be robust to low image resolution, occlusions, rotations, and lighting variation [38] that have been used in several UAV-related applications [9,39,40] (Figure 2). Other 2D markers such as ARTag [41] have been used for UAV localization [42]; however, in this work, we used AprilTag mainly due to the restrictions set by the aforementioned challenge. However, our approach can be easily adapted to work with other tags.

In both scenarios, the 25h9 AprilTags with dimensions of 6 cm × 6 cm are used to identify the victim, whereas the 16h9 and 36h11 tags with dimensions 40 cm × 40 cm are used to determine the landing platform. In the first scenario, the camera operates in 4 k @ 30 fps and 1080 p @ 60 fps, although in our application, the feed that was retrieved was 720 p @ 60 fps. In the second scenario, the majority of the smart-phone cameras record 1080 p @ 30 fps videos, although several devices support 1080 p @ 60 fps.

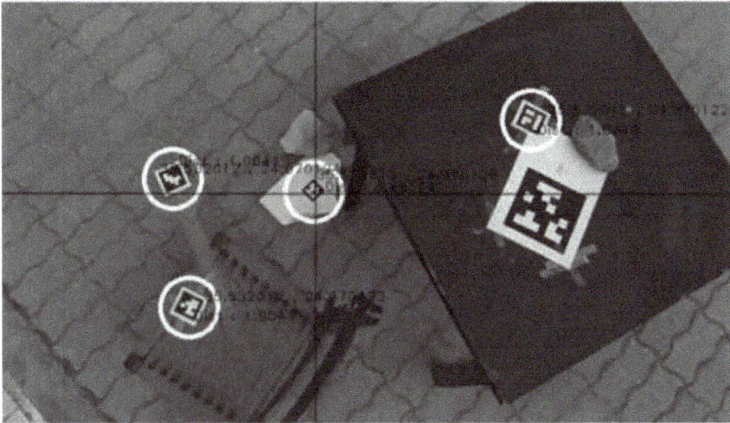

Figure 2. Multiple AprilTag detection, identification and geo-location. The large AprilTag on the right is not detected as its identity is not included in the survivor IDs.

The software was developed within the Robotic Operating System (ROS) environment using OpenCV functions. In the first scenario, the DJI libraries for the connection with the drone and camera were utilized, while in the second scenario, the application was implemented on the Android OS.

As described above, the current implementation refers mainly to multi-copters, the majority of which navigate under a tilted fashion, making landing while moving, a non-trivial task. Furthermore, the landing area is considered comparable to the UAV size (i.e., car roof-top), making experimentation and testing quite challenging.

In order for a user to be able to use the described capabilities, an Android application is implemented (Figure 3, down-left). Through this application, the end user can mark a survey area on a map. The UAV takes off from the moving platform, scans the full area, identifies the required targets; while upon completion, it seeks the landing platform. Using the camera feed, it tries to detect the landing target while approaching the GPS coordinates of the platform. Once the platform has been detected, the drone follows and starts descending towards the platform's landing pad.

Figure 3. Scanning and sequential AprilTag detection, identification and geo-location. The mobile application screenshot shows survivor coordinates and IDs.

3. Search, Detection and Geo-Location Implementation and Results

3.1. Target Recognition of Survivor AprilTags

As described by the implementation scenario, the UAV, after taking off from the moving platform, navigates towards the search area, where markers in the form of AprilTags are scattered to represent victims (true IDs) and non-victims (false IDs). According to the AprilTag size and placement, flying height, gimbal speed and camera mode (resolution vs. frame rate), the UAV adjusts its speed to accomplish a safe frame acquisition. Each acquired frame by the camera is forwarded to the embedded processor, where the dedicated ROS node processes it in order to detect the AprilTag.

We focused our efforts on implementing a high frame rate detection approach (i.e., ~30 fps), enabling the survivor detection in real-time. Thus, several computational approaches were undertaken (Figure 4):

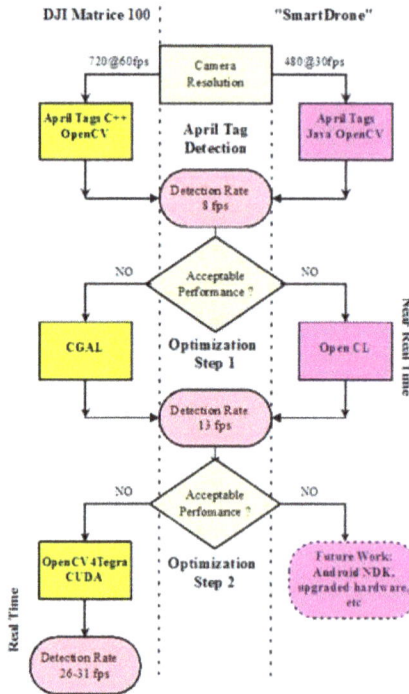

Figure 4. AprilTag detection flowchart for the two settings (DJI Matrice 100 and custom "smart-drone"). With the first setting and after two optimization steps, a detection rate of 26–31 fps was accomplished, whereas the second solution accomplishes a detection rate of 13 fps.

Initially, the AprilTags C++ library [43] that is available in open source under GNU LPGL v2.1 [44] was employed. This detection approach was successful in identifying and locating the AprilTags but within a low frame rate, mainly due to its poor computational performance.

The second detection approach included the OpenCV implementation based on the solution provided in [44]. Under this implementation, AprilTag detection and recognition was accomplished under a rate of about eight frames per second.

Although the frame rate was greatly improved, in order to further improve the performance of the algorithm, we used the CGal-based approach available in [45] to detect quads. This approach gave a great performance upgrade, resulting in 13 frames per second, and the results seemed to be near real-time.

To further enhance the system's performance, we parallelized the process as Graphics Processing Unit (GPU) functions. This seems to be viable, as the detection process described in [38] does not involve sophisticated algorithms that demand large local memory [46]. A similar approach was proposed for QR tags in [47]. Therefore, we used the OpenCV4Tegra [48] framework that permits performance of all OpenCV functions in parallel as GPU functions. However, some of the functions described in [38] are computationally expensive, and, furthermore, they are not available in OpenCV. For example, the "Find Union" function iteratively constructed disjoint subsets inside the initial pixel set of the image, resulting in a processing bottleneck. To overcome such computationally demanding tasks, the CUDA parallel computing platform [49] was employed along with appropriate modifications to the implementation.

The final AprilTag detection optimization was performed at a rate of 26–31 frames per second. In addition, 6 cm × 6 cm moving AprilTags were detected and identified at a distance of about 4 m, even at extreme visual angles (Figure 5). Single false positive detections occur rarely, while more than one "true positive" frame recognitions verify the AprilTag ID, making the overall false positive occasions obsolete. When the wind, UAV or survivor movements are considerable, some disruptions in the detection occur, yet they do not contaminate the real-time processing of the implementation. In any case, approaches such as the one presented in [50] for the control of the UAV velocity and attitude under various disturbances may be further incorporated into the overall system architecture.

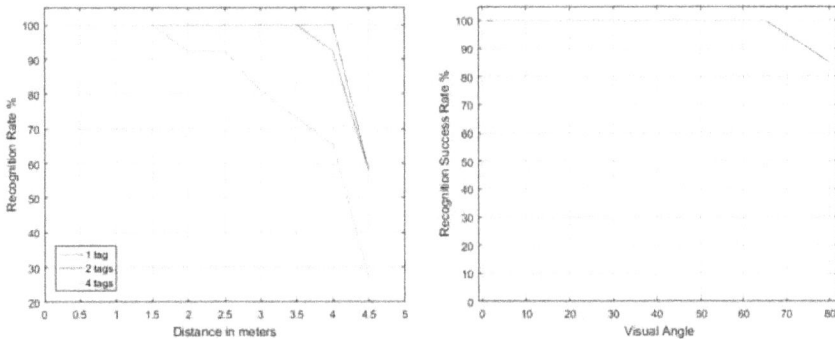

Figure 5. Left: the detection success rate for multiple (1, 2 and 4 AprilTags) targets and distances (UAV to target); **Right:** the detection success rate for single target per visual angle.

3.2. Geo-Location of Survivor AprilTags

The coordinates of the identified AprilTags are determined via the image geometry, gimbal orientation and the UAV GPS readings. However, as an object moves away from the principal point of the frame, large distortions occur in the image geometry. Thus, after the initial identification of the AprilTag, the gimbal moves to position it in the center of the camera's field-of-view (FOV), the principal point. The detection algorithm uses homography pose estimation to extract the translation and rotation of the AprilTag's center coordinates with respect to the image center, AprilTag size, and the camera's intrinsic and extrinsic characteristics. The latter are derived through dedicated camera calibration, performed either offline or even online using: (a) the DJI API (Application Programming Intreface) calls for acquiring camera data; and (b) well known implementations using OpenCV. The AprilTag coordinates are generally calculated relative to the camera center.

In order to provide geographical coordinates to the AprilTag, we have to take into account that three different coordinate systems exist: (1) the 3D UAV system; (2) the gimbal mounted under the frame that houses the camera; and (3) the 2D image coordinates. Thus, the problem is obtaining the world coordinates of the AprilTag marker given its image coordinates. Thus, the following approach was implemented to transform the camera's to the UAV's reference system: when a new frame is available for processing, the gimbal and compass values are recorded and their angular distance is calculated. Therefore, the target's distance and orientation with respect to the UAV can be computed, knowing the geometry of the UAV and its components. As a final step, the UAV's GPS records are used as inputs into the Vincenty's Direct Formulae [51] along with the target's distance and orientation provided in the previous step. These formulae deliver the target's GPS coordinates.

Furthermore, in the case of multiple AprilTag detection, the identification and localization initiates by sequentially detecting the closest to the principal point AprilTag, centering it through the camera-gimbal movement and providing its coordinates in real-time (Figure 2). Then, the specific AprilTag ID is excluded from further searching, and the next one is consecutively centered providing its coordinates. Thus, step by step, all AprilTags are identified only once.

3.3. Area Scanning Approach

The navigation approach in terms of the routing and scanning schedule is a key task that needs to be optimized. A first approach divides the search area in parallel adjacent flying strips, a well-known geometry-based photogrammetric task. Based on the camera's focal length, flying height, camera's pixel resolution (x and y), side-lap, angular FOV, target scaling factor, etc., the scan lines can be easily estimated.

The UAV initially retains a constant flight height, determined by the size of the AprilTags with the use of the barometer and sonars onboard. Occlusion and obstacles are undertaken through an obstacle avoidance system (e.g., MB1230 sonar sensor, MaxBotix, Brainerd, MN, USA,) or by determining a height map—elevation model of the area through a sensor system (e.g., DJI Guidance). According to the obstacle avoidance readings, the UAV adjusts its altitude not only to avoid the obstacles but also to lower its elevation and acquire a better view of occluded areas. Under the lowered or elevated positions, a full camera rotation is performed in order to include as many occluded AprilTags as possible for an optimized visibility scan (Figure 3). The rotation is facilitated by using both gimbal and drone attitude.

4. Takeoff and Landing Approach

Independent of the UAV setting used (i.e., DJI Matrice 100 or "smart-drone") takeoff is a straightforward task and no problems were identified since the flight controller, in both settings, is able to manage the stabilization issues that occur when taking off from a moving platform, tested up to 30 km/h (Figure 6). However, distinct procedures for landing have been taken into consideration.

Figure 6. UAV taking off of a moving vehicle.

4.1. High-Computational Processing Setting

In this case, when the landing target is stationary, the actual landing point is within a few centimeters (1–5 cm) from the expected one. When the landing is performed on a moving platform, three distinct approaches were implemented, one based on *follow*, one on *aggressive*, and a third on an integrated *hybrid* mode.

According to the *follow* mode, the UAV follows the moving platform from a safe distance and flying height according to the GPS readings of the platform, until the camera locks the AprilTag landing pattern (Figure 7). The approach of the UAV is gradual, meaning that it lowers its altitude, while the AprilTag remains targeted until it reaches the landing pad. If, for any reason, the AprilTag is lost for a predefined frame number, the UAV rises to retarget the AprilTag.

Figure 7. The *following* approach under extreme turning and speed sequences.

Under the *aggressive* approach, the UAV hovers over the area of interest and approximates the platform's relative angular position moving trajectory through the GPS and then uses the camera to target and lock in the moving platform. This approach uses real-time error correction and trajectory planning, achieved with the use of the OpenCV, ROS API, and DJI API calls. In this approach, the drone detects the vehicle, directly targets the theoretical predicted intersection between their courses, and follows the closest distance approach. The UAV then navigates in a direct approach towards the platform.

The *hybrid* approach takes advantage of the speed and optimization of the *aggressive* approach, until the platform is close enough to identify the landing AprilTag. Then, the landing procedure of the *follow* approach is performed. Thus, the UAV selects the shortest path to the target, and at the final stage of approach, tries to hover above the target and slowly descend into landing.

All approaches were tested under various platform speeds and conditions. We had successful touchdowns from stationary targets to moving targets, with speeds reaching 30 km/h. The moving target scenarios tested included both following a target and landing, and also locating a distant target, reaching it and then landing (Figure 8). In our testing, all landings were successful in approaching the platform; however, at high speeds and wide platform turning angles, the final 25 cm of descent resulted sometimes in less accurate landings and instability.

Therefore, depending on the moving platform type, when its speed reaches more than 20 km/h, a stabilizing practice should be devised in order to alleviate the gradual friction of the UAV landing legs that may lead to unstable landing or even crashing. Such implementations may include reconfigurable perching elements [52], electromagnets in landing legs or platform, fastening straps, retractable hooks, etc.

Figure 8. UAV landing on a moving platform.

4.2. Low-Computational Processing Setting

The previous section detailed the procedures for landing a UAV powered with a high-computational processing unit on a moving platform. In this section, we discuss the case of the smart-drone unit, where the obstacle avoidance and landing procedures need to be revisited.

In order to detect the landing base, real-time image processing takes place in the smart-drone's mobile device (Figure 9). More specifically, the acquired feed from the smart-phone camera is processed in real-time to detect the landing base. The base is designed as a discrete circle on a homogenous background that is colored in such a way that it can be easily distinguished from the surroundings. From an algorithmic point of view, each acquired still frame from the image stream is transformed to HSV (Hue, Saturation, Value), and the contours that are relative to the target's specific HSV values are calculated. After this filtering, the circle areas in the image are determined through the Hough circles function, so that that the base can be identified. In order to estimate the UAV distance from the landing base, the pose of the target circle needs to be determined. When facing the target at a 90° angle, the distance is estimated by calculating the target size in pixels and referencing it to its calibrated values. When facing from a different angle, we calculate the homography and extract the rotation and translation of the target. Thus, while the exact position of the object into the 3D space and its projection to a 2D surface are determined, we estimate the distance by considering the pixel area of the target.

Figure 9. Smart-drone prototype and corresponding landing pad.

The processing is undertaken by the smart-phone processor, and flight correction data are forwarded to the flight controller in order to change its position to new relative (x, y, z) coordinates and descend at a given rate, until the UAV reaches the landing pad.

5. Discussion and Conclusions

In this study, we presented a UAV related scenario representing an abundance of applications characterized by search, detect, geo-locate and automatic landing components. The key function of our approach is the UAV computational autonomy, which adjusts its attitude according to the specific environment. The underlying application adheres to a Search and Rescue paradigm, where the UAV takes off, scans a predefined area, geo-locates possible survivor targets, and, upon completion, returns to land on a moving platform. In order to support computational autonomy and perform "expensive" real-time tasks, we integrated high and relatively low-computational embedded systems.

Both implementations were able to identify the targets and proceeded successfully to the landing platforms. The first approach, however, proved more reliable, as it was able to keep on tracking and approaching in real-time, fast moving landing targets. The low-computational smart-drone approach seems very promising, but still needs improvement in the stabilization system of the phone's camera. Specifically, a gimbal-like mechanism needs to be constructed specifically for the mobile phone.

The challenging tasks of the described scenario include the real-time target detection and identification, and the follow, approach and landing implementations in the moving platform. Real time target detection was accomplished using a GPU parallelization, in order to achieve 30 fps processing speed. The significance of the processing frame rate is justified due to the high speed landing platforms, under which our implementation succeeded with following and landing upon. In several other studies, either the platform velocity was low, or the size of the moving platform was considerably large in order to enable a safe landing [32–35]. To the best of our knowledge, there are only a few studies that propose techniques for automated UAV landing on fast moving (i.e., >20 km/h) cars [35]. Even then, the platform speed facilitates landing of airplane-like UAVs, since it actually provides a moving air runway.

For multi-copters and relatively small landing areas, our implementation provides a novel approach scheme. Thus, our "aggressive" approach for the multi-copters is used to provide a robust and direct way for the UAV to reach the landing target. Furthermore, the usual "follow" mode used in recreation or most of the commercial follow-me approaches remains dangerous when occluded, and vegetated areas are evident since the UAV will adjust its attitude relative to the moving target under a lagged approach. The "aggressive" approach, when utilized after the landing target camera locking, ensures an obstacle-free corridor towards a safe landing.

Another subtle detail of our approach is the real-time geo-location capability that provides geographic coordinates under an optimized implementation. Since camera distortion may provide inaccurate location estimations, principal point targeting, pose extraction and calibration are used to provide more accurate coordinates. Geo-location is crucial especially in SAR applications in large regions, such as the ocean, where multiple people may be in danger. The exact location of multiple people is not achieved by today's means, since the common practice involves merely visual inspection, especially when UAVs fly relatively high. Under our approach, the authorities along with the rescue and recovery teams, may receive, in real-time, correct coordinates for multiple moving people, all at once, assisting with precise and timely rescue.

It is also noted that the whole methodology is not limited in SAR applications, but seamlessly reflects other detection scenarios including environmental monitoring and pattern recognition (overgrazing, deforestation, etc.), vegetation identification for endangered species geo-location, animal observation, protection from poaching, etc.

When the moving platform speed is more than 20 km/h, a stabilizing system should be used to protect the UAV from flipping or crashing. Thus, for secure landing, electromagnets in landing legs or platform, suction systems or servo-actuated hooks are some solutions that currently are being tested by our group.

Supplementary Materials: A "demonstration video" is available online at https://youtu.be/uj1OQxMQetA.

Acknowledgments: We would like to acknowledge DJI who under the DJI–Ford 2016 challenge offered free of charge the equipment of the Matrice 100, Manifold, X3 Zenmuse camera, Guidance, replacement parts and the challenge tasks.

Author Contributions: Sarantis Kyritsis, Angelos Antonopoulos, Theofilos Chanialakis and Emmanouel Stefanakis were the main developers of the implementations, and, along with Achilles Tripolitsiotis, performed the tests. Christos Linardos, as a certified airplane pilot, designed the landing approach scenario, and Achilles Tripolitsiotis, along with Sarantis Kyritsis and Panagiotis Partsinevelos, who served as the coordinator of the research group, wrote the paper.

Conflicts of Interest: The authors declare no conflict of interest. The founding sponsors had no role in the design of the study; in the collection, analyses, or interpretation of data; in the writing of the manuscript, and in the decision to publish the results.

References

1. GrandViewResearch. Available online: http://www.grandviewresearch.com/industry-analysis/commercial-uav-market (accessed on 15 August 2016).
2. Austin, R. Unmanned Aircraft Systems: UAVs design, development and deployment. In *Aerospace Series*; Moir, I., Seabridge, A., Langton, R., Eds.; Wiley and Sons: Chichester, UK, 2010.
3. Ham, Y.; Ham, K.K.; Lin, J.J.; Golparvar-Fard, M. Visual monitoring of civil infrastructure systems via camera-equipped unmanned aerial vehicles (UAVs): A review of related works. *Vis. Eng.* **2016**, *4*, 1. [CrossRef]
4. Chan, B.; Guan, H.; Blumenstein, M. Towards UAV-based bridge inspection systems: A review and an application perspective. *Struct. Monit. Maint.* **2015**, *2*, 283–300. [CrossRef]
5. Eschmann, C.; Kuo, C.M.; Kuo, C.H.; Boller, C. High-resolution multisensor infrastructure inspection with unmanned aircraft systems. *ISPRS* **2013**, *XL-1/W2*, 125–129. [CrossRef]
6. Máthé, K.; Busoniu, L. Vision and control for UAVs: A survey of general methods and of inexpensive platforms for infrastructure inspection. *Sensors* **2015**, *15*, 14887–14916. [CrossRef] [PubMed]
7. Bryson, M.; Sukkarieh, S. Inertial sensor-based simultaneous localization and mapping for UAVs. In *Handbook of Unmanned Aerial Vehicle*; Valavanis, K., Vachtsevanos, G., Eds.; Springer: Dordrecht, The Netherlands, 2015; pp. 401–431.
8. Dijkshoorn, N. Simultaneous Localization and Mapping with the AR. Drone. Master's Thesis, Universiteit van Amsterdam: Amsterdam, The Netherlands, 14 July 2012.
9. Li, C. *Simultaneous Localization and Mapping Implementations via Probabilistic Programming with Unmanned Aerial Vehicles*; Research Project Report, Massachusetts Institute of Technology: Boston, MA, USA, 28 July 2015.
10. Dehghan, S.M.M.; Moradi, H. SLAM-inspired simultaneous localization of UAV and RF sources with unknown transmitted power. *Trans. Inst. Meas. Control* **2016**, *38*, 895–907. [CrossRef]
11. Gruz, G.; Encarnação, P. Obstacle avoidance for unmanned aerial vehicles. *J. Intell. Robot. Syst.* **2012**, *65*, 203–217. [CrossRef]
12. Gageik, N.; Benz, P.; Montenegro, S. Obstacle detection and collision avoidance for a UAV with complementary low-cost sensors. *IEEE Access* **2015**, *3*, 599–609. [CrossRef]
13. Call, B. Obstacle Avoidance for Unmanned Air Vehicles. Master's Thesis, Brigham Young University, Provo, UT, USA, 11 December 2006.
14. Barry, A. High-Speed Autonomous Obstacle Avoidance with Pushbroom Stereo. Ph.D. Thesis, Massachusetts Institute of Technology, Cambridge, MA, USA, February 2016.
15. Partsinevelos, P.; Agadakos, I.; Athanasiou, V.; Papaefstathiou, I.; Mertikas, S.; Kyritsis, S.; Tripolitsiotis, A.; Zervos, P. On-board computational efficiency in real-time UAV embedded terrain reconstruction. *Geophys. Res. Abstr.* **2014**, *16*, 9837.
16. Bulatov, D.; Solbrig, P.; Gross, H.; Wernerus, P.; Repasi, E.; Heipke, C. Context-based urban terrain reconstruction from UAV-videos for geoinformation applications. In Proceedings of the Conference on Unmanned Aerial Vehicle in Geomatics, Zurich, Switzerland, 14–16 September 2011.

17. Gottlieb, Y.; Shima, T. UAVs task and motion planning in the presence of obstacles and prioritized targets. *Sensors* **2015**, *15*, 29734–29764. [CrossRef] [PubMed]
18. Witayangkurn, A.; Nagai, M.; Honda, K.; Dailey, M.; Shibasaki, R. Real-time monitoring system using unmanned aerial vehicle integrated with sensor observation service. In Proceedings of the Conference on Unmanned Aerial Vehicle in Geomatics, Zurich, Switzerland, 14–16 September 2011.
19. Nagai, M.; Witayangkurn, A.; Honda, K.; Shibasaki, R. UAV-based sensor web monitoring system. *Int. J. Navig. Obs.* **2012**, *2012*. [CrossRef]
20. Bravo, R.; Leiras, A. Literature review of the application of UAVs in humanitarian relief. In Proceedings of the XXXV Encontro Nacional de Engenharia de Producao, Fortaleza, Brazil, 13–16 October 2015.
21. Baiocchi, V.; Dominici, D.; Milone, M.V.; Mormile, M. Development of a software to plan UAVs stereoscopic flight: An application on post earthquake scenario in L'Aquila city. In *Computational Science and its Applications-ICCSA 2013*; Murgante, B., Misra, S., Carlini, M., Torre, C.M., Nguyen, H.Q., Taniar, D., Apduhan, B.O., Gervasi, O., Eds.; Springer: Berlin, Germany, 2013; pp. 150–165.
22. Yeong, S.P.; King, L.M.; Dol, S.S. A review on marine search and rescue operations using unmanned aerial vehicles. *Int. J. Mech. Aerosp. Ind. Mech. Manuf. Eng.* **2015**, *9*, 396–399.
23. Waharte, S.; Trigoni, N. Supporting search and rescue operation with UAVs. In Proceedings of the 2010 International Conference on Emerging Security Technologies (EST), Canterbury, UK, 6–7 September 2010; pp. 142–147.
24. Herissé, B.; Hamel, T.; Mahony, R.; Russotto, F.X. Landing a VTOL unmanned aerial vehicle on a moving platform using optical flow. *IEEE Trans. Robot.* **2012**, *28*, 1–13. [CrossRef]
25. Sapiralli, S. Vision-based autonomous landing of an helicopter on a moving target. In Proceedings of the AIAA Guidance, Navigation, and Control Conference, Chicago, IL, USA, 10–13 August 2011.
26. Araar, O.; Aouf, N.; Vitanov, I. Vision based autonomous landing of multirotor UAV on moving platform. *J. Intell. Robot. Syst.* **2016**, 1–16. [CrossRef]
27. Wenzel, K.E.; Masselli, A.; Zell, A. Automatic take off, tracking and landing of a miniature UAV on a moving carrier vehicle. *J. Intell. Robot. Syst.* **2011**, *61*, 221–238. [CrossRef]
28. Alarcon, F.; Santamaria, D.; Viguria, A. UAV helicopter relative state estimation for autonomous landing on platforms in a GPS-denied scenario. *IFAC-PapersOnLine* **2015**, *48*, 37–42. [CrossRef]
29. Oh, S.; Pathak, K.; Agrawal, S. Autonomous helicopter landing on a moving platform using a tether. In Proceedings of the 2005 IEEE International Conference on Robotics and Automation, Barcelona, Spain, 18–22 April 2005; pp. 3960–3965.
30. Wu, C.; Song, D.; Qi, J.; Han, J. Autonomous landing of an unmanned helicopter on a moving platform based on LP path planning. In Proceedings of the 8th International Conference on Intelligent Unmanned Systems (ICIUS 2012), Singapore, 22–24 October 2012; pp. 978–981.
31. Kim, J.; Jung, Y.; Lee, D.; Shim, D.H. Outdoor autonomous landing on a moving platform for quadrotors using an omnidirectional camera. In Proceedings of the International Conference on Unmanned Aircraft Systems (ICUAS 2014), Orlando, FL, USA, 27–30 May 2014.
32. Xu, G.; Zhang, Y.; Ji, S.; Cheng, Y.; Tian, Y. Research on computer vision-based for UAV autonomous landing on a ship. *Pattern Recognit. Lett.* **2009**, *30*, 600–605. [CrossRef]
33. Sanchez-Lopez, J.L.; Pestana, J.; Saripalli, S.; Campoy, P. An approach toward visual autonomous ship board landing of a VTOL UAV. *J. Intell. Robot. Syst.* **2014**, *74*, 113–127. [CrossRef]
34. Fourie, K.C. The Autonomous Landing of an Unmanned Helicopter on a Moving Platform. Master's Thesis, Stellenbosch University, Stellenbosch, South Africa, December 2015.
35. Frølich, M. Automatic Ship Landing System for Fixed-Wing UAV. Master's Thesis, Norwegian University of Science and Technology, Trondheim, Norway, June 2015.
36. DLR. Available online: http://www.dlr.de/dlr/en/desktopdefault.aspx/tabid-10081/151_read-16413/#/gallery/21679 (accessed on 20 March 2016).
37. Robotic Trends. Available online: http://www.roboticstrends.com/photo/meet_the_6_finalists_of_the_drones_for_good_award/2 (accessed on 25 February 2016).
38. Olson, E. AprilTag: A robust and flexible visual fiducial system. In Proceedings of the 2011 IEEE International Conference on Robotics and Automation, Shanghai, China, 9–13 May 2011; pp. 3400–3407.
39. Chaves, S.M.; Wolcott, R.W.; Eustice, R.M. NEEC Research: Toward GPS-denied landing of unmanned aerial vehicles on ships at sea. *Naval Eng. J.* **2015**, *127*, 23–35.

40. Ling, K. Precision Landing of a Quadrotor UAV on a Moving Target Using Low-Cost Sensors. Master's Thesis, University of Waterloo, Waterloo, ON, Canada, 16 September 2014.

41. Fiala, M. ARTag, a fiducial marker system using digital techniques. In Proceedings of the 2005 IEEE Computer Society Conference on Computer Vision and Pattern Recognition (CVPR '05), Washington, DC, USA, 20–25 June 2005; Volume 2, pp. 590–596.

42. Jayatilleke, L.; Zhang, N. Landmark-based localization for unmanned aerial vehicles. In Proceedings of the 2013 IEEE International Systems Conference (SysCon 2013), Orlando, FL, USA, 15–18 April 2013.

43. Kaess, M. AprilTags C++ Library. Available online: http://people.csail.mit.edu/kaess/apriltags/ (accessed on 10 April 2016).

44. GNU Operating System. Available online: https://www.gnu.org/licenses/old-licenses/lgpl-2.1.en.html (accessed on 10 April 2016).

45. Swatbotics. Available online: https://github.com/swatbotics/apriltags-cpp (accessed on 10 April 2016).

46. Asano, S.; Maruyama, T.; Yamaguchi, Y. Performance comparison of FPGA, GPU and CPU in image processing. In Proceedings of the 2009 Field Programmable Logic and Applications (FPL 2009), Prague, Czech Republic, 31 August–2 September 2009; pp. 126–131.

47. Sörös, G. GPU-Accelerated Joint 1D and 2D Barcode Localization on Smartphones. In Proceedings of the 39th International Conference on Acoustics, Speech, and Signal Processing (ICASSP 2014), Florence, Italy, 4–9 May 2014; pp. 5095–5099.

48. NVIDIA. Linux for Tegra R24.2. Available online: https://developer.nvidia.com/embedded/linux-tegra (accessed on 10 April 2016).

49. NVIDIA. CUDA. Available online: http://www.nvidia.com/object/cuda_home_new.html (accessed on 10 April 2016).

50. Xiong, H.; Yuan, R.; Yi, J.; Fan, G.; Jing, F. Disturbance rejection in UAV's velocity and attitude control: Problems and Solutions. In Proceedings of the 30th Chinese Control Conference, Yantai, China, 22–24 July 2011; pp. 6293–6298.

51. Vincenty, T. Direct and inverse solutions of geodesics on the ellipsoid with application of nested equations. *Surv. Rev.* **1975**, *23*, 88–93. [CrossRef]

52. Erbil, M.A.; Prior, S.D.; Keane, A.J. Design optimisation of a reconfigurable perching element for vertical take-off and landing unmanned aerial vehicles. *Int. J. Micro Air Veh.* **2013**, *53*, 207–228. [CrossRef]

Article

Development and Testing of a Two-UAV Communication Relay System

Boyang Li, Yifan Jiang, Jingxuan Sun, Lingfeng Cai and Chih-Yung Wen *

Department of Mechanical Engineering, The Hong Kong Polytechnic University, Hong Kong, China;
boyang.li@connect.polyu.hk (B.L.); jiang.uhrmacher@connect.polyu.hk (Y.J.);
jingxuan.j.sun@connect.polyu.hk (J.S.); 11811766d@connect.polyu.hk (L.C.)
* Correspondence: cywen@polyu.edu.hk; Tel.: +852-2766-6644

Academic Editors: Felipe Gonzalez Toro and Antonios Tsourdos
Received: 22 July 2016; Accepted: 10 October 2016; Published: 13 October 2016

Abstract: In the development of beyond-line-of-sight (BLOS) Unmanned Aerial Vehicle (UAV) systems, communication between the UAVs and the ground control station (GCS) is of critical importance. The commonly used economical wireless modules are restricted by the short communication range and are easily blocked by obstacles. The use of a communication relay system provides a practical way to solve these problems, improving the performance of UAV communication in BLOS and cross-obstacle operations. In this study, a communication relay system, in which a quadrotor was used to relay radio communication for another quadrotor was developed and tested. First, the UAVs used as the airborne platform were constructed, and the hardware for the communication relay system was selected and built up. Second, a set of software programs and protocol for autonomous mission control, communication relay control, and ground control were developed. Finally, the system was fully integrated into the airborne platform and tested both indoor and in-flight. The Received Signal Strength Indication (RSSI) and noise value in two typical application scenarios were recorded. The test results demonstrated the ability of this system to extend the communication range and build communication over obstacles. This system also shows the feasibility to coordinate multiple UAVs' communication with the same relay structure.

Keywords: two-UAV system; communication relay; mission control software; Ground Control Station

1. Introduction

In recent years, as a result of decreased weights and sizes, reduced costs, and increased functionalities of different sensors and components, Unmanned Aerial Vehicles (UAVs) are becoming increasingly popular in a wide range of domestic applications (photography, surveillance, environment monitoring, search-and-rescue, etc.) [1–3]. As the market develops and the variety of applications expands, more requirements are now being imposed on extending the flight range and increasing the adaptability to the complex missions of UAV systems. In missions such are surveillance and search-and-rescue, UAV systems are usually required to operate in complicated environments while still being able to maintain uninterrupted communication with the ground control station for the purpose of reporting, monitoring, and control. For example, in the mountainous or high-density urban regions, the communication between UAVs and the ground control station can be easily interrupted, causing the ground control station to lose real-time data feedback from the UAVs, leading to mission failure. In addition, if the mission requires long flight range, the communication might also be interrupted by the distance.

Commercially available yet economical UAV communication solutions, however, have a number of limitations that potentially hinder their widespread application. First, the communication range of major civil UAV communication solutions is comparatively short. Telemetry modules, such as 3DR

Sik2, XBee, and other Wi-Fi modules, which are common options for civil UAVs, have communication ranges that are limited to a few kilometers. Second, it is difficult to establish a stable data link when obstacles such as trees, tall buildings, or mountains separate the UAVs and GSC.

In situations discussed above, it can be difficult to complete the mission with only one UAV deployed, while keeping cost and system complexity requirements met [4–7]. To harness the advantages of UAVs in these situations and minimize the drawbacks at the same time, an alternative solution is to deploy a multi-UAV system, which could utilize the inter-connectivity among multiple UAVs to maintain uninterrupted communication between every UAV and the ground control station. In this study, a UAV communication relay solution was developed, which uses relay and routing to extend the communication range and bypass obstacles at low cost. The objective was to develop and test a proof-of-concept two-UAV system that demonstrated the ability to relay radio communication. This system only consisted of two UAVs, but its design could accommodate the addition of more UAVs. The system uses one UAV as communication relay point and enables the other UAV to operate in areas where direct communication with ground control cannot be established. This configuration allows communication to be established across obstacles or over a distance exceeding the range of the onboard radio transceiver, both of which will be demonstrated in field tests later. The typical application scenario of this system is shown in Figure 1 schematically.

Figure 1. Application scenarios of UAV communication relay system.

The paper is organized as follows. Section 2 introduced the related works. Section 3 introduces the hardware configurations of the multiple UAVs communication relay system. Section 4 discusses the software aspects of the system. In Section 5, both the indoor and outdoor tests methods and results are presented. A conclusion of the work is given in Section 6.

2. Related Works

Numerous researches have been performed in UAV communication relay scheme. In the conceptual study of UAV based military communication relays, Pinkney et al. [8] described a multi-year airborne communication range extension program plan which was primarily used to support the Battlefield Information Transmission System. The main objective of the program was to provide the beyond line of sight (BLOS) communications capabilities without the help of scarce satellite resources. Their range of coverage can vary from 50 to 100 miles for UAVs of interest. Rangel et al. [9] developed a multipurpose aerial platform using a UAV for tactical surveillance of maritime vessels. They concluded that the communication relay using a UAV system is a low-cost application which could provide air dominance with satisfied communication range with lost cost resources. In Orfanus et al. [10], the use of the self-organizing paradigm to design efficient UAV relay networks to support military operations was proposed.

Abundant previous studies have focused on the algorithms for vehicle position assignment and network configuration. Choi et al. [11] developed guidance laws to optimize the position of a

Micro-UAV which relayed communication and video signals when the other UAV is out of radio contact range with the base. Zhan et al. [12] investigated a communication system in which UAVs were used as relays between ground terminals and a network base station. Their work mainly focused on developing the algorithm for optimizing the performance of the relay system. In [13,14], Cetin and coworkers proposed a novel dynamic approach to the relay chain concept to maintain communication between vehicles in the long-range communication relay infrastructure. Their path planning method was based on the artificial potential field.

Many other researchers used simulation methods to propose and test the relay systems. Hauert et al. [15] conducted 2D simulations of the design on a swarm system which does not make use of positioning information or stigmergy for application in aerial communication relay. Jiang and Swindlehurst [16] simulated the dynamic UAV relay positioning for multiple mobile ground nodes and multi-antenna UAVs. In [17], a relay network composed of UAVs keeping the Wireless Sensor Network connected was proposed. They also used cooperative multiple input multiple output (MIMO) techniques to support communication. Ben Moussa et al. [18] provided a simulation-based comparison between two strategies for an autonomous mobile aerial node communication relay between two mobile BLOS ground nodes. Lee et al. [19] proposed an airborne relay-based regional positioning system to address pseudolite system limitations. They conducted simulations and demonstrated that the proposed system guarantees a higher accuracy than an airborne-based pseudolite system.

For experimental studies, most of the previous works were based on IEEE 802.11b (Wi-Fi) standard. Brown et al. [20,21] implemented a wireless mobile ad hoc network with radio nodes mounted at fixed sites, ground vehicles, and UAVs. One of their scenarios was to extend small UAVs' operational scope and range. They measured detailed data on network throughput, delay, range, and connectivity under different operating regimes and showed that the UAV has longer range and better connectivity with the ground nodes. Jimenez-Pacheco et al. [22] also built a mobile ad hoc network (MANET) of light UAVs. The system has implemented a line of sight dynamic routing in the network. Morgenthaler et al. [23] described the implementation and characterization of a mobile ad hoc network for light flying robots. Their UAVNet prototype was able to autonomously interconnect two end systems by the airborne relay. Asadpour et al. [24,25] conducted simulation and experimental study of image relay transmission for UAVs. Yanmaz et al. [26] proposed a simple antenna extension to 802.11 devices for aerial UAV nodes. They tested the performance of their system and showed that 12 Mbps communication can be achieved at around 300 m. Yanmaz et al. [27] also did a follow up work on a two-UAV network using 802.11a. In this work, they showed that high throughout can be achieved using two-hop communications. Johansen et al. [28] described the field experiment with a fixed wing UAV acted as a wireless relay for Underwater Vehicle. They only tested the download data rates. Notably, although the Wi-Fi network is convenient to build multiple access points, its communication range is highly restricted and a large number of nodes are needed for long range applications. One of the advantages of a Wi-Fi network is its high throughput. However, for UAV telemetry communication, the distance is a more important requirement than the transmission speed.

For experimental works without the use of Wi-Fi network, Guo et al. [29] analyzed the use of small UAVs for 3G cellular network relay. They tested the upload and download throughputs and the ping time in both urban and rural environments. Deng et al. [30] developed the UAVs and the communication relay system for power line inspection. Their system consisted of multi-platform UAVs and multi-model communication system for image/video transmission. However, they did not test the communication range or quality of their system. The only one work used 433/915 MHz as the communication frequency was Willink et al. [31] who measured the low-altitude air-to-ground channel at 915 MHz. They mainly focused on studying the spatial diversity to and from the airborne node and also measured time delay and received power.

For all above-mentioned works, The UAVs in their systems did not exchange messages with the GCS by the relayed communication, which means the UAVs were used as carriers to realize communication relay for other systems rather than acting as the active parts in the relay systems. Therefore, the applications would be constrained if UAVs lost connection with the GCS. The highlights of our work are the building of an interface and protocol with the flight controller and direct monitoring and control of the UAVs with relayed communication. This study provides a viable and economic way, with detailed hardware and software methods described, to construct a prototype of the UAV communication relay system.

3. Hardware Platform

In this section, the hardware construction of the miniature two-UAV system is introduced. A miniature UAV relay system with the ability to relay radio communications was built. The hardware mainly consisted of three communication nodes: a GCS; a UAV that relays radio communication, which was named "Mom"; and a second UAV named "Son", whose communication with the GCS was relayed through the UAV "Mom". The schematic of the major hardware platform for the "Mom and Son" relay system is illustrated in Figure 2 and the detailed communication subsystem will be introduced in Section 3.3 later.

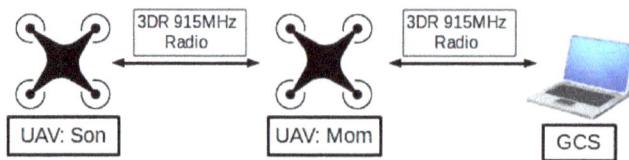

Figure 2. Whole architecture of the communication relay system.

3.1. System Architecture

The relay system can be conceptualized as the following major functional blocks, with the corresponding hardware architecture shown in Figure 3. The figure mainly describes the main components in UAV "Mom" while all other UAVs have the same hardware architecture.

- **Aircraft**: This is the body of the UAV, usually consisting of structural components, power system, and battery. In the test system, the aircraft were constructed using commercially available frame kits. Both of the UAVs, i.e., "Mom" and "Son", were selected as quadrotors for easily launch and recovery ability during the development period. Communication relay systems, however, are not limited to quadrotors and can be integrated into any desired airborne platform such as fixed wing UAV for long endurance and distance relay service.
- **Flight Controller**: This includes main flight control boards as well as the necessary sensors, that together provide attitude and position control for the UAVs. In this test system, 3DR Pixhawk kit [32] was selected as the main flight controller. The controller has an internal IMU, barometer, external GPS, and a compass for attitude and position measure. It also includes safety switch, buzzer, LEDs, and different kinds of interfaces combined to provide flight control.
- **Communication Module**: This functional block directly handles wireless radio communications, and consists of radio transceivers and antennas. The paired modules will connect automatically and transmit data instructed by the mission controller.
- **Mission Controller**: This functional block is a microcomputer running Linux operation system and mainly responsible for the following functions: (1) conducting communication flow control, message routing, and message error checking; (2) monitoring UAV flight conditions and reporting to GCS; and (3) translating mission commands from GCS to low-level instructions, which are directly supplied to the flight controller.
- **Ground Control Station**: This functional block refers to a Windows-based laptop running the GCS software to send commands to all the UAV nodes and monitor the status of all UAVs.
- **Emergency Remote Controller**: The remote controller is connected to the flight controller via 2.4 GHz wireless communication. In normal operation, all the commands and status can be

set by the GSC point. This device has switch to directly takeover the control of UAVs in case of emergency situation in the debugging process.

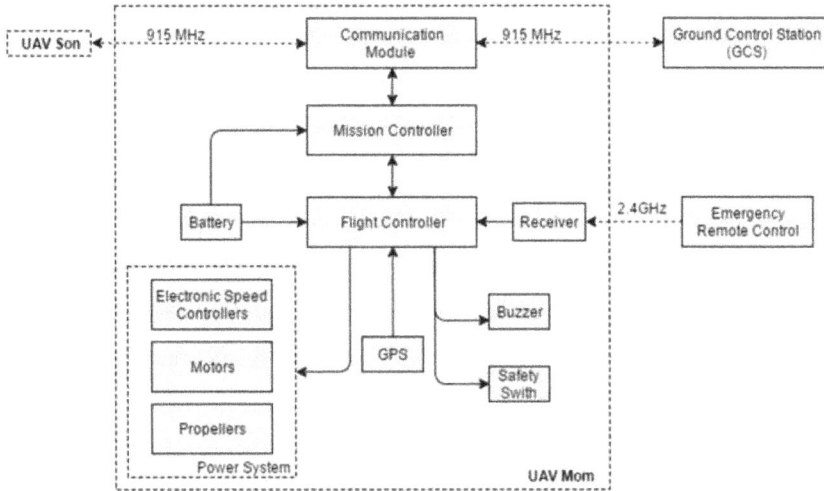

Figure 3. System hardware architecture.

3.2. Communication Subsystem

To construct the communication subsystem, a number of commonly used wireless communication technologies were considered as the possible candidate: IEEE 802.11n Wi-Fi, 3G/4G-LTE mobile network, and 433 MHz/915 MHz radio. Wi-Fi can provide a reliable, high-speed connection, but has a comparatively small range. Typical Wi-Fi routers have an effective range of fewer than 100 m, rendering them unsuitable for long-range outdoor missions. The applicability of the 3G/4G-LTE mobile network relies heavily on the strength of mobile phone signal, which varies greatly in different locations, and is not always available in remote areas where surveillance and rescue missions are likely to take place. A 433 MHz/915 MHz radio can be operated without a license in many parts of the world, and can be implemented with easily accessible, low-cost transceivers (for example, 3DR Radio Telemetry as discussed below). A range of a few kilometers could be achieved. The radio can also be deployed at any desirable site without the need for supporting infrastructure. These features give 433 MHz/915 MHz radio advantages in various outdoor missions. The frequency options, 433 MHz or 915 MHz, can be selected to ensure compliance with local radio regulations.

Based on the discussion above, to establish communication between the UAVs and GCS and to facilitate the communication relay system, 3DR 915 MHz Radio Telemetry (3DR Radio) [33] was selected as our radio transceiver hardware. 3DR Radio, the built-in processor of which runs open-source SiK firmware [34], has the following useful features that enable communication relay:

- Built-in time division multiplexing (TDM) support.
- Supports frequency hopping spread spectrum (FHSS) among a configurable number of channels (configured to 50 channels in this study). Its frequency hopping sequence is user-configurable.

Figure 4 illustrates the radio communication relay setup. Four 3DR Radios were used, numbered T1-1, T1-2, T2-1, and T2-2. Table 1 gives the details of the configuration and installation of the telemetries. T1-1 and T1-2 carried communication between GCS and UAV "Mom" exclusively, whereas T2-1 and T2-2 carried communication between UAV "Mom" and UAV "Son".

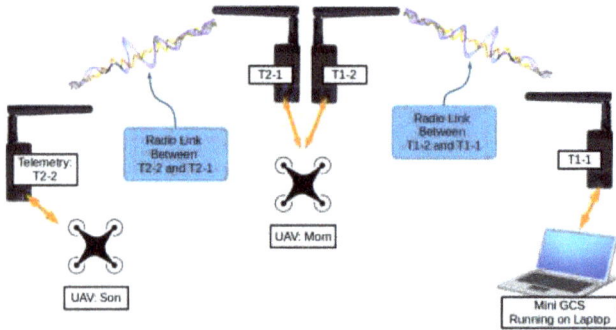

Figure 4. Radio communication relay setup.

Table 1. Telemetry Installation and Settings.

3DR Radio	Installed at	Frequency Hopping Setting
T1-1	GCS	hopping sequence 1
T1-2	UAV "Mom"	hopping sequence 1
T2-1	UAV "Mom"	hopping sequence 2
T2-2	UAV "Son"	hopping sequence 2

A combination of TDM and FHSS techniques was used to ensure that the presence of multiple radio telemetries did not lead to radio-frequency interference. By default, telemetries Tn-1 and Tn-2 (where $n = 1, 2 \dots$) formed one radio link between each another. On power-up, Tn-1 and Tn-2 first searched for synchronization signals from each another; these signals were used to synchronize the internal clocks of both telemetries. Once the clocks were synchronized, the telemetries then started a TDM process, in which Tn-1 and Tn-2 alternately transmitted data over radio, and listened to incoming data when not transmitting. The alternation of a transmission window occurred at a fixed frequency, and the time between consecutive alternations was referred to as the turnover time. The selection of the turnover time depends on the time required to transmit messages. On one hand, the turnover time should be long enough to allow an adequate number of messages to be transmitted in a single transmission window. On the other hand, the turnover time should not be too long, otherwise, the other telemetry would have to wait longer before it can transmit its own messages, causing delays in information update, and compromising the synchronization accuracy between telemetries. The value of the turnover time could be tuned based on the above-mentioned principles. In this research, the turnover time was set to 100 ms, which ensured stable operation of the communication relay function. Values ranging from 90 ms to 140 ms have been tested and could provide similar results. This process is illustrated in Figures 5 and 6.

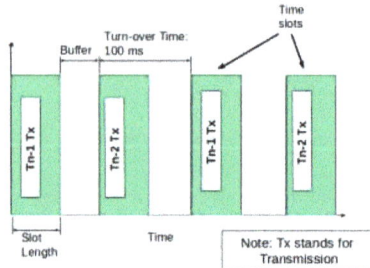

Figure 5. Time division multiplexing (TDM) working principle.

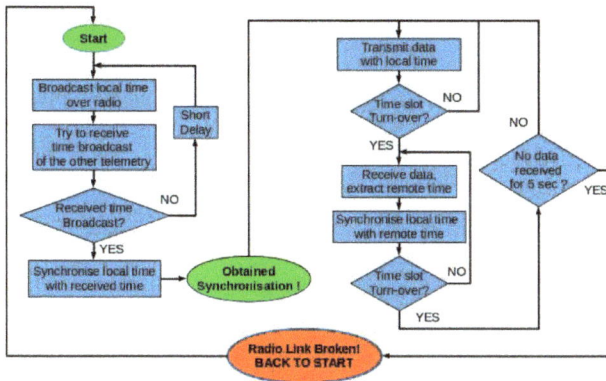

Figure 6. TDM process flow chart.

In addition to the TDM process, the transmission of telemetries Tn-1 and Tn-2 were frequency-hopped. The telemetries transmitted in one of the 50 sub-channels that were obtained by dividing the 915 MHz band into 50 equal portions. On the occurrence of the alternation of a transmission window, both Tn-1 and Tn-2 simultaneously hopped to a new sub-channel, which was determined by a random channel hopping sequence shared between Tn-1 and Tn-2. As shown in Table 1, T1-1 and T1-2 were configured to use a different hopping sequence than T2-1 and T2-2. Therefore, although multiple devices might be transmitting radio waves at the same time, the sub-channels in which they are transmitting are different, as illustrated in Figure 7.

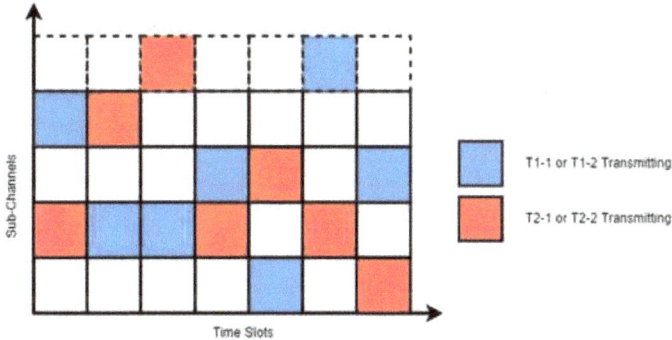

Figure 7. Frequency hopping of telemetries.

3.3. Mission Control Subsystem

The control of key onboard functions: communication relay, flight status monitoring, and mission control, is centrally managed by the mission control subsystem, which UAV "Mom" and "Son" both carry onboard. The onboard mission control subsystem is mainly in charge of the following functions:

- Executing radio communication frame processing and relay control.
- Decoding commands contained in frames, and translating the commands into low-level instructions to the UAV's flight controller.
- Monitoring UAV flight status, and reporting to the GCS on a regular basis.
- Handling ad hoc functions that specific missions necessitate, for example, image processing, dynamics flight route planning, smart target identification, etc.

The above-mentioned functions require the UAVs to be equipped with an on-board computer with adequate processing power. In this research, Raspberry Pi 2 Module B (Rpi 2B) was chosen as the onboard computer for both UAVs because of its relatively small size, strong processing power, and abundant IO port. Figure 8 shows the on-board computer module, with Rpi 2B mounted on top.

Figure 8. Onboard computer module.

The onboard computer needs to frequently exchange data with and acquire information from other onboard devices: radio telemetries, the flight controller, the GPS module, and various onboard sensors. The onboard computer communicates with the peripheral devices through USB ports. Rpi 2B readily supports USB protocol and has four USB ports available. Most peripheral devices, however, utilize a low-level communication method named Universal Asynchronous Receiver and Transmitter (UART). Therefore, a protocol conversion module based on CP2102 microchip was wired between the USB ports of Rpi 2B and the peripheral devices, which handles the conversion between the USB protocol and the UART protocol. Figure 9 shows the hardware connection diagrams of the onboard computer module with other subsystems. The Rpi 2B of "Son" was connected to two USB devices which are T2-2 and the flight controller. The Rpi 2B of "Mom" was connected to three USB devices including T1-2, T2-1, and the flight controller.

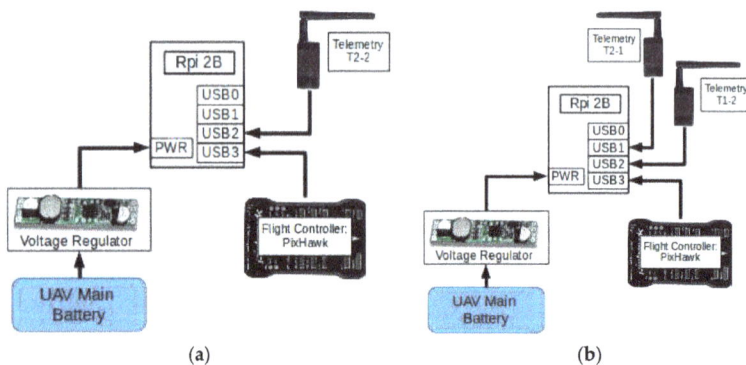

Figure 9. Hardware connection diagram of (**a**) UAV "Son" and (**b**) UAV "Mom".

The onboard computer module is powered by the UAV's main battery and is switched on when the UAV is powered up. After power-up, the onboard computer module then automatically executes a self-initialization program, which configures necessary system parameters and starts up a set of software programs. The programs then run on the onboard computer module and execute functions such as communication relay control and mission control. Details about software structure and realization of the communication relay subsystem will be discussed in Section 4.

4. Software Development

The software developed in this project includes mission control software running on the mission control board and Mini GCS running on the GCS laptop. The mission control software is in charge of analyzing the target node as well as the command contents of each message it receives. If the ID of the messages represents other UAV nodes, the software will bypass this message to the corresponding targets. If the ID of the message corresponds to the receiving UAV itself, the software will further resolve the command data in the message and send it to the flight controller. The Mini GCS was developed to better coordinate the multiple UAV system. The condition of each UAV will be displayed on the screen including the flight mode, the GPS position, the attitude etc. The Mini GCS was also used to send commands to each UAV node by the user friendly Graphic User Interface (GUI).

4.1. Data Framing for Communication Relay

To facilitate the communication relay, the messages and data to be exchanged between the UAVs and GCS were grouped into frames, the format of which is shown in Figure 10. Each communication node ("Mom", "Son", and GCS) was given a unique ID. The header of the frame contained two fields, "Source ID" and "Target ID". "Source ID" indicated the node from which the frame was transmitted, and "Target ID" indicated the target receiver of this frame. This mechanism enabled the identification of the sender and receiver of a data frame, which was important to the realization of frame routing.

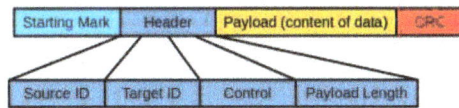

Figure 10. Format of frame.

4.2. Mission Control Software

The mission control software, which runs on the UAV's on-board computer, consists of the following seven modules, and the software's block diagram is shown in Figure 11.

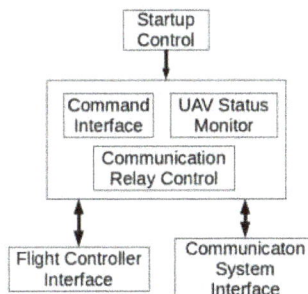

Figure 11. Block diagram of mission control software.

1. Flight Controller Interface: manages direct communication with Pixhawk flight controller over USB port.
2. UAV Status Monitor: monitors flight and mission parameters of the UAV, and reports them regularly to the GCS.

3. Command Interface: translates user commands into low-level instructions to the flight controller; records and reports command execution status.
4. Communication System Interface: manages low-level message traffic between mission control subsystem and ration telemetries; executes message framing, frame reception, and error checking.
5. Communication Relay Control: controls the relay of radio communication.
6. Startup Control: bootstraps the whole mission control software system during UAV power-up process.

4.3. Software Realization of Relay Process

The presence of a relay implies that the relaying node (UAV "Mom" in this case) needs to handle a frame routing. When processing received frames, the UAVs made use of the "Source ID" and "Target ID" fields to filter and/or route the frames in the following manner.

*   UAV "Son" is the terminal receiver, which only receives frames over the relay from "Mom." Therefore, "Son" only saves for further processing frames whose "Target ID" is equal to the ID of "Son" itself.
*   UAV "Mom" functions as a communication relay node, but it also receives control commands from GCS. Therefore, "Mom" will save for further processing frames whose "Target ID" is equal to the ID of "Mom" itself, and will re-transmit frames targeted at "Son" over Radio T2-1.

The above-mentioned frame routing logic is elaborated in the following pseudocode (Algorithms 1 and 2) and flow chart in Figure 12.

Algorithm 1: Frame Processing Logic, UAV Son

For (true) do:
    if (frame receive from T2-2) do:
        if (frame_target_id == id_son) do:
            save frame for local processing.
        else do:
            discard frame.
        end
    end
end

Algorithm 2: Frame Processing Logic, UAV Mom

For (true) do:
    if (frame receive from T1-2) do:
        if (frame_target_id == id_mom) do:
            save frame for local processing.
        else if (frame_target_id == id_son) do:
            send frame out from T2-1.
        end
    else if (frame received from T2-1) do:
        if (frame_target_id == id_mom) do:
            save frame for local processing.
        else if (frame_target_id == id_GCS) do:
            send frame out from T1-1.
        end
    end
end

The last field of the frame contained a four-byte cyclic redundancy checking (CRC) code, which was computed using the IEEE 802.3 32-bit CRC polynomial. This CRC code was important because it helps verify data frame integrity and protect the system from the influence of error-corrupted data or messages. Error-corrupted data or messages could be dangerous, as they might be mistakenly understood by the UAV's mission control subsystem as wrong commands, resulting in unpredictable and potentially dangerous behavior.

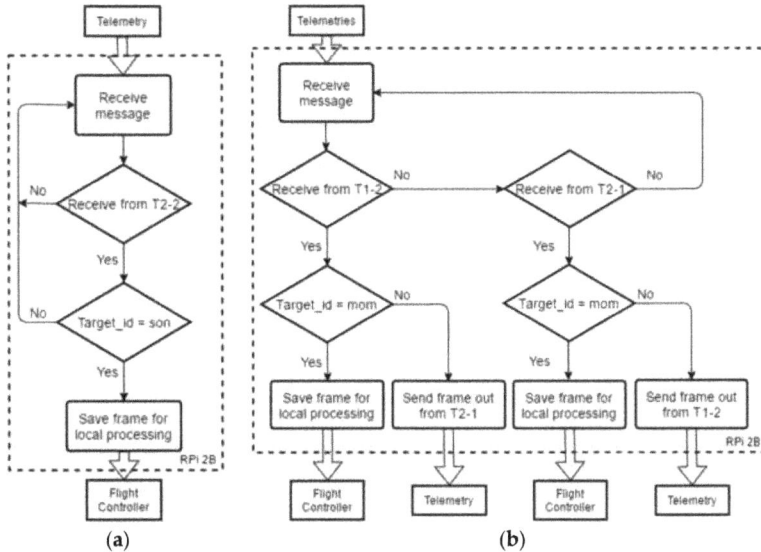

Figure 12. Flow chart of mission control software for (**a**) UAV "Son" and (**b**) UAV "Mom".

4.4. Mini GCS

A ground control software, named "Mini GCS", was developed in-house. Mini GCS is an integrated UAV control interface that is compatible with Windows 7, 8, and 10 systems. The functions of Mini GCS include handling data exchanges with UAVs, displaying information about the UAVs, and providing an interface for users to send commands to the UAVs. It allows users to monitor the status of both UAVs, and to control the mission of each UAV from a single, integrated interface, which makes it a crucial communication node in the communication relay network. Mini GCS was implemented using C#, and the graphic user interface (GUI) was constructed using Microsoft Windows .NET framework. Because the main focus of this work is to demonstrate a signal relay system for the UAV applications, details of this Mini GCS software are not emphasized in the manuscript and the source code is included in the section of Supplementary Materials for reference.

Figure 13 shows a screenshot of Mini GCS. A number of features are available.

- **UAV Status Display:** displays to users a selected set of important information about each UAV, including communication status, GPS status, flight mode, altitude, speed etc.
- **UAV Command Line:** a command line interface that allows users to send specific commands to any UAV nodes in the network such as takeoff, land, return to launch, etc.
- **UAV Position Display:** displays the real-time position and heading of both UAVs in a map of the mission area.

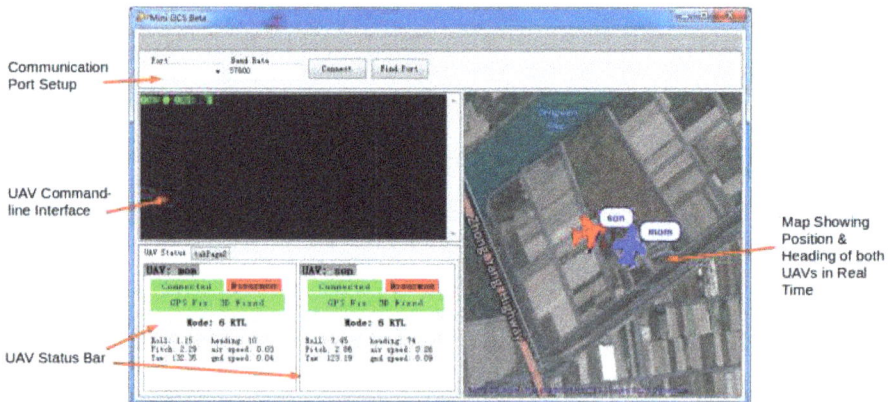

Figure 13. Graphic user interface (GUI) of Mini GCS.

5. Performance Tests

To verify the performance of the system, a series of indoor and outdoor tests were carried out. First, the data loss rate and bit error rate (BER) were measured in the laboratory environment to choose the suitable data transmission rate for the relay system. Then, for outdoor tests, Received Signal Strength Indicator (RSSI), a measurement of the power present in a received radio signal, was measured. The RSSI value scales as $1.9\times$ the dBm signal strength, plus an offset, which means the RSSI is proportional to the dBm [35]. We tested both UAVs in two typical application scenerios in Hong Kong: one was to cross obstacles and the other was to extend the communication range. Finally, a BLOS demonstration was conducted to show the rundown and results of a whole application process in Taiwan. Notably, when the transmission power of the telemetry module is set to the maximum value, which is 20 dBm (100 mW), the typical range achieved using standard configuration and antenna is around 500 m. Due to the stringent flight regulation and the area constraint of the test site in Hong Kong, it is almost impossible for the authors to test the beyond-line-of-sight flight and the maximum power relay which would exceed the allowed flight area. Therefore, to demonstrate the relay function in Hong Kong, the authors have adjusted the transmission power of the radio to its minimum value (1/10 of the max.) so that the radios will disconnect at about 100 m. In this way, the relay methods can be tested in a relatively small area, otherwise, the telemetry modules will always be connected. In missions requiring long distance communication, the transmission power can be easily turned to a maximum value to get the best communication range.

5.1. Indoor Tests

To investigate the reliability of our radio communication relay, the data loss rate and the BER were first tested indoors. Verifying the data loss rate and BER in an indoor environment under different data rates gave us knowledge about the usable bandwidth of the system and helped us estimate the maximum data rate that can be used in the flight tests. The GCS was configured to transfer data at a fixed data rate, and the statistics of the received data at UAV "Son" were measured. Two aspects of the data transfer were studied: the data loss rate, which is the percentage of data that was lost during transmission (data that was transmitted from GCS but was not received by UAV "Son"); and the BER, which was the percentage of data that was received by "Son" but did not pass the CRC test.

Tests were conducted at four different data rates: 4 kilobits per second (kbps), 8 kbps, 12.8 kbps, and 16 kbps. Figure 14 shows the setup of the test. The percentages of data loss and BER with respect to the data rate are shown in Figure 15.

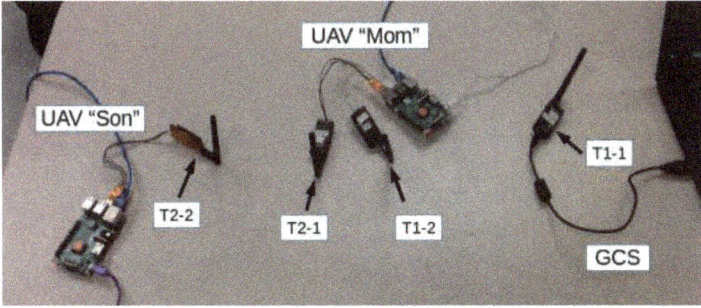

Figure 14. Radio communication relay testing setup.

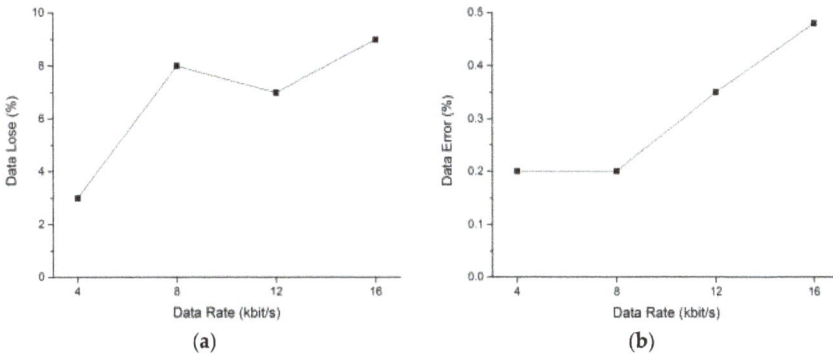

Figure 15. (**a**) Percentage of data loss and (**b**) bit error rate with respect to data rate.

The tests indicated that when the data rate was equal to or higher than 4 kbps, the data lost percentage rose from 3% to 9% and the data error rate rose from 0.2% to 0.5%, approximately. A higher data rate resulted in both more data loss and a higher chance of data transmission errors. Andre et al. [7] estimated that the throughput of typical UAV mission commands is less than 10 kbps. In our tests, the current used data rate for a two-UAV system was about 8 kbps. Then, for communication relay applications with this system and configuration, a data rate of 8 kbps is recommended to ensure an acceptable compromised data lose and error rate.

5.2. Cross-Obstacle Test

The outdoor flight tests were first conducted in the International Model Aviation Center of the Hong Kong Model Engineering Club (22°24′58.1″ N 114°02′35.4″ E). Figure 16 shows the Google map of the test site and final positions of the UAVs in the test. In the simulated cross-obstacle communication relay test, the UAV "Son" was designed to conduct a mission that takes off from the "Home Position", flies over a part of the forest with tall trees, and then lands on the road across the forest (UAV "Son" in Figure). The tall trees would block the communication signal from GCS to UAV "Son". Then the UAV "Mom" took off and loitered at the UAV "Mom" position, which was above the forest. The UAV "Son" would recover the communication with home GSC with the help of relay point UAV "Mom". The relative positions of all the communication nodes can be more clearly conceived in the schematic shown in Figure 17.

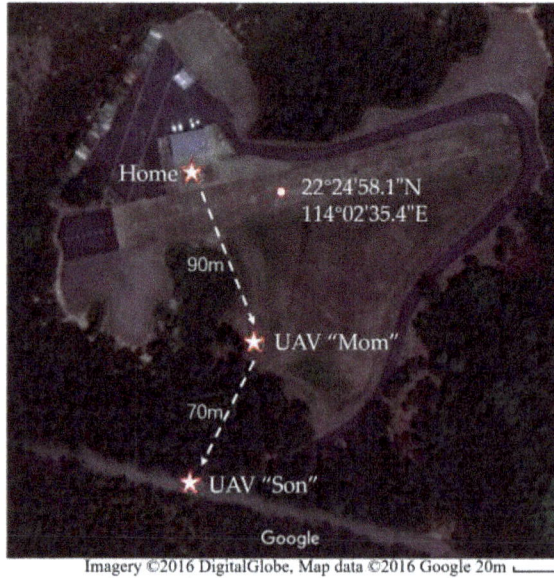

Figure 16. Final positions of nodes in the Google map in the cross-obstacle tests test in Hong Kong.

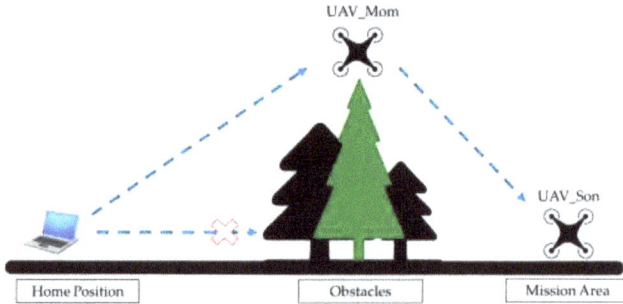

Figure 17. Schematic of cross-obstacle communication relay flight test.

RSSI and noise of both UAV "Son" and UAV "Mom" were recorded and shown in Figure 18. RSSI can reflect the strength of wireless signal received by the UAV "Mom" and UAV "Son". After several tests, we found that only when RSSI value is 20 scale (about 7 dBm) higher than the average noise value can the wireless modules keep a stable connection. Therefore, a dashed blue line is added to each figure to indicator when the connection is lost or unstable. The test started at time 0 s, when UAV "Son" took off and flew to the mission area. The duration of the whole test mission was about 10 min. Since 120 s, the UAV "Son" reached the area behind the tall trees so the RSSI decreased significantly until it reached the noise level and the connection with the GCS was lost. At this moment, the UAV "Mom" took off and flew to the relay point. Before 120 s, UAV "Mom" stayed at home near the GSC so the RSSI remained at a high value. After take off, the RSSI of UAV "Mom" decreased but still remained much higher than the noise, which means that the UAV "Mom" kept connecting with the GSC. During the 180 to 420 s, the UAV "Son" flew over the forest and landed on the road. In the Figure 17a, the RSSI fell to the same scale with noise value. In this period, the UAV "Son" remained disconnected with GCS. After the time 420s, the UAV "Mom" flew close to UAV "Son" and relayed the communication to help UAV "Son" reconnect with the GCS. Then the RSSI value of "Son" recovered. The UAV "Mom" continually flied to UAV "Son" (away from the GCS) for a while which caused further increase of RSSI of "Son" and decrease of "Mon" at the end period.

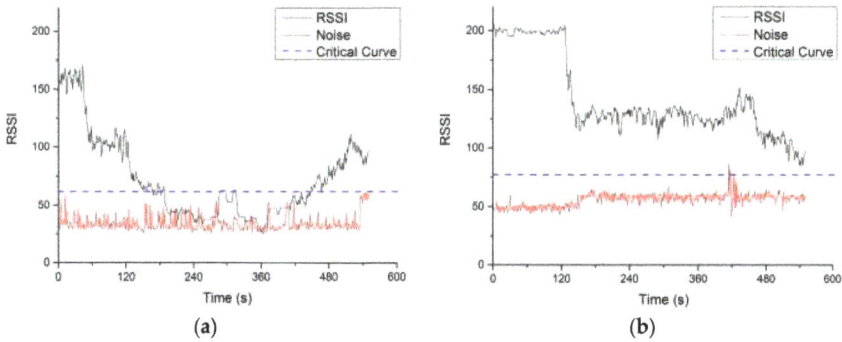

Figure 18. Received Signal Strength Indicator (RSSI), noise and critical curve of both (**a**) UAV "Son" and (**b**) UAV "Mom".

5.3. Extend Distance Test

. In the communication distance extension tests in Hong Kong, the UAV "Son" was firstly given a mission to fly away as far as possible until the connection was lost. After recording the lost position, which is about 100 m, the UAV "Son" continued to fly away for about 60 m and stayed at the position, waiting for the UAV "Mom" to relay the communication. Thereafter, the UAV "Mom" took off and flew to the direction of UAV "Son" until the connection was rebuild. The final positions of UAV nodes in this test is shown in Figure 19. With the help of a relay node, a 160 m total connection was achieved.

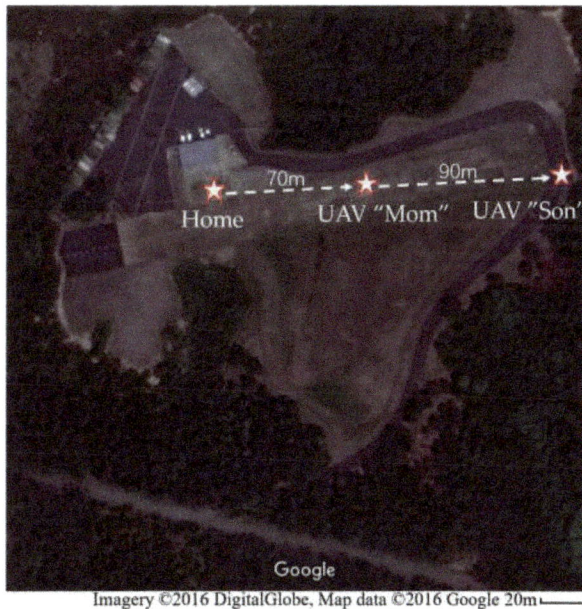

Figure 19. Final positions of nodes in the Google map in the extended distance test in Hong Kong.

The RSSI, noise and critical curves during the full test are shown in Figure 20. As is similar with the cross obstacle test, the communication between UAV "Son" and GCS was first lost at about 180 s and finally reconnected at about 300 s when the UAV "Mom" takeoff and flew near to UAV "Son". The drop of RSSI of UAV "Mom" was due to its takeoff and flying away from home position.

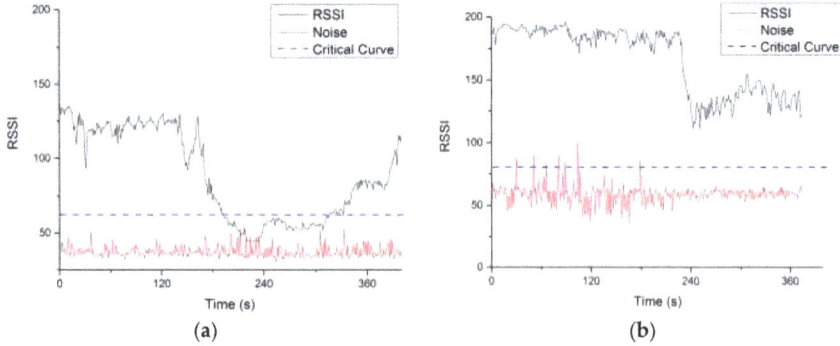

Figure 20. RSSI, noise, and critical curve of both (**a**) UAV "Son" and (**b**) UAV "Mom".

5.4. Beyond-Line-of-Sight Tests

To further verify the reliability of the communication relay system, including mini GCS, in a beyond-line-of-sight missions, outdoor flight tests were conducted in Xigang flight site, Tainan City, Taiwan (23°6′21.5″ N, 120°12′36.0″ E). In this test, the transmission power of the telemetry modules were turned to the maximum value to get the longest communication range. The Google map of the test site and final positions of the UAVs in the test are shown in Figure 21.

Figure 21. Final positions of nodes in the Google map in the extended distance test in Taiwan.

The rundown of the typical flight mission is given below:

1. All UAVs ("Mom" and "Son" in this case) are powered up.
2. Start GCS software "Mini GCS".

3. Wait for communication relay network to start. The startup process is automatic and normally takes less than three minutes.

4. Using "Mini GCS", command "Son" to take off and start the execution of a preprogrammed mission, which brings "Son" to a destination far enough for the radio communication between "Son" and "Mini GCS" to break.

5. Wait until the communication between "Son" and "Mini GCS" breaks due to distance.

6. Using "Mini GCS", command "Mom" to take off and move into the proximity of the destination area of "Son". "Mom" automatically performs a communication relay when it moves toward "Son" and reconnects.

The rundown of the test is shown as a series of pictures in Figure 22. A video clip demonstrating the complete test procedure is given in the Supplementary Material. In the video demonstration, UAV "Son" started a pre-defined mission. The mission is to fly away for about 600 m until the UAV was BLOS in Figure 22c. When the connection between "Son" and GCS was lost (indicated by the green "Connected" icon changing to the red "Disconnected"), "Mom" was deployed to a designated relaying position (about 400 m away from home position). Once "Mom" reached the relaying position, communication between "Son" and GCS was re-established automatically and was not lost again before the end of the mission.

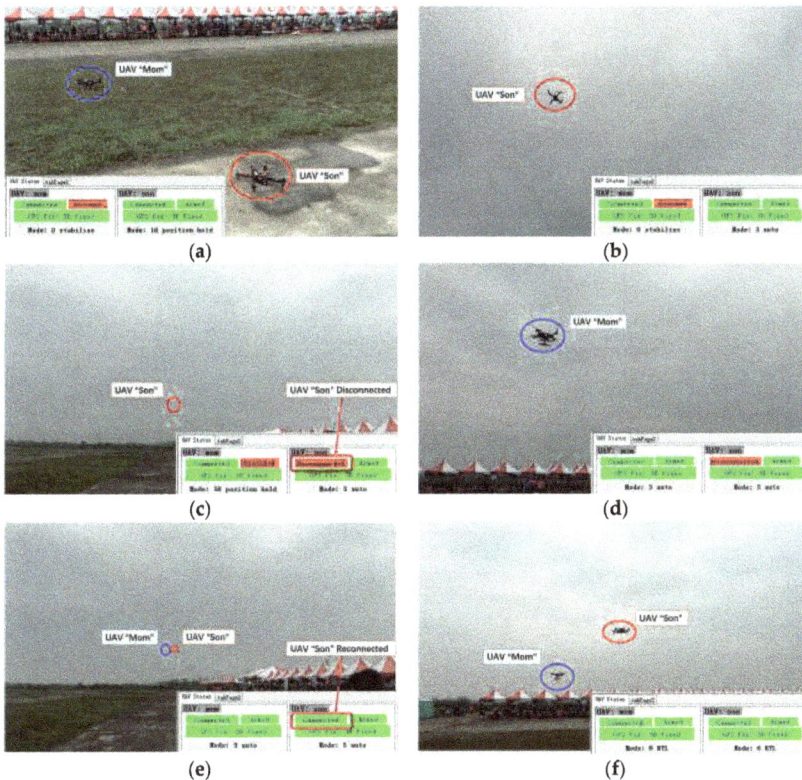

Figure 22. A rundown of flight test (**a**) Pre-takeoff setup of flight test; (**b**) UAV "Son" starts the mission; (**c**) "Son" is out of radio range and communication with it is lost; (**d**) UAV "Mom" is deployed to the relay position; (**e**) Communication with "Son" is re-established after being relayed through "Mom"; (**f**) "Mom" and "Son" return to launch position; the mission is completed.

Eight flight tests were conducted. In all of the tests, the system was able to re-establish communication between "Son" and "Mini GCS" through the relay of "Mom". When "Mom" reached the designated relaying position, the communication connection between "Son" and "Mini GCS" was adequately stable for mission operations. The successful rate of the communication re-establishment is 100%. The tests demonstrated that: (1) the system is capable of maintaining uninterrupted communication between GCS and an end node once the relaying node is properly deployed; and (2) connectivity between GCS and the end node can be automatically re-established when the relaying node reaches the proper relaying position, even if it has been previously lost due to an inadequate relay.

6. Conclusions and Future Development

In this study, major attention was given to implementing a functional and reliable proof-of-concept UAV communication relay system that uses commercially available hardware components combined with customized software. In the prototype system, two quadrotors, each mounted with a Pixhawk flight controller and a Raspberry Pi 2 Model B microcomputer were used as the aerial platforms. The onboard relay control protocol and program were developed to realize UAV communication extension and cross-obstacle communication. The GSC software was also developed to facilitate the two-UAV relay operations. A series of indoor and outdoor tests were carried out. The indoor tests aimed to determine the suitable data transmission rate, while the outdoor tests demonstrated reliable communication performance in two proposed application scenarios in Hong Kong and one BLOS mission simulation in Taiwan. All the outdoor tests showed the system's capability of maintaining good quality real-time communication through relays, and automatically recovering from communication losses.

The current work focused only on communication relay through a single relaying node (UAV "Mom"). Relays through multiple relaying nodes, although feasible, will be conducted in the near future. The current configuration of the system allows the system to be extended by adding more relaying nodes. A possible configuration is illustrated in Figure 23, in which multiple relaying nodes ("Mom 1" to "Mom N") are connected in series, forming a bidirectional relay system. The current design of the data frame permits, at most, 13 relaying nodes between "Son" and the GCS. When the format of the data frames was designed, four binary bits were assigned to represent the "ID" of the UAV. With four bits to use, at most 16 unique IDs could be represented, among which UAV "Son", GCS, and a special broadcasting ID occupied three places. Therefore, there were 13 unique IDs left, which could then be taken up by 13 multiple UAV "Moms", theoretically. However, the maximum number of relaying nodes in practice may also be subjected to the constraint of the data rate. With more UAVs in the system, more information will be exchanged. If the data rate is kept unchanged, communication between UAVs and GCS might experience longer delays.

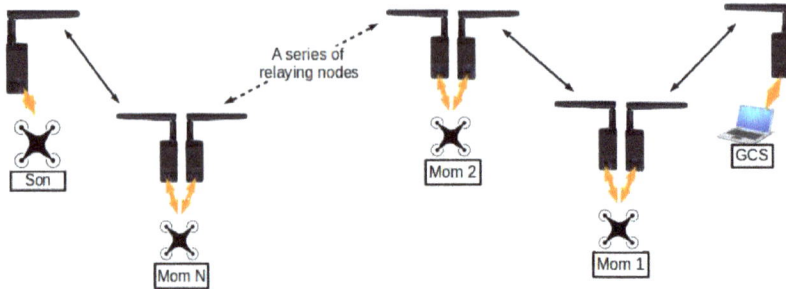

Figure 23. Extending the relay system by adding relaying nodes.

Supplementary Materials: The following is available online at https://www.youtube.com/watch?v=HxN0oafNmzw, Video S1: Multiple UAVs Telecommunication Demonstration. https://github.com/YifanJiangPolyU/UAV-communication-relay, Source code 1: Mission control software source code. https://github.com/HKPolyU-UAV/Mini_GCS, Source code 2: Mini-GSC source code.

Acknowledgments: This work is sponsored by Innovation and Technology Commission, Hong Kong under Contract No. ITS/334/15FP. We also appreciate Hiu Fung Wan for his help in mechanical parts of this work.

Author Contributions: Y.J. and J.S. conceived the relay system. B.L. built up the hardware platform. Y.J. developed the protocol and onboard software. L.C. developed the G.C.S. software. B.L. and Y.J. performed the tests and wrote the paper. C.W. was in charge of the whole project management.

Conflicts of Interest: The authors declare no conflict of interest.

Abbreviations

The following abbreviations are used in this manuscript:

BLOS	Beyond-Line-of-Sight
UAV	Unmanned Aerial Vehicle
GCS	Ground Control Station
RSSI	Received Signal Strength Indication
FHSS	Frequency-Hopping Spread Spectrum
TDM	Time Division Multiplex
UART	Universal Asynchronous Receiver and Transmitter
CRC	Cyclic Redundancy Checking
BER	Bit Error Rate
GUI	Graphic User Interface

References

1. Nonami, K. Prospect and recent research & development for civil use autonomous unmanned aircraft as UAV and MAV. *J. Syst. Des. Dyn.* **2007**, *1*, 120–128.
2. Francis, M.S. Unmanned air systems: Challenge and opportunity. *J. Aircr.* **2012**, *49*, 1652–1665. [CrossRef]
3. Colomina, I.; Molina, P. Unmanned aerial systems for photogrammetry and remote sensing: A review. *ISPRS J. Photogramm. Remote Sens.* **2014**, *92*, 79–97. [CrossRef]
4. Bekmezci, İ.; Sahingoz, O.K.; Temel, Ş. Flying ad hoc networks (fanets): A survey. *Ad Hoc Netw.* **2013**, *11*, 1254–1270. [CrossRef]
5. Arena, A.; Chowdhary, G.; Conner, J.; Gaeta, R.; Jacob, J.; Kidd, J. Development of an Unmanned Aircraft Systems (UAS) Option at the Graduate Level. In Proceedings of the 51st AIAA Aerospace Sciences Meeting Including the New Horizons Forum and Aerospace Exposition, Grapevine, TX, USA, 7–10 January 2013.
6. Gupta, L.; Jain, R.; Vaszkun, G. Survey of important issues in UAV communication networks. *IEEE Commun. Surv. Tutor.* **2016**, *18*, 1123–1152. [CrossRef]
7. Andre, T.; Hummel, K.A.; Schoellig, A.P.; Yanmaz, E.; Asadpour, M.; Bettstetter, C.; Grippa, P.; Hellwagner, H.; Sand, S.; Zhang, S.W. Application-driven design of aerial communication networks. *IEEE Commun. Mag.* **2014**, *52*, 128–136. [CrossRef]
8. Pinkney, F.J.; Hampel, D.; DiPierro, S. Unmanned Aerial Vehicle (UAV) Communications Relay. In Proceedings of the Milcom 96, McLean, VA, USA, 21–24 October 1996; Volumes 1–3, pp. 47–51.
9. Rangel, R.K.; Kienitz, K.H.; de Oliveira, C.A.; Brandao, M.P. Development of a Multipurpose Tactical Surveillance System Using UAV'S. In Proceedings of the 2014 IEEE Aerospace Conference, Big Sky, MT, USA, 1–8 March 2014; pp. 1–9.
10. Orfanus, D.; de Freitas, E.P.; Eliassen, F. Self-organization as a supporting paradigm for military UAV relay networks. *IEEE Commun. Lett.* **2016**, *20*, 804–807. [CrossRef]
11. Choi, Y.; Pachter, M.; Jacques, D. Optimal Relay UAV Guidance—A New Differential Game. In Proceedings of the 50th IEEE Conference on Decision and Control and European Control Conference (CDC-ECC), Orlando, FL, USA, 12–15 December 2011; pp. 1024–1029.
12. Zhan, P.C.; Yu, K.; Swindlehurst, A.L. Wireless relay communications with unmanned aerial vehicles: Performance and optimization. *IEEE Trans. Aerosp. Electron. Syst.* **2011**, *47*, 2068–2085. [CrossRef]

13. Cetin, O.; Zagli, I. Continuous airborne communication relay approach using unmanned aerial vehicles. *J. Intell. Robot. Syst.* **2012**, *65*, 549–562. [CrossRef]

14. Cetin, O.; Zagli, I.; Yilmaz, G. Establishing obstacle and collision free communication relay for UAVs with artificial potential fields. *J. Intell. Robot. Syst.* **2013**, *69*, 361–372. [CrossRef]

15. Hauert, S.; Zufferey, J.-C.; Floreano, D. Evolved swarming without positioning information: An application in aerial communication relay. *Auton. Robot.* **2009**, *26*, 21–32. [CrossRef]

16. Jiang, F.; Swindlehurst, A.L. Dynamic UAV Relay Positioning for the Ground-to-Air Uplink. In Proceedings of the 2010 IEEE Globecom Workshops, Miami, FL, USA, 5–10 December 2010; pp. 1766–1770.

17. Marinho, M.A.M.; de Freitas, E.P.; da Costa, J.P.C.L.; de Almeida, A.L.F.; de Sousa, R.T. Using Cooperative Mimo Techniques and UAV Relay Networks to Support Connectivity in Sparse Wireless Sensor Networks. In Proceedings of the 2013 International Conference on Computing, Management and Telecommunications (Commantel), Ho Chi Minh City, Vietnam, 21–24 January 2013; pp. 49–54.

18. Ben Moussa, C.; Gagne, S.; Akhrif, O.; Gagnon, F. Aerial Mast vs. Aerial Bridge Autonomous UAV Relay: A Simulation-Based Comparison. In Proceedings of the International Conference on New Technologies, Dubai, United Arab Emirates, 30 March–2 April 2014; pp. 3118–3123.

19. Lee, K.; Noh, H.; Lim, J. Airborne relay-based regional positioning system. *Sensors* **2015**, *15*, 12682–12699. [CrossRef] [PubMed]

20. Brown, T.X.; Argrow, B.; Dixon, C.; Doshi, S.; Thekkekunnel, R.-G.; Henkel, D. Ad Hoc UAV Ground Network (Augnet). In Proceedings of the AIAA 3rd Unmanned Unlimited Technical Conference, Chicago, IL, USA, 20–23 September 2004; pp. 1–11.

21. Brown, T.X.; Doshi, S.; Jadhav, S.; Himmelstein, J. Test Bed for a Wireless Network on Small UAVs. In Proceedings of the AIAA 3rd Unmanned Unlimited Technical Conference, Chicago, IL, USA, 20–23 September 2004; pp. 20–23.

22. Jimenez-Pacheco, A.; Bouhired, D.; Gasser, Y.; Zufferey, J.C.; Floreano, D.; Rimoldi, B. Implementation of a Wireless Mesh Network of Ultra Light MAVs with Dynamic Routing. In Proceedings of the IEEE Globe Work, Anaheim, CA, USA, 3–7 December 2012; pp. 1591–1596.

23. Morgenthaler, S.; Braun, T.; Zhao, Z.L.; Staub, T.; Anwander, M. Uavnet: A Mobile Wireless Mesh Network Using Unmanned Aerial Vehicles. In Proceedings of the IEEE Globe Work, Anaheim, CA, USA, 3–7 December 2012; pp. 1603–1608.

24. Asadpour, M.; Giustiniano, D.; Hummel, K.A. From Ground to Aerial Communication: Dissecting WLAN 802.11N for the Drones. In Proceedings of the WiNTECH '13 8th ACM International Workshop on Wireless Network Testbeds, Experimental Evaluation & Characterization, Miami, FL, USA, 30 September–4 October 2013; pp. 25–32.

25. Asadpour, M.; Giustiniano, D.; Hummel, K.A.; Heimlicher, S.; Egli, S. Now or Later?—Delaying Data Transfer in Time-Critical Aerial Communication. In Proceedings of the 2013 ACM International Conference on Emerging Networking Experiments and Technologies (Conext '13), New York, NY, USA, 9–12 December 2013; pp. 127–132.

26. Yanmaz, E.; Kuschnig, R.; Bettstetter, C. Achieving Air-Ground Communications in 802.11 Networks with Three-Dimensional Aerial Mobility. In Proceedings of the IEEE Infocom Ser, Turin, Italy, 14–19 April 2013; pp. 120–124.

27. Yanmaz, E.; Hayat, S.; Scherer, J.; Bettstetter, C. Experimental Performance Analysis of Two-Hop Aerial 802.11 Networks. In Proceedings of the 2014 IEEE Wireless Communications and Networking Conference (WCNC), Istanbul, Turkey, 6–9 April 2014; pp. 3118–3123.

28. Johansen, T.A.; Zolich, A.; Hansen, T.; Sørense, A.J. Unmanned Aerial Vehicle as Communication Relay for Autonomous Underwater Vehicle—Field Tests. In Proceedings of the Globecom 2014 Workshop—Wireless Networking and Control for Unmanned Autonomous Vehicles, Austin, TX, USA, 8–12 December 2014.

29. Guo, W.; Devine, C.; Wang, S. Performance Analysis of Micro Unmanned Airborne Communication Relays for Cellular Networks. In Proceedings of the 9th International Symposium on Communication Systems, Networks & Digital Signal Processing (CSNDSP), Manchester, UK, 23–25 July 2014; pp. 658–663.

30. Deng, C.; Wang, S.; Huang, Z.; Tan, Z.; Liu, J. Unmanned aerial vehicles for power line inspection: A cooperative way in platforms and communications. *J. Commun.* **2014**, *9*, 687–692. [CrossRef]

31. Willink, T.J.; Squires, C.C.; Colman, G.W.K.; Muccio, M.T. Measurement and characterization of low-altitude air-to-ground mimo channels. *IEEE Trans. Veh. Technol.* **2016**, *65*, 2637–2648. [CrossRef]

32. 3DR Pixhawk. Available online: https://store.3dr.com/products/3dr-pixhawk (accessed on 30 June 2016).
33. Sik Telemetry Radio. Available online: http://ardupilot.org/copter/docs/common-sik-telemetry-radio.html (accessed on 30 June 3016).
34. Dronecode/Sik. Available online: https://github.com/Dronecode/SiK (accessed on 30 June 2016).
35. Sik Radio—Advanced Configuration. Available online: http://ardupilot.org/copter/docs/common-3dr-radio-advanced-configuration-and-technical-information.html#common-3dr-radio-advanced-configuration-and-technical-information (accessed on 15 August 2016).

![sensors](sensors logo) *sensors*

MDPI

Article

A Novel Method for Vertical Acceleration Noise Suppression of a Thrust-Vectored VTOL UAV

Huanyu Li [1], Linfeng Wu [1], Yingjie Li [1] ,Chunwen Li [1] and Hangyu Li [2,*]

[1] Department of Automation, Tsinghua University, Beijing 100084, China;
lihuanyu12@mails.tsinghua.edu.cn (H.L.); wulf13@mails.tsinghua.edu.cn (L.W.);
liyj08@mails.tsinghua.edu.cn (Y.L.); lcw@mail.tsinghua.edu.cn (C.L.)

[2] Department of Energy Resources Engineering, Stanford University, Stanford, CA 94305, USA

* Correspondence: hangyuli@stanford.edu or lihangyupku@gmail.com; Tel.: +1-650-804-2820

Academic Editors: Felipe Gonzalez Toro and Antonios Tsourdos
Received: 7 September 2016; Accepted: 29 November 2016; Published: 2 December 2016

Abstract: Acceleration is of great importance in motion control for unmanned aerial vehicles (UAVs), especially during the takeoff and landing stages. However, the measured acceleration is inevitably polluted by severe noise. Therefore, a proper noise suppression procedure is required. This paper presents a novel method to reduce the noise in the measured vertical acceleration for a thrust-vectored tail-sitter vertical takeoff and landing (VTOL) UAV. In the new procedure, a Kalman filter is first applied to estimate the UAV mass by using the information in the vertical thrust and measured acceleration. The UAV mass is then used to compute an estimate of UAV vertical acceleration. The estimated acceleration is finally fused with the measured acceleration to obtain the minimum variance estimate of vertical acceleration. By doing this, the new approach incorporates the thrust information into the acceleration estimate. The method is applied to the data measured in a VTOL UAV takeoff experiment. Two other denoising approaches developed by former researchers are also tested for comparison. The results demonstrate that the new method is able to suppress the acceleration noise substantially. It also maintains the real-time performance in the final estimated acceleration, which is not seen in the former denoising approaches. The acceleration treated with the new method can be readily used in the motion control applications for UAVs to achieve improved accuracy.

Keywords: data fusion; noise suppression; Kalman filter; acceleration; VTOL UAV; thrust-vectoring

1. Introduction

Vertical takeoff and landing (VTOL) unmanned aerial vehicles (UAV) have attracted the attention of many researchers [1,2]. For VTOL UAVs, the vertical motion control design is of great importance, especially during the takeoff and landing stages [3,4]. This is because a motion control failure can result in severe damage to the UAV. To achieve well-performed vertical motion control, high-quality vertical acceleration feedback should be used because it enables the controllers to take action before the unexpected dynamic manifests itself in velocity and height [5–7].

However, the acceleration measurement can be severely polluted by noise in real applications, which is a major challenge of acceleration feedback control [8]. A lot of research has been done to reduce the acceleration noise.

Sun et al. developed an infinite impulse response (IIR) digital lowpass filter to remove the high frequency noise in acceleration measurement [9]. The parameters of the lowpass filter were adopted according to the characteristics of the acceleration noise. An experiment was conducted to verify the performance of the lowpass filter. Because of its simplicity, a lowpass filter has been widely used in denoising applications [10]. Lu et al. developed a procedure which consists of an IIR lowpass filter and a Kalman filter to attenuate the influence of sensor noise and external disturbance [11]. The lowpass filter is applied at first to eliminate the high frequency noise of the measured acceleration. The Kalman filter is then used to further reduce the noise. The noise reduction performance was verified by experiment. The standard deviation of acceleration noise was reduced from 0.33 m/s^2 to 0.02 m/s^2. In addition, El-Sheimy et al. introduced the wavelet filter to reduce the acceleration noise [12]. The six-level wavelet decomposition is used to eliminate the high frequency noise from the low frequency signal. A significant improvement in the quality of the acceleration signal was achieved. The standard deviation of the acceleration noise was reduced by about 99%.

Though the above methods are able to reduce acceleration noise effectively, these studies were all taken in a constrained environment in which the acceleration is very stable with only small variations. For acceleration, which varies significantly with time, the noise reduction becomes much more challenging. Kownacki investigated such problems using the Kalman filter, though the model was built with the assumption of constant acceleration [13]. A conflict between the output signal noise level and filter response rate was discussed. This discussion indicates that the filter response rate has to be sacrificed in order to achieve effective noise suppression performance. In fact, this dilemma also exists in all of the above noise reduction methods. Hebbale and Ghoneim exploited constant jerk (the derivative of acceleration) model in the design of a Kalman filter to reduce the acceleration noise [14]. Nevertheless, this method also suffers from the time delay just like the other approaches.

The above-mentioned methods have been used for UAV acceleration noise reduction. For example, Abellanosa et al. used lowpass filter to suppress the acceleration noise for position estimation of a Quadcopter UAV [15]. Quadri et al. applied a Kalman filter to reduce acceleration noise for attitude estimation of a multi-rotor UAV [16]. Related work can also be found in [17,18]. In those studies, the accelerations were mainly used for position and attitude estimation for UAVs. The presence of time delay in the existing denoising methods prevents the usage of acceleration feedback control for UAVs.

To overcome the problems in the previous denoising methods, we develop a new frame for acceleration noise suppression of a thrust-vectored UAV. Instead of the direct filter used in previous work, the new procedure combines the thrust information with the accelerometer measurement to obtain a more accurate acceleration estimate. The new method is able to suppress the acceleration noise without introducing time delay. A thrust-vectored tail-sitter UAV takeoff experiment is conducted to verify the effectiveness of our method. To the best of our knowledge, this is the first attempt to estimate UAV acceleration in this way.

This paper is organized as follows. In Section 2, the configuration of the thrust-vectored VTOL UAV and the experiment platform is introduced. Section 3 presents the detailed analysis of the statistical characteristics of the acceleration noise, which is the prerequisite for the noise suppression approach. In Section 4, we build an engine thrust model and a thrust vector model based on ground experiments which are used to calculate the vertical thrust when UAV is flying. The noise reduction procedure is then described in Section 5, which includes a Kalman filter to estimate UAV mass and a data fusion step to obtain the minimum variance estimate of acceleration. In Section 6, we present the results for the UAV takeoff experiment. Two existing methods are tested as well for comparison purpose. A section to summarize the work is finally presented.

2. UAV Configuration and Experiment Platform

Tsinghua University has developed a new prototype tail-sitter VTOL UAV, which is equipped with a thrust-vectoring engine system as shown in Figure 1. The UAV has a length of 2 m, a wingspan of 1.6 m, and the gross weight of the UAV (including the thrust vectoring system) is about 24 kg. The thrust vectoring system includes two micro turbine engines with independent vectoring nozzles. The turbine engine used is Jetcat P200. P200 is a single-spool axial flow turbine engine with a net weight of 2.4 kg and a maximum thrust of 230 N which corresponds to a maximum rotor speed of 112,000 revolutions per minute (RPM). With two turbine engines, this UAV has a maximum thrust of 460 N which enables an approximately 46 kg maximum takeoff mass. The detailed technical parameters of the engine are available in [19].

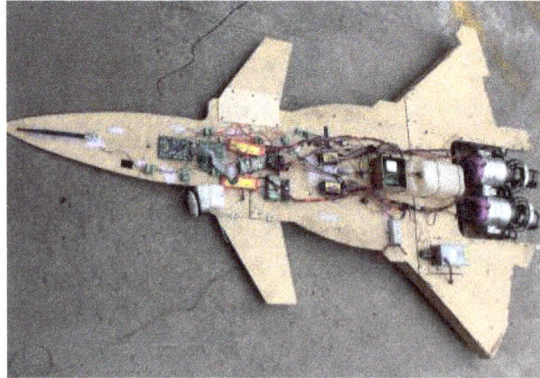

Figure 1. The thrust-vectored tail-sitter unmanned aerial vehicle (UAV).

The electronic system of the UAV includes a microcomputer, an inertial measurement unit (IMU), a Hall sensor, a laser range finder and other auxiliary communication and measurement instruments. The microcomputer board with STM32F407 microprocessor is used to perform data processing and give commands to the thrust vectoring system in order to keep the UAV stable. The IMU (ADIS16488A from Analog Devices) provides the attitude and acceleration information to the microcomputer with a sample frequency of 25 Hz. It is installed in the center of gravity of the UAV to prevent the rotation of the UAV body from disturbing acceleration measurement. The Hall sensor is used to measure the rotor speed with very high accuracy (±10 RPM for rotor speed at over 85,000 RPM). The laser range finder (DLSB-15 from Dimetix) with a sample frequency of 5 HZ and a measurement error within ±1.5 mm is used to measure the height from the ground for the UAV.

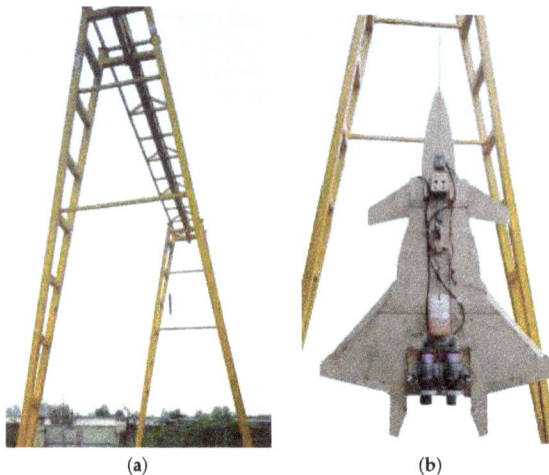

Figure 2. The experiment platform. (**a**) Gantry crane; (**b**) Suspended UAV.

To ensure the safety of the experiment (for both the staff and UAV body) and also for the convenience of the experiment, the UAV does not directly take off and land on the ground. Similarly to the UAV VTOL experiment study in [20], it is hung with a cable under a gantry crane which is shown in Figure 2a. The height of the gantry crane is 7.2 m, and the width is 6.8 m. To avoid the ground effect influencing the thrust performance, the tail of the UAV is hung at about 1.5 m high. As mentioned previously the UAV itself has a length of 2 m, therefore the operating window for the UAV is about 3.5 m. During the experiment, the UAV height is monitored closely to ensure it does not hit the crane. The UAV hanging under the crane is shown in Figure 2b. The experiment is conducted in calm days with wind speed less than 1 km/h to minimize the external disturbance.

3. Analysis of Acceleration Noise

The acceleration measured with IMU is inevitably polluted by noise. The noise can be caused by many reasons, for example, the external disturbances (such as wind), and the vibration of the turbine engines. To develop a noise reduction procedure, the characteristics of the noise need firstly to be investigated.

In this work, only the vertical acceleration is analyzed because it is of greater importance for UAV motion control. The vertical acceleration is not directly measured in the strap-down inertial system. Instead, the acceleration is measured in the UAV body coordinate system. This measured acceleration needs to be transformed from the UAV body coordinate system to the inertial coordinate system. Previous work on coordinate transformation for tail-sitter UAVs has been done by other researchers [21]. It is directly used in this paper without repetitive description. In this work, all the acceleration data presented is in the inertial coordinate system (after transformation).

The measurements of UAV vertical acceleration at three different engine states are shown in Figure 3. Figure 3a shows the acceleration before the engines start, Figure 3b shows the acceleration when the engines work at an idle speed (33,000 RPM), while Figure 3c shows the acceleration when the engines work close to the takeoff speed (85,000 RPM). It should be noted that the engine system does not reach the takeoff thrust at the rotor speed of 85,000 RPM, therefore the UAV remains stationary for all three states.

Because the IMU measures the proper acceleration (physical acceleration) instead of coordinate acceleration (rate of change of velocity), the measured acceleration should be the gravitational acceleration when the UAV stays stationary. As shown in Figure 3a the acceleration measurement is very close to 9.8 m/s^2 with small-scale noise. In contrast, the acceleration noise becomes more significant after the engines start and it grows as the rotor speed increases from 33,000 to 85,000 RPM. This is because of the vibration in the engines due to the rotation of the rotor and the disturbance of the gas flow. Note that the rotor speed of 85,000 RPM is close to the critical rotor speed for UAV takeoff.

The mean and variance for each of the three acceleration measurements are shown in Table 1. It can be seen that the mean values are very close to each other. The slight difference between them is because of the finite amount of data measured during the experiments. The largest difference is only 0.009 m/s^2, which is negligible compared with the gravitational acceleration. In this work, gravitational acceleration is adopted as 9.797 m/s^2 which is the average of the three mean values in Table 1. In contrast, the computed variance values differ from each other significantly. When engines are shut down (before start), the variance is very small and it increases dramatically when engine rotor speed is of 33,000 RPM or 85,000 RPM. This is consistent with the observation in Figure 3, since the variance in acceleration represents the intensity of the noise.

Table 1. Mean and variance of acceleration measurement.

Engine States	Mean (m/s^2)	Variance (m^2/s^4)
Before start	9.801	0.00022
Idle speed (33,000 RPM)	9.792	0.193
Rotor speed (85,000 RPM)	9.797	0.351

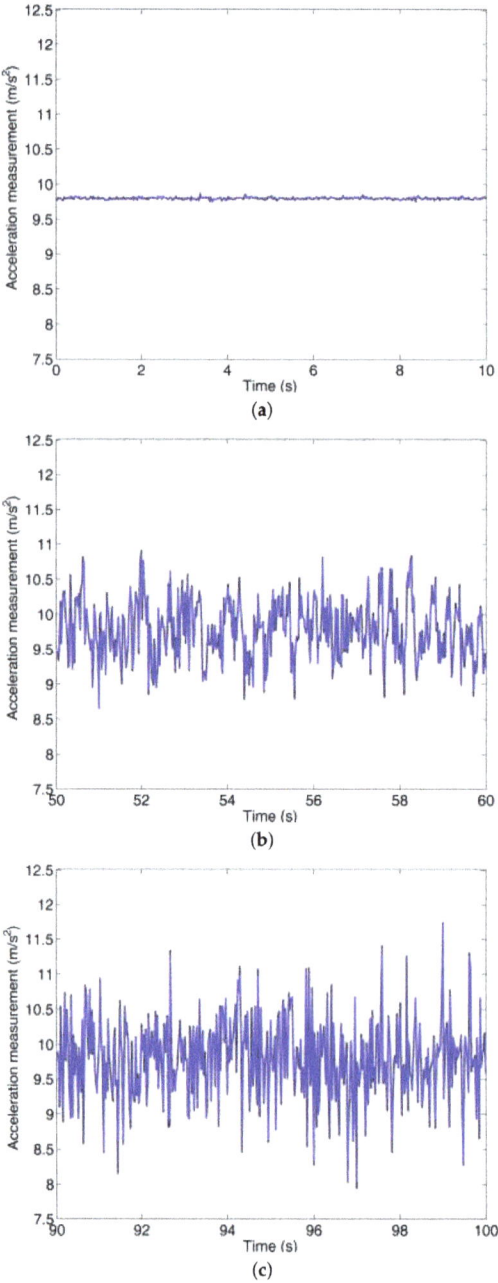

Figure 3. Vertical acceleration measurements at different engine states. (**a**) Vertical acceleration measurement before the engines start; (**b**) Vertical acceleration measurement when the engines work at idle (33,000 RPM); (**c**) Vertical acceleration measurement when the engines work close to takeoff (85,000 RPM).

Many filtration and data fusion algorithms have intrinsic assumptions on the statistical characteristics of the noise. More detailed investigation of the noise is therefore needed. In this work, the acceleration noise in measured data is defined as:

$$w = a_m - \bar{a}_m \tag{1}$$

where w represents the noise, a_m is the measured acceleration and $\bar{a}_m$ is the mean of the measured acceleration.

We first compute the normalized autocorrelation, designated $R(k)$, of the acceleration noise using Equation (2):

$$R(k) = \frac{1}{N\sigma^2} \sum_{i=1}^{N} w(i)w(i+k) \tag{2}$$

where N is the number of sampled data points, σ^2 is the variance of the noise, $w(i)$ and $w(i+k)$ are the noise values at sampled data points i and $i+k$ respectively, and k is the lag of the autocorrelation. The acceleration noise for engine rotor speed of 85,000 RPM (as shown in Figure 3c) is used for analysis because it corresponds to a state of greater importance (near the takeoff speed).

The result of Equation (2) is shown in Figure 4 with the x-axis indicating the lag and the y-axis indicating the normalized autocorrelation of the sampled data. According to Figure 4, the autocorrelation of the acceleration noise can be regarded as an impulse function with high fidelity, which indicates that the acceleration noise can be treated as white noise.

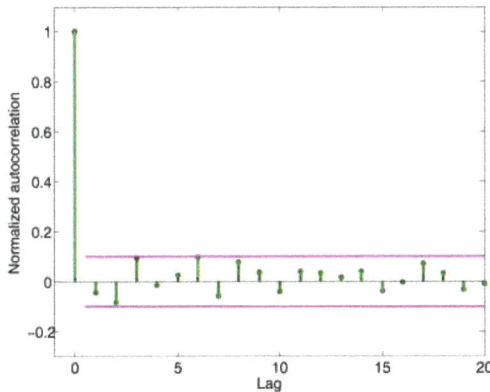

Figure 4. Normalized autocorrelation of the noise in measured acceleration.

In addition, Figure 5 shows the histogram of the acceleration noise with a normal distribution curve fit (purple line). It can be seen that the normal distribution fits the experimental results very well, which suggests that the noise is of the Gaussian type.

In addition to the above analysis, Woodman [22] and Titterton [23] also stated that it is standard practice to assume that random errors follow a Gaussian distribution when modelling errors in acceleration measurement. Therefore, it is reasonable to assume the accelerometer measurement noise is a white Gaussian noise. This assumption is widely used in the literature, for example, in [11,13,16] mentioned before. The same assumption is used in this work as well.

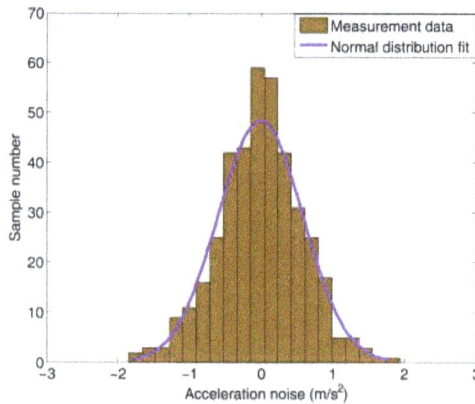

Figure 5. Histogram of the acceleration noise and normal distribution curve fit.

4. Modeling of Engine Thrust System

During the takeoff and landing stages, the aerodynamic forces can be ignored because of the low air speed relative to the UAV. The engine thrust is the only non-gravitational force exerts on the UAV. The UAV motion dynamics in the vertical direction can therefore be written as:

$$F_v - mg = ma \tag{3}$$

where F_v denotes the vertical thrust of the engine system, m is the mass of the UAV, g is the gravitational acceleration, and a is the vertical coordinate acceleration with the positive direction upward. By introducing the proper acceleration a_p, Equation (3) can be rewritten as:

$$a_p = g + a = F_v/m \tag{4}$$

This proportional relationship (Newton's Second Law) between the proper acceleration and the vertical thrust is demonstrated in Equation (4). It inspires us to incorporate the thrust information into the acceleration estimation. However, because the engine thrust cannot be directly measured during flight, mathematical models are needed to calculate the thrust. In this section, we develop models to compute the thrust and thrust deflection angle based on measurements that can be obtained when the UAV is flying.

Note that the noise reduction method developed in this work (presented in Section 5) is independent of the engine thrust system model built here. The method can be readily used to reduce acceleration noise for other types of UAV platforms with different engines for which different thrust system models are needed. The thrust system model described in this section is for Jetcat P200 turbine engine used in our UAV platform.

4.1. Engine Thrust Model

Previous work has demonstrated that a single input single output model can be used to characterize the relationship between the thrust and rotor speed for micro turbine engines [24]. In this work, a similar relationship is built to model the P200 turbine engine. Since the engine thrust cannot be measured during flight, the ground experiments are conducted.

Figure 6 shows the settings for engine thrust ground experiment. A turbine engine and a vectoring nozzle are fixed on a precise balance which can measure the triaxial forces. The detailed descriptions for the turbine engine ground experiment can be found at [24]. For the thrust experiment, the deflection angle of the vectoring nozzle is set to be zero, meaning the nozzle axis is aligned with the engine axis. During the experiment, the rotor speed is fixed at a constant value and the corresponding thrust is measured at the steady state. The rotor speed is then adjusted to different values and the thrust measurements are taken to build the model.

Figure 6. Engine thrust ground experiment platform.

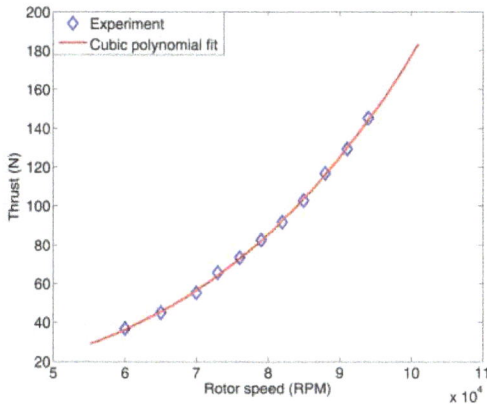

Figure 7. Relationship between the rotor speed and engine thrust in steady state conditions.

The thrust experiment results for a single P200 engine at steady state are shown in Figure 7 as the blue points for rotor speed ranges from 60,000 RPM to 94,000 RPM. A cubic polynomial function is used to fit the experimental data, and it gives us the red curve as shown in Figure 7. The mathematical expression for the cubic polynomial relationship is written as:

$$F_t = 439.61\Omega^3 - 499.46\Omega^2 + 293.67\Omega - 55.26 \tag{5}$$

where F_t represents the engine thrust with the unit of Newton (N) and Ω represents the rotor speed with the unit of 10^5 RPM. It can be seen that the polynomial function matches the experiment data closely. The mean of the regression residuals is -1.2×10^{-4} N and the standard deviation of the residuals is 0.67 N.

The above polynomial relationship is validated by comparing the experimental thrust with the computed thrust (using Equation (5)). The comparison is shown in Table 2 for the rotor speeds of 96,000, 98,000 and 100,000 RPM which are not used in obtaining Equation (5). The relative error, designated E_F, is computed as:

$$E_F = \frac{F_{tm} - F_{te}}{F_{te}} \tag{6}$$

where F_{tm} is the engine thrust computed using the model, and F_{te} is the engine thrust measured in the experiment. The close agreement between the measured and computed thrusts and the very small (0.3%) relative errors shown in Table 2 demonstrate high accuracy of the thrust model. Different experimental data (rather than the blue points shown in Figure 7) can be used to fit the cubic equation as long as the relative errors shown in Table 2 are within an acceptable range (0.5% used in this work). Detailed discussion about the cubic thrust model can be found in an earlier work by Tsinghua University [24].

Table 2. Comparison of the experimental thrust and the modeled thrust.

Rotor Speed (RPM)	96,000	98,000	100,000
Experimental thrust (N)	154.8	167.3	177.1
Computed thrust (N)	154.6	166.8	177.4
Relative error (%)	−0.13	−0.30	0.17

For real problems, the engines typically operate with time-varying rotor speed, especially during the takeoff and landing stages. Therefore, the polynomial thrust model needs to be validated by a dynamic thrust experiment as well. The time-varying rotor speed as shown in Figure 8a is used in the dynamic experiment. The range for this time-varying rotor speed is from 45,000 to 95,000 RPM which covers the rotor speed for UAV takeoff and landing. The corresponding thrust computed using the polynomial model is then compared with the thrust measured during the experiment in Figure 8b. A close match is observed which again confirms the accuracy and effectiveness of the thrust model.

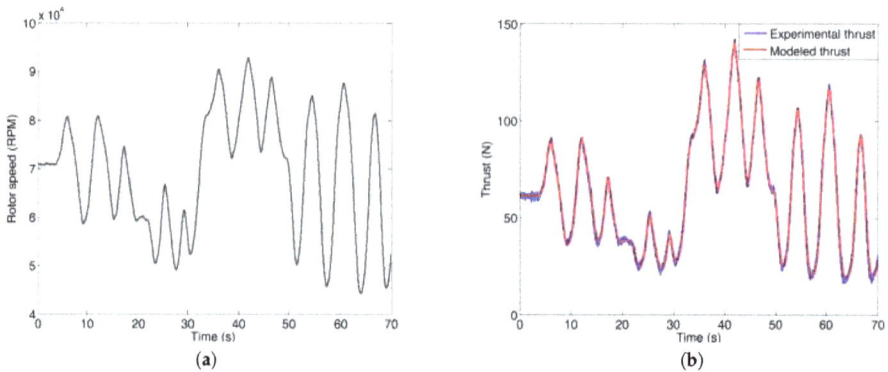

Figure 8. Thrust model validation using time-varying rotor speed. (a) Time-varying rotor speed; (b) Measured thrust and modeled thrust for time-varying rotor speed.

As mentioned before, the engine thrust cannot be measured during flight, but the rotor speed can be measured by Hall sensor with very high accuracy. The measured rotor speed will serve as the input in the engine thrust model (Equation (5)) to compute the thrust.

4.2. Thrust Vector Model

During flight, the attitude of the UAV is controlled by the engine thrust vectoring system to ensure the stability of the UAV body (with the UAV body nose pointing up vertically). The thrust vectoring system generates triaxial moments by deflecting the vectoring nozzles. Consequently, the engine thrust is not exactly aligned with the engine axis due to the nozzle deflection. The engine thrust model (Equation (5)) developed in the previous section is used to compute the overall thrust. To obtain the vertical component of the overall thrust, the deflection angle of the engine thrust is needed.

We are not able to measure the deflection angle of the thrust directly when the UAV is flying. However, the deflection angle is controlled by the microcomputer and the command of the deflection angle can be recorded. The thrust vectoring system is well calibrated to ensure the thrust deflection angle is the same as the deflection command. It is validated by performing a thrust vectoring experiment on the ground.

The same experimental settings shown in Figure 6 are used. The engine rotor speed for this experiment is fixed at 84,140 RPM which corresponds to a total thrust of 100 N. Since the nozzle is axisymmetric, only the experimental results for yaw movement are presented. During the experiment, the sinusoidal deflection angle command with an amplitude of 10 degree (shown in Figure 9 as the purple curve) is used and the corresponding thrust vector is measured. The deflection angle of the thrust, designated α_F, is calculated using Equation (7) as:

$$\alpha_F = tan^{-1} \frac{F_{ty}^b}{F_{tx}^b} \qquad (7)$$

where F_{tx}^b and F_{ty}^b are the measured thrusts in the x and y directions in the UAV body coordinate system. The UAV coordinate is defined as: the x-axis points out of the tailsitter nose, the y-axis points out the right wing and the z-axis points out the bottom of the fuselage.

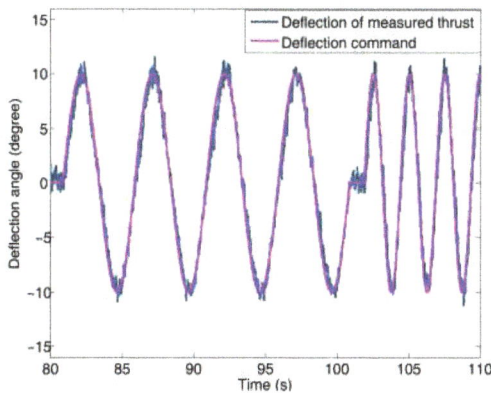

Figure 9. Comparison of the deflection angle command and the actual deflection angle of thrust.

The actual thrust deflection angle (α_F) derived from the thrust vector measurement using Equation (7) is compared with the deflection command in Figure 9 using the blue curve. A close match between the two curves is observed. This suggests that the deflection command can be used as the actual thrust deflection angle with sufficient accuracy. Therefore, the thrust vector model can be written as:

$$\begin{pmatrix} F_{tx}^b \\ F_{ty}^b \\ F_{tz}^b \end{pmatrix} = \frac{F_t}{\sqrt{1 + tan^2\alpha + tan^2\beta}} \begin{pmatrix} 1 \\ tan\beta \\ tan\alpha \end{pmatrix} \tag{8}$$

where α and β are the pitch and yaw deflection commands, F_t is the total engine thrust which is estimated using engine thrust model described in the previous section, F_{tx}^b, F_{ty}^b and F_{tz}^b are the engine thrusts in the x, y and z directions respectively. Note that the x, y and z directions are in the UAV coordinate system, not in the inertial coordinate system.

By combining the engine thrust model (Equation (5)) and thrust vector model (Equation (8)), we are able to obtain the thrust vector using the measured rotor speed and the deflection commands. The coordinate transformation is again used to obtain the vertical thrust, which is indicated by F_{vc}.

Note that there are two engines in the propulsion system. During the flight, the two engines receive the same rotor speed command, which means they have the same overall thrust. But the deflection angles are controlled independently. Therefore, the vertical thrusts generated by the two engines, designated F_{vc1} and F_{vc2}, need to be computed separately. The total vertical thrust of the engine system is the sum of F_{vc1} and F_{vc2}:

$$F_{vc} = F_{vc1} + F_{vc2} \tag{9}$$

Though the modeled results are of good quality, small modeling errors are inevitable. In addition, the measurement noise is also present. To incorporate these effects into the vertical thrust computation, we assume the error of the computed vertical thrust is white Gaussian noise. The genuine vertical thrust is thus expressed as:

$$F_v = F_{vc} + w_F \tag{10}$$

where F_v is the genuine vertical thrust, F_{vc} is the computed vertical thrust, and w_F is the white Gaussian noise for vertical thrust with the variance of σ_F^2.

5. Methodology for Acceleration Noise Suppression

By defining $\lambda = \frac{1}{m}$ in which m denotes the UAV mass, we can rewrite Equation (4) as:

$$a_m = \lambda F_v + w_m \tag{11}$$

where F_v is the vertical thrust force and $a_m = a + g + w_m$ represents the measurement of the IMU with w_m is the vertical acceleration noise which is white Gaussian noise with a variance of $\sigma_{a_m}^2$. By combining with Equation (10), the above equation can be rewritten as:

$$\begin{aligned} a_m &= \lambda(F_{vc} + w_F) + w_m \\ &= \lambda F_{vc} + (\lambda w_F + w_m) \end{aligned} \tag{12}$$

Since λ is a constant (the inverse of the UAV mass), λw_F is white Gaussian noise with the variance of $\lambda^2 \sigma_F^2$.

We then define a new term, w, as:

$$w = \lambda w_F + w_m \tag{13}$$

where w represents the overall noise which combines the measured acceleration noise and the thrust modeling error. Because both λw_F and w_m are white Gaussian noise, w is white Gaussian noise as well. In addition, λw_F is independent of w_m due to the independence between w_F and w_m. According to the above analysis, w can be treated as white Gaussian noise with the variance, designated σ^2, written as:

$$\sigma^2 = \lambda^2 \sigma_F^2 + \sigma_{a_m}^2 \tag{14}$$

We take λ as the state variable, a_m as the observation variable and F_{vc} as the input parameter, the discrete-time state space model can be written as:

$$\begin{cases} \lambda(k+1) = \lambda(k) \\ a_m(k) = \lambda(k)F_{vc}(k) + w(k) \end{cases} \tag{15}$$

The standard Kalman filter can be used to estimate the λ by combining the computed thrust and measured acceleration information. The updating process is as follows:

$$\begin{cases} K_\lambda(k) = P_\lambda(k-1)F_{vc}(k)[F_{vc}(k)P_\lambda(k-1)F_{vc}(K) + \sigma^2(k)]^{-1} \\ P_\lambda(k) = [1 - K_\lambda(k)F_{vc}(k)]P_\lambda(k-1) \\ \hat{\lambda}(k) = \hat{\lambda}(k-1) + K_\lambda(k)[a_m(k) - F_{vc}(k)\hat{\lambda}(k-1)] \end{cases} \tag{16}$$

where $\hat{\lambda}(k-1)$ and $\hat{\lambda}(k)$ are the estimates of λ at sampled data points $k-1$ and k respectively, $K_\lambda(k)$ is the Kalman filter gain, and $P_\lambda(k)$ is the variance of $\hat{\lambda}(k)$. Note that the variance of $w(k)$, designated $\sigma^2(k)$, is estimated as follows in the calculation process:

$$\sigma^2(k) = \hat{\lambda}^2(k-1)\sigma_F^2 + \sigma_{a_m}^2 \tag{17}$$

Here we assume the variance of the measured acceleration noise during UAV takeoff and landing is the same as the acceleration variance when the engine rotor speed is at 85,000 RPM. This is because that the rotor speed of 85,000 RPM is very close to speed of the engines during takeoff and landing. Therefore, $\sigma_{a_m} = 0.351$ (from Table 1) is used in this work. The variance of vertical thrust error is assumed to be $\sigma_F^2 = 10\,\text{N}^2$.

With the estimated λ and calculated vertical thrust F_{vc}, vertical acceleration can be obtained as:

$$a_c(k) = \hat{\lambda}(k)F_{vc}(k) \tag{18}$$

where $a_c(k)$ is the calculated acceleration at sampled data point k. By combing with Equation (10), the above equation can be rewritten as:

$$\begin{aligned} a_c(k) &= \hat{\lambda}(k)[F_v(k) - w_F(k)] \\ &= \hat{\lambda}(k)F_v(k) - \hat{\lambda}(k)w_F(k) \end{aligned} \tag{19}$$

We then investigate the expectation of $a_c(k)$ as:

$$\begin{aligned} E\{a_c(k)\} &= E\{\hat{\lambda}(k)F_v(k)\} - E\{\hat{\lambda}(k)w_F(k)\} \\ &= \lambda F_v(k) - \rho_{\hat{\lambda}(k),w_F(k)}\sigma_{\hat{\lambda}(k)}\sigma_F \end{aligned} \tag{20}$$

where $\rho_{\hat{\lambda}(k),w_F(k)}$ represents the correlation coefficient between $\hat{\lambda}(k)$ and $w_F(k)$, and the derivation of $\rho_{\hat{\lambda}(k),w_F(k)}$ is shown in Equation (A1). Note that the absolute value of the correlation coefficient does not exceed 1. Because $\sigma_{\hat{\lambda}(k)}$ approaches zero as k tends to infinity, the expectation of $a_c(k)$ approaches the genuine vertical acceleration, $\lambda F_v(k)$, as k tends to infinity according to Equation (20). Therefore $a_c(k)$ is the asymptotically unbiased estimate of the vertical acceleration.

According to Equation (19), the variance of $a_c(k)$ is then given by:

$$\sigma_{a_c}^2(k) = F_v^2(k)\sigma_{\hat{\lambda}(k)}^2 + \hat{\lambda}(k)^2\sigma_F^2 - 2\rho_{\hat{\lambda}(k),w_F(k)}\hat{\lambda}(k)F_v(k)\sigma_{\hat{\lambda}(k)}\sigma_F \tag{21}$$

To obtain the minimum variance estimate of the vertical acceleration, the calculated acceleration $a_c(k)$ and the measured acceleration $a_m(k)$ are fused as :

$$a_f(k) = \eta(k)a_c(k) + [1 - \eta(k)]a_m(k) \tag{22}$$

where $a_f(k)$ is the acceleration fusion result, $\eta(k)$ is the weighting coefficient with $0 < \eta(k) < 1$. It is obvious that $a_f(k)$ is asymptotically unbiased due to the asymptotically unbiasedness of $a_c(k)$ and the unbiasedness of $a_m(k)$. The variance of $a_f(k)$ is given by:

$$\sigma^2_{a_f(k)} = \eta(k)^2\sigma^2_{a_c(k)} + [1 - \eta(k)]^2\sigma^2_{a_m} + 2\eta(k)[1 - \eta(k)]\rho_{a_c(k),a_m(k)}\sigma_{a_c(k)}\sigma_{a_m} \tag{23}$$

where $\rho_{a_c(k),a_m(k)}$ is the correlation coefficient between $a_c(k)$ and $a_m(k)$ which is derived in Equation (A2). The variable $\eta(k)$ is adopted as follows to achieve the minimum variance estimate:

$$\eta(k) = \frac{\sigma^2_{a_m} - \rho_{a_c(k),a_m(k)}\sigma_{a_c(k)}\sigma_{a_m}}{\sigma^2_{a_c(k)} + \sigma^2_{a_m} - 2\rho_{a_c(k),a_m(k)}\sigma_{a_c(k)}\sigma_{a_m}} \tag{24}$$

The flowchart of the noise reduction procedure developed in this work is shown in Figure 10. At first, the engine system model (including engine thrust model and thrust vector model in Section 4) needs to be built. It is used to obtain the UAV vertical thrust when it is in flight. Then the Kalman filter (Equation (16)) is applied to estimate UAV mass by combining the information of engine vertical thrust and the IMU measured acceleration. The estimated UAV mass can then be used together with the vertical thrust to compute acceleration using Newton's Second Law. A data fusion step (Equation (22)) is finally applied to obtain the minimum variance estimate of the UAV acceleration. This overall procedure can be used for any type of UAVs for noise reduction purpose. The only change will be in the engine system model if the method is applied to different UAV platforms.

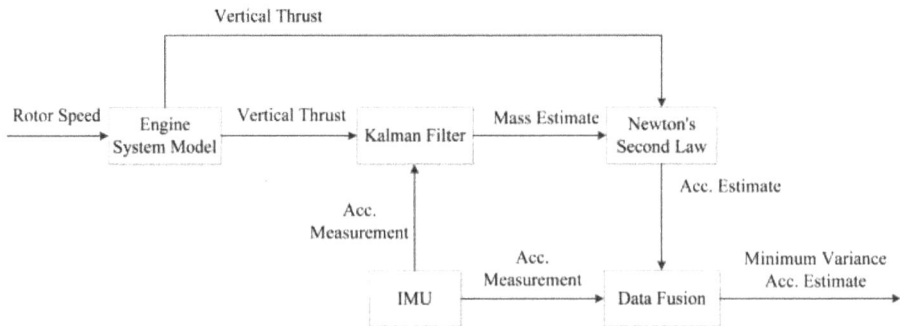

Figure 10. Flowchart of the noise reduction procedure developed in this work.

6. Experiment Result

To assess the performance of the new method for UAV vertical acceleration noise suppression, a vertical takeoff experiment is conducted. The data record began at the time when the UAV has positive (upward) elevation change which is indicated by the height measurement from the laser range finder as shown in Figure 11a. The UAV ascended from the initial position of about 1.5 m to around 3.2 m at $t = 2.8$ s. This corresponds to a maximum height gain of about 1.7 m. After reaching the highest point, the UAV started to descend, though it had not returned to its original position for the period that we measured. The rotor speed measured during the experiment is shown in Figure 11b. As can be seen, the rotor speed increased for $t < 1$ s, and dropped quickly during $1 < t < 2$ s. The rotor speed reached a stable value around 86,000 RPM eventually. The corresponding vertical acceleration measurement from IMU is shown in Figure 11c. The same as in Figure 3, the acceleration measurement is heavily polluted by the noise, though the similar trend with the rotor speed is observed.

We then subtract the gravitational acceleration from the IMU measured acceleration shown in Figure 11c. This gives us the coordinate acceleration (rate of change of velocity) which is displayed in Figure 11d. It can be clearly seen that this UAV experiment includes both the upward acceleration period (positive coordinate acceleration) and downward acceleration period (negative coordinate acceleration). Therefore, it is sufficient to validate the noise reduction method developed in this work.

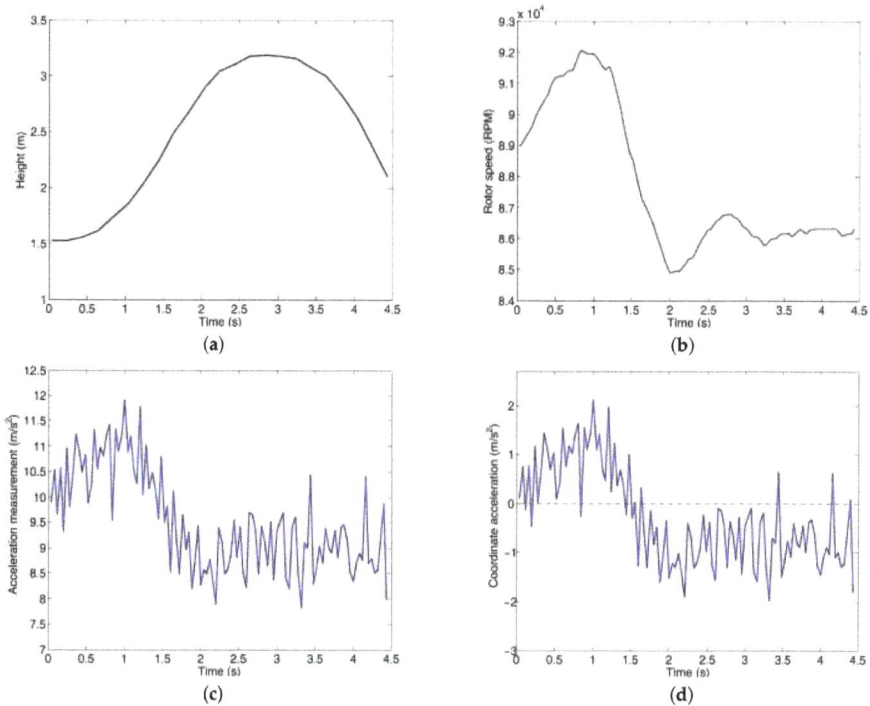

Figure 11. Experiment results. (**a**) Experiment height measurement; (**b**) Experiment rotor speed; (**c**) Experiment acceleration measurement; (**d**) Coordinate acceleration (computed from Figure 11c).

We apply the engine thrust model and thrust vector model developed in Section 3 to the experimental data. The calculated vertical thrust is shown in Figure 12. It is then used to estimate λ using the Kalman filter (Equation (16)). As shown in Figure 13, the value of λ quickly converges to 0.0404 kg^{-1}, which means that the mass of the UAV $(1/\lambda)$ is about 24.75 kg. Note that, the variable λ is unknown during flight, though it can be considered as a constant within a short period of time (for example, during the takeoff and landing stages). This is because that the UAV mass is impacted by the fuel consumption, which is unknown when the UAV is flying. Therefore, the Kalman filter (Equation (16)) is required to estimate λ.

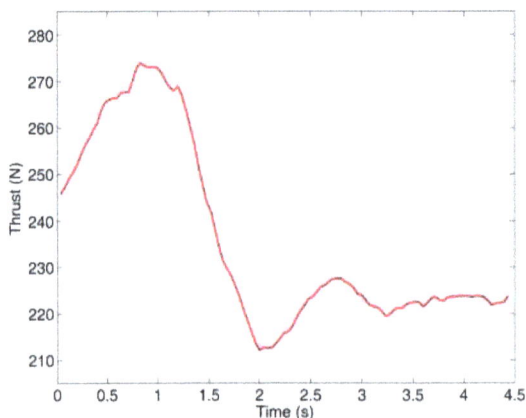

Figure 12. Calculated vertical thrust.

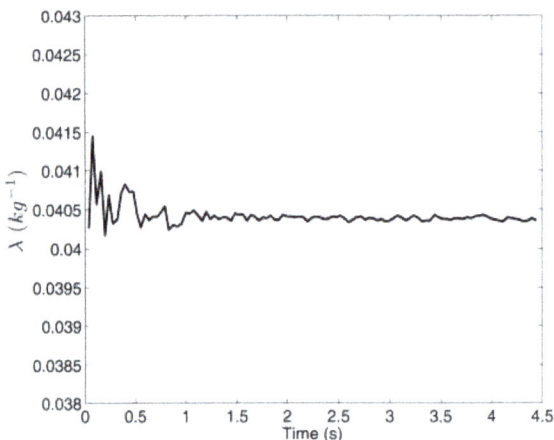

Figure 13. Estimate of λ.

The final acceleration fusion result is shown in Figure 14 as the red line. For comparison purpose, the measured vertical acceleration is shown again in Figure 14. It can be seen clearly that the acceleration fusion result suppresses the noise in the measured acceleration and meanwhile preserves the trend in measurements. To quantify that, we calculate the variance of the acceleration computed with data fusion method using Equation (23). The variance is shown in Figure 15. Initially the variance is of similar magnitude as the variance of the measured acceleration (as shown in Table 1). It drops quickly as more data is used in the fusion calculation and becomes stable at a value close to $0.02 \text{ m}^2/\text{s}^4$ after about 2 s.

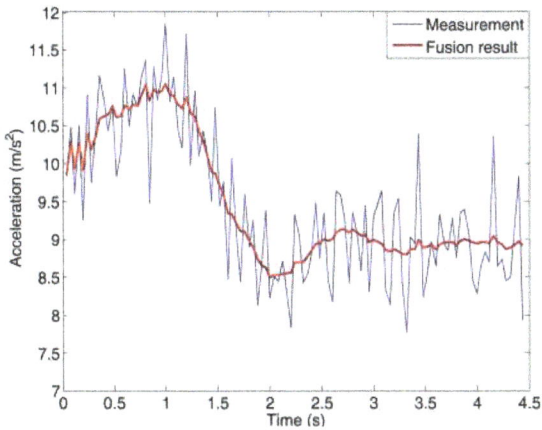

Figure 14. Acceleration fusion result.

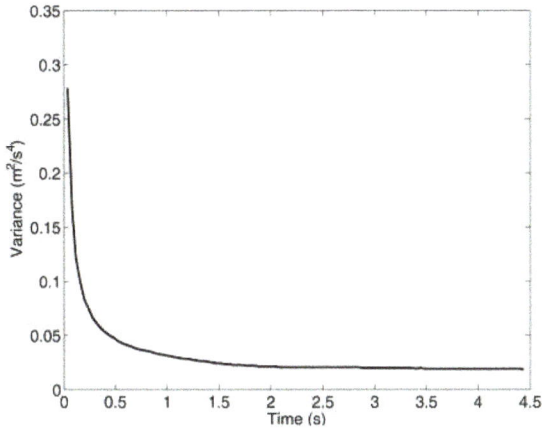

Figure 15. Variance of computed acceleration.

For the purpose of comparison, an infinite impulse response (IIR) lowpass filter developed in [9] and the Kalman filter developed in [13] are also used to denoise the measured acceleration. The parameters of the IIR filter and the Kalman filter are adopted to achieve similar denoising performance to the data fusion method developed in this work.

The IIR lowpass filter is shown as follows:

$$a_l(k) = b_0 a_m(k) + b_1 a_m(k-1) + b_2 a_m(k-2) + b_3 a_m(k-3) + b_4 a_m(k-4) + b_5 a_m(k-5) \\ - c_1 a_l(k-1) - c_2 a_l(k-2) - c_3 a_l(k-3) - c_4 a_l(k-4) - c_5 a_l(k-5) \tag{25}$$

where $a_l(k)$ and $a_m(k)$ are the filtered and measured accelerations for sampled data point k, respectively. The coefficients used here are:

$$b_0 = 0.0013, b_1 = 0.0064, b_2 = 0.0128, b_3 = 0.0128, b_4 = 0.0064, b_5 = 0.0013$$
$$c_1 = -2.9754, c_2 = 3.8060, c_3 = -2.5453, c_4 = 0.8811, c_5 = -0.1254$$

The state space model used in the Kalman filter developed in [13] is shown as:

$$\begin{cases} a_s(k+1) = a_s(k) + w_s(k) \\ a_m(k) = a_s(k) + w_m(k) \end{cases} \tag{26}$$

where $a_k(k)$ and $a_k(k+1)$ are the acceleration states at sampled point k and $k+1$, $w_s(k)$ is the process noise, $a_m(k)$ is the observed acceleration with observation noise indicated by $w_m(k)$. Note that $w_s(k)$ and $w_m(k)$ are mutually independent white Gaussian noise. Their variances used here are adopted as $\sigma_{w_s}^2 = 0.01$ and $\sigma_{w_m}^2 = 0.351$, respectively.

The comparison is shown in Figure 16. At a very early time value ($t < 0.2$ s), the results using the Kalman filter and the new data fusion method are very close. As more data is included into the calculation, they start to deviate and an obvious time delay is seen in the Kalman filter result for $t < 2.5$ s. In contrast, the lowpass result differs from the fusion result from the very beginning, and the time delay in the lowpass result is very obvious as well. At the late stage ($t > 2.5$ s), when the acceleration becomes relatively stable (due to stable engine rotor speed shown in Figure 11b), the results using the three approaches become close.

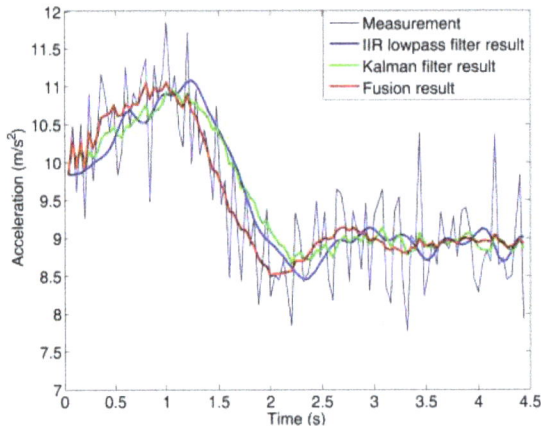

Figure 16. Acceleration fusion result compared with the IIR lowpass filter and Kalman filter results.

The difference between each of the three computed accelerations and the measured acceleration, designated a_d, is computed as:

$$a_d = a_c - a_m \tag{27}$$

where a_c is the computed acceleration, and a_m is the measured acceleration. The acceleration differences are shown in Figuire 17. Dashed lines are used for Kalman filter and data fusion results for better visualization. We can see that the acceleration differences are of high amplitude due to the significant noise in the measured acceleration.

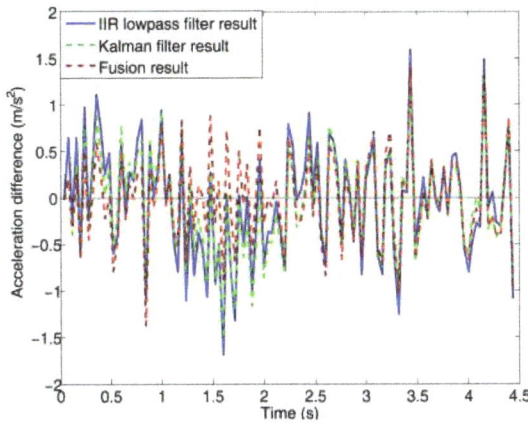

Figure 17. Acceleration differences.

A nine-order mean filter is used to extract the low frequency component of the acceleration differences shown in Figure 17. The nine-order mean filter is expressed as:

$$\bar{a}_d(k) = \frac{1}{9} \sum_{i=k-4}^{k+4} a_d(i), \quad \text{for } 4 < k < N - 4 \tag{28}$$

where $\bar{a}_d(k)$ is the low frequency component result at sampled data point k. The results are shown in Figure 18, from which we can clearly see the difference between the new method and the two existing methods. The lowpass filter and Kalman filter provide similar results and both of them have apparent non-zero values in their low frequency components (greater than zero for $t < 0.9$ s and less than zero during $1 < t < 2.2$ s). This is because of the time delay introduced by the filtering process. The time delay is due to the weighting of the past time acceleration measurements in the filtered results in both the lowpass and Kalman filters. In contrast, the results using the new method closely distribute around zero without any obvious non-zero trends. This is because that the new method uses the past time information only for the estimation of the variable λ. As shown in Figure 13, the λ converges very quickly to a constant value during the takeoff stage. Therefore, the new method does not introduce any time delay in acceleration computation.

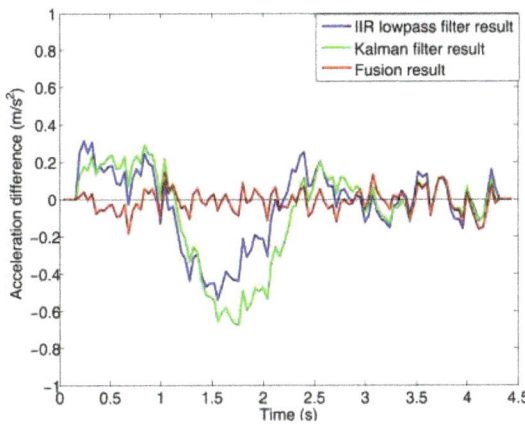

Figure 18. Low frequency component of acceleration differences.

Note that the time delay can be reduced for lowpass filter and Kalman filter by adjusting their parameters, but the noise suppression performance of the two methods will be severely degraded. Compared with the lowpass filter and Kalman filter, the newly developed method aided with the thrust information have much better real time performance with strong ability of noise suppression.

7. Conclusions

In this work, we developed a novel method to reduce the noise in acceleration measurement for a thrust-vectored VTOL UAV. The method combines the information in engine thrust with the information in the measured acceleration. To do that, a Kalman filter is first applied to estimate the UAV mass, which is used to compute acceleration together with the vertical thrust. This estimated acceleration is further fused with the measured acceleration to obtain the minimum variance acceleration estimate.

To test the effectiveness of the newly developed approach, a vertical takeoff experiment for the UAV was performed. The new method was used to compute the vertical acceleration using the data collected from the experiment. The results demonstrated that the new method has very good performance in terms of noise reduction. The variance of the vertical acceleration is reduced by about 95%. In addition, the acceleration calculated using the new approach shows no time delay, which enables its usage in acceleration feedback control for UAV takeoff and landing (or for any other situations with fast-changing accelerations). In contrast, two existing approaches based on IIR and Kalman filter suffer from significant time delay which limits their usability for real problems, although they can achieve similar denoising performance.

In addition, we developed numerical models for engine thrust and thrust vector which are important components in the entire noise suppression procedure. A polynomial function is used to characterize the relationship between the rotor speed and the engine thrust. Though the model was built based on steady-state ground experiment, it is able to predict accurately the thrust in dynamic condition (with time-varying rotor speed). The thrust deflection angle was found to be consistent with the deflection angle command, which is then used to compute the engine thrust in the vertical direction.

We analyzed the statistical characteristics of the acceleration noise. It was found that the amplitude of the noise grows with the increasing rotor speed. We also demonstrated that the acceleration noise can be treated as white Gaussian noise based on the autocorrelation and histogram of the sampled noise data.

The noise suppression method developed here can be readily applied into the acceleration feedback motion control for VTOL UAVs during a critical period, for example, takeoff and landing stages. Improved precision in motion control is expected, which will be tested in future work.

Acknowledgments: The authors thank Chao Zhang for his experiment advices.

Author Contributions: The method was proposed by Huanyu Li. Huanyu Li and Linfeng Wu designed the experiment. The experiments were performed by Huanyu Li, Linfeng Wu, Yingjie Li. Data analysis was done by Huanyu Li and Hangyu Li, and the paper was mainly written by them as well. Chunwen Li supervised the work. All authors contributed to writing the article.

Conflicts of Interest: The authors declare no conflict of interest.

Abbreviations

The following abbreviations are used in this manuscript:

UAV	Unmanned aerial vehicle
VTOL	Vertical takeoff and landing
IMU	Inertial measurement unit
RPM	Revolutions per minute
IIR	Infinite impulse response

Appendix A

$$
\begin{aligned}
\rho_{\hat{\lambda}(k),w_F(k)} &= E\{w_F(k)(\hat{\lambda}(k) - \lambda)\}/(\sigma_{\hat{\lambda}(k)}\sigma_F) \\
&= E\{w_F(k)[\hat{\lambda}(k-1) + K_\lambda(k)(a_m(k) - F_{vc}(k)\hat{\lambda}(k-1)) - \lambda]\}/(\sigma_{\hat{\lambda}(k)}\sigma_F) \\
&= E\{w_F(k)[\hat{\lambda}(k-1) + K_\lambda(k)(\lambda F_{vc}(k) + \lambda w_F(k) + w_m(k) - F_{vc}(k)\hat{\lambda}(k-1)) - \lambda]\}/(\sigma_{\hat{\lambda}(k)}\sigma_F) \\
&= E\{w_F(k)[(\lambda - \hat{\lambda}(k-1))(K_\lambda(k)F_{vc}(k) - 1) + w_F(k)K_\lambda(k)(\lambda w_F(k) + w_m(k))]\}/(\sigma_{\hat{\lambda}(k)}\sigma_F) \\
&= E\{K_\lambda(k)w_F(k)[\lambda w_F(k) + w_m(k)]\}/(\sigma_{\hat{\lambda}(k)}\sigma_F) \\
&= K_\lambda(k)\lambda\sigma_F/\sigma_{\hat{\lambda}(k)}
\end{aligned}
\tag{A1}
$$

$$
\begin{aligned}
\rho_{a_c(k),a_m(k)} &= E\{w_m(k)[\hat{\lambda}(k)w_F(k) + \hat{\lambda}(k)F_v(k) - \lambda F_v(k)]\}/(\sigma_{a_c(k)}\sigma_{a_m}) \\
&= E\{\hat{\lambda}(k)w_m(k)w_F(k) + w_m(k)(\hat{\lambda}(k) - \lambda)F_v(k)\}/(\sigma_{a_c(k)}\sigma_{a_m}) \\
&= F_v(k)E\{w_m(k)[\hat{\lambda}(k) - \lambda]\}/(\sigma_{a_c(k)}\sigma_{a_m}) \\
&= F_v(k)E\{w_m(k)[\hat{\lambda}(k-1) + K_\lambda(k)(a_m(k) - F_{vc}(k)\hat{\lambda}(k-1)) - \lambda]\}/(\sigma_{a_c(k)}\sigma_{a_m}) \\
&= F_v(k)E\{w_m(k)[(\lambda - \hat{\lambda}(k-1))(K_\lambda(k)F_{vc}(k) - 1) + K_\lambda(k)w_m(k) + K_\lambda\lambda w_F(k)]\}/(\sigma_{a_c(k)}\sigma_{a_m}) \\
&= F_v(k)K_\lambda\sigma_{a_m}/\sigma_{a_c(k)}
\end{aligned}
\tag{A2}
$$

References

1. Johnson, E.N.; Turbe, M.A. Modeling, control, and flight testing of a small-ducted fan aircraft. *J. Guid. Control Dyn.* **2006**, *29*, 769–779.
2. Kubo, D.; Suzuki, S. Tail-sitter vertical takeoff and landing unmanned aerial vehicle: Transitional flight analysis. *J. Aircr.* **2008**, *45*, 292–297.
3. Tomic, T.; Schmid, K.; Lutz, P.; Domel, A.; Kassecker, M.; Mair, E.; Grixa, I.L.; Ruess, E.; Suppa, M.; Burschka, D. Toward a fully autonomous UAV: Research platform for indoor and outdoor urban search and rescue. *IEEE Robot. Autom. Mag.* **2012**, *19*, 46–56.
4. Lei, X.; Li, J. An adaptive altitude information fusion method for autonomous landing processes of small unmanned aerial rotorcraft. *Sensors* **2012**, *12*, 13212–13224.
5. Peddle, I.K. *Acceleration Based Manoeuvre Flight Control System for Unmanned Aerial Vehicles*; Stellenbosch University: Stellenbosch, South Africa, 2008.
6. Boyle, D.P.; Chamitoff, G.E. Autonomous maneuver tracking for self-piloted vehicles. *J. Guid. Control Dyn.* **1999**, *22*, 58–67.
7. Wang, X. Takeoff/landing control based on acceleration measurements for VTOL aircraft. *J. Frankl. Inst.* **2013**, *350*, 3045–3063.
8. Gan, Y.; Sui, L.; Wu, J.; Wang, B.; Zhang, Q.; Xiao, G. An EMD threshold de-noising method for inertial sensors. *Measurement* **2014**, *49*, 34–41.
9. Sun, F.; Sun, W. Mooring alignment for marine SINS using the digital filter. *Measurement* **2010**, *43*, 1489–1494.
10. Parks, T.W.; Burrus, C.S. *Digital Filter Design*; Wiley-Interscience: Hoboken, NJ, USA, 1987.
11. Lu, S.; Xie, L.; Chen, J. New techniques for initial alignment of strapdown inertial navigation system. *J. Frankl. Inst.* **2009**, *346*, 1021–1037.
12. El-Sheimy, N.; Nassar, S.; Noureldin, A. Wavelet de-noising for IMU alignment. *IEEE Aerosp. Electron. Syst. Mag.* **2004**, *19*, 32–39.
13. Kownacki, C. Optimization approach to adapt Kalman filters for the real-time application of accelerometer and gyroscope signals' filtering. *Dig. Signal Process.* **2011**, *21*, 131–140.
14. Hebbale, K.V.; Ghoneim, Y.A. A speed and acceleration estimation algorithm for powertrain control. In Proceedings of the 1991 American Control Conference, Boston, MA, USA, 26–28 June 1991; pp. 415–420.
15. Abellanosa, C.B.; Lugpatan, R.P.J.; Pascua, D.A.D. Position Estimation using Inertial Measurement Unit (IMU) on a Quadcopter in an Enclosed Environment. *Int. J. Comput. Commun. Instrum. Eng.* **2016**, *3*, 332–336.
16. Quadri, S.A.; Sidek, O.; Bin Abdullah, A. A Study of State Estimation Algorithms in an OktoKopter. *Int. J. u- e-Serv. Sci. Technol.* **2014**, *7*, 247–266.

17. Edu, I.R.; Adochiei, F.C.; Grigorie, T.L.; Botez, R.M. Tuning of a Wavelet Filter for Miniature Accelerometers Denoising based Joint Symbolic Dynamics (JSD) Method. *INCAS Bull.* **2015**, *7*, 71–81.

18. Nebylov, A.; Sukrit, S.; Arifuddin, F. Perspectives for Development of an Autonomous and Intelligent WIG-Craft and Its Peculiar Control Problems. In Proceedings of the 2009 IFAC "AGNFC" Workshop Proceedings, Samara, Russia, 30 June–2 July 2011; pp. 1–6.

19. Li, H.; Wu, L.; Li, Y.; Li, C. Identification of turbine engine dynamics with the governor in the loop. In Proceedings of the 2016 IEEE International Conference on Systems, Man, and Cybernetics (SMC), Budapest, Hungary, 9–12 October 2016.

20. Stone, R.H.; Anderson, P.; Hutchison, C.; Tsai, A.; Gibbens, P.; Wong, K.C. Flight testing of the t-wing tail-sitter unmanned air vehicle. *J. Aircr.* **2008**, *45*, 673–685.

21. Beach, J.M.; Argyle, M.E.; McLain, T.W.; Beard, R.W.; Morris, S. Tailsitter heading estimation using a magnetometer. In Proceedings of the 2014 American Control Conference, Portland, OR, USA, 4–6 June 2014; pp. 91–96.

22. Woodman, O.J. *An Introduction to Inertial Navigation*; Technical Reports; UCAMCL-TR-696; University of Cambridge, Computer Laboratory: Cambridge, UK, 2007.

23. Titterton, D.; Weston, J. *Strapdown Inertial Navigation Technology*, 2nd ed.; The Institution of Engineering and Technology: London, UK, 2004.

24. Yang, J.; Zhu, J. Dynamic modelling of a small scale turbojet engine. In Proceedings of the 2015 European Control Conference (ECC), Linz, Austria, 15–17 July 2015; pp. 2750–2755.

sensors

MDPI

Article

Cubature Information SMC-PHD for Multi-Target Tracking

Zhe Liu [1,2], Zulin Wang [1,3] and Mai Xu [1,*]

[1]   School of Electronic and Information Engineering, Beihang University, Beijing 100191, China;
      liuzhe201@buaa.edu.cn (Z.L.); wzulin@buaa.edu.cn (Z.W.)
[2]   School of Information and Communication Engineering, North University of China, Taiyuan 030051, China
[3]   Collaborative Innovation Center of Geospatial Technology, 129 Luoyu Road, Wuhan 430079, China
*   Correspondence: maixu@buaa.edu.cn; Tel.: +86-10-8231-5461

Academic Editors: Felipe Gonzalez Toro and Antonios Tsourdos
Received: 6 March 2016; Accepted: 1 May 2016; Published: 9 May 2016

Abstract: In multi-target tracking, the key problem lies in estimating the number and states of individual targets, in which the challenge is the time-varying multi-target numbers and states. Recently, several multi-target tracking approaches, based on the sequential Monte Carlo probability hypothesis density (SMC-PHD) filter, have been presented to solve such a problem. However, most of these approaches select the transition density as the importance sampling (IS) function, which is inefficient in a nonlinear scenario. To enhance the performance of the conventional SMC-PHD filter, we propose in this paper two approaches using the cubature information filter (CIF) for multi-target tracking. More specifically, we first apply the posterior intensity as the IS function. Then, we propose to utilize the CIF algorithm with a gating method to calculate the IS function, namely CISMC-PHD approach. Meanwhile, a fast implementation of the CISMC-PHD approach is proposed, which clusters the particles into several groups according to the Gaussian mixture components. With the constructed components, the IS function is approximated instead of particles. As a result, the computational complexity of the CISMC-PHD approach can be significantly reduced. The simulation results demonstrate the effectiveness of our approaches.

Keywords: Sequential monte carlo; probability hypothesis density; importance sampling; cubature information filter; Gaussian mixture

PACS: J0101

1. Introduction

1.1. Background

Multi-target tracking refers to sequential approximation of the states (positions, velocities, *etc.*), and the number of individual targets. It has been widely used in ground-moving-target tracking [1], visual tracking [2], and distribution fusion [3]. In multi-target tracking, both the state and observation sets of targets are time-varying. In practice, the associations between state and observation sets are always unknown, thus posing a challenge for multi-target tracking. The conventional approaches, such as the nearest-neighbour Kalman filter (NNKF) [4], Extended Kalman Filter (EKF) [5,6], multiple hypothesis tracking (MHT) [7], and joint probabilistic data association (JPDA) [8], are used to formulate the explicit associations between states and observations of targets. However, caused by the targets appearing and disappearing, the state and observation sets of the multi-target are uncertain. Such uncertainty costs high complexity in the conventional approaches on constructing the association between them. Several approaches have emerged to improve the performance of conventional approaches in terms of tracking accuracy [9–13].

Recently, filters based on random finite sets (RFS) have been developed as alternative frameworks of the traditional algorithms [14–17] to estimate the multi-target number and states. In RFS formulations, both the multi-target state and observation sets are modelled as random finite sets. Given such models, the multi-target tracking problem is formulated under the Bayesian framework [14]. Since the optimal RFS-based Bayesian filter is computationally intractable, the probability hypothesis density (PHD) filter has been proposed as a sub-optimal Bayesian filter for multi-target tracking [15–17]. By propagating the first order moment (namely the intensity), the PHD filter can save much computational complexity in comparison to the optimal RFS-based Bayesian filter. Moreover, it avoids the combinatorial problem arising from data association, which is the bottleneck for conventional multi-target tracking approaches to estimate multi-target number and states.

However, the difficulty of the PHD filter is that it is intractable to derive close-form solutions to the PHD filter formulations. To solve such difficulties, Vo *et al.* [18] proposed the bootstrap SMC-PHD (BSMC-PHD) filter, which selects the transition density as the IS function. The modified versions of BSMC-PHD filter have emerged in [19–21]. As for those BSMC-PHD filters, the IS function selection simplifies the weight computation in the prediction stage of the SMC-PHD filter. In practice, such a selection leads to a few particles with large-valued weights in the update stage, when targets have nonlinear motion [22].

To improve the tracking performance of the conventional SMC-PHD filter, several efficient approaches have been presented. Morelande *et al.* proposed the Rao-Blackwellised SMC-PHD (RB-SMC-PHD) filter in [23]. The RB-SMC-PHD filter utilizes several auxiliary variables to define the IS function, thus making it suitable for the intensity approximation of the SMC-PHD filter with conditionally linear Gaussian models. Motivated by the auxiliary particle filter, Whiteley *et al.* proposed the auxiliary particle PHD (APHD) filter in [24], in which the auxiliary variable is pre-selected to minimize the variance of the IS weights. Although the APHD filter enhances the efficiency of the SMC-PHD filter in nonlinear scenario, it yields poor performance in case of severe nonlinearities or high process noise [22]. Inspired by the unscented particle filter, Yoon *et al.* [25] utilized the unscented information filter (UIF) to design the IS function of SMC-PHD (called USMC-PHD filter). Since such a design takes the current observations of targets into account, it is more stable than the BSMC-PHD filter. However, the drawback of this filter is that its performance has been influenced by the selection of the sigma-points of UIF. Ristic *et al.* [20] proposed a novel state estimation method (called IBSMC-PHD), rather than partitioning particles in *ad-hoc* manner. Their method groups the particles in the update stage, thus enjoying the computational efficiency.

1.2. Our Work and Contributions

In this paper, we propose a cubature information SMC-PHD (CISMC-PHD) and its fast implementation (F-CISMC-PHD) approaches, which can be used to estimate the time-varying number and states of multi-target. As aforementioned, the disadvantage of conventional SMC-PHD filters is the tracking inefficiency in nonlinear scenario. To avoid such inefficiency, our CISMC-PHD approach applies the posterior intensity as the IS function. With such a selection, the current observations can be incorporated into the IS function design. Then, we utilize the cubature information filter (CIF) [26] with a gating method to calculate the IS function. Benefitting from tracking accuracy of CIF in high dimensional nonlinear case, the CISMC-PHD approach is capable of estimating the time-varying number and states of targets. To avoid initializing birth intensity in whole state space, a birth intensity initialization method is proposed for our CISMC-PHD approach. At last, we present the F-CISMC-PHD approach to reduce the computational complexity by considering groups of particles as Gaussian mixture components. These components are applied to approximate the IS functions instead of particles of the CISMC-PHD approach. Since the number of Gaussian components is much less than that of particles, the computational complexity can be greatly reduced. The main contributions of our work are listed as follows.

(1) We propose a novel IS function approximating method, which utilizes the CIF with a gating method to enhance the estimation accuracy of the SMC-PHD filter. Specifically, first, the posterior intensity is applied as the IS function of our CISMC-PHD approach, in order to incorporate the current observation set into the IS function approximation. Then, the gating method is integrated into the update step of the CIF for approximating the IS function. Benefitting from the most recent success of CIF in nonlinear state estimation, the tracking performance of the proposed CISMC-PHD approach is significantly enhanced.
(2) We develop a method to initialize the birth intensity for the next tracking recursion. Since the current estimated targets (*i.e.*, current survival targets) are not possible to be the birth targets at the next recursion, the observations of estimated targets are removed from the current observation set. Then, using an unbiased model, the remaining observations are mapped to state space for the birth intensity initialization. As such, the birth intensity can be adaptively initialized, making the target tracking more accurate and stable.
(3) We develop a fast version of the CISMC-PHD approach (namely F-CISMC-PHD). We first consider each group of particles as a Gaussian mixture component. Then these components are used to approximate the IS functions of the CISMC-PHD approach. With the approximated IS functions, the particles can be sampled from these components for the intensity prediction and update steps. As a result, the computational complexity of the proposed CISMC-PHD approach can be significantly reduced.

The rest of this paper is organized as follows. In Section 2, a brief overview of the SMC-PHD filter is provided. Section 3 proposes our CISMC-PHD approach. A fast implementation of CISMC-PHD approach is presented in Section 4. Simulation results are demonstrated in Section 5, and Section 6 concludes this paper.

2. A Brief Overview of The SMC-PHD Filter

In this section, we review the basic idea of the SMC-PHD filter in detail. The main notations used in this section are defined as follows.

$\mathbf{x}_k$	The state of a dynamic target at time k
$\mathbf{z}_k$	The observation of a dynamic target at time k
$\gamma_k(\cdot)$	The intensity of birth target at time k
L_k	The number of the survival particles at time k
J_k	The number of the birth particles at time k
$p_s(\cdot)$	The survival probability of target
$p_d(\cdot)$	The detected probability of target
$\pi(\cdot,\cdot)$	The IS function of birth intensity
$q(\cdot|\cdot)$	The IS function of survival intensity

The SMC-PHD filter, motivated by the particle filter, is a sequential implementation of the PHD filter. In the SMC-PHD filter, the posterior intensity can be represented by a set of random samples of state vector $\mathbf{x}_k$ with associated weights, which are usually called particles. By substituting these particles into the recursion of the PHD filter, the multi-dimensional integrals can be replaced by summations of the particles, which is computationally tractable.

More specifically, we define the particle set at time $k-1$ as $\{\mathbf{x}_{k-1}^{(i)}, w_{k-1|k-1}^{(i)}\}_{i=1}^{L_{k-1}}$, where $\mathbf{x}_{k-1}^{(i)}$ and $w_{k-1|k-1}^{(i)}$ are the state and weight of the *i*-th particle at time $k-1$, respectively. The posterior intensity at time $k-1$ can be modelled by

$$D_{k-1|k-1}(\mathbf{x}_{k-1}|\mathbf{Z}_{1:k-1}) = \sum_{i=1}^{L_{k-1}} w_{k-1|k-1}^{(i)} \cdot \delta(\mathbf{x} - \mathbf{x}_{k-1}^{(i)}) \tag{1}$$

where $\mathbf{Z}_{1:k-1}$ is the multi-target observations from time 1 to $k-1$, and $\delta(\cdot)$ is the Dirac Delta function. Notice that $\mathbf{Z}_k = \{\mathbf{z}_{k,1}, \mathbf{z}_{k,2}, \ldots, \mathbf{z}_{k,M}\}$. Given $D_{k-1|k-1}(\mathbf{x}_{k-1}|\mathbf{Z}_{1:k-1})$, the implementation of the SMC-PHD filter consists of the *prediction* and *update* stages.

Prediction: We first denote IS functions for the survival and birth targets as $q(\mathbf{x}_k^{(i)}|\mathbf{x}_{k-1}^{(i)}, \mathbf{Z}_k)$ and $\pi(\mathbf{x}_k^{(i)}, \mathbf{Z}_k)$, respectively. Then, according to [18], the predicted intensity can be formulated by these IS functions,

$$D_{k|k-1}(\mathbf{x}_k|\mathbf{Z}_{1:k-1}) = \sum_{i=1}^{L_{k-1}+J_k} w_{k|k-1}^{(i)} \cdot \delta(\mathbf{x} - \mathbf{x}_k^{(i)}) \tag{2}$$

where

$$\mathbf{x}_k^{(i)} \sim \begin{cases} q(\mathbf{x}_k^{(i)}|\mathbf{x}_{k-1}^{(i)}, \mathbf{Z}_k) & i = 1, \ldots, L_{k-1} \\ \pi(\mathbf{x}_k^{(i)}, \mathbf{Z}_k) & i = L_{k-1}+1, \ldots, L_{k-1}+J_k \end{cases} \tag{3}$$

$$w_{k|k-1}^{(i)} = \begin{cases} \dfrac{p_s(\mathbf{x}_{k-1}^{(i)}) \cdot f(\mathbf{x}_k^{(i)}|\mathbf{x}_{k-1}^{(i)})}{q(\mathbf{x}_k^{(i)}|\mathbf{x}_{k-1}^{(i)}, \mathbf{Z}_k)} & i = 1, 2, \ldots, L_{k-1} \\ \dfrac{\gamma(\mathbf{x}_k^{(i)})}{J_k \cdot \pi(\mathbf{x}_k^{(i)}, \mathbf{Z}_k)} & i = L_{k-1}+1, \ldots, L_{k-1}+J_k \end{cases} \tag{4}$$

In Equation (4), J_k is calculated by $J_k = \nu \int \gamma(\mathbf{x})d\mathbf{x}$, where ν is the particle number of each birth target. Equations (2)–(4) can be then used to predict the states and weights of the particles.

Update: In this stage, we obtain the posterior intensity by updating Equation (2). Then, we have the following update strategy,

$$D_{k|k}(\mathbf{x}_k|\mathbf{Z}_{1:k}) = \sum_{i=1}^{L_{k-1}+J_k} w_{k|k}^{(i)} \delta(\mathbf{x} - \mathbf{x}_k^{(i)}) \tag{5}$$

where

$$w_{k|k}^{(i)} = \left(1 - p_d(\mathbf{x}_k^{(i)}) + \sum_{\mathbf{z} \in \mathbf{Z}_k} \frac{p_d(\mathbf{x}_k^{(i)})g_k(\mathbf{z}|\mathbf{x}_k^{(i)})}{\kappa(\mathbf{z}) + C_z}\right) w_{k|k-1}^{(i)} \tag{6}$$

$\kappa(\cdot)$ denotes the clutter intensity, and

$$C_z = \sum_{i=1}^{L_{k-1}+J_k} p_d(\mathbf{x}_k^{(i)})g_k(\mathbf{z}|\mathbf{x}_{k,i})w_{k|k-1} \tag{7}$$

Equations (2)–(6) include the main procedure of SMC-PHD at one recursion. Commonly, to avoid the degeneracy of particles, the resampling strategy is utilized to resample particle set $\{\mathbf{x}_k^{(i)}, w_{k|k}^{(i)}\}_{i=1}^{L_{k-1}+J_k}$. After resampling, we can use the clustering methods to extract number and states of targets.

3. The CISMC-PHD Approach for Multi-Target Tracking

In this section, we present our CISMC-PHD approach. To be more specific, we propose a novel IS function approximation algorithm incorporating the CIF and a gating method in Section 3.1. Then, Section 3.2 develops a method to initialize the birth intensity. Finally, Section 3.3 introduces the state extraction method for state estimation.

The nonlinear dynamic model of the target with state $\mathbf{x}_k$ at time k is given as follows,

$$\text{Process model:} \quad \mathbf{x}_k = \phi(\mathbf{x}_{k-1}) + \mathbf{v}_{k-1} \tag{8}$$

$$\text{Observation model:} \quad \mathbf{z}_k = \varphi(\mathbf{x}_k) + \mathbf{w}_k \tag{9}$$

where $\phi(\cdot)$ is the state transition function, and $\varphi(\cdot)$ denotes the relationship between state and observation. $\mathbf{v}_{k-1}$ and $\mathbf{w}_k$ are the process and observation noises at time $k-1$ and k, respectively. Both $\mathbf{v}_{k-1}$ and $\mathbf{w}_k$ are assumed to be Gaussian noises with zero means, and their covariances are denoted as

$\mathbf{Q}_{k-1}$ and $\mathbf{R}_k$. According to this model, the transition density $f(\mathbf{x}_k|\mathbf{x}_{k-1})$ and likelihood $g_k(\mathbf{z}_k|\mathbf{x}_k)$ are subject to Gaussian distributions.

3.1. The IS Function Approximation Algorithm

As mentioned in Section 2, most of the conventional SMC-PHD filters utilize the transition density function as the IS functions, resulting in great tracking error for targets with nonlinear dynamics. A novel IS function approximation algorithm, incorporating the CIF with a gating method, is presented to improve the tracking accuracy.

In our approach, we select the IS functions of Equations (3) and (4) as

$$\text{Survival IS:} \quad q(\mathbf{x}_k^{(i)}|\mathbf{x}_{k-1}^{(i)}, \mathbf{Z}_k) = N(\mathbf{x}_k^{(i)}; \mathbf{m}_{k,s}^{(i)}, \mathbf{P}_{k,s}^{(i)}) \tag{10}$$

$$\text{Birth IS:} \quad \pi(\mathbf{x}_k^{(i)}|\mathbf{Z}_k) = N(\mathbf{x}_k^{(i)}; \mathbf{m}_{k,b}^{(i)}, \mathbf{P}_{k,b}^{(i)}) \tag{11}$$

where $\mathbf{m}_{k,s}^{(\cdot)}$ and $\mathbf{m}_{k,b}^{(\cdot)}$ are means of the survival and birth particles, respectively. $\mathbf{P}_{k,s}^{(\cdot)}$ and $\mathbf{P}_{k,b}^{(\cdot)}$ denotes the corresponding covariances of them.

Then, the problem of IS function design can be reduced to calculate $\mathbf{m}_k^{(i)}$ and $\mathbf{P}_k^{(i)}$. Now, we discuss on how to calculate them. For simplicity, they are replaced by $\mathbf{m}_k$ and $\mathbf{P}_k$, respectively. Here we use the CIF and gating methods to estimate them.

Before introducing the CIF method, we review the cubature rules [27]. The cubature rules are used to approximate the Gaussian weight integral. Assuming $c(\mathbf{x})$ is a function on the n-dimension $\mathbb{R}^n$, its Gaussian weight integral can be approximated by

$$I_N(c) = \int_{\mathbb{R}^n} c(\mathbf{x})N(\mathbf{x}; \mathbf{m}, \mathbf{P}) \approx \frac{1}{2n} \sum_{j=1}^{2n} c(\mathbf{m} + \sqrt{\mathbf{P}}\alpha_j) \tag{12}$$

where

$$\alpha_j = \sqrt{n}[1]_j, \, j = 1, 2, \ldots, 2n \tag{13}$$

and $[1]_j$ is the j-th vector of the set

$$\left\{ \begin{bmatrix} 1 \\ 0 \\ \vdots \\ 0 \end{bmatrix}, \cdots, \begin{bmatrix} 0 \\ 0 \\ \vdots \\ 1 \end{bmatrix}, \begin{bmatrix} -1 \\ 0 \\ \vdots \\ 0 \end{bmatrix}, \begin{bmatrix} 0 \\ 0 \\ \vdots \\ -1 \end{bmatrix} \right\}$$

According to Equation (12), the cubature rules can be used to compute the multi-dimension integrals in the prediction and update steps of the CIF method.

Prediction: In this step, we first predict the state $\mathbf{m}_{k|k-1}$ and covariance $\mathbf{P}_{k|k-1}$ according to cubature rules. Then, the predicted information state vector $\mathbf{y}_{k|k-1}$ and matrix $\mathbf{Y}_{k|k-1}$ are estimated for the *update* step.

Let $\mathbf{m}_{k-1}$ and $\mathbf{P}_{k-1}$ be the previous state and covariance, respectively. According to Equations (12) and (13), the j-th cubature point $\chi_{k-1,j}$ can be estimated by

$$\chi_{k-1,j} = \sqrt{\mathbf{P}_{k-1}}\alpha_j + \mathbf{m}_{k-1} \tag{14}$$

Then, we can calculate $\mathbf{m}_{k|k-1}$ and $\mathbf{P}_{k|k-1}$ using the following formulations:

$$\mathbf{m}_{k|k-1} = \frac{1}{2n}\sum_{j=1}^{2n}\chi^*_{k-1,j} \tag{15}$$

$$\mathbf{P}_{k|k-1} = \frac{1}{2n}\sum_{j=1}^{2n}\chi^*_{k-1,j}(\chi^*_{k-1,j})^T - \mathbf{m}_{k|k-1}(\mathbf{m}_{k|k-1})^T$$
$$+\mathbf{Q}_{k-1} \tag{16}$$

where $(\cdot)^T$ is the transpose operator, and

$$\chi^*_{k-1,j} = \phi(\chi_{k-1,j}) \tag{17}$$

Given Equations (15) and (16), the information forms of $\mathbf{m}_{k|k-1}$ and $\mathbf{P}_{k|k-1}$ are represented [28] by

$$\mathbf{y}_{k|k-1} = \mathbf{Y}_{k|k-1}\mathbf{m}_{k|k-1} \tag{18}$$

and

$$\mathbf{Y}_{k|k-1} = (\mathbf{P}_{k|k-1})^{-1} \tag{19}$$

where $\mathbf{y}_{k|k-1}$ and $\mathbf{Y}_{k|k-1}$ are the information state and matrix, respectively.

Update: We use the observation set $\mathbf{Z}_k$ to update the predicted $\mathbf{y}_{k|k-1}$ and $\mathbf{Y}_{k|k-1}$ in the current step. In order to construct the associations between the observation set $\mathbf{Z}_k$ and predicted observation $\mathbf{z}_{k|k-1}$, a gating method is applied to extract the associated observations. With the extracted observations, we can finally obtain $\mathbf{m}_k$ and covariance $\mathbf{P}_k$.

We denote $\mathbf{z}_{k|k-1}$ as the predicted observation, computed by

$$\mathbf{z}_{k|k-1} = \frac{1}{2n}\sum_{j=1}^{2n}\chi^*_{k|k-1,j} \tag{20}$$

where

$$\chi^*_{k|k-1,j} = \varphi(\chi_{k|k-1,j}) \tag{21}$$

and

$$\chi_{k|k-1,j} = \sqrt{\mathbf{P}_{k|k-1}}\alpha_j + \mathbf{m}_{k|k-1} \tag{22}$$

Utilizing the predicted observation $\mathbf{z}_{k|k-1}$ of Equation (20), the error cross covariance matrix of state and observation can be evaluated by

$$\mathbf{P}^{mz}_{k|k-1} = \frac{1}{2n}\sum_{j=1}^{2n}(\chi_{k|k-1,j} - \mathbf{m}_{k|k-1})(\chi^*_{k|k-1,j} - \mathbf{z}_{k|k-1})^T$$
$$= \frac{1}{2n}\sum_{j=1}^{2n}\chi_{k|k-1,j}(\chi^*_{k|k-1,j})^T - \mathbf{m}_{k|k-1}(\mathbf{z}_{k|k-1})^T \tag{23}$$

With the above obtained parameters, we can calculate the state contribution and its corresponding information matrix as

$$\mathbf{i}_{k,j} = \mathbf{Y}_{k|k-1}\mathbf{P}^{mz}_{k|k-1}\mathbf{R}_k^{-1}(\mu_j + (\mathbf{Y}_{k|k-1}\mathbf{P}^{mz}_{k|k-1})^T\mathbf{m}_{k|k-1}) \tag{24}$$

and

$$\mathbf{I}_k = \mathbf{Y}_{k|k-1}\mathbf{P}^{mz}_{k|k-1}\mathbf{R}_k^{-1}(\mathbf{Y}_{k|k-1}\mathbf{P}^{mz}_{k|k-1})^T \tag{25}$$

where μ_j is the innovation of the j-th observation z_j ($z_j \in Z_k$), expressed by

$$\mu_j = z_j - z_{k|k-1} \tag{26}$$

In our scenario, z_j is a two-dimension vector, and μ_j follows a two-dimension Gaussian distribution.

In practice, the observation set may contain large clutters. The existence of these clutters cannot only degenerate the estimation accuracy, but also increase the computational complexity. Recently, several gating technologies have been proposed to remove the clutters from the observation set [29,30]. Inspired by [30], we utilize the gating technology to reduce the influence of clutters.

Intuitively, observations far away from the predicted observation are subject to be generated by clutters. These observations must be removed from the observation set. With the gating technology, the left observations can be represented by

$$\hat{Z}_k = \{z_{k,j} | \mu_j^T (P_{k|k-1}^{zz})^{-1} \mu_j < \sqrt{T_h}\} \quad z_{k,j} \in Z_k \tag{27}$$

where $P_{k|k-1}^{zz}$ is the covariance matrix of the predicted observation $z_{k|k-1}$, and $(\cdot)^{-1}$ is the matrix inversion. T_h is the threshold of the gate. According to Equation (27), the innovation μ_j follows the Chi-square distribution. Thus, T_h can be determined by the dimension of μ_j and association probability. Commonly, the square root of T_h is known as the number of Sigma. Literature [30] proved that the number of Sigma gates ranging from 3 to 5 (corresponding to $T_h = 9 - 25$) can guarantee the true observation lying inside the gate with "enough" probability (≥ 0.971), when the dimension of μ_j is less than three. In this paper, we select the number of Sigma gates being to 4 (corresponding to $T_h = 16$). When the dimension of μ_j is less than three, such a selection guarantee that the association probability ≥ 0.998.

Then, we concentrate on computing $P_{k|k-1}^{zz}$ of Equation (27). Let $P_{k|k-1}^{mz}$ be the cross covariance matrix between observation and state space. According to the linear error propagating of [31], $P_{k|k-1}^{mz}$ can be rewritten as

$$P_{k|k-1}^{mz} \simeq P_{k|k-1} H_k^T \tag{28}$$

where H_k is the linearized matrix.

Obviously, H_k can be approximated by

$$H_k \simeq (P_{k|k-1}^{mz})^T P_{k|k-1}^{-1} \tag{29}$$

With the achieved H_k, according to [32], $P_{k|k-1}^{zz}$ can be calculated by

$$P_{k|k-1}^{zz} = H_k P_{k|k-1} H_k^T + R_k \tag{30}$$

Substituting the achieve $P_{k|k-1}^{zz}$ into Equation (27), $\hat{Z}_k$ can be extracted from the current observation set Z_k.

With the extracted observation set $\hat{Z}_k$, the information state vector y_k and matrix Y_k are represented as:

$$y_k = y_{k|k-1} + \sum_{j=1}^{|\hat{Z}_k|} i_{k,j} \tag{31}$$

$$Y_k = Y_{k|k-1} + \sum_{j=1}^{|\hat{Z}_k|} I_{k,j} \tag{32}$$

Given information state $\mathbf{y}_k$ and matrix $\mathbf{Y}_k$, posterior state $\mathbf{m}_k$ can be reconstructed based on Equation (18):

$$\mathbf{m}_k = \mathbf{Y}_k^{-1}\mathbf{y}_k \tag{33}$$

Moreover, the posterior covariance $\mathbf{P}_k$ is recovered based on Equation (19):

$$\mathbf{P}_k = (\mathbf{Y}_k)^{-1} \tag{34}$$

If there is no observation that lies inside the gate ($\hat{\mathbf{Z}}_k = \varnothing$), we approximate $\mathbf{m}_k$ and $\mathbf{P}_k$ in the following,

$$\mathbf{m}_k = \mathbf{m}_{k|k-1} \tag{35}$$
$$\mathbf{P}_k = \mathbf{P}_{k|k-1} \tag{36}$$

Substituting the above obtained $\mathbf{m}_k$ and $\mathbf{P}_k$ into Equations (10) and (11), we can approximate the IS functions of survival and birth targets for our CISMC-PHD approach.

3.2. The Birth Intensity Initialization Method

According to Equation (2), the birth intensity has large effect on the posterior intensity estimation. Targets may "born at anywhere" of the state space. In other words, birth intensity $\gamma(\mathbf{x})$ may cover the whole state space, which is rather exhaustive [25]. To avoid such a disadvantage, observation-driven birth intensity initiation methods were proposed [20,33,34]. Inspired by these methods, an adaptive birth intensity initialization method is proposed for the CISMC-PHD approach. Instead of initializing birth intensity across the whole state space, the proposed method of CISMC-PHD approach utilizes the current observations and estimated targets to initialize the birth intensity at the next recursion. Compared with the conventional SMC-PHD filters, our method can initialize the birth intensity without knowing it as a prior.

The implementation of our method consists of two steps. First, in order to initialize the birth intensity, we remove observations generated by the estimated targets That is because the current survival targets cannot be new-born targets at the next recursion. Second, we use the remaining observations to estimate the birth target components, which can be used to calculate the birth intensity. With these two steps, the birth intensity can be initialized for the next recursion.

Step1. Remove observations generated by the estimated targets.

In the basic PHD filter, it is assumed that each target can yield at most one observation [35]. According to this assumption, each target has one and only one corresponding observation. Influenced by the noises and clutters, the observation generated by the target may appear around the target. In other words, observations around the target has the large probability to be generated by the same target. Therefore, the birth target state set can be estimated by removing states of estimated targets from the multi-target state.

Here, we adopt the bearing and range tracking model [36] to illustrate the birth intensity initialization method of our CISMC-PHD approach. Let $\mathbf{x}_{k,i}^e$ be the state of the i-th target in the estimate state set $\mathbf{X}_k^e$. $\mathbf{x}_{k,i}^e$ consists of position and velocity, while $\mathbf{z}_{k,j}$ ($\mathbf{z}_{k,j} \in \mathbf{Z}_k$) consists of the bearing angle and range. We define the distance between $\mathbf{x}_{k,i}^e$ and $\mathbf{z}_{k,j}$ ($\mathbf{z}_{k,j} \in \mathbf{Z}_k$) as

$$d_{i,j} = |(\mathbf{z}_{k,j} - \varphi(\mathbf{x}_{k,i}^e))_r| \tag{37}$$

where $(\cdot)_r$ denotes the range-dimension element, and $|\cdot|$ is the absolution value.

We follow the way of [30] to select the certain threshold for Equation (37),

$$d_{i,j} < l \cdot \sigma_r \tag{38}$$

where σ_r is the error of the range-dimension (known as a prior). l is the confidence level, commonly selected from $l = 3 \sim 5$. Here, we use $l = 3$, which can guarantee that the associated probability equals to 0.997.

With Equations (37) and (38), we can remove the observations associated with the estimated targets. Let $\tilde{\mathbf{Z}}_k$ be the observations of birth targets, the removing procedure is summarized in Table 1.

Table 1. Removing observations generated by the estimated targets.

–   **Input:** Observation set $\mathbf{Z}_k$, and estimated state set $\mathbf{X}_k^e$
–   **Output:** Observation set of birth targets $\tilde{\mathbf{Z}}_k$
-   **For:** $j = 1, 2, \cdots, \lvert \mathbf{X}_k^e \rvert$,
1   Compute the distance $d_{i,j}$ between $\mathbf{x}_{k,i}^e$ ($\mathbf{x}_{k,i}^e \in \mathbf{X}_k^e$) and $\mathbf{z}_{k,j}$ for each $\mathbf{z}_{k,i} \in \mathbf{Z}_k$ by Equation (37).
2   Extract $\mathbf{z}_{k,i}$ satisfying $i = \{i \lvert d_{i,j} <= 3 \cdot \sigma_r\}$.
3   Remove $\mathbf{z}_{k,i}$ from $\mathbf{Z}_k$ to obtain $\tilde{\mathbf{Z}}_k$.
-   **End For**
-   **Return** $\tilde{\mathbf{Z}}_k$.

Step2. Estimate the birth target components.

Once $\tilde{\mathbf{Z}}_k$ is obtained, we turn to estimate the birth target components (the mean of the i-th target state vector $\mathbf{m}_{k,b}^{(i)}$ and its corresponding covariance $\mathbf{P}_{k,b}^{(i)}$) by the unbiased model of [37].

Let $\tilde{\mathbf{z}}_{k,i} \in \tilde{\mathbf{Z}}_k$, we map $\tilde{\mathbf{z}}_{k,i}$ into state space denoted by $\tilde{\mathbf{z}}_{k,i}^c = [p_{k,i}^x, p_{k,i}^y]^T$. $p_{k,i}^x$ and $p_{k,i}^y$ can be computed by $p_{k,i}^x = \beta_\theta^{-1} r_{k,i} \cos \theta_{k,i}$ and $p_{k,i}^y = \beta_\theta^{-1} r_{k,i} \sin \theta_{k,i}$. $\beta_\theta = \sigma_\theta$ is a biased comparison factor, where σ_θ, as a prior, is the error of bearing angle $\theta_{k,i}$. According to [37], $\mathbf{m}_k^{(i)}$, the mean of the i-th birth target state, can be estimated as

$$\mathbf{m}_k^{(i)} = [p_{k,i}^x, 0, p_{k,i}^y, 0, 0]^T \tag{39}$$

The covariance can be approximated by

$$\mathbf{P}_k^{(i)} = \begin{bmatrix} \sigma_{xx} & 0 & \sigma_{xy} & 0 & 0 \\ 0 & \sigma_v & 0 & 0 & 0 \\ \sigma_{yy} & 0 & \sigma_{xy} & 0 & 0 \\ 0 & 0 & 0 & \sigma_v^2 & 0 \\ 0 & 0 & 0 & 0 & \sigma_\theta^2 \end{bmatrix} \tag{40}$$

where σ_v, as a prior, is the standard deviation of velocity. In Equation (40), the following exists,

$$\begin{cases} \sigma_{xx} = & (\beta_\theta^{-2} - 2)(\tilde{r}_{k,i})^2 \cos^2(\theta_{k,i}) + 0.5((\tilde{r}_{k,i})^2 \\ & + \sigma_r^2)(1 + \beta_\theta^4 \cos(2\theta_{k,i})) \\ \sigma_{xy} = & (\beta_\theta^{-2} - 2)(\tilde{r}_{k,i})^2 \cos(\theta_{k,i}) \sin(\theta_{k,i}) + 0.5((\tilde{r}_{k,i})^2 \\ & + \sigma_r^2)(1 + \beta_\theta^4 \cos(2\theta_{k,i})) \\ \sigma_{yy} = & (\beta_\theta^{-2} - 2)(\tilde{r}_{k,i})^2 \sin^2(\theta_{k,i}) + 0.5((\tilde{r}_{k,i})^2 \\ & + \sigma_r^2)(1 - \beta_\theta^4 \cos(2\theta_{k,i})) \end{cases} \tag{41}$$

Finally, we can construct the new-born targets as $\{\mathbf{m}_{k,i}, \mathbf{P}_{k,i}\}_{i=1}^{N_{k,b}}$, where $N_{k,b} = |\check{\mathbf{Z}}_k^c|$ is considered as the number of birth targets. We use Equation (11) to sample states of birth particles. The weights of these birth particles are initialized with the same values, $w_{k,b}^{(i)} = \frac{\int \gamma(\mathbf{x})d\mathbf{x}}{N \cdot N_{k,b}}$, where N is the number particles for each target, and $\gamma(\mathbf{x})$ is defined in Section 2. On this basis, these birth particles become survival particles at time $k + 1$. That is to say, the IS functions of these particles at time $k + 1$ can be computed by the method of Section 3.1. Notice that the new-born target in this section may contain clutters, and these clutters can be removed in the resampling step of the CISMC-PHD approach.

The proposed initialization method may cause overestimation of targets. To overcome the issue of overestimation, some advanced methods, such as [33,34], *etc.*, may be incorporated for initialization of our approach. It is an interesting future work.

3.3. State Estimation

In multi-target tracking, it is rather important to estimate the target number and states. As for the state estimation, clustering methods, are commonly used in SMC-PHD filters [16,18]. However, they are subject to biased estimation [21]. Ristic *et al.* [21] proposed an method that clusters the particles into several groups at the *update* stage. In this paper, we intend to adopt the method of [38] for state estimation, which is an improved method of [21]. There are also several alternative methods, such as Zhao's method [39] and MEAP method [40], which have better estimation performance.

According to Equation (6), the updated weight w_k^i of the i-th particle consists of two parts,

$$w_k^{(i,j)} = \begin{cases} (1 - p_d(\mathbf{x}_k))w_{k|k-1}^{(i)} & j = 0 \\ \frac{p_d(\mathbf{x}_k^{(i)})g_k(\mathbf{z}_{k,j}|\mathbf{x}_k^{(i)})}{\kappa(\mathbf{z}_{k,j}) + C_{z_{k,j}}} w_{k|k-1}^i & j = 1, \dots, |\hat{\mathbf{Z}}_k| \end{cases} \tag{42}$$

In Equation (42), $j = 0$ denotes that there is no observation, and $C_{z_{k,j}}$ can be computed by Equation (7). For state estimation, we aggregate W^j of particle weights corresponding to observation $\mathbf{z}_{k,j}$,

$$W^j = \sum_{i=1}^{L_{k-1}+J_k} w_k^{(i,j)} \quad j = 0, \dots, |\hat{\mathbf{Z}}_k| \tag{43}$$

According to Equations (42) and (43), if $z_{k,j}$ is generated by the clutter, then the likelihood $g_k(\mathbf{z}_{k,j}|\mathbf{x}_k^{(i)})$ may be small, leading to low value of W^j. However, if $z_{k,j}$ is generated by the target, then W^j may be large due to the large value of $g_k(\mathbf{z}_{k,j}|\mathbf{x}_k^{(i)})$. Thus, setting certain threshold W_{th} for W^j, we can assign particles $\{\mathbf{x}_{k,j}^{(i)}, w_k^{(i,j)}\}$ that satisfy $W^j > W_{th}$ to the j-th target. In this paper, we set $W_{th} = 0.5$, the same as [21]. Then, $\mathbf{x}_{k,j}$ and $\mathbf{P}_{k,j}$ can be calculated in the following

$$\mathbf{x}_{k,j} = \frac{1}{W^j} \sum_{i=1}^{L_{k-1}+J_k} w_k^{i,j} \mathbf{x}_{k,j}^{(i)} \tag{44}$$

$$\mathbf{P}_{k,j} = \frac{1}{W^j} \sum_{i=1}^{L_{k-1}+J_k} w_k^{i,j} (\mathbf{x}_{k,j}^{(i)} - \mathbf{x}_{k,j})(\mathbf{x}_{k,j}^{(i)} - \mathbf{x}_{k,j}) \tag{45}$$

Given Equations (44) and (45), the states of multi-target can be finally output. We summarize our approach at time k in Table 2.

Table 2. The CISMC-PHD filter at time k.

- **Input:** Birth particle set $\{x_{k-1,b}^{(i)}, \mathbf{P}_{k-1,b}^{(i)}, w_{k-1,b}^{(i)}\}_{i=1}^{J_{k-1}}$, survival particle set $\{x_{k-1,s}^{(i)}, \mathbf{P}_{k-1,s}^{(i)}, w_{k-1,s}^{(i)}\}_{i=1}^{L_{k-1}}$, and current observation set $\mathbf{Z}_k$
- **Output:** Target number M_k, and estimated state set $\mathbf{X}_k^e$

1. Calculate the IS function $q(\mathbf{x}_k^{(i)}|\mathbf{x}_{k-1,s}^{(i)}, \mathbf{Z}_{1:k})$ of Equation (10) by Equations (33) and (34), and draw particles $\{\mathbf{x}_{k|k}^{(i)}, w_{k|k-1}^{(i)}\}_{i=1}^{L_{k-1}}$ for survival targets by Equations (3) and (4), where $\mathbf{m}_{k-1,s}^{(i)} = \mathbf{x}_{k-1,s}^{(i)}$.

2. Approximate the IS function $q(\mathbf{x}_{k,b}^{(i)}|\mathbf{x}_{k-1,b}^{(i)}, \mathbf{Z}_{1:k})$ of Equation (10) using Equations (33) and (34), and draw particle set $\{\mathbf{x}_{k|k}^{(i)}, w_{k|k-1}^{(i)}\}_{i=L_{k-1}+1}^{L_{k-1}+J_{k-1}}$ for birth particles by Equations (3) and (4), where $\mathbf{m}_{k-1,b}^{(i)} = \mathbf{x}_{k-1,b}^{(i)}$.

3. Calculate $w_{k|k}^{(i)}$ by Equation (6) for resampling and $w_{k|k}^{(i,j)}$ by Equation (42) for estimating, using the particle set $\{\mathbf{x}_{k|k}^{(i)}, w_{k|k-1}^{(i)}\}_{i=1}^{L_{k-1}+J_{k-1}}$.

4. Compute W^j by Equation (43) with $w_{k|k}^{(i,j)}$ calculated by step 3 and assign particles into the corresponding group by W_{th} to estimate the target states and number.

5. Estimate the state set $\mathbf{X}_k^e$ by Equation (44) with $w_{k|k}^{(i,j)}$ and $\mathbf{x}_{k|k,j}^{(i)}$, where $w_{k|k}^{(i,j)}$ and $\mathbf{x}_{k|k,j}^{(i)}$ belong to the j−th group of step 4. In addition, target number can be approximated by $M_k = |\mathbf{X}_k^e|$.

6. Resample particles $\{\mathbf{x}_{k|k}^{(i)}, \mathbf{P}_{k|k}^{(i)}, w_{k|k}^{(i)}\}_{i=1}^{L_{k-1}+J_k}$ to obtain $\{\mathbf{x}_{k,s}^{(i)}, \mathbf{P}_{k,s}^{(i)}, w_{k,s}^{(i)}\}_{i=1}^{L_k}$ for the next recursion, where $L_k = [\sum_{i=1}^{L_{k-1}+J_k} w_{k|k}^{(i)}]$, and $[\cdot]$ denotes the nearest integer.

7. Remove the observations of survival targets using Table 1, and estimate the birth components by Equations (39) and (40) to obtain the birth component set $\mathbf{B}_k$.

8. Draw particle from Equation (11) to obtain the birth particles $\{x_{k,b}^{(i)}, \mathbf{P}_{k,b}^{(i)}, w_{k,b}^{(i)}\}_{i=1}^{J_k}$ when $\mathbf{B}_k$ is given.

9. Return M_k and $\mathbf{X}_k^e$.

---

4. A Fast Approach For The CISMC-PHD Filter

In this section, we focus on reducing the computational complexity of our CISMC-PHD approach. Section 4.1 presents a fast implementation of the CISMC-PHD approach, namely the F-CISMC-PHD approach. Then, we analyze the computational complexity of the CISMC-PHD and F-CISMC-PHD approaches in Section 4.2. The framework of the improved approach is illustrated in Figure 1.

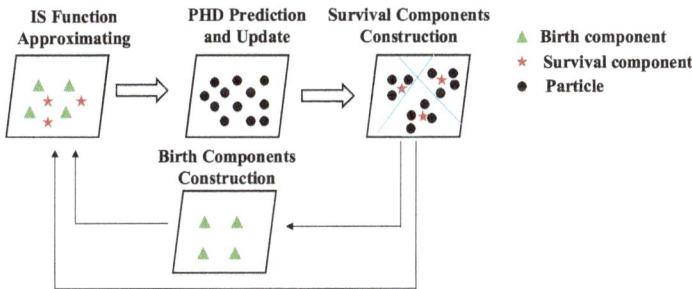

Figure 1. Framework of the F-CISMC-PHD approach. In the *IS function Approximating* step, we utilize the survival and birth components to approximate the IS functions. Then the predicted particles are generated and updated in the *PHD prediction and update* step to achieve the posterior particles. By clustering the particles into several groups, the Gaussian Mixture components (namely survival components) can be constructed in the *survival components construction* step. Meanwhile, we also apply the survival components to estimate the birth components in the *birth components construction* step. These components are used to approximate the IS functions in the next iteration.

4.1. The Fast CISMC-PHD Filter

In this section, we introduce the F-CISMC-PHD approach, which can save the computational time of the CISMC-PHD approach. Inspired by [41], we consider the particle groups of targets as the Gaussian mixture components. Recall that $N_{k-1,b}$ is the number of birth components. $\mathbf{m}_{k-1,b}^{(i)}$ and $\mathbf{P}_{k-1,b}^{(i)}$ denote the mean and covariance of the i-th birth component, respectively. The posterior intensity of Equation (1) can be approximated by

$$D_{k-1|k-1}(\mathbf{x}_{k-1}|\mathbf{Z}_{1:k-1}) \approx \sum_{i=1}^{N_{k-1,s}} G_{k-1,s}^{(i)} N(\mathbf{x}_{k-1};\mathbf{m}_{k-1,b}^{(i)},\mathbf{P}_{k-1,b}^{(i)}) + \sum_{i=1}^{N_{k-1,b}} G_{k-1,b}^{(i)} N(\mathbf{x}_{k-1};\mathbf{m}_{k-1}^{(i)},\mathbf{P}_{k-1}^{(i)}) \quad (46)$$

where $\mathbf{m}_{k-1,s}^{(i)}$ and $\mathbf{P}_{k-1,s}^{(i)}$ are the mean and covariance of the i-th survival component, respectively. $G_{k-1,s}^{(i)}$ is its corresponding weight, $N_{k-1,s}$ is the number of survival components, and $G_{k-1,b}^{(i)}$ is the weight of the i-th birth component. Here, $G_{k-1,b}^{(i)} = \frac{\int \gamma(\mathbf{x})d\mathbf{x}}{N_{k-1,b}}$.

Commonly, birth components at time $k-1$ become survival components at time k. We combine birth and survival components into one set. That is $\{G_{k-1}^{(i)},\mathbf{m}_{k-1}^{(i)},\mathbf{P}_{k-1}^{(i)}\}_{i=1}^{N_{k-1,b}+N_{k-1,s}} = \{G_{k-1,b}^{(i)},\mathbf{m}_{k-1,b}^{(i)},\mathbf{P}_{k-1,b}^{(i)}\}_{i=1}^{N_{k-1,b}} \cup \{G_{k-1,s}^{(i)},\mathbf{m}_{k-1,s}^{(i)},\mathbf{P}_{k-1,s}^{(i)}\}_{i=1}^{N_{k-1,s}}$, where $G_{k-1}^{(i)},\mathbf{m}_{k-1}^{(i)},\mathbf{P}_{k-1}^{(i)}$ denote the weight, mean and covariance of the i-th combined component.

With the combined components, we use the CIF method of Section 3.1 to approximate the IS function of each component. Here, we utilize $q(\mathbf{x}|\mathbf{m}_{k-1}^{(i)},\mathbf{Z}_k)$ to represent the IS function of the i-th component. On this basis, the j-th predicted particle, which is generated by the i-th component, can be represented by

$$\mathbf{x}_k^{(j)} \sim q(\mathbf{x}_k^{(j)}|\mathbf{m}_{k-1}^{(i)},\mathbf{Z}_k) \quad j = \sum_{a=1}^{i-1}\lfloor G_{k-1}^{(a)} \cdot N\rfloor + 1, \sum_{a=1}^{i-1}\lfloor G_{k-1}^{(a)} \cdot N\rfloor + 2,\ldots,\sum_{a=1}^{i}\lfloor G_{k-1}^{(a)} \cdot N\rfloor \quad (47)$$

$$w_{k|k-1}^{(j)} = \frac{p_s(\mathbf{x}_{k-1}^{(i)})\cdot f(\mathbf{x}_k^{(j)}|\mathbf{x}_{k-1}^{(i)})}{q(\mathbf{x}_k^{(j)}|\mathbf{m}_{k-1}^{(i)},\mathbf{Z}_k)} \cdot \frac{G_{k-1}^{(a)}}{\lfloor G_{k-1}^{(a)}N\rfloor} \quad j = \sum_{a=1}^{i-1}\lfloor G_{k-1}^{(a)} \cdot N\rfloor + 1, \sum_{a=1}^{i-1}\lfloor G_{k-1}^{(a)} \cdot N\rfloor + 2,\ldots,\sum_{a=1}^{i}\lfloor G_{k-1}^{(a)} \cdot N\rfloor \quad (48)$$

where $f(\mathbf{x}_k^{(j)}|\mathbf{x}_{k-1}^{(i)}) = N(\mathbf{x}_k^{(j)};\mathbf{m}_{k|k-1}^{(i)},\mathbf{P}_{k|k-1}^{(i)})$. In addition, $\mathbf{m}_{k|k-1}^{(i)}$ and $\mathbf{P}_{k|k-1}^{(i)}$ are the predicted mean and covariance of i-th component, respectively, which can be computed by Equations (15) and (16). $\lfloor \cdot \rfloor$ denotes the nearest floor integer, and $\lfloor G_{k-1}^{(i)} \cdot N\rfloor$ is the number of particles generated by the i-th component. According to Equations (47) and (48), the number of predicted particles is $N_k = \sum_{a=1}^{N_{k,s}+N_{k,b}}\lfloor G_{k-1}^{(a)} \cdot N\rfloor$.

Then, the predicted weight of Equation (48) is substituted into Equation (42) for particle grouping and state estimation. Recall that, weight W^j of Equation (43) can be used to assign the particles into the j-th group, where the mean $\mathbf{x}_{k,j}$ and covariance $\mathbf{P}_{k,j}$ can be calculated by Equations (44) and (45) in Section 3.3. Given W^j, $\mathbf{x}_{k,j}$ and $\mathbf{P}_{k,j}$, we model the j-th group as a Gaussian mixture component $\{G_k^{(j)},\mathbf{x}_{k,j},\mathbf{P}_{k,j}\}$, where we set $G_k^{(j)} = \frac{W_j}{\max_j W_j}$. With such a selection, it can guarantee that the groups with large W^j have enough number of sampling particles. Note that the group with small W^j has the large probability to be generated by the clutter, and such a group should be neglected to avoid the waste of computational time. We set a certain threshold T_g for $\{G_k^{(j)}|G_k^{(j)} > T_g\}$, subject to $T_g \cdot N \ll N$. In this paper, we set $T_g = 0.1$. The construction procedure of the target components is summarized in Table 3.

The target components of Table 3 refer to the survival target components. The birth components and target state estimation can be achieved by Sections 3.2 and 3.3, respectively.

Table 3. The construction of target components.

- **Input:** The group weight W^j, particle $\{\mathbf{x}_{k,j}^{(i)}, w_k^{i,j}\}$, $,j = 0,1,\ldots,|\hat{\mathbf{Z}}_k|$, and $i = 1,2,\ldots,L_{k-1} + J_k$
- **Output:** The target components $\{G_k^{(a)}, \mathbf{m}_{k,a}, \mathbf{P}_{k,p}\}_{a=1}^{N_{s,k}}$.

1 Initiate the weight of the target component $G_k^{(1)} = 0, N_{s,k} = 0$.
2 Normalize W^j by $\bar{W}^j = \frac{W^j}{\max_j W^j}$
3 **for:** $j = 0,1,\ldots,|\mathbf{Z}_k|$
   if $\bar{W}^j > T_g$
      - Add the number of target components $N_{s,k} = N_{s,k} + 1$.
      - Substitute $\{w_k^{(i,j)}\}$ and $\{\mathbf{x}_k^{(i)}\}$ into Equation (44) to approximate the mean $\mathbf{x}_{k,N_{s,k}}$ and covariance $\mathbf{P}_{k,N_{s,k}}$.
      - Save the weight of the current component as $G_k^{(N_{s,k})} = \bar{W}^j$.
4 Return the target components $\{G_k^{(a)}, \mathbf{m}_{k,a}, \mathbf{P}_{k,p}\}_{a=1}^{N_{s,k}}$.

4.2. Computational Complexity

In this section, we analyze the computational complexity of the CISMC-PHD, F-CISMC-PHD and conventional SMC-PHD approaches. For justice, we adopt the same state estimation and birth target initialization methods for the three approaches. The computational complexity of the three approaches on state estimating and birth target initializing is the same, when they have same particle numbers and observations. Thus, it can be neglected for the computational complexity comparing. In addition, we select the multinomial resampling method as the resampling methods of the CISMC-PHD and conventional SMC-PHD approaches. The particle numbers of these approaches are equal to N_p.

We begin with the computational complexity analysis of the CISMC-PHD approach. Incorporating the CIF and gating methods into the SMC-PHD approach, the CISMC-PHD approach can achieve a good estimation accuracy in nonlinear target tracking. As mentioned in Section 3.1, each particle is applied to approximate the IS functions. The computational complexity of the CIF and gating method per particle is nearly $O(n_d^3)$, where n_d is the dimension of the particle state. The computational complexity of the CIF-based IS functions is $O(N_p \cdot n_d^3)$. Besides, the computational complexity of the PHD update step is $O(N_p \cdot M)$, where M is the number of observations. In addition, the resampling step of our CISMC-PHD approach consumes $O(N_p)$ computational complexity. Thus, the computational complexity of our CISMC-PHD approach in total is $O(N_p \cdot n_d^3 + N_p \cdot M + N_p)$.

Then, we turn to the analysis of the computational complexity of the F-CISMC-PHD approach. Since the F-CISMC-PHD approach adopts the target components to compute the CIF-based IS functions, the computational complexity of the CIF-based IS functions is $O(N_G \cdot n_d^3)$, where N_G is the number of target components, $N_G \ll N_p$. Assuming that in the PHD prediction step, N_G target components generate N_p particles. The computational complexity of the PHD update step is $O(N_p \cdot M)$. Besides, the Gaussian target component forming consumes $O(M + 1)$. Combining the computational complexity of these steps together, the computational complexity of the F-CISMC-PHD approach is $O(N_G \cdot n_d^3 + N_p \cdot M + M + 1)$. In practice, $M \ll N_p$, the computational complexity of F-CISMC-PHD filter is much less than the CISMC-PHD filter.

The conventional SMC-PHD approach uses the transitional density as the IS function. Compared with the CISMC-PHD and F-CISMC-PHD approaches, it does not need to the IS function computing. Hence, the computational complexity of the conventional SMC-PHD approach can be approximated as $O(N_p \cdot M + N_p)$, where $O(N_p \cdot M)$ and $O(N_p)$ are the computational complexity of the update and resampling steps. Obviously, $O(N_p \cdot M + N_p) < O(N_G \cdot n_d^3 + N_p \cdot M + M + 1) < O(N_p \cdot n_d^3 + N_p \cdot M + N_p)$. Thus, the computational complexity of the conventional SMC-PHD approach is smaller than the CISMC-PHD and F-CISMC-PHD approaches.

With the above discussion, we can conclude that the conventional SMC-PHD approach has the lowest computational complexity, and the CISMC-PHD approach has the highest computational complexity. However, the estimation accuracy of the conventional SMC-PHD approach is the lowest, and the estimation accuracy of the CISMC-PHD approach is the highest. The F-CISMC-PHD approach can make a trade-off between the computational complexity and estimation accuracy. Such a conclusion can be observed in Section 5.

5. Simulation Results

In this section, we validate the tracking performance of the proposed CISMC-PHD and F-CISMC-PHD approaches. In Section 5.1, a nonlinear simulation scenario composed of five targets is constructed. Then, Section 5.2 compares the estimation results of the IBSMC-PHD [20], proposed CISMC-PHD and F-CISMC-PHD approaches, in terms of the optimal subpattern assignment (OSPA) metric [42] and Root Mean Square Error (RMSE). At last, we compare the simulation results of all three approaches with different numbers of clutters and detection probabilities, to validate the effectiveness of our approaches.

5.1. Simulation Scenarios

In our simulations, we use the nonlinear scenario, the same as [36]. Let $\mathbf{x}$ be the target states, represented by $\mathbf{x} = [p_x, v_x, p_y, v_y, \alpha]^T$. In this paper, (p_x, p_y) is the position, (v_x, v_y) is the velocity, and α is the turn rate. With the above definitions, we model the nonlinear dynamic equation as

$$
\mathbf{x}_k =
\begin{bmatrix}
1 & \frac{\sin(\alpha_{k-1}T)}{\alpha_{k-1}} & 0 & -\frac{1-\cos(\alpha_{k-1}T)}{\alpha_{k-1}} & 0 \\
0 & \cos(\alpha_{k-1}T) & 0 & -\sin(\alpha_{k-1}T) & 0 \\
0 & \frac{1-\cos(\alpha_{k-1}T)}{\alpha_{k-1}} & 1 & \frac{\sin(\alpha_{k-1}T)}{\alpha_{k-1}} & 0 \\
0 & \sin(\alpha_{k-1}T) & 0 & \cos(\alpha_{k-1}T) & 0 \\
0 & 0 & 0 & 0 & 1
\end{bmatrix} \mathbf{x}_{k-1}
$$
$$
+
\begin{bmatrix}
\frac{T^2}{2} & 0 & 0 \\
T & 0 & 0 \\
0 & \frac{T^2}{2} & 0 \\
0 & T & 0 \\
0 & 0 & 1
\end{bmatrix} \varepsilon_{k-1}
\tag{49}
$$

where $T = 1$ second (s) is the sampling interval. In addition, ε_{k-1} is the noise, defined by $\varepsilon_{k-1} \sim N(\varepsilon_{k-1}; 0, \mathbf{Q})$. We denote $\mathbf{Q} = diag(\sigma_{x,\varepsilon}^2, \sigma_{y,\varepsilon}^2, \sigma_\alpha^2)$ as covariance matrices of ε_{k-1}. In this paper, we set $\sigma_{x,\varepsilon} = \sigma_{y,\varepsilon} = 1$ meter/second^2 (m/s^2), $\sigma_\alpha = \pi/180$ rad. The initial states, appearing times, and disappearing times of targets are listed in Table 4.

Table 4. Initial states of the targets.

Target	State	Appearing(s)	Disappearing(s)
1	$[320, 5, 320, 5, 0]$	1	40
2	$[400, -5, 400, 5, 0]$	8	50
3	$[375, 5, 375, -5, 0]$	25	70
4	$[400, 5, 325, -5, 0]$	59	70
5	$[325, -5, 375, 5, 0]$	59	70

Besides, the observation model is given by

$$\mathbf{z}_k = \begin{bmatrix} \arctan\left(\frac{p_y}{p_x}\right) \\ \sqrt{p_x^2 + p_y^2} \end{bmatrix} + \boldsymbol{\eta}_k \tag{50}$$

where $\boldsymbol{\eta}_k$ is the observation noise defined by $\boldsymbol{\eta}_k \sim N(\boldsymbol{\eta}_k; 0, \mathbf{R})$, and $\mathbf{R} = diag(\sigma_\theta^2, \sigma_r^2)$ is the covariance. Here, we set $\sigma_\theta = \pi/180$ rad and $\sigma_r = 1$ meter (m). We also assume that clutters are uniformly distributed in the detection region, where the angle range is $(0, \pi/2)$ rad, and distance range is $(0, 1000)$ m. Trajectories of the five targets in both scenarios are shown in Figure 2, where the clutter number is set to be 10 for each scenario.

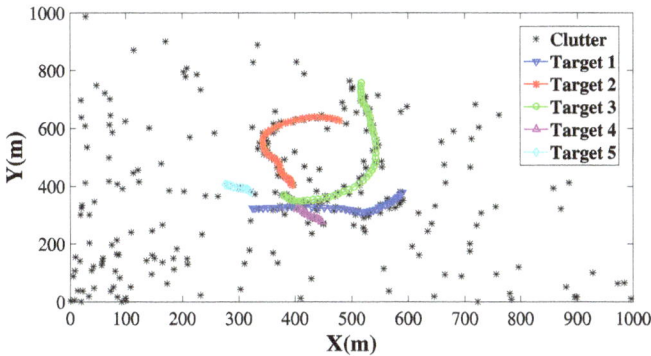

Figure 2. Ground-truth trajectories of five targets with the clutter number setting to be 10. The target trajectories are depicted by circle-solid lines with different colors, while the asterisks denote clutters.

For parameters, we set the gating threshold $T_h = 16$ according to [32], the probability of detection and survival are $p_d(\mathbf{x}_k) = 0.95$, and $p_s(\mathbf{x}_k) = 0.99$, in accordance with [36]. The particle numbers for each birth target and survival targets are 5 and 100, respectively. All of the simulations are run in a computer with MATLAB 2015a, and i5 3.2 GHz processor with 4GB RAM.

5.2. Comparison of Estimation Accuracy on Certain Number of Clutters

To compare the estimation accuracy, we adopt the first order OSPA and RMSE as the metric. Here, we discuss the first order OSPA distance in the following. Let $\mathbf{X} = \{\mathbf{x}_1, \ldots, \mathbf{x}_n\}$ and $\mathbf{Y} = \{\mathbf{y}_1, \ldots, \mathbf{y}_n\}$ be two RFSs, where m and n are numbers of elements in $\mathbf{X}$ and $\mathbf{Y}$, respectively. Supposing that Ω_n represents the set of permutations of $\{1, 2, \ldots, n\}$, the first order OSPA metric can be rewritten by

$$\bar{d}_p^c(\mathbf{X}, \mathbf{Y}) = \frac{1}{n}\left(\min_{\varsigma \in \Omega_n} \sum_{i=1}^{m} d^c(\mathbf{x}_i, \mathbf{y}_{\varsigma(i)})^p + c^p(n - m)\right) \tag{51}$$

where $d^c(\mathbf{x}, \mathbf{y}) = \min(c, d(\mathbf{x}, \mathbf{y}))$, $c > 0$ is a cut-off factor, and $d(\mathbf{x}, \mathbf{y})$ is the distance between $\mathbf{x}$ and $\mathbf{y}$. In this paper, we set $c = 150$ in accordance with [43], and use the Euclidean distance to compute $d(\mathbf{x}, \mathbf{y})$.

Then, we conducted 500 Monte Carlo runs for multi-target tracking with the IBSMC-PHD, our CISMC-PHD, and F-CISMC-PHD approaches. The estimated trajectories of all three approaches are demonstrated in Figure 3. We can observe that most of estimated points are covered with the true trajectories in Figures 3b,c, while most of points are not covered with the ground-truth trajectories.

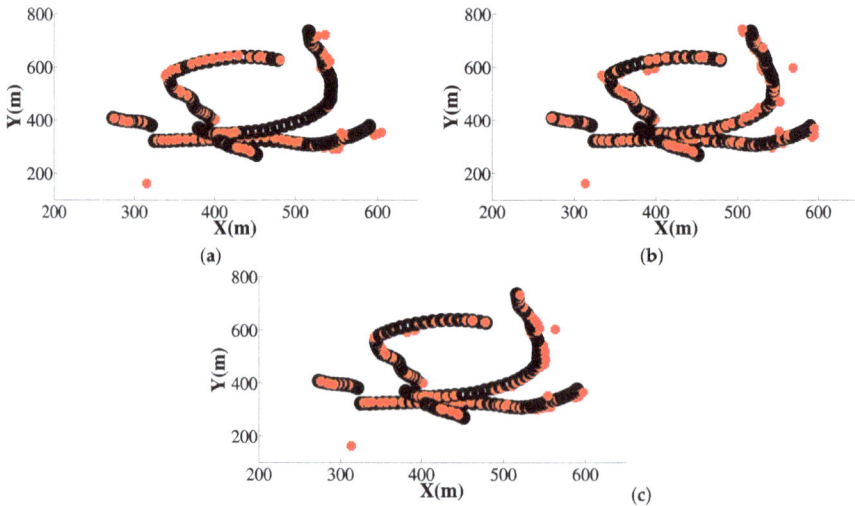

Figure 3. Estimated trajectories of three approaches with clutter number being 10. The estimated trajectories are represented with the red (light) points, while the true trajectories are with the black (dark) solid line. (**a**) Trajectories of IBSMC-PHD; (**b**) Trajectories of CISMC-PHD; (**c**) Trajectories of F-CISMC-PHD.

Furthermore, Figure 4 depicts the OSPA distances of all three approaches. In these figures, the OSPA distances of the proposed CISMC-PHD and F-CISMC-PHD approaches are smaller than the BSMC-PHD approach. Note that large OSPA distance denotes large tracking error. Thus, the proposed CISMC-PHD and F-CISMC-PHD approaches have the smaller tracking error than the IBSMC-PHD approach. We can also observe that the OSPA distances of the proposed CISMC-PHD and F-CISMC-PHD approaches have four peaks at time 2, 9, 26, and 60. That is to say, targets may appear at these times.

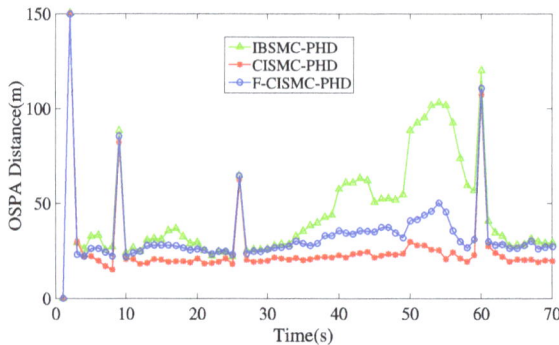

Figure 4. OSPA distances of the IBSMC-PHD, CISMC-PHD and F-CISMC-PHD approaches with clutter number being 10.

We also plot the estimated numbers and corresponding RMSEs of all three approaches in Figure 5. As seen from Figure 5a, the estimated number of the proposed CISMC-PHD approach is closest to the ground truth among three approaches, thus enjoying the lowest RMSE in Figure 5b. In addition, the numerical results of the averaged OSPA distances and RMSEs are listed in Table 5, demonstrating that our CISMC-PHD approach achieves the best estimation on numbers and states. According to Table 5, the F-CIMS-PHD approach can make a compromise between the computational time and estimation accuracy.

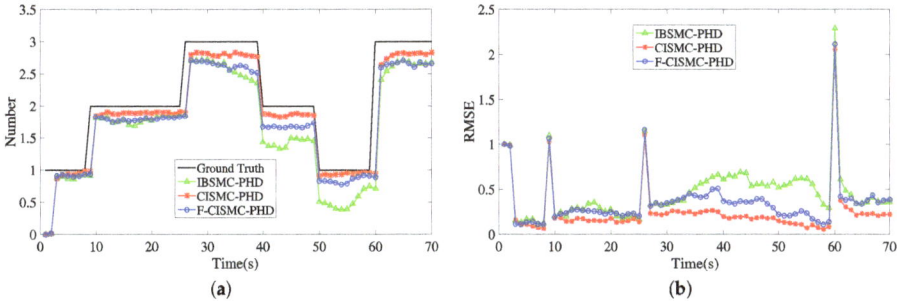

Figure 5. Estimated numbers and RMSEs of IBSMC-PHD, CISMC-PHD and F-CISMC-PHD approaches with clutter number being 10. (**a**) Estimated numbers of the three approaches; (**b**) RMSEs of the three approaches.

Table 5. Averaged Estimation Errors and Computational Times per 100 particles.

Approaches	OSPA (m)	RMSE	Time (s)
IBSMC-PHD	46.20	0.44	0.02
CISMC-PHD	25.48	0.24	0.2
F-CISMC-PHD	31.93	0.30	0.06

5.3. Comparison of Estimation Accuracy on Various Numbers of Clutters

To validate the influence of clutters on multi-target tracking, the IBSMC-PHD, CISMC-PHD and F-CISMC-PHD approaches were implemented with 500 Monte Carlo simulations alongside the clutter number changing from 1 to 30. The results are illustrated in Figure 6a,b. From this figure, we can see that the averaged OSPA distances of all three approaches are enhanced, when the clutter number increases from 1 to 30. Among these approaches, the CISMC-PHD approach has the smallest averaged OSPA distance and RMSE.

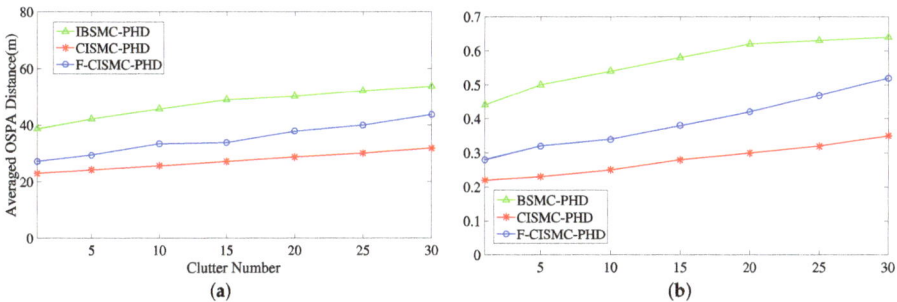

Figure 6. Estimated error of the IBSMC-PHD, CISMC-PHD and F-CISMC-PHD approaches along with the clutter number changing from 1 to 30. (**a**) Averaged OSPA distances; (**b**) Averaged RMSEs of estimated number.

5.4. Comparison of Estimation Accuracy over Different Probabilities of Detection

In this section, we compare the estimation accuracy at different detection probabilities (varying from 0.92 to 0.98). Here, the clutter number is chosen to be 10, and 500 Monte Carlo simulations are run for the comparison. Table 6 reports the OSPA distances and RMSEs of the IBSMC-PHD, CISMC-PHD and F-CISMC-PHD approaches.

Table 6. Tracking Performance over different detection probabilities.

	$p_d = 0.92$			$p_d = 0.94$		
	IBSMC-PHD	**CISMC-PHD**	**F-CISMC-PHD**	**IBSMC-PHD**	**CISMC-PHD**	**F-CISMC-PHD**
OSPA(m)	52.13	34.40	40.21	43.55	27.39	29.82
RMSE	0.54	0.37	0.47	0.47	0.32	0.37
	$p_d = 0.96$			$p_d = 0.98$		
	IBSMC-PHD	**CISMC-PHD**	**F-CISMC-PHD**	**IBSMC-PHD**	**CISMC-PHD**	**F-CISMC-PHD**
OSPA(m)	43.55	27.39	29.82	39.35	23.20	23.02
RMSE	0.42	0.27	0.31	0.35	0.22	0.24

From Table 6, we can observe that the estimated accuracy of the F-CISMC-PHD approach get close to the CISMC-PHD approach, when the probabilities of detection increase from 0.92 to 0.98. Thus, the F-CISMC-PHD approach is suitable for the high probabilities of detection.

5.5. Comparison of Estimation Accuracy at Challenging Nonlinear Scenarios

In this section, we compare the estimation accuracy of the IBSMC-PHD, CISMC-PHD and F-CISMC-PHD approaches at challenging nonlinear scenarios. For the challenging nonlinear scenarios, the standard deviation σ_θ varies from $\frac{1.5\pi}{180}$ to $\frac{3\pi}{180}$. We implement each approach with 500 Monte Carlo simulations.

The averaged accuracy, evaluated by OSPA and RMSE, is listed in Table 7. Table 7 indicates that the estimation accuracy of all three approaches decreases, when σ_θ increases from $\frac{1.5\pi}{180}$ to $\frac{3\pi}{180}$. Compared with the IBSMC-PHD approach, the OSPA distances and RMSEs of the CISMC-PHD and F-CISMC-PHD approaches are smaller at all four values of σ_θ. It means that the estimation accuracy of the CISMC-PHD and F-CISMC-PHD approaches is more stable than the IBSMC-PHD approach in the challenging nonlinear scenarios.

Table 7. Tracking Performance over different σ_θ.

	$\sigma_\theta = \frac{1.5\pi}{180}$			$\sigma_\theta = \frac{2\pi}{180}$		
	IBSMC-PHD	**CISMC-PHD**	**F-CISMC-PHD**	**IBSMC-PHD**	**CISMC-PHD**	**F-CISMC-PHD**
OSPA(m)	49.62	27.32	29.66	50.40	27.64	29.99
RMSE	0.43	0.33	0.37	0.47	0.34	0.38
	$\sigma_\theta = \frac{2.5\pi}{180}$			$\sigma_\theta = \frac{3\pi}{180}$		
	IBSMC-PHD	**CISMC-PHD**	**F-CISMC-PHD**	**IBSMC-PHD**	**CISMC-PHD**	**F-CISMC-PHD**
OSPA(m)	51.04	27.43	30.74	52.56	28.02	31.3
RMSE	0.50	0.35	0.39	0.45	0.35	0.39

6. Conclusions

In this paper, we have proposed the CISMC-PHD and F-CISMC-PHD approaches, which can estimate the time-varying number and states of multi-target nonlinear tracking. In our CISMC-PHD approach, a novel IS function approximation method is presented, which incorporates a gating method into the CIF method. To initiate the birth intensity of the next recursion, we use the current observations

Sensors **2016**, *16*, 653

and estimated states to estimate the birth target components. In addition, we also present a fast implementation of the CISMC-PHD approach, namely F-CISMC-PHD, to reduce the time complexity of the CISMC-PHD approach. By clustering the particles into several groups, the target components can be obtained by representing the groups as Gaussian mixture components. Utilizing these components to approximating the IS functions, the computational time can be reduced magnificently. The simulation results demonstrate that the proposed CISMC-PHD and F-CISMC-PHD approaches outperform the conventional BSMC-PHD approach.

This paper concentrates on improving efficiency and accuracy of the conventional SMC-PHD filter. It simply utilizes the multinomial resampling method as the resampling method. Other resampling methods may be integrated in the CISMC-PHD approach for the future work. Besides, the F-CISMC-PHD approach is only suitable for the high probability of detection and small number of clutters. Study on improving the tracking performance of the F-CISMC-PHD approach at low probability of detection and large number of clutters may be seen as another direction of the future work.

Acknowledgments: This work was supported by the National Nature Science Foundation of China projects under Grants 61471022 and 61573037, and the Fok Ying Tung Education Foundation under Grant 151061.

Author Contributions: Zhe Liu provided insights in formulating the ideas, performed the simulations, and analyzed the simulation results. Zulin Wang provided some insights on motivation and basic idea in introduction. Mai Xu offered some insights on complexity analysis and technical derivations. Zhe Liu wrote the paper. Mai Xu and Zulin Wang carefully revised the paper.

Conflicts of Interest: The authors declare no conflict of interest.

References

1. Yu, H.; Meier, K.; Argyle, M.; Beard, R.W. Moving-Target Tracking in Single-Channel Wide-Beam SAR. *IEEE/ASME Trans. Mech.* **2015**, *20*, 541–552.
2. Wu, J.; Hu, S.; Wang, Y. Adaptive multifeature visual tracking in a probability-hypothesis-density filtering framework. *Signal Process.* **2003**, *93*, 2915–2926.
3. Uney, M.; Clark, D.E.; Julier, S.J. Distributed Fusion of PHD Filters via Exponential Mixture Densities. *IEEE J. Sel. Top. Signal Process.* **2013**, *7*, 521–531.
4. Pulford, G. Taxonomy of multiple target tracking methods. *IEE Proc. Radar Sonar Navig.* **2005**, *152*, 291–304.
5. Yang, B.; Xu, G.; Jin, J.; Zhou, Y. Comparison on EKF and UKF for geomagnetic attitude estimation of LEO satellites. *Chin. Space Sci. Technol.* **2012**, *32*, 23–30.
6. Yang, B.; He, F.; Jin, J.; Xiong, H.; Xu, G. DOA estimation for attitude determination on communication satellites. *Chin. J. Aeronaut.* **2014**, *27*, 670–677.
7. Blackman, S. Multiple Hypothesis Tracking for Multiple Target Tracking. *IEEE Trans. Aerosp. Electron. Syst.* **2004**, *19*, 5–18.
8. Bar-Shalom, Y.; Li, X.R. *Multitarget-Multisensor Tracking: Principles and Techniques*; YBS Publishing: Storrs, CT, USA, 1995.
9. Roecker, J. Suboptimal Joint Probabilistic Data Association. *IEEE Trans. Aerosp. Electron. Syst.* **1994**, *30*, 504–510.
10. Bar-Shalom, Y.; Li, X.R.; Kirubarajan, T. *Estimation with Applications to Tracking and Navigation: Theory Algorithms and Software*; John Wiley & Sons: New York, NY, USA, 2001.
11. Puranik, S.; Tugnait, J.K. Tracking of Multiple Maneuvering Targets using Multiscan JPDA and IMM Filtering. *IEEE Trans. Aerosp. Electron. Syst.* **2007**, *43*, 23–34.
12. Musicki, D.; Scala, B.L. Multi-target Tracking in Clutter without Measurement Assignment. *IEEE Trans. Aerosp. Electron. Syst.* **2008**, *44*, 887–896.
13. Yang, J.; Ji, H.; Fan, Z. Probability hypothesis density filter based on strong tracking MIE for multiple maneuvering target tracking. *Int. J. Control Autom. Syst.* **2013**, *11*, 306–316.
14. Mahler, R. Engineering statistics for multi-object tracking. In Proceedings of the 2001 IEEE Workshop on Multi-Object Tracking, Vancouver, BC, Canada, 8–8 July 2001; p. 53–60.

15. Mahler, R. Multitarget Bayes filtering via first-order multitarget moments. *IEEE Trans. Aerosp. Electron. Syst.* **2003**, *39*, 1152–1178.

16. Mahler, R. A Survey of PHD Filter and CPHD Filter Implementations. In Proceedings of the Signal Processing, Sensor Fusion, and Target Recognition XVI, Orlando, FL, USA, 9–11 April 2007.

17. Mahler, R. "Statistics 101" for multisensor, multitarget data fusion. *IEEE Aerosp. Electron. Syst. Mag.* **2004**, *19*, 53–64.

18. Vo, B.N.; Singh, S.; Doucet, A. Sequential Monte Carlo Methods for Multitarget Filtering with Random Finite Sets. *IEEE Trans. Aerosp. Electron. Syst.* **2005**, *41*, 1224–1245.

19. Mahler, R. *Statistical Multisource-Multitarget Information Fusion*; Artech House: Norwood, MA, UK, 2007; Volume 685.

20. Ristic, B.; Clark, D.; Vo, B.-N.; Vo, B.-T. Adaptive Target Birth Intensity for PHD and CPHD Filters. *IEEE Trans. Aerosp. Electron. Syst.* **2012**, *48*, 1656–1668.

21. Ristic, B.; Clark, D.; Vo, B.N.; Vo, B.T. Improved SMC implementation of the PHD filter. In Proceedings of the 2010 13th Conference on Information Fusion (FUSION), Edinburgh, UK, 26–29 July 2010; pp. 1–8.

22. Candy, J.V. *Bayesian Signal Processing: Classical, Modern and Particle Filtering Methods*; John Wiley & Sons: Hoboken, NJ, USA, 2011.

23. Morelande, M. A sequential Monte Carlo method for PHD approximation with conditionally linear/Gaussian models. In Proceedings of the 2010 13th Conference on Information Fusion (FUSION), Edinburgh, UK, 26–29 July 2010; pp. 1–7.

24. Whiteley, N.; Sumeetpal, S.; Godsill, S. Auxiliary Particle Implementation of Probability Hypothesis Density Filter. *IEEE Trans. Aerosp. Electron. Syst.* **2010**, *46*, 1437–1454.

25. Yoon, J.H.; Kim, D.U.; Yoon, K.J. Efficient importance sampling function design for sequential Monte Carlo PHD filter. *Signal Process.* **2012**, *92*, 2315–2321.

26. Chandra, K.P.B.; Gu, D.W.; Postlethwaite, I. Square Root Cubature Information Filter. *IEEE Sens. J.* **2013**, *13*, 750–758.

27. Arasaratnam, I.; Haykin, S. Cubature Kalman Filters. *IEEE Trans. Autom. Control* **2009**, *54*, 1254–1269.

28. Mutambara, A.G.O. *Decentralized Estimation and Control for Multi-Sensor Systems*, 1st ed.; CRC press: Boca Raton, FL, USA, 1998.

29. Bailey, T.; Upcroft, B.; Durrant-Whyte, H. Validation gating for non-linear non-Gaussian target tracking. In Proceedings of the 2006 9th International Conference on Information Fusion, Florence, Italy, 26–29 July 2006.

30. Li, T.; Sun, S.; Sattar, T. High-speed Sigma-gating SMC-PHD filter. *Signal Process.* **2013**, *93*, 2586–2593.

31. Sibley, G.; Sukhatme, G.; Matthies, L. The Iterated Sigma Point Kalman Filter with Applications to Long Range Stereo. *Robot. Sci. Syst.* **2006**, *8*, 235–244.

32. Zhang, H.; Jing, Z.; Hu, S. Gaussian mixture CPHD filter with gating technique. *Signal Process.* **2010**, *89*, 1521–1530.

33. Reuter, S.; Meissner, D.; Wilking, B.; Dietmayer, K. Cardinality balanced multi-target multi-Bernoulli filtering using adaptive birth distributions. In Proceedings of the 2013 16th International Conference on Information Fusion (FUSION), Istanbul, Turkey, 9–12 July 2013; pp. 1608–1615.

34. Li, T.; Sun, S.; Corchado, J.M.; Siyau, M.F. Random finite set-based Bayesian filters using magnitude-adaptive target birth intensity. In Proceedings of the 2014 17th International Conference on Information Fusion (FUSION), Salamanca, Spana, 7–10 July 2014; pp. 1–8.

35. Mahler, R. "Statistics 102" for Multisource-Multitarget Detection and Tracking. *IEEE J. Sel. Top. Signal Process.* **2013**, *7*, 376–389.

36. Vo, B.N.; Ma, W.K. The Gaussian Mixture Probability Hypothesis Density Filter. *IEEE Trans. Signal Process.* **2006**, *54*, 4091–4104.

37. Mo, L.; Song, X.; Zhou, Y.; Kang, S.; Bar-Shalom, Y. Unbiased converted measurements for tracking. *IEEE Trans. Aerosp. Electron. Syst.* **1998**, *34*, 1023–1027.

38. Schikora, M.; Koch, W.; Streit, R.; Cremers, D. A sequential Monte Carlo method for multi-target tracking with the intensity filter. In *Advances in Intelligent Signal Processing and Data Mining*; Springer: Berlin, Germany, 2013; pp. 55–87.

39. Zhao, L.; Ma, P.; Su, X.; Zhang, H. A new multi-target state estimation algorithm for PHD particle filter. In Proceedings of the 2010 13th Conference on Information Fusion (FUSION), Edinburgh, UK, 26–29 July 2010; pp. 1–8.

40. Li, T.; Sun, S.; Bolić, M.; Corchado, J.M. Algorithm design for parallel implementation of the SMC-PHD filter. *Signal Process.* **2016**, *119*, 115–127.

41. Van der Merwe, R.; Wan, E. Gaussian mixture sigma-point particle filters for sequential probabilistic inference in dynamic state-space models. In Proceedings of the 2003 IEEE International Conference on Acoustics, Speech, and Signal Processing, (ICASSP'03), Hong Kong, China, 6–10 April 2003; pp. 701–704.

42. Schuhmacher, D.; Vo, B.T.; Vo, B.N. A Consistent Metric for Performance Evaluation of Multi-Object Filters. *IEEE Trans. Signal Process.* **2008**, *56*, 3447–3457.

43. Yoon, J.H.; Kim, D.U.; Yoon, K.J. Gaussian mixture importance sampling function for unscented SMC-PHD filter. *Signal Process.* **2013**, *93*, 2664–2670.

sensors

MDPI

Article

Epipolar Rectification with Minimum Perspective Distortion for Oblique Images

Jianchen Liu [1], Bingxuan Guo [2,*], Wanshou Jiang [2], Weishu Gong [3] and Xiongwu Xiao [2]

[1]   School of Remote Sensing and Information Engineering, Wuhan University, 129 Luoyu Road, Wuhan 430079, China; liujianchen@whu.edu.cn

[2]   State key Laboratory for Information Engineering in Surveying, Mapping and Remote Sensing, Wuhan University, 129 Luoyu Road, Wuhan 430079, China; jws@whu.edu.cn (W.J.); xwxiao@whu.edu.cn (X.X.)

[3]   Department of Geographical Sciences, University of Maryland, 2181 Samuel J. LeFrak Hall, 7251 Preinkert Drive, College Park, MD 20742, USA; weishugong@gmail.com

*   Correspondence: mobilemapuav@163.com or 00201550@whu.edu.cn; Tel.: +86-155-2718-6115

Academic Editors: Felipe Gonzalez Toro and Antonios Tsourdos
Received: 14 September 2016; Accepted: 3 November 2016; Published: 7 November 2016

Abstract: Epipolar rectification is of great importance for 3D modeling by using UAV (Unmanned Aerial Vehicle) images; however, the existing methods seldom consider the perspective distortion relative to surface planes. Therefore, an algorithm for the rectification of oblique images is proposed and implemented in detail. The basic principle is to minimize the rectified images' perspective distortion relative to the reference planes. First, this minimization problem is formulated as a cost function that is constructed by the tangent value of angle deformation; second, it provides a great deal of flexibility on using different reference planes, such as roofs and the façades of buildings, to generate rectified images. Furthermore, a reasonable scale is acquired according to the dihedral angle between the rectified image plane and the original image plane. The low-quality regions of oblique images are cropped out according to the distortion size. Experimental results revealed that the proposed rectification method can result in improved matching precision (Semi-global dense matching). The matching precision is increased by about 30% for roofs and increased by just 1% for façades, while the façades are not parallel to the baseline. In another designed experiment, the selected façades are parallel to the baseline, the matching precision has a great improvement for façades, by an average of 22%. This fully proves our proposed algorithm that elimination of perspective distortion on rectified images can significantly improve the accuracy of dense matching.

Keywords: epipolar rectification; oblique images; UAV images; minimum perspective distortion; 3D reconstruction

1. Introduction

Aerial oblique imagery has become an important source for acquiring information about urban areas because of their visualization, high efficiency and wide application in domains such as 3D modeling, large-scale mapping and emergency relief planning. An important characteristic of oblique images is the big tilt angles [1], and they usually contain large perspective distortions relative to the surfaces. This large distortion reduces the image correlations and makes dense image matching more difficult, so traditional techniques usually perform poorly on oblique images. However, the precise 3D reconstruction tasks require an accurate dense disparity map, e.g., using a SGM (Semi-global Matching) based stereo method [2], therefore, epipolar rectification is a necessary initial step for 3D modeling [3]. To guarantee completeness, robustness and precision, image rectification for the purpose of 3D reconstruction should take the perspective distortion into account.

This paper is inspired by the fact that epipolar rectification can minimize perspective distortion; thus, features can be matched very accurately by correlation and an accurate dense disparity map can be generated [4]. Each urban scene usually contains multiple surface planes, and these planes can be grouped according to their feature directions (horizontal or vertical). The epipolar rectification should minimize the distortion of planes in the same group to create an exact match for those planes. The epipolar rectification principle of minimum perspective distortion relative to the original image is desirable in the vast majority of cases; however, rectified images with a minimum perspective distortion relative to surface planes are also useful in certain circumstances. For example, in oblique photogrammetry, rectified horizontal images are usually used to reconstruct the roofs of buildings, meanwhile rectified vertical images are expected to generate more accurate depth maps of vertical planes such as the façades of buildings.

Many epipolar rectification algorithms have been proposed, and they can generally be categorized into linear transformation and non-linear transformation algorithms. The former algorithms transform an image from one plane to another plane, making the corresponding epipolar lines coincide with the scan lines [5]. The linear approaches' advantages are that it is mathematically simple, fast and preserves image features. In contrast, non-linear approaches typically use Bresenham's algorithm [6] to extract pixels along epipolar lines, thus avoiding most of the problems that linear rectification approaches have, e.g., generating unbounded, large or badly warped images. Two similar methods [7,8] involve parameterizing the image with polar coordinates (around the epipoles). These methods have two important features: they can address epipoles located in the images, and they can reduce the matching ambiguity to half the epipolar lines. Instead of resampling the epipolar lines on the original image planes, an attractive method proposed by [9] employs a rectification process based on a cylinder rather than a plane. However, this method is relatively complex and requires large numbers of calculations in three-dimensional space. Another non-linear transformation method expressed in a paper by [10] proposes an accurate computation method based on rectification of spherical-camera images via resampling the same longitude line. However, this method is suitable only for spherical panoramic images.

Because linear transformation is simple and intuitive, this article focuses on the homography based method. Due to the redundant degrees of freedom, the solution to rectification is not unique and can actually lead to undesirable distortions. The distortion constraint leads to reduced degrees of freedom for homographies in solving the rectification problem. The first work on using a distortion constraint was performed by [11], followed by [12]. They suggest using the transformation that minimizes the range of disparity between the two images, i.e., the distance between the rectified corresponding points. In their state-of-the-art methods [13,14], the authors attempt to make the effects of rectification "as affine as possible" over the area of the images. In papers by [15], a different distortion criterion consists of preserving the sampling of the original images. The method proposed in the paper [16] uses the squared Sampson error for the constrained rectification. The algorithm [17] is decomposed into three steps: the first and second step involve making the image plane parallel to the baseline, while the third crucial step minimizes the distortion relative to the original images. None of the above methods mention the fact that there is no agreement on what the distortion criterion should be, and they all require corresponding points. In a paper by [18], the authors aim to eliminate relative distortion between the rectified images by choosing the reference plane in the scene. However, only the reference plane and the planes that are parallel to the reference plane have no relative distortion and their method still requires the corresponding points. In the case of calibrated cameras, the rotation matrix for rectified images is determined directly. The method [19] determines the rectified image plane according to the baseline and the optical axis of the old left matrix; however, in that case, the distortion for one of the rectified images is small, but the distortion of the other may be larger in the oblique photography case. The algorithm expressed in the paper [20] improves the preceding algorithm by making the distortion of the two rectified images small relative to the original images. However, these methods still do not consider distortion relative to surfaces in object space.

Unlike the methods described above, which reduce distortion by explicitly minimizing an empirical measure, the proposed approach is to minimize a cost function that is constructed by the tangent value of angle deformation. In this manner, the rectified images will have smallest perspective distortion for some surface planes and features can be matched quite accurately by correlation. In addition, the homography based method may yield very large images or cannot rectify at all. These issues can be solved by the scope constraint which can also crop the low-quality regions of oblique images.

In this paper, we investigated the rectification method of minimum perspective distortion by taking into account surface planes, such as original image planes, roofs and the façades of buildings. The method is flexible in order to generate rectified images with respect to different reference planes. The remainder of this paper is organized as follows. The innovative rectification algorithms and their distortion constraints are presented in detail in Section 2. The performance of the proposed methods and the quantitative evaluation of the matching results are subsequently evaluated in Section 3. Finally, concluding remarks are provided.

2. Methodology

2.1. Algorithm Principle

There is little difference between computer vision (CV) and photogrammetry (DP) in terms of definitions of projective geometry. The projective matrix P generated by both the computer vision and photogrammetry definitions is the same. However, the expressions of the camera matrix K and the rotation matrix R are different. This is because they define the camera coordinate frame differently [21], which is shown in Figure 1.

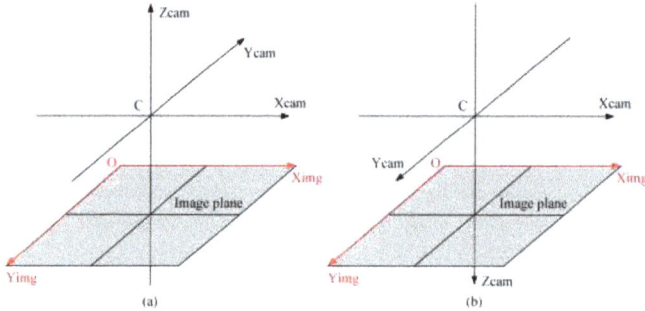

Figure 1. The camera coordinate frames for (**a**) photogrammetry definition; (**b**) computer vision definition.

The origin of coordinates C and coordinate axes $X_{cam}, Y_{cam}, Z_{cam}$ constitute the camera coordinate frame. The image coordinate system is consisted of the origin of coordinates O and coordinate axes X_{img}, Y_{img}. From Figure 1, we can see that the camera coordinate frames are both right hand Euclidean coordinate systems. However, the image plane is $Z = f$ in computer vision and $Z = -f$ in photogrammetry, where $f > 0$. The relationship between R_{CV} and R_{DP} is shown below:

$$R_{CV} = \begin{bmatrix} 1 & 0 & 0 \\ 0 & -1 & 0 \\ 0 & 0 & -1 \end{bmatrix} R_{DP} \tag{1}$$

As per the different camera coordinate frames, the camera calibration matrix K can be respectively denoted as:

$$K_{CV} = \begin{bmatrix} f_x & 0 & x_0 \\ 0 & f_y & y_0 \\ 0 & 0 & 1 \end{bmatrix} K_{DP} = \begin{bmatrix} -f_x & 0 & x_0 \\ 0 & f_y & y_0 \\ 0 & 0 & 1 \end{bmatrix} K_{CV} = -K_{DP} \begin{bmatrix} 1 & 0 & 0 \\ 0 & -1 & 0 \\ 0 & 0 & -1 \end{bmatrix} \qquad (2)$$

where f_x, f_y represents the focal length of the camera in terms of pixel dimensions in the x and y direction, respectively. The expression (x_0, y_0) is the principal point in terms of pixel dimensions. In this article, the symbols K and R refer to the definition used by the photogrammetry field. Thus, the direction of the image plane and the Z axis of the camera coordinate frame defined in photogrammetry are in accord. This selection is more convenient for the subsequent rectifying transformation when considering the perspective distortion relative to reference planes.

2.1.1. Homographic Transformation

Epipolar rectification can be viewed as the process of transforming the epipolar geometry of a pair of images into a canonical form. It can be accomplished by applying a homographic matrix to each image that maps the original image to a predetermined plane. Let H and H' be the homographic matrix to be applied to images I and I', respectively. Also, let $p \in I$ and $p' \in I'$ be a pair of corresponding points. The camera matrix K_{rec} and the rotation matrix R_{rec} can be generated by the algorithm proposed in this paper, while the symbols R and K refer to the original image. Considering the rectified image points $p_{rec} \in I_{rec}$, $p'_{rec} \in I'_{rec}$, the transformation can be defined as:

$$\begin{aligned} p_{rec} &= Hp \\ p'_{rec} &= H'p' \end{aligned} \qquad (3)$$

$$\begin{aligned} H &= K_{rec}R_{rec}R^T K^{-1} \\ H' &= K'_{rec}R'_{rec}R'^T K'^{-1} \end{aligned} \qquad (4)$$

However, there are countless types of transformation matrices H that meet the above conditions of the solution. Moreover, poor choices for H and H' can result in rectified images that are dramatically changed in scale or severely distorted. Therefore, rectified image planes should be selected according to the criteria of minimum perspective distortion, and it will be discussed in the next section.

2.1.2. Minimizing Perspective Distortion

The angle deformation always exists in the perspective transformation from a reference configuration to a current configuration. The scale of rectified images can be determined by the focal length, and it does not affect the angle deformation. In the process of rectification, a method is developed to minimize the tangent value of angle deformation. The subscript L and R denote the left and right images respectively in the following of the paper. Here, it can be defined as:

$$\varepsilon = (\omega_L)^2 + (\omega_R)^2 \qquad (5)$$

where ω is the tangent value of angle deformation. The result can be determined by minimizing the squared error ε. The angle deformation presents a notable positive correlation with the rotation angle, i.e., the dihedral angle between the rectified image plane and its reference plane (original image plane or surface plane). It is easy to discuss the characteristics of angle deformation by decomposing it into two directions: the rotation direction and its perpendicular direction. A line in reference plane which is perpendicular to the rotation direction has no angle deformation, while the angle deformation of a line that is parallel to the rotation direction could not be ignored. The relationship of the rotation angle θ with the tangent value of angle deformation is given below:

$$\omega = -\sin(\theta) \cdot b \qquad (6)$$

in which b determines the position of a line in the reference plane. Thus, Equation (5) can be rewritten as:

$$\varepsilon = (\sin(\theta_L))^2 + (\sin(\theta_R))^2 \tag{7}$$

According to the principle of epipolar rectification, the two rectified image planes (I_{rec} and I'_{rec}) must be corrected to be coplanar, and must both be parallel to the baseline (B). Thus, the direction Z of the rectified image plane is constrained to be perpendicular to the baseline, which lies in a plane A perpendicular to the baseline. In the case that left and right reference planes are different, the rectification of minimum perspective distortion is illustrated in Figure 2. N_L and N_R are the directions of reference planes. Their projections on a plane A are N'_L and N'_R. The α in Figure 2 denotes the angle between N'_R and Z. Thus θ_L and θ_R can be expressed respectively by a function that takes one parameter α, and these expressions can be easily derived by the analytic geometry. Furthermore, the direction Z of rectified image plane is determined by one parameter α. The solution is to minimize the squared error ε by gradient descent method. In the case that left and right reference planes are the same, the direction Z of rectified image plane is the projection of the direction vector N onto the plane A.

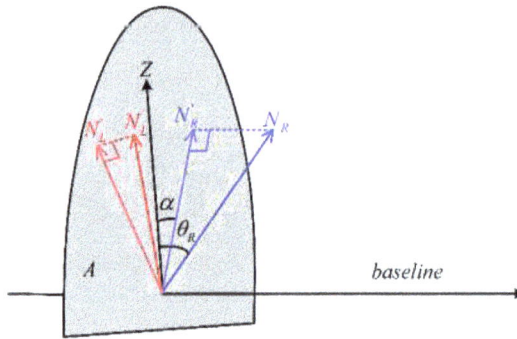

Figure 2. The rectification of minimum perspective distortion.

2.2. Rectification Algorithm

2.2.1. R Matrix of Rectified Image

After expressing the observational coordinate axes of the camera coordinate frame numerically as three unit vectors (e_0, e_1, e_2) in the world coordinate system, together they comprise the rows of the rotation matrix R (world to camera). The rectified images with respect to different reference planes are controlled by the R matrix. The R matrix calculation is simple and flexible as explained in the following sections.

Basic Rectification

A minimum distortion rectification relative to the original image planes is discussed first and it can be applied to a variety of cases. To carry out this method, it is important to construct a triple of mutually orthogonal unit vectors (e_1, e_2, e_3). The first vector e_1 can be given by the baseline. Because the baseline is parallel to the rectified image plane and the epipolar line is horizontal, vector e_1 coincides with the direction of the baseline. C_1, C_2 are the camera station coordinates and e_1 can be deduced as:

$$e_1 = \frac{B}{||B||}, \quad B = C_2 - C_1 \tag{8}$$

The two constraints on the second vector, e_2, are that it must be orthogonal to e_1 and that the perspective distortion relative to the original images must both be minimal. To achieve these, it should compute and normalize the cross product of e_1 with e_{temp}, which is the direction Z of rectified image plane (see Section 2.1.2). It can be expressed as:

$$e_2 = \frac{e_{temp} \times e_1}{|| e_{temp} \times e_1 ||} \tag{9}$$

The third unit vector is unambiguously determined as:

$$e_3 = e_1 \times e_2 \tag{10}$$

Together, they comprise the rows of the rotation matrix R, which is defined as:

$$R_{rec} = \begin{bmatrix} e_1 \\ e_2 \\ e_3 \end{bmatrix} \tag{11}$$

Thus, the rectified camera coordinate frames are defined by getting the R matrix of the rectified images. Noting that the left and right R matrices are same.

Horizontal or Vertical Rectification

When the image models are absolutely oriented, horizontally or vertically rectified images can be generated. At the same time, it can minimize the perspective distortion relative to horizontal or vertical planes, making the result conducive for image-matching purposes for regular buildings. Because the baseline is not absolutely horizontal, the way to minimize the perspective distortion relative to horizontal planes is to generate the rectified images that are closest to the horizontal plane. However, absolutely rectified vertical images can be generated according to Section 2.1.2. The computational process is similar to the above procedures. There is only a slight difference in the definition of e_{temp}. For horizontal images, it is defined as follows:

$$e_{temp} = \begin{bmatrix} 0 & 0 & 1 \end{bmatrix} \tag{12}$$

When vertical images are needed, the e_{temp} should meet the following constraints:

1. $e_{temp} = \begin{bmatrix} x & y & 0 \end{bmatrix}$;
2. e_{temp} must be orthogonal to the baseline;
3. e_{temp} should be consistent with the two direction vectors of the original images' optical axes, i.e., $e_{temp} \cdot R_3 > 0$, $e_{temp} \cdot R_3' > 0$.

General Rectification

When the images models are relatively oriented or when a non-horizontal or non-vertical plane exists in the world coordinates, the direction of the rectified image planes should be closest to the direction of the plane to minimize the distortion with respect to that plane. The computational process is similar to the above two procedures, requiring only a small difference in the definition of e_{temp}. Given a plane expressed as $\begin{bmatrix} a & b & c & d \end{bmatrix}$, its normal form is expressed as $\begin{bmatrix} a & b & c \end{bmatrix}$, which is consistent with the two direction vectors of the original images' optical axes. Then, e_{temp} is determined as:

$$e_{temp} = \begin{bmatrix} a & b & c \end{bmatrix} \tag{13}$$

From the above discussion, it is easy to obtain various rectified images with different distortion characteristics by defining different R matrices, which works because the definition of the R matrix is flexible.

2.2.2. Camera Matrix of Rectified Image

The scale of both rectified images can be adjusted by setting a suitable focal length. As we know, the focal lengths of rectified images have the same value. The most commonly used method, shown in Figure 3a, is to set the focal length the same as the original images; however, in that case, the rectified images will be larger than the original images. Note that although the resolution is higher than in the original images, it is meaningless. Our method, shown in Figure 3b, is to keep the principle point of the original image unchanged during the perspective transformation.

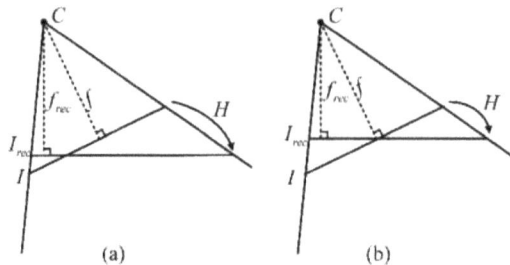

Figure 3. The two methods for setting focal length: (**a**) the focal length unchanged; (**b**) the principle point unchanged.

In Figure 3a, the rectified images will be larger than the original images. In particular, as the rotation angle between the optical axes of the original and rectified images grows larger, the rectified image size becomes significantly bigger. In contrast, in Figure 3b, the rectified image size would not be significantly different from the original image. The proposed method may result in rectified images in which some part of the image is compressed and the other part is stretched compared to the original images. However, the average resolution remains almost unchanged from the original images. In this paper, the focal length is defined as:

$$
\begin{aligned}
f_1 &= f \cdot R_3 \cdot e_3 \\
f_2 &= f' \cdot R_3' \cdot e_3 \\
f_{rec} &= \min(f_1, f_2)
\end{aligned}
\tag{14}
$$

After getting the R matrix and the focal length of the rectified images, it is easy to obtain the K matrices of the rectified images. According to Section 2.1.1, the H matrices can be calculated to rectify the original images to rectified images.

2.3. Distortion Constraints

2.3.1. Distortion Coordinate Frame

To better express the character of distortion relative to the original images, a distortion coordinate frame is defined. The optical axis of the rectified image is the Z_{dis} axis of the distortion coordinate frame, i.e., the third row of the rectified rotation matrix R_{rec}. The X_{dis} and Y_{dis} axes of this coordinate frame can be obtained by the cross products. The Y_{dis} axis must be orthogonal to the two optical axes of the original image and rectified image, and can be expressed as:

$$
Y_{dis} = R_3 \times R_{rec3}
\tag{15}
$$

The third unit vector is unambiguously determined as:

$$X_{dis} = Y_{dis} \times Z_{dis} \tag{16}$$

The three unit vectors of $X_{dis}, Y_{dis}, Z_{dis}$ form a rotation matrix R_{dis}.

2.3.2. Characteristics of Distortion

This section focus on the distortion of rectified image relative to the original image. It is easy to discuss the characteristics of distortion in the distortion coordinate frame. The distortion within an image line that is parallel to the Y_{dis} axis has the same size, while the distortion within an image line that is parallel to the X_{dis} axis gradually becomes larger along the positive direction of the X_{dis} axis. The size of distortion (denoted as t) is the ratio between the size of a point in the original image and the size of its corresponding point in the referencing image. It is derived from the projection geometry and shown in Equation (17), a schematic diagram is introduced in Figure 4:

$$t = \cos\alpha(\cos\alpha - \sin\alpha\cot\theta)^2 \tag{17}$$

where α is the angle between the Z axes of the original and the rectified camera coordinate frames. Assume that the field of view (FOV) of the original image is π, although the FOV is usually less than that value in actuality. For the rectified image, the valid FOV range is (α, π) and $\theta \in (\alpha, \pi)$. The θ in Figure 4 denotes the angle from the directional vector X_{dis} to the ray of light. The size of the distortion is closely related to the angle α. Its characteristics are illustrated in Figure 5.

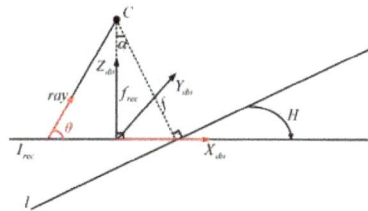

Figure 4. The distortion coordinate frame.

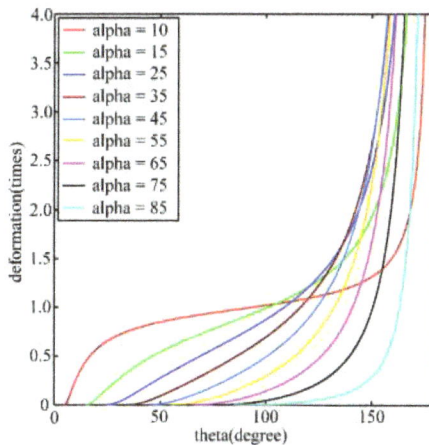

Figure 5. The curves of distortion relative to original image.

Figure 5 shows nine curves that correspond to different α values. The horizontal axis represents the ray direction θ, and the vertical axis represents the image distortion relative to original image. Through the distribution curve of distortion, a curve that is far away from the line $y = 1$ corresponds to an image with large distortion. It can be observed that the greater the angle α is, the greater the corresponding distortion is. Moreover, the distortion near the image edges is greater than the distortion near the image center. For the images used in this paper, the tilt angle of oblique images is approximately 45°, and the field of view is approximately 35°–50°, so rectified images typically do not have such large distortions.

2.3.3. Constraint Method

The size of the maximum distortion can constrain the scope of the image and can be applied to the following two aspects: constraining the unbounded images and getting the highest quality image region. When generating the rectified horizontal images, distortion constraints can remove the image region with the smaller base-to-height ratio and the image areas that are likely to be blurred due to atmospheric influence. In oblique aerial photography, the upper part of the image is the region with the smaller base-to-height ratio and is also highly likely to be affected by air quality, making it blurry.

If the rotation angle α is large and the FOV of original image is also large, it is likely to generate an unbounded image. This phenomenon is most likely to appear in close range photogrammetry and oblique photogrammetry. Given the threshold (the size of maximum distortion) T, thus Equation (17) can be rewritten as:

$$\tan\theta = \frac{\sin\alpha}{\cos\alpha - \sqrt{\frac{T}{\cos\alpha}}} \tag{18}$$

Equation (18) can provide the result of calculating the desirable image region. Then, translating the coordinates of the constrained image scope from the distortion coordinate frame to the rectified camera coordinate frame $(R_{dis} \rightarrow R_{rec})$. Finally, solving for the intersection area of the constrained image scope (solved by Equation (18) under a threshold T) and the original image's projective scope (calculated by Equation (3) using image 4 corner points) in the rectified image plane for the final scope of the rectified image.

3. Experimental Evaluations

3.1. Performance of Rectification

The presented approach is tested with oblique images captured by the SWDC-5 aerial multi-angle photography system. This system is composed of five large format digital cameras with one vertical angle and four tilt angles. The image size of the five cameras is 8176 × 6132, and the pixel size is 6 μ. The angles of the four tilt cameras are 45° relative to the vertical camera. The focal length of the tilt cameras is 80 mm, while the focal length of the vertical camera is 50 mm. The relative height of flight is 1000 m possessing a GSD (Ground Sampling Distance) of 12 cm. The side and forward overlapping rates are 50% and 80% respectively. The coordinates are recorded in the WGS-84 coordinate system. Oblique photography captures more information, including the façade textures of the buildings, which can be used to create a more realistic appearance in 3D urban scene modelling.

When reconstructing 3D architectures from oblique images, calibration is mandatory in practice and can be achieved in many situations and by several algorithms [22,23]. Given a pair of stereo oriented images, the corresponding P (projection matrix), or the intrinsic parameters of each camera and the extrinsic parameters of the images, it is straightforward to define a rectifying transformation. Meanwhile, it is needed to minimize the perspective distortion according to the above methods. In this article, the lens distortions are not considered and have already been removed in the experimental data.

Figure 6a,b shows the original image pair captured from Wuhan City (China) in which the red lines are epipolar lines. In the sub-region pair in Figure 6c,d, the roofs are shown with perspective distortion, which is especially apparent in Figure 6e,f with the building façades. Examples of rectified image pairs illustrate basic rectification (Figure 7a,b), horizontal rectification (Figure 8), vertical rectification (Figure 9) and scope-constrained rectification (Figure 7c,d). Figure 7a,b shows the rectified image pair with minimum distortion properties relative to the original images, i.e., the changes to the optical axis are minimal. It is apparent that the epipolar lines (red lines) are horizontal in the rectified images and that the corresponding lines are in nearly the same vertical position. Figure 8a,b shows a horizontally rectified image pair while Figure 8c,d shows their sub-regions in which the roofs (red areas) are similar and without distortion, i.e., the disparities are close to a constant. There is no distortion for the horizontal objects projected into the horizontally rectified images, but absolutely rectified horizontal images do not exist for the non-horizontal baseline. Although there is a slight distortion for the horizontal objects in this type of rectification, the distortion is minimal and can be ignored for oblique aerial photography. Using a small adjustment, images without the distortion of horizontal objects can be achieved by setting different focal lengths and making the rectified image plane absolutely horizontal. However, the method cannot generate rectified images in this way. Figure 9 clarifies the concepts of vertically rectified images. In that figure, the vertical lines of façades are still vertical in the vertically rectified images as shown in Figure 9c,d compared to Figure 6e,f. Typically, there is no distortion for façades in the vertical direction, while in the horizontal direction, scale distortion is inevitable unless the façades are all parallel to the vertically rectified image plane. From these results, the façades can be considered for flight course planning. Figure 7c,d shows the rectification result under the scope constraint. Due to the large tilt angle, the image regions that have a smaller base-to-height ratio are removed as can be observed in Figure 7c,d compared to Figure 8a,b. This method can also be used to constrain the unbounded rectified image area, which happens when the tilt angle is large and the field of view of is also large, as in oblique photogrammetry.

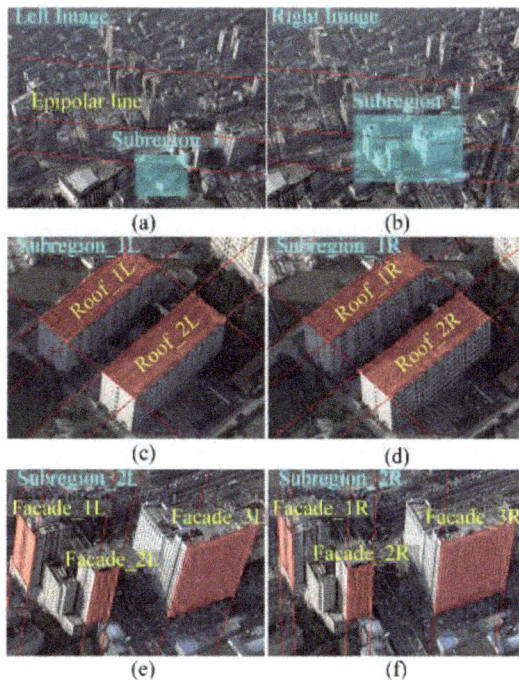

Figure 6. The original image pair and its sub-regions: the red lines in (**a**,**b**) are the epipolar lines in the original image; the red areas in (**c**,**d**) are the horizontal roofs projected in the original images; the red areas in (**e**,**f**) are vertical façades projected in the original images.

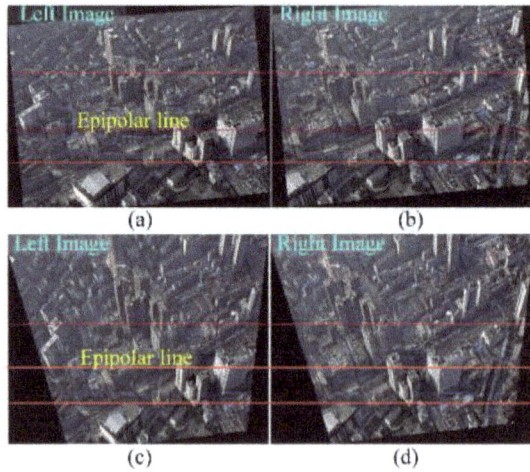

Figure 7. The result of epipolar rectification: (**a**,**b**) show the rectified image pair with the minimum distortion relative to original images; (**c**,**d**) show the horizontally rectified image pair with a scope constraint.

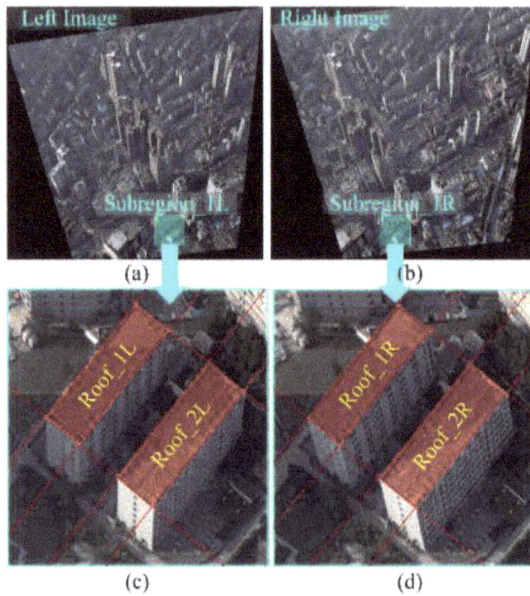

Figure 8. A rectified horizontal image pair (**a**,**b**); and their sub-regions (**c**,**d**), in which the shapes of horizontal planes projected in the rectified images are similar.

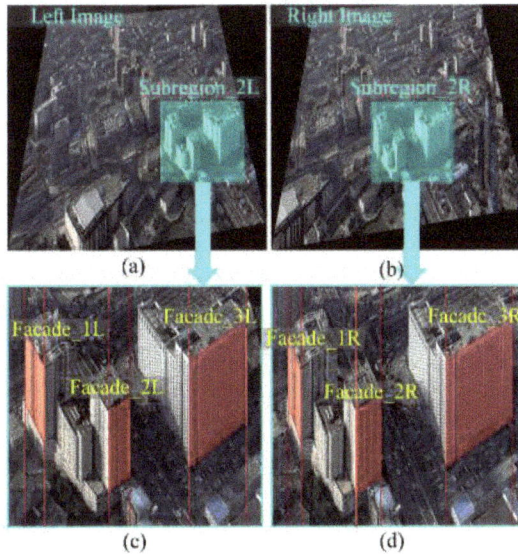

Figure 9. The rectified vertical image pair (**a**,**b**); and their sub-regions (**c**,**d**), in which the vertical lines projected into the rectified images are true vertical lines.

3.2. Quantitative Evaluation of the Matching Results

In this section, the matching results of commonly used rectifications [19,20] and the proposed rectification are comparatively analyzed. These two commonly used rectifications are very similar to the proposed basic rectification. In addition, there are small differences for their dense matching results, which can be ignored for the dataset used in this paper. However, the commonly used methods do not consider the perspective distortion relative to surfaces in object space. Therefore, there are dramatic differences compared to the proposed horizontal and vertical rectification. The following quantitative analysis shows the superiority of the proposed rectification method.

Here, we select horizontal roofs and vertical façades (red areas shown in Figure 10) to evaluate the matching precision influenced by the distortions. Three sets of rectified images pairs were matched by the tSGM algorithm [24] and the resulting depth maps were evaluated quantitatively. For the horizontal rectification, the roofs appear to be without perspective distortion, however, the distortions of façades are not eliminated. In contrast, the distortions of façades are minimized in the vertical rectification, which is opposite to the roofs. For the commonly used rectification, the roofs and the façades are both with geometry distortion. All dense image matches were carried out on full resolution imagery. For comparison purposes, the resulting depth maps have been transformed from rectified images to original images.

The matching results of horizontal objects are compared in Figure 10d–f, showing that the densities of point clouds in the roof areas are not the same, especially in the black area. The horizontally rectified image (matching result shown in Figure 10d) generates more points than the vertical rectification (matching result shown in Figure 10e) and commonly used rectification (matching result shown in Figure 10f) for the roofs. The matching result for the façades in Figure 10a–c shows that the vertical rectification is the best as expected. Vertical rectification is more convenient for matching façades and generates a denser set of points than the other rectifications. To differentiate them more convincingly, the result of a quantitative analysis is first shown in Table 1. The influence of deformation on matching can be analyzed from two aspects: the percentage of valid pixels and the precision. The former means the percentage of generated depth pixels within an area. In the latter case, the RMSE (Root Mean

Square Error) of plane fitting is used to scale the precision of image matching. From Table 1, we can see that due to the reduced size of the distortions, the density of point clouds increases from 98.89% to 99.15%, and the RMSE is reduced from 15.2 cm to 10.1 cm for Roof 1 (the red area in Figure 10). Similar changes occurred for Roof 2. Because of the high matching result from tSGM, the precision is obviously improved, though there is no significant improvement in the integrity of roofs. Using the same analytical method for façades, both increased integrity and precision for vertical rectification are shown in Table 2, but not as obviously as for the roofs. This is probably due to the fact that the façades are not parallel to the baseline, and the scale deformation in the horizontal direction still exists in vertical rectification. Nevertheless, the rectification methods proposed in this paper can improve the matching precision. The matching results of horizontal and vertical rectifications are compared in Table 3. It shows that the matching results of horizontal objects in horizontal rectification are better than that of the vertical rectification, while there are almost completely opposite conclusions for façades. In the case of roofs, experimental data also show that the matching results of horizontal rectification perform the best for precision, because the distortion of roofs in horizontal rectification is smallest in these three situations. A similar conclusion can be drawn for façades.

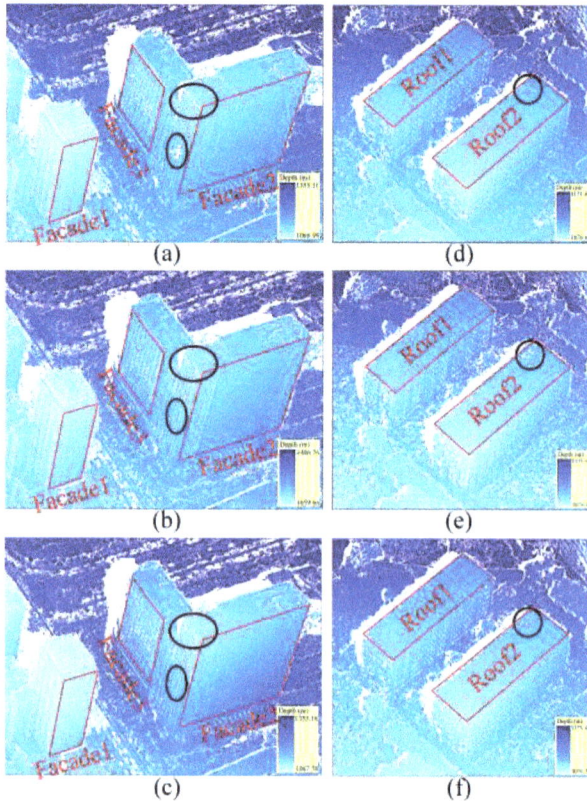

Figure 10. A comparison of matching results for roofs and façades: (**a,d**) the matching result of rectified horizontal images; (**b,e**) the matching result of rectified vertical images; (**c,f**) the matching result of commonly used rectification images.

Table 1. A comparison of matching results for horizontally rectified images and commonly used rectification images.

Objects		Horizontal Rectification		Commonly Used Methods	
		Integrity (%)	Precision (RMSE)	Integrity (%)	Precision (RMSE)
Horizontal Objects	Roof 1	99.15%	10.1 cm	98.89%	15.2 cm
	Roof 2	98.86%	18.1 cm	98.55%	21.5 cm

Table 2. A comparison of matching results for vertically rectified images and commonly used rectification images.

Objects		Commonly Used Methods		Vertical Rectification	
		Integrity (%)	Precision (RMSE)	Integrity (%)	Precision (RMSE)
Vertical Façades	Façade 1	92.12%	53.3 cm	92.47%	52.5 cm
	Façade 2	97.45%	24.7 cm	98.52%	24.5cm
	Façade 3	79.28%	27.1 cm	82.40%	25.6 cm

Table 3. A comparison of matching results for horizontally rectified images and vertically rectified images.

Objects		Horizontal Rectification		Vertical Rectification	
		Integrity (%)	Precision (RMSE)	Integrity (%)	Precision (RMSE)
Horizontal Objects	Roof 1	99.15%	10.1 cm	97.94%	22.4 cm
	Roof 2	98.86%	18.1 cm	97.31%	27.2 cm
Vertical Façades	Façade 1	91.44%	56.6 cm	92.47%	52.5 cm
	Façade 2	97.18%	30.4 cm	98.52%	24.5cm
	Façade 3	78.63%	27.6 cm	82.40%	25.6 cm

3.3. Robustness Evaluation

We choose another set of data captured from Nanchang City (China) to evaluate the robustness of the proposed algorithm. The image data is captured by a multi-angle oblique photography system composed of five large format digital cameras: one vertical angle and four tilt angles. The image size of vertical view is 9334 × 6000 and the image size of tilt views is 7312 × 5474. The pixel size is 6 μ. The angle of four tilt cameras is 45° relative to the vertical camera. The focal length of tilt cameras is 80 mm and that of vertical camera is 50 mm. The relative height of flight is 1000 m. The distance of adjacent strips is 500 m and that of adjacent images within the same strip is 200 m.

Three trips with a total of 210 images are used in the experiment and cover an area of 7 km². Coverage area is a city district, and there are a lot of horizontal and vertical planes. Images are processed through bundle adjustment, automatic DEM (Digital Elevation Model) extraction and orthoimage production steps with the GodWork software package (version 2.0), which has been developed by Wuhan University (Wuhan, China). We choose 12 façades and 12 horizontal planes (including roofs, playgrounds and roads) within the coverage area, which are shown in Figure 11. The selected planes are evaluated by using different oriented image pairs.

We use the same method as mentioned in Section 3.2 to evaluate the matching precision influenced by the distortions. Table 4 shows the evaluation results of the horizontal planes. It shows that the matching precision is increased by about 33% for horizontal planes. The analysis results are in agreement with the Section 3.2. From Table 5, we can see that the matching precision in vertical rectification also shows a great improvement for façades, by an average of 22%. The analysis results are significantly different from Section 3.2. This is within our expectations. Because the flight direction is from east to west, it is easier to select a number of façades parallel to the baseline. Thus, there is no distortion for the façades projected into the vertically rectified images. This fully proves our hypothesis that perspective distortion has a great influence on matching. Elimination of perspective distortion on rectified images can significantly improve the accuracy of dense matching.

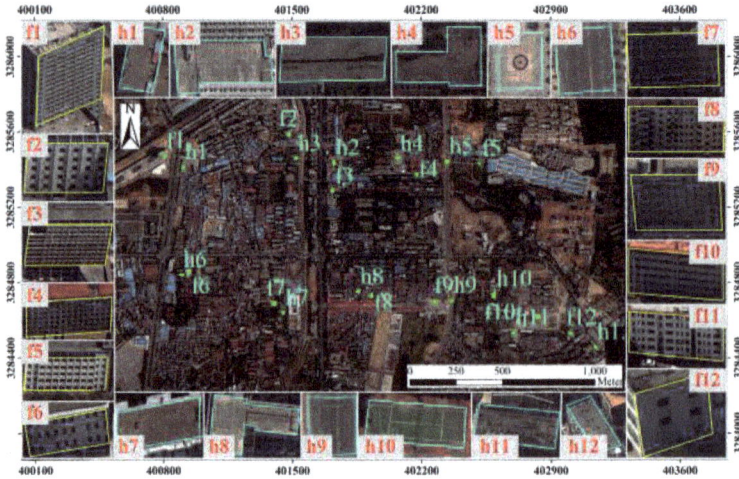

Figure 11. The illustration of coverage area and selected test planes. We choose 12 façades and 12 horizontal planes (including roofs, playgrounds and roads) within the coverage area. In the orthoimages, the plane positions are marked. The selected planes are shown in the surroundings.

Table 4. A comparison of matching results for horizontally rectified images and commonly used rectification images.

Objects	Commonly Used Methods Precision (RMSE)	Horizontal Rectification Precision (RMSE)	Improvement (%)
h1	20.02 cm	15.30 cm	23.56%
h2	15.81 cm	9.57 cm	39.46%
h3	14.98 cm	8.14 cm	45.69%
h4	10.08 cm	6.55 cm	35.00%
h5	39.75 cm	25.59 cm	35.62%
h6	8.92 cm	6.34 cm	28.91%
h7	20.50 cm	13.78 cm	32.80%
h8	13.82 cm	7.64 cm	44.75%
h9	37.63 cm	27.41 cm	27.16%
h10	33.30 cm	20.93 cm	37.15%
h11	17.99 cm	11.26 cm	37.41%
h12	28.20 cm	24.95 cm	11.54%

Table 5. A comparison of matching results for vertically rectified images and commonly used rectification images.

Objects	Commonly Used Methods Precision (RMSE)	Vertical Rectification Precision (RMSE)	Improvement (%)
f1	17.18 cm	15.05 cm	12.42%
f2	22.18 cm	18.25 cm	17.72%
f3	32.88 cm	25.12 cm	23.58%
f4	44.15 cm	36.67 cm	16.94%
f5	17.02 cm	8.95 cm	47.39%
f6	30.72 cm	22.92 cm	25.38%

Table 5. *Cont.*

Objects	Commonly Used Methods Precision (RMSE)	Vertical Rectification Precision (RMSE)	Improvement (%)
f7	23.71 cm	14.46 cm	39.02%
f8	39.87 cm	35.09 cm	12.00%
f9	27.83 cm	21.18 cm	23.90%
f10	41.20 cm	32.79 cm	20.40%
f11	44.21 cm	32.42 cm	26.68%
f12	50.84 cm	47.12 cm	7.31%

4. Conclusions

Epipolar rectification does not usually take into account the distortions of surface planes and the quality of original images. Therefore, a new rectification algorithm for aerial oblique images is proposed that minimizes the distortion of surface planes. The method is based on the minimization of a cost function that is constructed by the tangent value of angle deformation. In addition, a scope-constrained rectification is proposed to solve the problems of unbounded rectified images and crop out the low-quality areas of oblique images. Although the method proposed in this paper seems simple, it addresses epipolar rectification of oblique images in a flexible manner and solves many practical problems in oblique image matching.

The proposed strategy of epipolar rectification leads to depth maps with greater numbers of valid pixels and increased precision by minimizing the perspective distortion. The experiments have confirmed that the matching precision for horizontal objects can be significantly improved by using the proposed rectification method (increased by about 30%). This improvement is attributed to the fact that the horizontal objects appear to be without distortions in the horizontal rectification. However, the distortions of façades have not been completely eliminated, and scale deformation in the horizontal direction is inevitable unless the façades are parallel to the baseline. Therefore, the façade directions should be considered for flight course planning. In a second set of data, the flight direction is from east to west, and most of the visible façades are parallel to the baseline. In this condition, the matching precision shows a great improvement for façades, by an average of 22%. This fully proves that perspective distortion has a great influence on matching. Elimination of perspective distortion on rectified images can significantly improve the accuracy of dense matching. Furthermore, a better result could be achieved by integrating two depth maps of horizontal rectification and vertical rectification in 3D modeling.

Acknowledgments: This work was supported in part by the National Key Research and Development Program of China (earth observation and navigation part): High availability and high precision indoor intelligent hybrid positioning and indoor GIS technology (No. 2016YFB0502200) and in part by the Fundamental Research Funds for the Central Universities under Grant 2014213020203.

Author Contributions: All authors made great contributions to the work. Jianchen Liu and Bingxuan Guo conceived this research. Jianchen Liu performed the experiments and wrote the paper. Bingxuan Guo analyzed data, provided advice and modified the paper. Wanshou Jiang, Weishu Gong and Xiongwu Xiao put forward valuable suggestions and help to revise the paper.

Conflicts of Interest: The authors declare no conflict of interest.

References

1. Peters, J. A look at making real-imagery 3D scenes (texture mapping with nadir and oblique aerial imagery). *Photogramm. Eng. Remote Sens.* **2015**, *81*, 535–536. [CrossRef]

2. Heiko, H. Stereo processing by semiglobal matching and mutual information. *IEEE Trans. Pattern Anal. Mach. Intell.* **2008**, *30*, 328–341.

3. Shahbazi, M.; Sohn, G.; Théau, J.; Menard, P. Development and evaluation of a UAV-photogrammetry system for precise 3D environmental modeling. *Sensors* **2015**, *15*, 27493–27524. [CrossRef] [PubMed]

4. Kim, J.-I.; Kim, T. Comparison of computer vision and photogrammetric approaches for epipolar resampling of image sequence. *Sensors* **2016**, *16*, 412. [CrossRef] [PubMed]

5. Sun, C. Trinocular stereo image rectification in closed-form only using fundamental matrices. In Proceedings of the 2013 20th IEEE International Conference on Image Processing (ICIP), Melbourne, Victoria, Australia, 15–18 September 2013; pp. 2212–2216.

6. Chen, Z.; Wu, C.; Tsui, H.T. A new image rectification algorithm. *Pattern Recognit. Lett.* **2003**, *24*, 251–260. [CrossRef]

7. Oram, D. Rectification for any epipolar geometry. In Proceedings of the British Machine Vision Conference, Cardiff, UK, 2–5 September 2002.

8. Pollefeys, M.; Koch, R.; Van Gool, L. A simple and efficient rectification method for general motion. In Proceedings of the Seventh IEEE International Conference on Computer Vision, Corfu, Greece, 20–27 September 1999; pp. 496–501.

9. Roy, S.; Meunier, J.; Cox, I.J. Cylindrical rectification to minimize epipolar distortion. In Proceedings of 1997 IEEE Computer Society Conference on Computer Vision and Pattern Recognition (CVPR'97), San Juan, Puerto Rico, 17–19 June 1997; p. 393.

10. Fujiki, J.; Torii, A.; Akaho, S. *Epipolar Geometry via Rectification of Spherical Images*; Springer: Berlin/Heidelberg, Germany, 2007; pp. 461–471.

11. Hartley, R.; Gupta, R. Computing matched-epipolar projections. In Proceedings of the IEEE Computer Society Conference on Computer Vision and Pattern Recognition, New York, NY, USA, 15–17 June 1993; pp. 549–555.

12. Hartley, R.I. Theory and practice of projective rectification. *Int. J. Comput. Vis.* **1999**, *35*, 1–16. [CrossRef]

13. Loop, C.; Zhang, Z. *Computing Rectifying Homographies for Stereo Vision*; Microsoft Research Technical Report MSR-TR-99-21; Institute of Electrical and Electronics Engineers, Inc.: New York, NY, USA, 1999.

14. Wu, H.H.P.; Yu, Y.H. Projective rectification with reduced geometric distortion for stereo vision and stereoscopic video. *J. Intell. Robot. Syst.* **2005**, *42*, 71–94. [CrossRef]

15. Mallon, J.; Whelan, P.F. Projective rectification from the fundamental matrix. *Image Vis. Comput.* **2005**, *23*, 643–650. [CrossRef]

16. Fusiello, A.; Irsara, L. Quasi-euclidean epipolar rectification of uncalibrated images. *Mach. Vis. Appl.* **2011**, *22*, 663–670. [CrossRef]

17. Monasse, P.; Morel, J.M.; Tang, Z. *Three-Step Image Rectification*; BMVA Press: Durham, UK, 2010; pp. 1–10.

18. Al-Zahrani, A.; Ipson, S.S.; Haigh, J.G.B. Applications of a direct algorithm for the rectification of uncalibrated images. *Inf. Sci.* **2004**, *160*, 53–71. [CrossRef]

19. Fusiello, A.; Trucco, E.; Verri, A. A compact algorithm for rectification of stereo pairs. *Mach. Vis. Appl.* **2000**, *12*, 16–22. [CrossRef]

20. Kang, Y.S.; Ho, Y.S. *Efficient Stereo Image Rectification Method Using Horizontal Baseline*; Springer: Berlin/Heidelberg, Germany, 2011; pp. 301–310.

21. Wang, H.; Shen, S.; Lu, X. Comparison of the camera calibration between photogrammetry and computer vision. In Proceedings of the International Conference on System Science and Engineering, Dalian, China, 30 June–2 July 2012; pp. 358–362.

22. Kedzierski, M.; Delis, P. Fast orientation of video images of buildings acquired from a UAV without stabilization. *Sensors* **2016**, *16*, 951. [CrossRef] [PubMed]

23. Triggs, B.; Mclauchlan, P.F.; Hartley, R.I.; Fitzgibbon, A.W. *Bundle Adjustment—A Modern Synthesis*; Springer: Berlin/Heidelberg, Germany, 1999; pp. 298–372.

24. Rothermel, M.; Wenzel, K.; Fritsch, D.; Haala, N. Sure: Photogrammetric surface reconstruction from imagery. In Proceedings of the LC3D Workshop, Berlin, Germany, 4–5 December 2012.

sensors

MDPI

Article

A Novel System for Correction of Relative Angular Displacement between Airborne Platform and UAV in Target Localization

Chenglong Liu [1,2], Jinghong Liu [1,*], Yueming Song [1] and Huaidan Liang [1,2]

[1]   Changchun Institute of Optics, Fine Mechanics and Physics, Chinese Academy of Sciences,
     Changchun 130033, China; liuchenglong14@mails.ucas.ac.cn (C.L.); songyueming@ciomp.ac.cn (Y.S.);
     lianghuaidan14@mails.ucas.ac.cn (H.L.)
[2]   University of Chinese Academy of Sciences, Beijing 100049, China
*   Correspondence: liujinghong@ciomp.ac.cn; Tel.: +86-431-8617-6159

Academic Editors: Felipe Gonzalez Toro and Antonios Tsourdos
Received: 30 December 2016; Accepted: 27 February 2017; Published: 4 March 2017

Abstract: This paper provides a system and method for correction of relative angular displacements between an Unmanned Aerial Vehicle (UAV) and its onboard strap-down photoelectric platform to improve localization accuracy. Because the angular displacements have an influence on the final accuracy, by attaching a measuring system to the platform, the texture image of platform base bulkhead can be collected in a real-time manner. Through the image registration, the displacement vector of the platform relative to its bulkhead can be calculated to further determine angular displacements. After being decomposed and superposed on the three attitude angles of the UAV, the angular displacements can reduce the coordinate transformation errors and thus improve the localization accuracy. Even a simple kind of method can improve the localization accuracy by 14.3%.

Keywords: UAV; target localization; shock absorber; angular displacement; image registration; coordinate transformation

---

1. Introduction

Currently, enemy situation reconnaissance, target localization, directing and adjusting artillery fire, and other auxiliary functions are still the main UAV applications. With a very low safety risk [1], an operator can remotely operate an UAV to fly toward the target area in order to acquire the target's real-time image and location information, which can be sent to the control center for analysis and decision making by intelligence analysts and commanders. In several recent wars, UAV has played a key role in situations where it is used for real-time battlefield reconnaissance, for collecting and providing intelligence, and for providing accurate target information to facilitate firing. For civil use, such as search and rescue [2], target localization is also the important UAV applications.

The basic principle of current target localization is the R-θ (remove-angle) method [3–7], in which the distance (R) of the target relative to the UAV is determined by a laser rangefinder and the angle (θ) is determined by a series of sensors. Based on its own location (usually using the Earth coordinate system), the position of a target in the geodetic coordinate system can be obtained after a series of coordinate transformations. The common target localization process is usually done through the transformations among at least five coordinate systems, including the camera coordinate system C, UAV coordinate system (platform) B, UAV geographic coordinate system V, Earth-Centered Earth-Fixed coordinate system (ECEF) E (in line with the WGS-84 standard) and the geodetic coordinate system G (in line with the WGS-84 standard). The platform moves independently of the UAV, detects the target through rotating search, and outputs the azimuth and pitch information of the detected target relative to the UAV through continuous tracking [8]. Then these data are linked to the INS data of the UAV and converted into a geodetic coordinate system the same as the GPS standard.

Based on the combination of camera and laser range finder, this paper proposes a method to improve the accuracy of angles, which could improve the accuracy of single-point localization in real time. As the operating frequency of laser rangefinders in a single range-based localization task is quite low (usually at the Hz level), the system designed by this paper can be considered real-time provided that it can complete the ranging accuracy optimization in the range measurement period. In addition, this system takes up a small space, without a significant increase in cost. It can effectively improve the accuracy of one-time single-station localization and won't interfere with most of the multi-times multi-station methods, so it is quite universal. At present, researchers have used various methods to analyze the localization error and improve the localization accuracy. Currently, some people are studying how to improve the single-aircraft single-point localization accuracy, while others are using the methods such as flight course planning and multi-aircraft localization to realize accuracy improvement. Most of the work done by current researchers fails to consider the deviation angle of the platform relative to the UAV. What has been considered, if possible, is only the addition of a parallel translation or the analysis of error influence, which, however, was seldom quantified. Redding [3] considered the influence of wind resistance, but the angular displacements caused by UAV attitude change were left out of consideration. The system in this paper can correct the angular displacements between the UAV and its platform when the cause of angular displacement error doesn't need to be known. The method proposed by Pachter et al. [9] also needs to know the target elevation in advance. Chiang et al. [10] improved the localization accuracy through setting the ground control point. Yue et al. [11] proposed the use of a height-based Steepest Descent Method for single-aircraft localization optimization, which, however, needs time and fails to meet the real-time requirement. While studying the single-aircraft localization, some people have begun to explore multi-aircraft localization. Morbidi et al. [12] proposed the method of active target tracking and joint localization based on an UAV fleet to estimate the target position through a Kalman filter. Qu et al. [13] proposed a joint localization method based on the azimuth angles among various UAVs. These methods, which are based on sensor data fusion, involve the complex issue of fleet route control [14–16] and need coordination among multiple UAVs, thus resulting in a higher hardware cost, more task time in most of the cases, and a greater risk in case of emergency. These exceptions apart, methods based on Kalman filter, Recursive Least Squares (RLS) filter, nonlinear filter [3,6,17,18] and methods based on video sequence [19–24] are also proposed to estimate the location. Most of these use the same aircraft at different time to improve the accuracy, but the optimization process requires time, that cannot meet the real-time requirement. On the other hand, the self-localization method integrating IMU into the platform involves the issue of size restriction. For a smaller IMU, its accuracy can't meet the requirement easily so that an additional onboard IMU whose size is larger is needed, which will increase the load.

This paper is structured as follows: first, we build the target localization model. Then the principle and working process of the system designed in this paper are introduced in detail, followed by effectiveness analysis, experiments, simulation and verification, analysis of verification results, and finally a summary.

2. Methodology

The core process of the target localization method adopted by this paper, is shown in Figure 1.

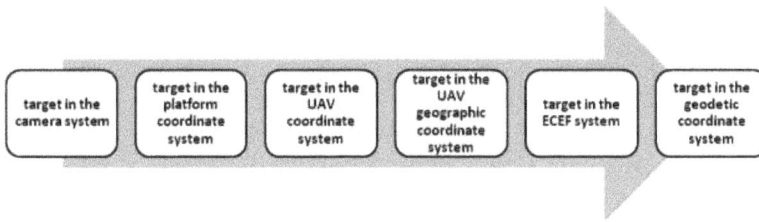

Figure 1. Localization process.

Compared with most of the target localization models [25], the main advantage of this method is the separation of the platform coordinate system from UAV coordinate system [26]. The addition of transformational matrix has effectively reduced the directional error. This is more evident during the high-pitch big-slope reconnaissance. Take a large UAV whose rising limit is 8 km as an example, In the case of vertical down-view, the localization error brought by 1 mrad of angle error is only 8 m, but during the oblique-view reconnaissance, the slope distance can easily reach 30 km, where the localization error brought by the same angle error is as big as 30 m. As shown in Figure 2, the bigger the slope distances from the platform to the target, the higher the directional requirement. The specific localization process is as follows.

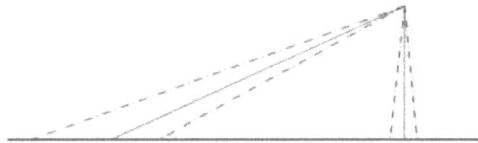

Figure 2. Different localization errors resulting from the same angle error.

2.1. Transformation from Camera Coordinate System C to Platform Coordinate System P

The relationship between the two coordinate systems is illustrated in Figure 3.

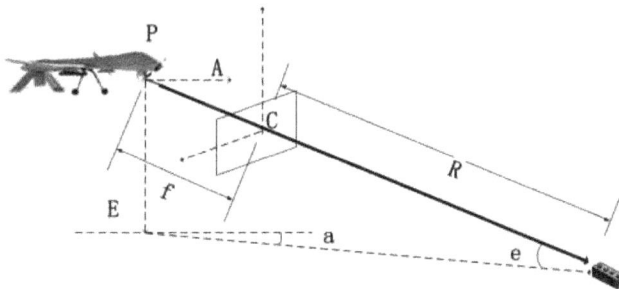

Figure 3. Relationship between camera coordinate system C and platform coordinate system P.

The homogeneous coordinates of a target in the camera coordinate system are:

$$[x_c \ y_c \ z_c \ 1]^T = [u \ v \ f \ 1]^T, \tag{1}$$

where u and v are the target's coordinates in the image (in pixel), and f is the current focal length of camera. Usually, when the UAV is detecting a target, the photoelectric platform will lock the detected target at the Field of View (FOV) center with multiple pixels and therefore the target can be considered at the image center. When the error inside camera coordinate system is ignored, the homogeneous coordinates of the target can be expressed as $[0\ 0\ f\ 1]^T$. A photoelectric platform is the camera carrier, which outputs the information on the angles a and e between Line of Sight (LOS) and the zero positions of two platform angles and measures the target distance R through a laser range finder. Since the platform uses the polar coordinate system, a coordinate transformation listed below is needed:

$$[x_p\ y_p\ z_p\ 1]^T = R \times Q_{pc}[0\ 0\ f\ 1]^T, \tag{2}$$

where Q_{pc} is the conversion matrix from the camera coordinate system C to the platform coordinate system P.

$$Q_{pc} = \begin{bmatrix} \cos a & 0 & \sin a & 0 \\ 0 & 1 & 0 & 0 \\ -\sin a & 0 & \cos a & 0 \\ 0 & 0 & 0 & 1 \end{bmatrix} \begin{bmatrix} 1 & 0 & 0 & 0 \\ 0 & \cos e & -\sin e & 0 \\ 0 & \sin e & \cos e & 0 \\ 0 & 0 & 0 & 1 \end{bmatrix}. \tag{3}$$

2.2. Transformation from Platform Coordinate System P to UAV Coordinate System B

This transformation process is the core of localization accuracy improvement over other methods [27]. Considering the demand for fast disconnection, the airborne platform is usually attached to the UAV in a strap-down manner. Most of the platforms have a shafting structure so that the onboard imaging systems (such as cameras and IR thermal imagers) can expand the reconnaissance field through rotation. Therefore, most of the platforms can be divided into two parts, namely the base and rotating part. The base is fixed to the UAV. The rotating part is linked, through shafting, to the base, putting the imaging system in motion to search for the target and locking the target to LOS via the servo system [28]. To improve the reconnaissance imaging quality, the base of an airborne platform often needs to be fixed to the UAV through a shock absorber, thus isolating part of high-frequency vibration and enhancing the stability of the platform itself, as show in Figure 4. However, the damping structure used by most of the shock absorber body is a flexible material, so in actual flight, the platform will produce a displacements relative to the UAV due to the influence of such factors as engine vibration, wind resistance, UAV attitude change and motion of platform. These displacements include monolithic translation, angular displacement and mixed displacement.

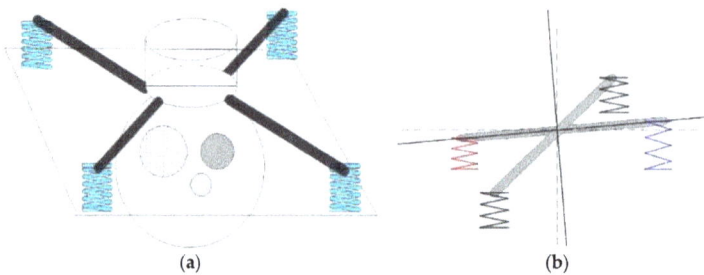

(a) (b)

Figure 4. (a) Position of shock absorber; (b) Deformation of shock absorber.

The target is usually locked in the center of the image (just like when you need to turn your head around to face a target and turn your eyes to the target so that you can look at it attentively) during flight. To determine the target angle we need to know the angle of the detected target relative to the platform, as output by the platform, and the angle of the platform itself relative to reference azimuth angle. Then the required angle can be obtained through transformation. For example, if a stone lies on the ground east-northeast of you, you need to know both the angle of the stone relative to your body and the angle of your body relative to the East or North.

The Inertial Measurement Unit (IMU) can determine the UAV attitude angles. As stated above, the size of IMU will be larger if we need a high-precision one. Accordingly, the IMU in most of the cases is installed not on the platform, but in other UAV compartments through the rigid connection due to the limitation of platform size. Otherwise, the platform will be too large to be installed.

Through the adjustment of airborne IMU angle during initial assembly, the reference 0° direction of the platform is considered consistent with the 0° direction of the UAV, but with the platform translation available, the 0° directions of platform might deflect. In this case, the 0° lines of the platform will move and/or rotate. Among all types of rotation, only the transition along the indication line has a negligible influence on localization accuracy. The non-parallel translation and rotation in any other direction, including the rotation around the zero line, will have a great influence on localization result. When the deflection angle is non-zero, the resolving based on the pitch and azimuth angles output by the platform is actually inaccurate [29]. That is, an error exists in the obtained target relative to the reference direction. For example, you think your eyes are 30° to the left in front of your body, but actually it is 29°. At this time, you need to have an error compensation factor (1°) to correct that. The transformation from the platform coordinate system P to the UAV coordinate system B is shown in Figure 5.

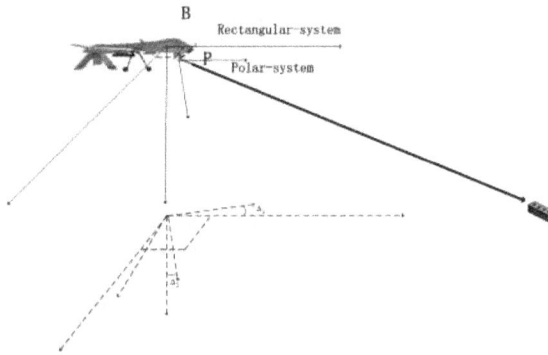

Figure 5. Relationship between platform coordinate system P and UAV coordinate system B.

To determine the transformation relation, the direct relative angular displacements Δ1 and Δ2 between platform and UAV must be determined first. For the concrete method, see the subsequent system description. Since this is a transformation from platform to UAV, converting the determined angular displacements directly into the UAV coordinate system can effectively reduce the workload. Then the angular displacements can be expressed as Δα, Δβ, Δγ and the transformation relation expressed as:

$$[x_b \ y_b \ z_b \ 1]^T = Q_{bp}[x_p \ y_p \ z_p \ 1]^T = R \times Q_{bp}Q_{pc}[0 \ 0 \ f \ 1]^T, \tag{4}$$

$$Q_{bp} = \begin{bmatrix} \cos \Delta\alpha & 0 & \sin \Delta\alpha & 0 \\ 0 & 1 & 0 & 0 \\ -\sin \Delta\alpha & 0 & \cos \Delta\alpha & 0 \\ 0 & 0 & 0 & 1 \end{bmatrix} \begin{bmatrix} 1 & 0 & 0 & 0 \\ 0 & \cos \Delta\beta & -\sin \Delta\beta & 0 \\ 0 & \sin \Delta\beta & \cos \Delta\beta & 0 \\ 0 & 0 & 0 & 1 \end{bmatrix} \begin{bmatrix} \cos \Delta\gamma & -\sin \Delta\gamma & 0 & 0 \\ \sin \Delta\gamma & \cos \Delta\gamma & 0 & 0 \\ 0 & 0 & 1 & 0 \\ 0 & 0 & 0 & 1 \end{bmatrix}, \tag{5}$$

where Q_{bp} is the conversion matrix from the platform coordinate system P to the UAV coordinate system B.

2.3. Transformation from UAV Coordinate System B to UAV Geographic Coordinate System V

The relationship between UAV coordinate system B and UAV geographic coordinate system V is illustrated in Figure 6.

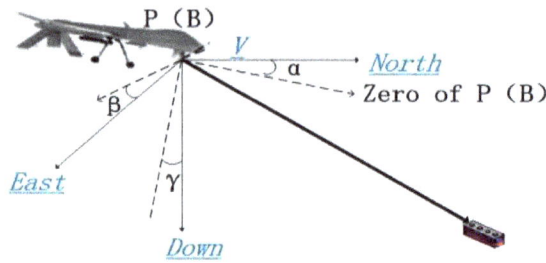

Figure 6. Relationship between UAV coordinate system B to UAV geographic coordinate system V.

The UAV geographic coordinate system V in this paper is defined as (North-East-Down) NED, of which the three axes are North Pole, due east and the earth's core respectively. The transformation relation in this process is:

$$[x_v \ y_v \ z_v \ 1]^T = Q_{vb}[x_b \ y_b \ z_b \ 1]^T, \tag{6}$$

$$Q_{vb} = \begin{bmatrix} \cos\gamma & -\sin\gamma & 0 & 0 \\ \sin\gamma & \cos\gamma & 0 & 0 \\ 0 & 0 & 1 & 0 \\ 0 & 0 & 0 & 1 \end{bmatrix} \begin{bmatrix} 1 & 0 & 0 & 0 \\ 0 & \cos\beta & -\sin\beta & 0 \\ 0 & \sin\beta & \cos\beta & 0 \\ 0 & 0 & 0 & 1 \end{bmatrix} \begin{bmatrix} \cos\alpha & 0 & -\sin\alpha & 0 \\ 0 & 1 & 0 & 0 \\ \sin\alpha & 0 & \cos\alpha & 0 \\ 0 & 0 & 0 & 1 \end{bmatrix}, \tag{7}$$

where Q_{vb} is the conversion matrix from the UAV coordinate system B to UAV geographic coordinate system V.

2.4. Transformation from UAV Geographic Coordinate System V to ECEF System E

The relationship between UAV geographic coordinate system V and ECEF system E is illustrated in Figure 7.

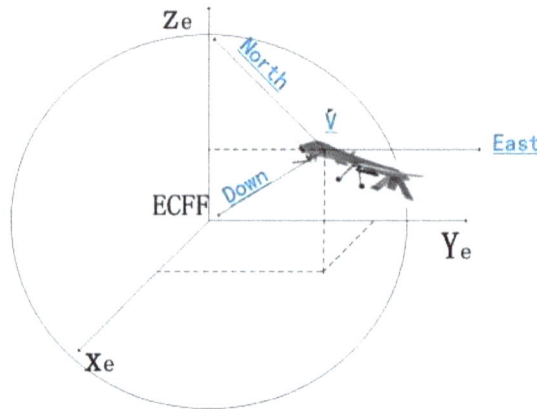

Figure 7. Relationship between UAV geographic coordinate system V and ECEF system E.

In the ECEF system, the origin is the center of mass of the earth, the axis z_e points from the Earth's spin axis to the North Pole, the axis x_e points to the intersection between prime meridian and equator, and all the three axes, namely y_e, x_e and z_e, jointly constitute a right-handed coordinate system. The transformation relation in this process is:

$$[x_e \ y_e \ z_e \ 1]^T = Q_{ev}[x_v \ y_v \ z_v \ 1]^T, \tag{8}$$

$$Q_{ev} = \begin{bmatrix} \cos\gamma & -\sin\gamma & 0 & 0 \\ \sin\gamma & \cos\gamma & 0 & 0 \\ 0 & 0 & 1 & 0 \\ 0 & 0 & 0 & 1 \end{bmatrix} \begin{bmatrix} 1 & 0 & 0 & 0 \\ 0 & 1 & 0 & 0 \\ 0 & 0 & 1 & (N+h) \\ 0 & 0 & 0 & 1 \end{bmatrix} \begin{bmatrix} \cos\alpha & 0 & -\sin\alpha & 0 \\ 0 & 1 & 0 & 0 \\ \sin\alpha & 0 & \cos\alpha & 0 \\ 0 & 0 & 0 & 1 \end{bmatrix}, \tag{9}$$

where Q_{ev} is the conversion matrix from the UAV geographic coordinate system V to the ECEF system E, and N is the radius of curvature in the prime vertical of the Earth. The semi-major axis of ellipsoid is a = 6,378,137 m, the semi-minor axis is b = 6,356,752 m, the first eccentricity of spheroid is:

$$e = \frac{\sqrt{a^2 - b^2}}{a}, \tag{10}$$

and the second eccentricity of the spheroid is:

$$e' = \frac{\sqrt{a^2 - b^2}}{b}, \tag{11}$$

By combining them with the current latitude, we can obtain:

$$N = \frac{a}{\sqrt{1 - e_e^2 \sin^2 M}}. \tag{12}$$

2.5. Transformation from ECEF Frame E to Geodetic Coordinate System G

The relationship between the ECEF frame E and geodetic coordinate system G is shown in Figure 8.

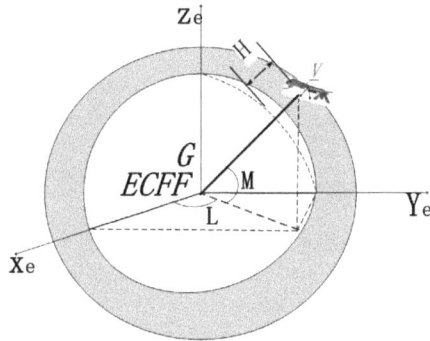

Figure 8. Relationship between ECEF system E and geodetic coordinate system G.

The geodetic coordinate system is spherical. In this frame, the origin is also the center of mass of the Earth, the axis z_e points from the Earth's spin axis to the North Pole, the axis x_e points to the intersection between prime meridian and equator, and all the three axes, namely y_e, x_e and z_e, jointly constitute a right-handed coordinate system. The transformation relation in this process is:

$$[x_g \ y_g \ z_g \ 1]^T = Q_{ge}[x_e \ y_e \ z_e \ 1]^T + [0 \ 0 \ Ne^2 \sin M \ 0]^T \tag{13}$$

$$Q_{ge} = \begin{bmatrix} 1 & 0 & 0 & 0 \\ 0 & \cos M & -\sin\gamma & 0 \\ 0 & \sin\gamma & \cos\gamma & 0 \\ 0 & 0 & 0 & 1 \end{bmatrix} \begin{bmatrix} \cos L & 0 & -\sin L & 0 \\ 0 & 1 & 0 & 0 \\ \sin L & 0 & \cos L & 0 \\ 0 & 0 & 0 & 1 \end{bmatrix} \begin{bmatrix} 1 & 0 & 0 & 0 \\ 0 & 1 & 0 & 0 \\ 0 & 0 & 1 & (N+h) \\ 0 & 0 & 0 & 1 \end{bmatrix} \tag{14}$$

where Q_{ge} is the conversion matrix from the ECFF system E to the geodetic coordinate system G.

A simpler calculation method is to obtain the ECEF system parameters at first, and then to calculate directly various parameters of geodetic coordinate system by using the following equation:

$$\begin{cases} M = \arctan(\frac{z_e + be'^2 \sin^3 U}{\sqrt{x_e^2 + y_e^2} - ae^2 \cos^3 U}) \\ L = \arctan\left(\frac{y_e}{x_e}\right) \\ H = \frac{\sqrt{x_e^2 + y_e^2}}{\cos M} - \frac{a}{\sqrt{1 - e^2 \sin^2 M}} \end{cases} , \tag{15}$$

$$U = \arctan(\frac{az_e}{b\sqrt{x_e^2 + y_e^2}}). \tag{16}$$

Thus, the target position (M, L, H) in the geodetic coordinate system G can be obtained. During the transformation, different kinds of errors are possible, which are summarized in Table 1.

Table 1. Sources of localization error.

Type of Function	Type of Subsystem	Influence Factor
Directional error of photoelectric platform	Optical system	Parallelism and conformance of optical axis
	Mechanical frame	Error in design and installation
	Control system	Error in stabilization and tracking system
	Others	Deformation, vibration, electromagnet interference, wear and tear etc.
Alignment error of photoelectric platform and INS	Installation alignment	Initial directional and horizontal leveling
	Shock absorber	High-frequency angular vibration and low-frequency shaking
Error in UAV (INS) motion parameter	Attitude measurement (INS)	UAV attitude measurement error
	Position measurement	UAV localization error
Range error	Laser range finder	Measurement error of range finder
Coordinate transformation error	Error in the transformations among different geodetic coordinate systems	

These errors have been studied by many scholars [18,30]. This paper only discusses the angular displacements between the platform and UAV, while processing other errors generally in the subsequent simulation.

3. System Composition and Working Principle

In this paper, the method to correct the relative angular displacement between the platform and the carrier UAV is to separate the platform coordinate system and the UAV coordinate system by adding a transformation matrix to reduce the angular displacement arising from the inconsistent deformation in shock absorber. Decomposing and superposing the angular displacements to the attitude angles of the UAV on the basis of the UAV can reduce the error in coordinate conversion and thus improve the localization accuracy [6].

In order to reduce the hardware requirements on the UAV, this paper assumes that the calculation is carried out by a command station located on the ground. The plane just needs to give its own position and the angle of the target, and the system designed in this paper is only used to measure the amount of angular displacement between the platform and the UAV. The localization process (including the coordinate transformation process) is carried out on the ground through the received data sent by the UAV. The advantage is that on the one hand the hardware consumption of the UAV much less, we can reduce the load weight; on the other hand, computers on the ground can achieve higher accuracy and faster speed.

3.1. System Composition

A combination of image registration and coordinate transformation is used to obtain the deflection angle of the platform relative to the UAV. The schematic diagram of the system is shown in Figure 9. For convenience of reading, only the front and rear bulkheads are shown, while the left and right bulkheads are omitted.

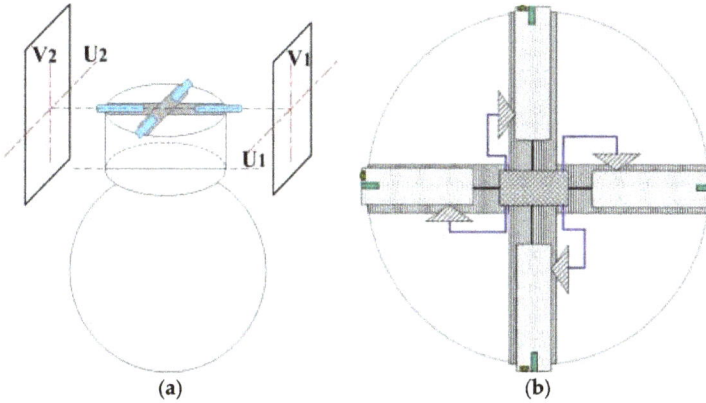

Figure 9. (**a**) System installation position; (**b**) Working principle diagram, top view.

The system for correcting the relative angular displacements between the UAV and its onboard platform includes a cruciform main frame, movable probes in the frame, a LED lighting system and a CMOS high-speed imaging system on each probe, a position measurement system, and an information processing system on the central PCB board. The cross frame is fixed on the top of the airborne platform base through a rigid connection, and its center is a PCB circuit board for processing information in real time.

Each probe is in the same plane and each of the outer ends of each probe is provided with an LED illumination system and a CMOS imaging system. Since the bulkhead is semi-closed and dimmed, the LED is used to illuminate the bulkhead area at which each probe is aimed. The imaging system is used for real-time high-frequency imaging of the bulkhead area.

The position measuring system is used for measuring the probe position in each arm and sending the position information to the information processing system, which calculates the total length of each arm according to the probe position information sent by position measurement system and at the same time, carries out registration and comparison with the initial image texture according to a series of bulkhead images obtained by the imaging system. After that, the displacement vectors in the image after registration could be obtained by comparing each center of the two images that has been encoded in a special way. The displacement of the probe relative to the initial position in the plane of the bulkhead can be obtained by a scale calculated following.

Through comprehensive analysis, the angular displacement of the cruciform frame relative to the UAV can be determined. Because the frame and the platform are rigidly connected, this angular displacement can be considered as the angular displacement of the platform relative to the UAV. In the subsequent coordinate conversion, the platform coordinate frame P is separated from the UAV coordinate system B, and a rotation matrix of the platform relative to the UAV is added. The coefficient of this matrix corrects the directional error of the target, thus improving the accuracy of target localization.

3.2. *The Specific Work Process*

3.2.1. Step 1: Calibrate the Initial Position of the Probes

The pattern is pre-printed on the four bulkheads of the payload capsule. When the zero installation line of the airborne platform coincides with the zero attitude angle of the UAV through adjustment, we fix the platform to the UAV and lift or lower it to its actual working position. Let the center of the frame vertically coincide with the geometric center of the polygon consisting of the connection lines of shock absorbers, and then fix the cruciform frame to the platform base. As the IMU and the platform have the same zero-line direction, it can be considered that there is only a parallel translation between platform coordinate system P and UAV coordinate system B at this time. This translation can be measured on the ground through side projection, but can be left out of consideration considering that the installation position of IMU is generally close to the platform, with only a small impact (on the order of cm) on final localization result. The probe was adjusted to be close to the four sides of the bulkhead and maintain a certain distance. The distance between the front and rear probes is denoted as l_{12}, and the distance between the left and right probes is l_{34}. Four illumination systems are then lit to illuminate the bulkhead walls that the probe is facing accordingly. The four CMOS imaging systems collect the image texture of the bulkhead walls and store them in the flash memory on the PCB. The four images at this time are referred to as reference images. Observe the top view of the UAV along the flight direction. The front and rear probes are defined as the front and rear probes, and the left and right probes are defined as the left probe and the right probe respectively. Take the center of the reference image of each bulkhead wall corresponding to each probe as the origin of a coordinate system, the vertical axis as the y axis, and the horizontal direction as the x axis. Four 2D rectangular coordinate systems are set on the plane of bulkhead walls, defined as $x_1O_1y_1$, $x_2O_2y_2$, $x_3O_3y_3$ and $x_4O_4y_4$. The significance of the bulkhead coordinate system is that the transformation matrix is obtained by differential calculation. And the bulkhead coordinate system does not appear in the subsequent coordinate transformation.

3.2.2. Step 2: Real Time Acquisition and Analysis of Displacement

During the flight, due to the effect of turning, bumps, resistance and other factors, the platform and the cross frame fixed on it would be displaced relative to the UAV. Each CMOS imaging system takes photos for the corresponding bulkhead wall at high frequency. The DSP chip in the information processing system matches and compares the texture of the current image with that of the reference image, and gives the displacement vector (in pixels) of the current image center relative to its original position. After that, we can obtain the displacement vector of every probe projection in the wall plane relative to its own reference origin through conversion using a factor λ. By combining the obtained displacement vector with the length of probe arm, the angular displacements of the frame relative to the UAV can be determined. The angular displacements include the displacements of pitch angle, roll angle and azimuth angle.

3.2.3. Step 3: Error Correction

The angular displacement of the frame relative to the UAV, as obtained by Step 2, will be transmitted to the receiver on the ground along with other date. The transformation matrix Q_{bp} is generated. Error correction is carried out during the localization process.

4. Validity Analysis

4.1. *The Shock Absorber*

The shock absorber used is shown in Figure 10. The black part is made from rubber. Figure 11 shows a side and top view of the displacement of the shock absorber. The radius of the model shock absorber is 1.25 cm. Its deformation in a plane will not exceed its radius, otherwise the shock absorber has been damaged. Here we take the limit value of 1 cm. As the platform is fixed to four shock absorbers, the final angular displacement T will not be large. The square edge of the stent structure is 25 cm. After projection in the plane, with $|X_1| \leq 1$ cm and $|X_2| \leq 1$ cm, the limit value of angular displacement:

$$\theta_{max} = \arctan\frac{|X_1| + |X_2|}{L} \leq 4.57°. \tag{17}$$

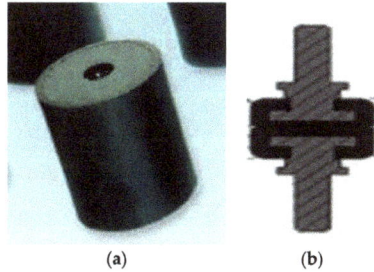

(a)   (b)

Figure 10. (**a**) Real shock absorber; (**b**) Structure of the shock absorber.

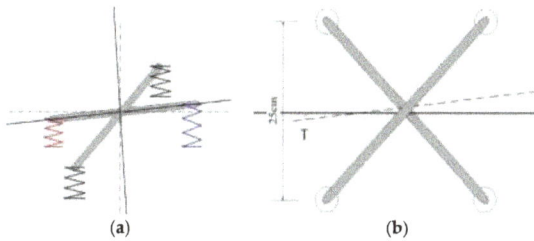

(a)   (b)

Figure 11.   (**a**) Side view of the shock absorber deformation; (**b**) top view of the shock absorber deformation.

Of course, this is a kind of very extreme situation. With the four shock absorbers working together, each single shock absorber will not deform to such a large extent.

After the introduction, verify the deformation and angular displacement. Figure 12 shows just the change in the length of absorber by comparing the length before and after the platform is placed on an inclined plane. It shows that there is indeed deformation.

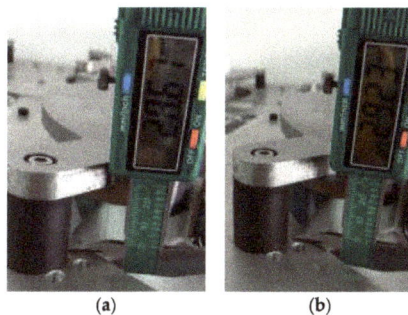

(a)   (b)

Figure 12. (**a**) Length of absorber before inclination; (**b**) Length of absorber after inclination.

The change in length has been 0.56 mm can be measured with the inclination angle of the inclined plane is merely 5°. According to the settings above (25 cm), the angular displacement can reach 0.1283°. It is easy to deduce that an angular displacement of 0.1° is common during the flight.

4.2. Image Registration

In the practical work, the image taken by the CMOS imaging system is sent to the information processing system. The DSP chip series [31] C64XX and SIFT algorithm [32] are used to calculate the number of pixels that shift between the reference image and the real-time image. Then, according to the distance between the imaging system and the bulkhead, the factor λ between the pixel and the actual distance is calculated.

As we adopt the pre-designed pattern, it is possible to control the size and kind of the pattern on the bulkhead. In view of the fact that the deformation of the shock absorber is not so great, a pattern having a size of 5 cm × 5 cm is enough. As the lens is close to the bulkhead, the distortion near the edge of the image caused by perspective projection will be large, so a small area in the center of the image is selected as the effective region to match, which can improve the registration speed and accuracy. Paste or print it on the area facing the probe on the bulkhead. We choose a pattern with simple feature, such as a variety of common graphics and directions inconsistent stripes (see Figure 13) to improve the efficiency of registration.

Figure 13. Pre-designed patterns.

The pattern can also ensure that a small area has a unique registration result, not mistakenly registered to other areas (for example, you would not match a small area that included both acute and right angles at the same time to another area that include both round and right angles), so it can ensure that any small area has a unique registration result.

4.3. The Transformation Matrix

By image registration, the offset is actually given in pixels. A scale factor λ must be converted to be the actual displacement m on the bulkhead. A schematic diagram of the pixel and actual displacement are shown in Figure 14. The conversion formula is:

$$\frac{l}{m} = \frac{f}{d} \tag{18}$$

where l is the distance on the CMOS sensor, m is the actual displacement size on the bulkhead, f is the focal length, d is the size of the lens on the probe from the bulkhead.

Take the pixel size of 5 µm as an example. If the focal length of the lens is 10 mm and the lens is 50 mm away from the wall, then the offset of a pixel corresponds to a displacement of 25 µm. That means the actual displacement is:

$$m = 25 \text{ µm} \times n \tag{19}$$

where n is the number of pixels offset.

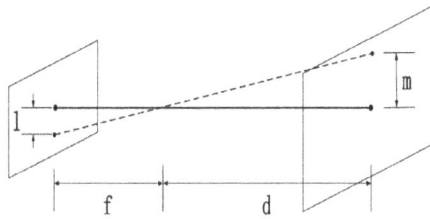

Figure 14. Schematic diagram of pixel and actual displacement.

Subsequent image offset calculation can continue to follow this scale factor. Then as long as the accuracy of image registration is 4 pixels, you can measure a displacement of 0.1 mm. The initial captured image is saved as a reference image, with its image center as the origin of the coordinates, after taking another image with displacement, the two images can be registered, we can see an offset has occurred in the image. This means that the lens has moved relative to its original position. A schematic diagram is illustrated in Figure 15.

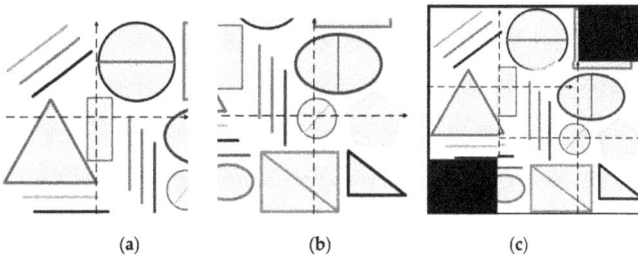

Figure 15. (**a**) Reference image; (**b**) displaced image; (**c**) image after registration.

When the imaging system moves in the direction of the probe, the real-time image is not on the same scale as the reference image. The registration process would adjust the image after zooming and rotation, and the final result refer to the reference image according to the principle of Scale Invariant Feature Transform (SIFT). Extracting the shape feature and scale in the moving image as invariants can ensure that the pixel-to-real distance scaling relationship is always valid [33].

After obtaining the displacement vector of the probe projecting in the plane of the bulkhead, the differential calculation is carried out to obtain the compensation matrix for correcting the error.

As defined above, the arm length between front probe and rear one is l_{12}, and the arm length between the left and right probe is l_{34}. The schematic for calculation of angular displacement size is shown in Figure 16. Pitch angle error is:

$$\Delta\alpha = \arctan\frac{x_1 - x_2}{l_{12}}. \tag{20}$$

Using the same method, we can calculate the roll angle error:

$$\Delta\beta = \arctan\frac{x_3 - x_4}{l_{34}}, \tag{21}$$

and yaw angle error:

$$\Delta\gamma = \arctan\frac{y_1 - y_2}{l_{12}}. \tag{22}$$

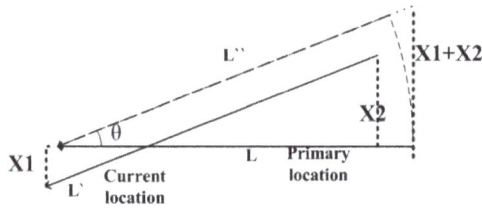

Figure 16. Schematic diagram of actual displacement conversion.

The transformation matrix used to compensate the offset angular displacement is then obtained:

$$Q_{bp} = \begin{bmatrix} \cos \Delta\alpha & 0 & \sin \Delta\alpha & 0 \\ 0 & 1 & 0 & 0 \\ -\sin \Delta\alpha & 0 & \cos \Delta\alpha & 0 \\ 0 & 0 & 0 & 1 \end{bmatrix} \begin{bmatrix} 1 & 0 & 0 & 0 \\ 0 & \cos \Delta\beta & -\sin \Delta\beta & 0 \\ 0 & \sin \Delta\beta & \cos \Delta\beta & 0 \\ 0 & 0 & 0 & 1 \end{bmatrix} \begin{bmatrix} \cos \Delta\gamma & -\sin \Delta\gamma & 0 & 0 \\ \sin \Delta\gamma & \cos \Delta\gamma & 0 & 0 \\ 0 & 0 & 1 & 0 \\ 0 & 0 & 0 & 1 \end{bmatrix}. \tag{23}$$

5. Experiments

5.1. The System

Limited to the experimental conditions and funding, this paper verified the system through a simulation of an actual flight in laboratory. The platform was fixed on the swing table shown in Figure 17, simulating the attitude change during flight.

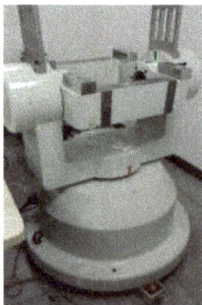

(a)

(b)

(c)

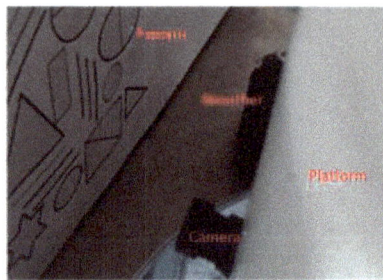

(d)

Figure 17. *Cont.*

(e)   (f)

Figure 17. (**a**) the swing table; (**b**) turntable; (**c**) the pattern picture pasted on the swing table; (**d**) position of each part; (**e**) the camera; (**f**) the DSP chip C64XX.

We use the multi-direction measuring turntable to measure angles under various attitudes After the vertical axis of the platform is tilted at various small angles with the horizontal plane, the error due to the deformation of the shock absorber can be measured because the center of gravity of the platform is far away from the base part in the vertical direction. The angle can reach $0.1°$ in the vertical direction. But the pitch angle can be set $0.05°$ for the upper limit value in the horizontal direction due to tangential deformation is small. Those parameters are set as the original error without optimization. According to the description above, the result of image registration will affect the measurement precision of probe displacement vector. At the same time, the registration time should be as short as possible to meet the real-time requirements [34]. In this paper, due to the use of a simple pattern, the registration algorithm used to be verified time-consuming in the 100-ms level. With the laser range finder limited in the level of several Hz, the solution can be considered real-time calculated. And by using the previous printed pattern, the image registration accuracy can reach the level of 2 pixels. According to above, 2 pixels correspond to the displacement of 0.05 mm.That means that we can measure an error of $0.01146°$. That is, the measurement accuracy can be $42''$. Compared with the original $0.1°$ error, the accuracy has been improved a lot.

5.2. Simulation

Simulation has been carried out in the localization process to prove it can improve the localization accuracy effectively. We used Monte Carlo method to simulate and analyze the target localization accuracy by using a simplified model. Taking into account the actual localization process, parameters that measured by the other sensors onboard are also with errors. So this simulation does not get rid of these errors except for the reference value. The parameters of each section are shown in Table 2. The errors were generated by the standard normal distribution function.

Table 2. Initial parameters and their error range.

Type	Focus	Laser Range	Azimuth Angle of Platform	Elevation Angle of Platform	Pitch	Roll	Yaw	Longitude	Latitude	Elevation of UAV
Reference	100 mm	12000 m	5°	−40°	0°	0°	0°	125.19°	43.54°	8000 m
Error range	5%	10 m	0.02865°	0.02865°	0.03°	0.03°	0.06°	0.0001°	0.0001°	10 m

At first, the angular displacement errors presented in this paper were not considered. The Monte Carlo method was used to generate 500 sets of localization parameters at random. The localization results are shown in Table 3 and Figure 18. The reference point was obtained with no angular displacement (set errors as $0°$). In order to reduce occasionality, we generate 500 sets of data and obtain the Root Mean Square Errors (RMSE) as the final result. Note that the process was done on the computer, because this system only needs to output a specific angular displacement, which conforms to the actual use of the entire UAV system environment. As mentioned earlier, the use of the computer on the ground will be faster and more accurate.

Table 3. Localization accuracy without angular displacement errors.

Configure	Longitude	Latitude	Elevation
Localization results	125.199925840984° E	43.6224190056630° N	293.18 m
Error (RMSE)	0.000170295986306348°	0.000119554074008099°	12.4911 m

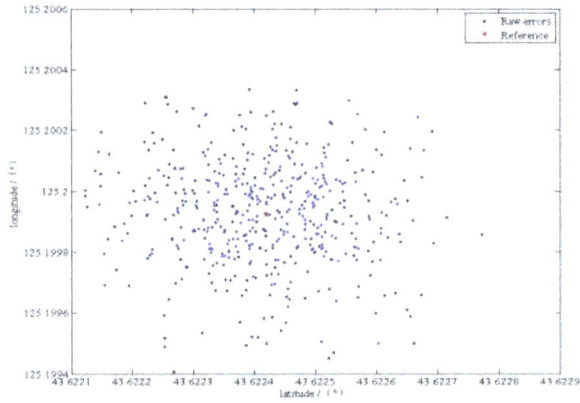

(a)

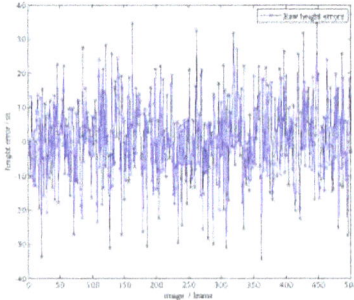

(b)

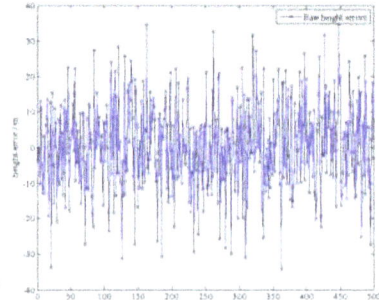

(c)

Figure 18. (a) The localization results in the plane without angular displacement errors; (b) the localization errors in the plane; (c) the localization errors in the elevation.

When taking the angular displacement errors shown in Table 4 between the platform and the carrier into account, the accuracy significantly decreased with the same other parameters. Results are shown in the Table 5 and Figure 19.

Table 4. The angular displacement errors without correction.

Angular Displacement Errors	Pitch	Roll	Yaw
Range	0.1°	0.1°	0.05°

Sensors **2017**, *17*, 510

Table 5. Results without correction.

Configure	Longitude	Latitude	Elevation
Error (RMSE)	0.000249456444498644°	0.000184640982639463°	21.53 m

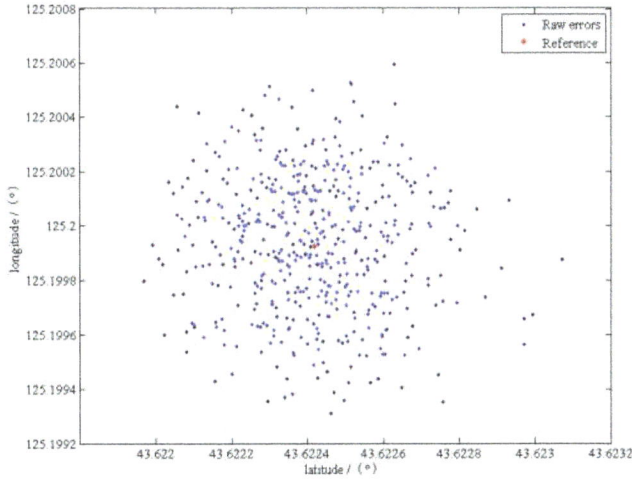

(a)

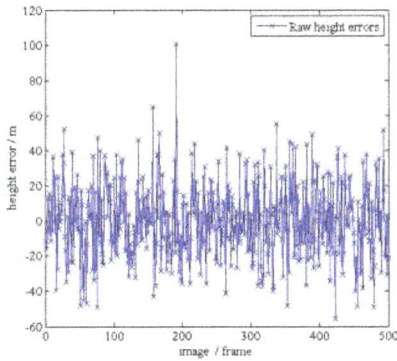

(b)

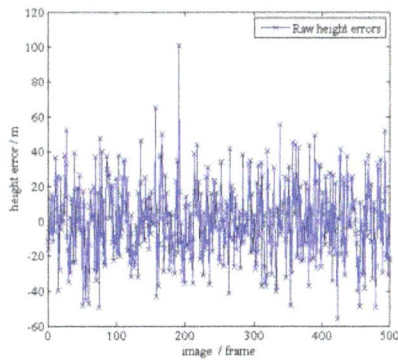

(c)

Figure 19. (a) The localization results in the plane with initial angular displacement errors; (b) the localization errors in the plane; (c) the localization errors in the elevation.

This is the case of a large angle to the horizontal direction. The angular displacement would impact greater if the angle is smaller [3]. The influence of angular displacement on the final localization results could be understood by comparison. When the UAV hovers above the target to reconnoiter it, the roll angle of UAV would be much larger. At this time, different grade deformations of shock absorber would greatly increase the inconsistency of the four distances and the center of gravity of platform is offset with the center of form. As a result, the angular displacement would be much greater than 0.1°. We adopted a conservative value when we carried out simulation in this paper. After correction of relative angular displacements between an UAV and its platform, we can obtain a series of more accurate localization results. As illustrated above, we can measure an error of 0.0115°, so we can set the angle errors to be 0.0115°, as shown in Table 6.

Table 6. The angular displacement errors with correction.

Angular Displacement Errors	Pitch	Roll	Yaw
Range	0.0115°	0.0115°	0.0115°

Results are shown in the Table 7 and Figure 20.

Table 7. Results with correction.

Configure	Pitch	Roll	Yaw
Error (RMSE)	0.000174500400122541°	0.000132910371952061°	14.3658 m

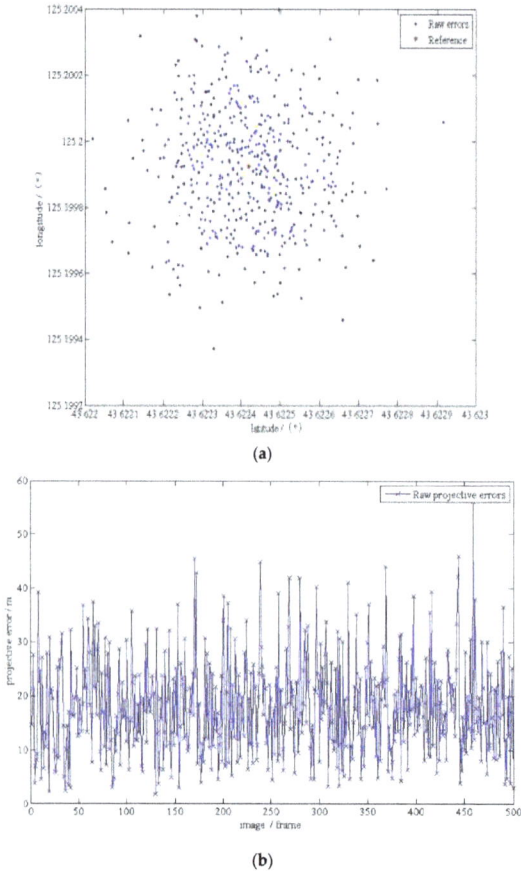

(a)

(b)

Figure 20. *Cont.*

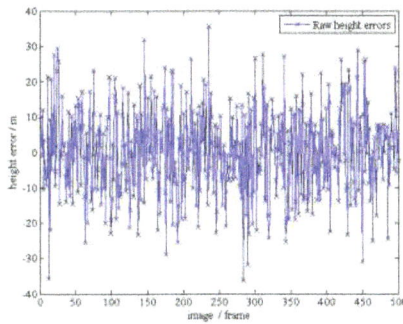

(c)

Figure 20. (a) The localization results in the plane with correction; (b) the localization errors in the plane; (c) the localization errors in the elevation.

We can obtain significant improvements by comparing the results. The platform was assumed to have an elevation angle of −40° (this elevation angle means the one that LOS of the platform relative to the target) in the simulation above. The influence of angular displacement on the final results can be seen by comparison. When the UAV reconnoiters target at a small elevation angle (set 0° in the horizontal plane), the target is almost in the front of the UAV. In this case, the directional errors—the angular errors—affect the final localization accuracy with a more significant factor.

Then we carried another simulation to analyze the effect of the system at different angles. We set the elevation angle at −60° to −20° as shown in Table 8 (0° in the horizontal plane) in the case of the remaining parameters of the same with Table 2. After obtaining the results without correction, we compare the results those have been corrected with them. The improvement can be obtained in this way:

$$I = \frac{|R - r|}{R},\tag{24}$$

where I represents the improvement; R represents the result without correction and r represents the results have been corrected.

Table 8. Improvement made by the system with different angle.

Current Angle	−60°	−50°	−40°	−30°	−25°	−20°
Improvement	39.12%	33.24%	31.13%	24.63%	18.56%	14.34%

Figure 21 shows the contribution of the angular displacement errors between the platform and the carrier UAV that have been corrected to final latitude errors (RMSE). The vertical axis represents the final latitude errors with a size of $10^{-5°}$.

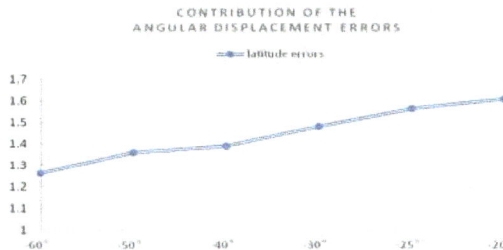

Figure 21. The contribution of the angular displacement errors to final latitude errors.

Figure 22 shows what percentage in the final latitude errors of the angular displacement errors are.

PERCENTAGE IN FINAL ERRORS

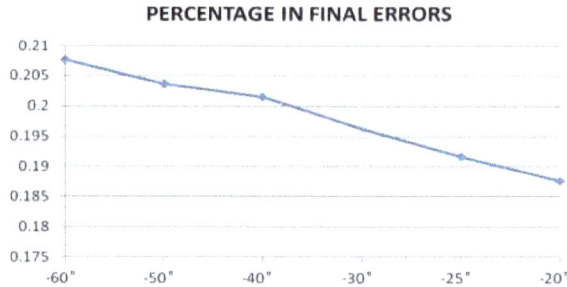

Figure 22. The percentage of the angular displacement errors in final latitude errors.

6. Discussion

We can conclude from Figure 20 that the effectiveness declines with the decreasing elevation angle, which has been illustrated in Figure 2. The percentage of the corrected errors in the final errors declines as the influence of other directing angle errors on the final result increases. Although the final localization accuracy declines significantly on the whole, we still achieve an improvement in accuracy by 14.34% as shown in Table 8. How to improve the accuracy on the whole with a small elevation angle should be considered in the future.

Simulation results show that the method proposed in this paper can effectively reduce the influence of angular displacement on the final localization results. However, the exploration must continue. Improving the accuracy of the system is still the most important. In this paper, the design of the system still has much room for improvement.

Firstly, in the equation of displacement measurement Equation (17), the resolution of measurement m can be smaller by reducing the distance between the probe and the bulkhead walls d and increasing the focal length f because the size of the pixel size l is limited by the CMOS sensor, which may not to be improved in a short period. However, it's easy to cause the probe to run into the bulkhead walls and get damaged in case of reducing d. Increasing the f will increasing the cost of the lens sharply. The specific control of d/f ratio should be studied in practice, but must not be blindly reduced.

Secondly, the registration accuracy of the system can be up to about 2 pixels currently. If the accuracy of a pixel can be improved, the accuracy of the measurement can be doubled. However, in order to meet the requirement of real-time calculation, the registration algorithm cannot be too complex, otherwise the time will be too long and that will increase the burden of calculation, and even result in data flow alternation errors. Taking the 2 Hz laser range finder as an example, this system needs to control the speed of the pattern registration within 500 ms, and better within 200 ms considering the clock signal synchronization, data transmission and other issues. A lot of efforts are still needed to study how to design a better and faster algorithm based on this system or to find the best predesigned pattern in order to improve the registration speed and accuracy.

7. Conclusions

The positive effect of the system lies in the fact that through measurement and calculation, a real-time angular displacement is obtained and then can be sent to the receiver, where the resulting angular displacement can serve as the error compensation item to be superposed with the current UAV attitude angle to obtain a more accurate LOS direction helpful for improving the localization accuracy. And this system has a small demand for space and hardware resources. The whole structure is compact, and low-cost, and can be installed in various forms. In terms of chip resources, the software is characterized by small calculation load, high real-time performance and effective correction. Earlier

Sensors **2017**, *17*, 510

calibration and alignment can be completed in dozens of minutes without adding too much workload. Through the image registration, the displacement in pixels is obtained, which can be proportionally converted into actual displacement, which, in turn, can be used to calculate the error of angular displacement between platform and UAV. With this system, real-time measurement can be taken in flight to improve the single-aircraft single-point localization accuracy, thus laying a good foundation for other improvement methods.

Acknowledgments: We acknowledge Academic Editor for his careful revision of the languages and grammatical structures in this article.

Author Contributions: Chenglong Liu, Jinghong Liu and Yueing Song conceived and designed the experiments; Chenglong Liu and Huaidan Liang performed the experiments; Jinghong Liu and Yueing Song contributed analysis tools; Chenglong Liu and Huaidan Liang wrote the paper.

Conflicts of Interest: The authors declare no conflict of interest.

References

1. United States Government Office of the Secretary of Defense. *Unmanned Aerial Vehicles Roadmap 2002–2027*; Office of the Secretary of Defense—Unmanned Aircraft & Drones: Washington, DC, USA, 2002.
2. Sun, J.; Li, B.; Jiang, Y.; Wen, C. A Camera-Based Target Detection and Positioning UAV System for Search and Rescue (SAR) Purposes. *Sensors* **2016**, *16*, 1778. [CrossRef] [PubMed]
3. Barber, D.B.; Redding, J.D.; McLain, T.W.; Beard, R.W.; Taylor, C.N. Vision-based Target Geo-location using a Fixed-wing Miniature Air Vehicle. *J. Intell. Robot. Syst.* **2006**, *47*, 361–382. [CrossRef]
4. Dobrokhodov, V.N.; Kaminer, I.I.; Jones, K.D.; Ghabcheloo, R. Vision-based tracking and motion estimation for moving targets using small uavs. In Proceedings of the 2006 American Control Conference, Minneapolis, MN, USA, 14–16 June 2006.
5. Monda, M.J.; Woolsey, C.A.; Reddy, C.K. Ground target localization and tracking in a riverine environment from a UAV with a gimbaled camera. In Proceedings of the AIAA Guidance, Navigation and Control Conference, Hilton Head, SC, USA, 18–21 August 2007; pp. 6747–6750.
6. Redding, J.; McLain, T.W.; Beard, R.W.; Taylor, C. Vision-based target localization from a fixed-wing miniature air vehicle. In Proceedings of the 2006 American Control Conference, Minneapolis, MN, USA, 14–16 June 2006; pp. 2862–2867.
7. Whang, H.; Dobrokhodov, V.N.; Kaminer, I.I.; Jones, K.D. On vision-based target tracking and range estimation for small UAVs. In Proceedings of the AIAA Guidance, Navigation and Control Conference, San Franscisco, CA, USA, 15–18 August 2005.
8. Ma, Y.; Soatto, S.; Kosecka, J.; Sastry, S.S. *An Invitation to 3-D Vision from Images to Geometric Models*; Springer: New York, NY, USA, 2012.
9. Pachter, M.; Ceccarelli, N.; Chandler, P.R. U.S. Air Force Institute of Technology, Vision-Based Target Geolocation Using Micro Air Vehicles. *J. Guidance Control Dyn.* **2008**, *31*, 597–615. [CrossRef]
10. Chiang, K.-W.; Tsai, M.-L.; Chu, C.-H. The Development of an UAV Borne Direct Georeferenced Photogrammetric Platform for Ground Control Point Free Applications. *Sensors* **2012**, *12*, 9161–9180. [CrossRef] [PubMed]
11. Yue, L.; Yang, C.Q.; Sheng, X.; Xi, H.Z. A fast target localization method with multi-point observation for a single UAV. In Proceedings of the 2016 28th Chinese Control and Decision Conference (CCDC), Yinchuan, China, 28–30 May 2016.
12. Morbidi, F.; Mariottini, G.L. Active target tracking and cooperative localization for teams of aerial vehicles. *IEEE Trans. Control Syst. Technol.* **2013**, *21*, 1694–1707. [CrossRef]
13. Qu, Y.; Wu, J.; Zhang, Y. Cooperative localization based on the azimuth angles among multiple UAVs. In Proceedings of the 2013 International Conference on Unmanned Aircraft Systems (ICUAS), Atlanta, GA, USA, 28–31 May 2013; pp. 818–823.
14. Sarunic, P.W.; Evans, R.J. Trajectory control of autonomous fixed-wing aircraft performing multiple target passive detection and tracking. In Proceedings of the 2010 Sixth International Conference on Intelligent Sensors, Networks and Information Processing (ISSNIP), Brisbane, Australia, 7–10 December 2010; pp. 169–174.

15. Rysdyk, R. UAV path following for constant line-of-sight. In Proceedings of the 2nd AIAA Unmanned Unlimited Systems, Technologies and Operations Aerospace, Land and Sea Conference, San Diego, CA, USA, September 2003.

16. Frew, E.; Rock, S. Trajectory Generation for Monocular-Vision Based Tracking of a Constant-Velocity Target. In Proceedings of the 2003 IEEE International Conference on Robotics and Automation, Taipei, Taiwan, 14–19 September 2003.

17. Grewal, M.S.; Henderson, V.D.; Miyasako, R.S. Application of Kalman filtering to the calibration and alignment of inertial navigation systems. *IEEE Trans. Autom. Control.* **1991**, *36*, 4–12. [CrossRef]

18. Dmitriyev, S.P.; Stepanov, O.A.; Shepel, S.V. Nonlinear filter methods application in INS alignment. *IEEE Trans. Aerosp. Electron. Syst.* **1997**, *33*, 260–271. [CrossRef]

19. Han, K.; DeSouza, G.N. Multiple Targets Geolocation using SIFT and Stereo Vision on Airborne Video Sequences. In Proceedings of the 2009 IEEE/RSJ International Conference on Intelligent Robots and Systems, St. Louis, MO, USA, 11–15 October 2009.

20. Conte, G.; Doherty, P. An Integrated UAV Navigation System Based on Aerial Image Matching. In Proceedings of the 2008 IEEE Aerospace Conference, 1–8 March 2008; pp. 1–10.

21. Schultz, H.; Hanson, A.; Riseman, E.; Stolle, F.; Zhu, Z. A system for real-time generation of geo-referenced terrain mod-els. In Proceedings of the SPIE Enabling Technologies for Law Enforcement, Boston, MA, USA, 6 November 2000.

22. Whitacre, W.; Campbell, M.; Wheeler, M.; Stevenson, D. Flight results from tracking ground targets using seascan UAVs with gimballing cameras. In Proceedings of the 2007 American Control Conference, New York, NY, USA, 9–13 July 2007.

23. Zhan, F.; Shen, H.; Wang, P.; Zhang, C. Precise ground target location of subsonic UAV by compensating delay of navigation information. *Opt. Precision Eng.* **2015**, *23*, 2506–2507. [CrossRef]

24. Cai, G.; Chen, B.M.; Lee, T.H. *Unmanned Rotorcraft Systems*; Springer: London, UK, 2011; pp. 223–254.

25. Zhou, Q.; Liu, J.; Wang, X. Automatic Correction of Geometric Distortion in Aerial Zoom Squint Imaging. *Opt. Precision Eng.* **2015**, *23*, 2927–2942. [CrossRef]

26. Wang, X.; Liu, J.; Zhou, Q. Real-Time Multi-Target Localization from Unmanned Aerial Vehicles. *Sensors* **2017**, *17*, 33. [CrossRef] [PubMed]

27. Miller, J.L.; Way, S.; Ellison, B.; June, C.A. Design challenges regarding highdefinition electro-optic/infrared stabilized imaging systems. *Optical Eng.* **2013**, *52*, 061310. [CrossRef]

28. Chen, V.C.; Miceli, W.J. The effect of roll, pitch and yaw motions on ISAR imaging. In Proceedings of the Radar Processing, Technology, and Applications IV, Denver, CO, USA, 18 July 1999; pp. 149–158.

29. Maier, A.; Kiesel, S.; Trommer, G.F. Performance analysis of federated filter for SAR/TRN/GPS/INS integration. *Gyroscopy Navig.* **2011**, *2*, 293–300. [CrossRef]

30. DeLima, P.; York, G.; Pack, D. Localization of ground targets using a flying sensor network. In Proceedings of the IEEE International Conference on Sensor Networks, Biquitous, and Trusworthy Computing, Taichung, Taiwan, 5–7 June 2006; pp. 194–199.

31. Texas Instruments Incorporated. *TMS320C6000 CPU and Instruction Set Reference Guide*; Texas Instruments Incorporated: Dallas, TX, USA, 1999.

32. Lowe, D.G. Distinctive image features from scale—Invariant key points. *Int. J. Comput. Vis.* **2004**, *60*, 91–110. [CrossRef]

33. Hu, M. Visual-pattern recognition by moment invariants. *Ire Trans. Inf. Theory* **1962**, *8*, 179–187.

34. Huang, D.; Wu, Z.; Liang, M.; Dong, Y. The Application of TMS320C64x DSP Assembly Language in Correlation Tracking Algorithms. In Proceedings of the 2010 3rd International Congress on Image and Signal Processing (CISP2010), Yantai, China, 16–18 October 2010.

sensors

MDPI

Article

Pedestrian Detection and Tracking from Low-Resolution Unmanned Aerial Vehicle Thermal Imagery

Yalong Ma [1,2], Xinkai Wu [1,2], Guizhen Yu [1,2,*], Yongzheng Xu [1,2] and Yunpeng Wang [1,2]

[1]   Beijing Key Laboratory for Cooperative Vehicle Infrastructure Systems and Safety Control, School of
     Transportation Science and Engineering, Beihang University, Beijing 100191, China;
     mayalong@buaa.edu.cn (Y.M.); xinkaiwu@buaa.edu.cn (X.W.); yongzhengxu@buaa.edu.cn (Y.X.);
     ypwang@buaa.edu.cn (Y.W.)
[2]   Jiangsu Province Collaborative Innovation Center of Modern Urban Traffic Technologies, SiPaiLou #2,
     Nanjing 210096, China
*   Correspondence: yugz@buaa.edu.cn; Tel.: +86-186-0101-2574

Academic Editor: Felipe Gonzalez Toro
Received: 20 January 2016; Accepted: 22 March 2016; Published: 26 March 2016

Abstract: Driven by the prominent thermal signature of humans and following the growing availability of unmanned aerial vehicles (UAVs), more and more research efforts have been focusing on the detection and tracking of pedestrians using thermal infrared images recorded from UAVs. However, pedestrian detection and tracking from the thermal images obtained from UAVs pose many challenges due to the low-resolution of imagery, platform motion, image instability and the relatively small size of the objects. This research tackles these challenges by proposing a pedestrian detection and tracking system. A two-stage blob-based approach is first developed for pedestrian detection. This approach first extracts pedestrian blobs using the regional gradient feature and geometric constraints filtering and then classifies the detected blobs by using a linear Support Vector Machine (SVM) with a hybrid descriptor, which sophisticatedly combines Histogram of Oriented Gradient (HOG) and Discrete Cosine Transform (DCT) features in order to achieve accurate detection. This research further proposes an approach for pedestrian tracking. This approach employs the feature tracker with the update of detected pedestrian location to track pedestrian objects from the registered videos and extracts the motion trajectory data. The proposed detection and tracking approaches have been evaluated by multiple different datasets, and the results illustrate the effectiveness of the proposed methods. This research is expected to significantly benefit many transportation applications, such as the multimodal traffic performance measure, pedestrian behavior study and pedestrian-vehicle crash analysis. Future work will focus on using fused thermal and visual images to further improve the detection efficiency and effectiveness.

Keywords: pedestrian detection; pedestrian tracking; aerial thermal image; video registration; unmanned aerial vehicle

1. Introduction

Pedestrian detection and tracking play an essential and significant role in diverse transportation applications, such as pedestrian-vehicle crash analysis, pedestrian facilities planning, the multimodal performance measure and pedestrian behavior study [1]. Many traditional methods for pedestrian detection and tracking use roadside surveillance cameras mounted on low-altitude light poles. For these methods, the coverage of the detection area is limited, and the installation is costly and time consuming. The unmanned aerial vehicles (UAVs) can be used as a high-altitude moving camera to cover a much wider area. Especially with the recent price drop of off-the-shelf UAV products, more and more researchers are exploring the potentials of using UAVs for pedestrian detection and tracking [2–4].

However, pedestrian detection from the images obtained from UAVs poses many challenges due to platform motion, image instability and the relatively small size of the objects. Depending on the flight altitude, camera orientation and illumination, the appearance of objects changes dramatically. This makes automatic object detection a challenging task. Using optical images, relatively larger objects, such as vehicles, can still be detected; but locating people presents an extremely difficult problem due to the small target size, shadow and low contrast of the target with the background or presence of clutter (see Figure 1a). To overcome this limitation, especially the potential optical camouflage, thermal imagery is employed, since the human thermal signature is somewhat more difficult to camouflage within the infrared spectrum [3]. However, even using thermal imagery, the detection of pedestrian trace is still a challenging problem due to the variability of human thermal signatures, which vary with meteorological conditions, environmental thermography and locales. As shown in Figure 1b, pedestrian objects in the thermal images appear as small hot blobs comprising only a few pixels and insufficient features, and non-pedestrian objects and backgrounds produce similar bright regions. All of these significantly downgrade the quality of the images and severely disturb accurate detection. In addition, the image instability caused by UAV jitter further increases the difficulty of detection.

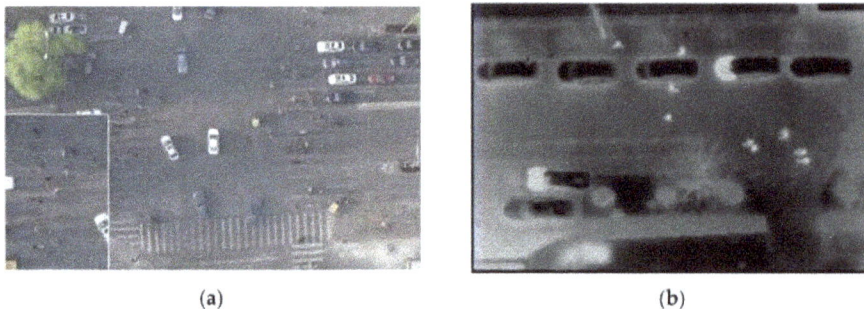

(a) (b)

Figure 1. (a) Aerial optical image and (b) thermal image.

This research aims to tackle the above challenges by proposing a pedestrian detection and tracking system using low-resolution thermal images recorded by the thermal infrared cameras installed in a UAV system. A two-stage approach, *i.e.*, Region of Interest (ROI) extraction and ROI classification, is first proposed for pedestrian detection. Given that the human body region is significantly brighter (or darker) than the background in the thermal images (see Figure 1b), a blob extraction method based on regional gradient features and geometric constraints is developed to extract bright or dark pedestrian candidate regions. The candidate regions are then fully examined by a linear Support Vector Machine (SVM) [5] classifier based on a weighted fusion of the Histogram of Oriented Gradient (HOG) [6] and Discrete Cosine Transform (DCT) descriptor for achieving accurate pedestrian detection. The experimental results illustrate the high accuracy of the proposed detection method.

With accurate pedestrian detection, this research then applies the pyramidal Lucas–Kanade method [7] to compute the local sparse optical flow, together with a secondary detection in the search region for correcting the drift, to track pedestrians. In order to eliminate the motion induced by UAVs, a semi-automatic airborne video registration method is applied before tracking. This semi-automatic registration method converts the spatiotemporal video into temporal information, thereby restoring the actual motion trajectory of pedestrians. The pedestrian velocity then can be derived from trajectory data.

The rest of the paper is organized as follows: Section 2 briefly reviews some work related to pedestrian detection and tracking in thermal images, followed by the methodological details of the proposed pedestrian detection and tracking system in Section 3. Section 4 presents a comprehensive evaluation of the proposed methods using diverse scenarios. Section 5 analyzes the pedestrian motion velocity. At the end, Section 6 concludes this paper with some remarks.

2. Related Work

2.1. Pedestrian Detection

Over the past two decades, considerable research efforts have been devoted to pedestrian detection in optical images [8,9], particularly using the HOG features first adopted by Dalal and Triggs. Concurrently, more and more research has focused on the effective detection of humans using thermal infrared images, mainly toward the applications of intelligent video surveillance [10–16] and Advanced Driver-Assistance System (ADAS) [17–19]. In [10], a two-stage template approach was proposed by employing a Contour Saliency Map (CSM) template with a background-subtraction method to obtain the location of a person, as well as examining the candidate regions using an AdaBoost classifier. In [11], the authors explored the human joint shape and appearance feature in infrared imagery. In their method, a layered representation was introduced to separate the foreground and background, and the shape and appearance cues were utilized for classification and localization, respectively. In [20], a DCT-based descriptor was used to construct the feature vector for person candidate regions, and a modified random naive Bayes classifier was applied to verify the person. The authors in [14] presented a background subtraction detection framework. The background image was extracted by various filtering and erasing of pedestrians, and pedestrians were segmented based on the size and histogram information. For a night-scene infrared video, the authors in [17] applied SVM classifiers using grayscale and binary features to recognize pedestrians from the candidate regions, and [18] proposed a tree-structured, two-stage Haar-like and HOG feature-based classifier to deal with the large variance in pedestrian size and appearance at night time for driver-assistance systems.

Very limited research has been conducted using images recorded by UAV systems for human detection, simply because the target size is too small in UAV images. The authors in [3] applied cascaded Haar classifiers with additional multivariate Gaussian shape matching for secondary confirmation to achieve people detection from UAV thermal images. The authors in [4] proposed a cascaded Haar feature-based body part detector by using the head, upper body and legs as the characteristic parts to generate individual body part classifiers. However, this method was evaluated by the images recorded by a stationary camera mounted on an elevated platform to replicate the UAV viewpoint. Considering that the human body parts in real UAV thermal images are not clear, the applications of this method are limited.

2.2. Pedestrian Tracking

Pedestrian tracking has been actively researched over the decades. Most of the existing methods are based on visual ground surveillance videos [12,21,22]. Considering the flexibility of aerial platforms and the large coverage of the scene from a top-down view, many researchers have begun to employ UAVs for object tracking. Our review here is only limited to approaches for pedestrian or object tracking with aerial platforms. The tracking of multiple objects from aerial platforms is a challenging topic because of the small size of objects, the low quality of images and the motion of the UAV. The authors in [23] combined background stabilization and layer representation to detect and track moving objects. They first applied a subtraction method to detect pedestrians, then employed the KLT (Kanade-Lucas-Tomasi) feature tracker and a post-linking process to track objects. The authors in [24] proposed a feature-based motion compensation algorithm, called image registration, to eliminate the motion of the aerial platform, and then adopted accumulative frame differencing to detect and track foreground objects. In [4], a particle filter-based approach was adopted to increase the identification and location accuracy of pedestrian objects. Furthermore, the authors in [25] explored a person tracking method for aerial thermal infrared videos using a combination of Harris corner detection and greedy correspondence matching. An OT-MACH (Optimal Trade-Off Maximum Average Correlation Height) filter is used in their method to determine whether a track is a person. However, when applying this method to aerial thermal images, the experiment results showed a poor performance due to the low quality of the images.

3. Pedestrian Detection and Tracking

3.1. Pedestrian Detection

A two-stage blob detection method is proposed in this research for pedestrian detection. The first stage is blob extraction. The proposed blob extraction approach applies the regional gradient feature and geometric constraint filtering to extract the precise pedestrian ROIs. Both bright and dark pedestrian regions can be detected. The second stage is blob classification. In this stage, a linear SVM classifier, which uses a weighted fusion of HOG and DCT descriptors as the feature vector, is utilized to verify the ROIs in order to achieve accurate detection. Figure 2 illustrates the basic workflow of the proposed detection framework.

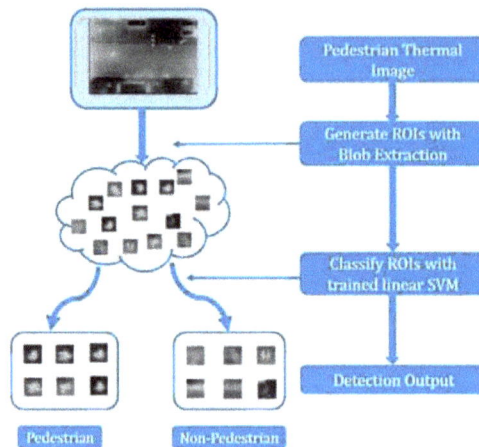

Figure 2. Pedestrian detection workflow.

3.1.1. Blob Extraction

Essentially, pedestrian detection is a typical binary classification problem [3,9,17,18,20]. The general classification procedure includes three phases: ROI selection, feature descriptor extraction and region classification. The selection of ROI is critical for the overall detection, because any actual object missed in the ROI extraction stage will not be detected in the subsequent classification stage. A number of approaches have been proposed for pedestrian ROI detection, such as background subtraction [4,12,14,15], intensity contrast [17,18,20,26] and CSM representation [10,16]. Background subtraction heavily relies on the stationary camera and can only segment the moving foreground objects, thus limiting its expansion to aerial platforms. Intensity contrast methods take advantage of the fact that humans appear brighter than other objects in thermal imagery and apply several methods, such as the adaptive threshold segmentation [17,18] and Maximally-Stable Extremal Regions (MSER) method [13], to select ROIs. However, these methods easily lead to the fragmentation of the whole body or merge the human region with the bright background regions, therefore likely resulting in the failure of classification. The CSM shows the strong and significant gradient differences between foreground and background; but the same pedestrian object may have different saliency levels in different locations in the aerial thermal images (*i.e.*, non-uniform brightness), therefore creating difficulties in detection.

To address the above problems and to take advantage of the fact that pedestrian objects appear blob shaped in aerial thermal top-views, a novel blob extraction approach, which uses the gray gradient feature of the region followed by geometric constraint filtering, is proposed to segment bright or black pedestrian regions.

Generally, pedestrians appear brighter (or darker) than the surrounding background in thermal imagery. Figure 3a presents an example of pedestrians from a top-view aerial thermal image. As shown in the corresponding 3D surface topographical picture (see Figure 3b; the height indicates the gray value of pixels), the edge of the human region indicates an obvious gradient feature. Therefore, the basic idea of the proposed blob extraction method is to first separate human regions and non-human regions using edge gradient features. During this process, the sliding window method is applied to scan the image and to find ROIs. Figure 4 presents the overall flowchart of blob extraction. Some critical techniques are explained in the following section.

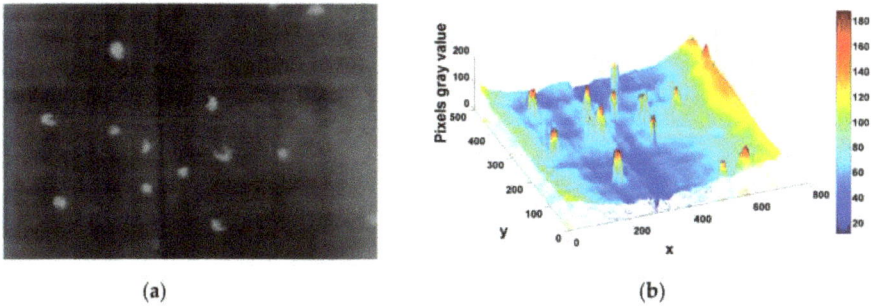

(a)   (b)

Figure 3. (**a**) Aerial thermal original image and (**b**) surface topographical picture.

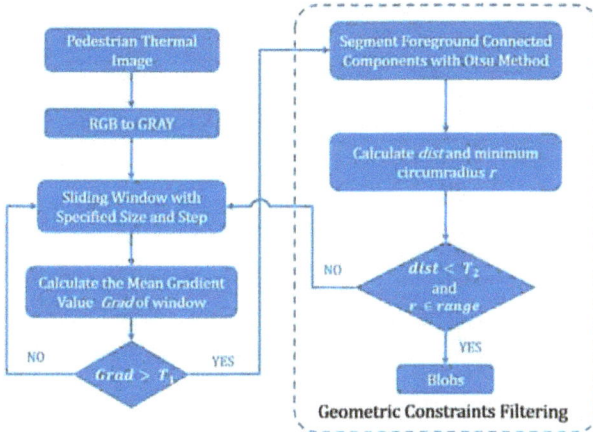

Figure 4. Blob extraction flow chart. T_1 is the predefined gradient threshold; T_2 is the predefined distance threshold; the range is the range of the minimum circumradius of a true pedestrian connected component; and *Grad* is the mean gradient value of local regions.

(1) Mean gradient value (*Grad*): First, each sliding window region is presented by a predetermined square matrix with $k \times k$ pixels (see Figure 5). To detect whether a local region is a pedestrian region, we simply need to check if the region is brighter (or darker) than the background. In other words, if a local region is a pedestrian region, the average gray value of inner rings should be larger (or lower) than the value of outer rings (see Figure 5). Note that due to the difficulty of detecting a pedestrian's

walking directions and walking states (e.g., arm swing and step type), we use the average gray value from all cells in a circle (*i.e.*, ring) in the matrix to represent the brightness level of the ring (see Figure 5). Then, the change of the brightness, *i.e.*, the gradient value G_i between the average gray values from the *i*-th and (*i* + 1)-th ring, can be calculated by the following equation (see Figure 5):

$$G_i = \overline{C}_{i+1} - \overline{C}_i \tag{1}$$

where $\overline{C}_i$ represents the average gray value from all cells in the *i*-th ring. Given the size of the square matrix of $k \times k$, the possible values of *i* are $i = \left[1, 2, \ldots, \dfrac{(k+1)}{2} - 1 \right]$. Note that when a pedestrian is colder than the environment, G_i is calculated by the following equation:

$$G_i = \overline{C}_i - \overline{C}_{i+1} \tag{2}$$

With G_i, the average gradient value of a local region, *i.e.*, *Grad*, is calculated by:

$$Grad = \frac{\sum_{i+1}^{n} G_i}{n} \tag{3}$$

where *n* represents the number of G_i in a local region. For the case presented in Figure 5, $n = (k+1)/2 - 1$.

After calculating *Grad*, a predefined gradient threshold T_1 is used to detect if the region is a pedestrian ROI. In practical implementations, the integral image technology [27] is utilized to accelerate the computation of the average gray value of every ring. The fixed size sliding window is adopted to scan the image, because all pedestrian objects have almost the same scale in top-view thermal imagery, therefore consuming less time than using multi-scale scanning. Note here that the size of sliding window is relative to the flight height. Additionally, with the flight height increasing, the size of the sliding window should be decreasing. If the flight height is known, the size of the sliding window could be selected based on the following empirical equation:

$$k = [(110 - h)/2]$$

where *k* (pixel) is the size of the sliding window and *h* (m) is the flight height. The notation of [x] indicates getting the largest integer less than or equal to x. Note that the flight height should be below 110 m.

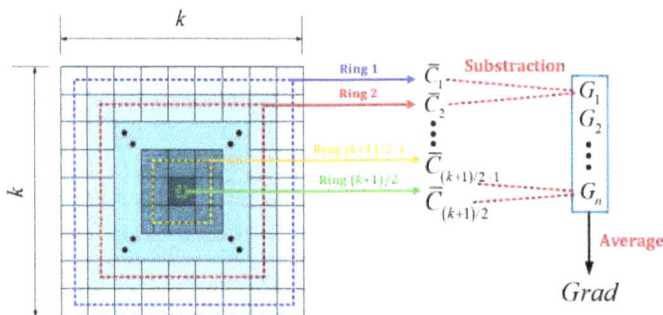

Figure 5. Illustration of *Grad* calculating process.

(2) Geometric constraints filtering: Although nearly all pedestrian ROIs can be detected through the abovementioned approach, a large number of false ROIs that have the same gradient features as the true ROIs will also be extracted because of the use of the mean gray value. To filter the false ROIs,

the Otsu threshold segmentation method [28] is employed to generate connected components for the pedestrian candidate in the binary image. The following two geometric constraints are used to refine the pedestrian candidate regions:

(1) Central location: According to the properties of a typical top-down view thermal image of a human, the distance between the centroid and center of the ROI should not be large. Therefore, the Euclidean distance between the centroid and center is calculated and then compared to a threshold (T_2) to filter ROIs (see Equation (4)):

$$\begin{cases} \text{Pedestrian ROI}: \text{ if } dist = \| P_m - P_0 \|_2 \leqslant T_2 \\ \text{Non} - \text{Pedestrian ROI}: \text{ if } dist = \| P_m - P_0 \|_2 > T_2 \end{cases} \tag{4}$$

where P_m and P_0 are the centroid and center of a square ROI, respectively.

This constraint eliminates most false pedestrian ROIs coming from the corners or edges of bright regions and the fragments of pedestrian blobs. Furthermore, this constraint ensures that the pedestrian is in the middle of the ROI, which is beneficial for subsequent classification.

(2) Minimum circumscribed circle: Because the scales of most of the pedestrian objects in top-view thermal images are similar, the pedestrian ROIs should not exceed a certain range. Therefore, by comparing the minimum circumradius r of a pedestrian connected component to a predefined *range*, we could eliminate too large or too small blobs and some rectangle regions. This method will further help filter the false ROIs.

Figure 6 illustrates two geometric constraints for both pedestrian and non-pedestrian ROIs. The example images are enlarged to 500 × 500 pixels for presenting the details. The first row represents the true pedestrian example, and other rows are non-pedestrian examples. From the figure, we can see that for non-pedestrian regions, either the distance between the centroid and center (*i.e.*, *dist*) is too large (see the case presented in the second row) or the circumscribed radius (*i.e.*, *r*) is too large (see the case presented in the third row). Therefore, by applying the above two filtering methods, most non-pedestrian regions can be eliminated, as shown in our testing presented in Section 4. In the practical extraction, for one blob, more than one window might be detected by the abovementioned procedures. We compute the average location for each group of windows. This reduces the subsequent classification times.

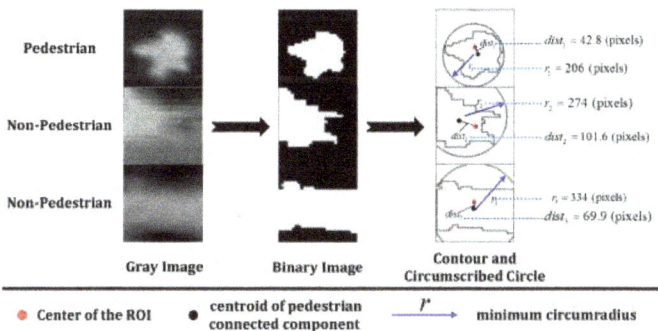

Figure 6. Illustration of geometric constraints. The first column is a gray image; the second column is the corresponding binary image; the third column is the image of the contour and minimum circumscribed circle: the red point is the center of the ROI; the black point is the centroid of the pedestrian connected component; *dist* is the Euclidean distance between the centroid and center; and *r* is the minimum circumradius represented by blue arrows; and the fourth column shows the examples of the values of *r* and *dist*.

3.1.2. Blob Classification

Blob extraction only detects pedestrian ROIs. To further classify the extracted blobs as pedestrians or non-pedestrians, a blob classification process is proposed here. A typical classification process consists of two steps: first, each blob is represented by a number of characteristics or denoted features; and second, some matching method is applied to compare the features of each blob with the features of the type of object that needs to be detected. In our research, due to the round shape of pedestrians in top-view thermal images, the pedestrian objects tend to be confused with other light sources, such as road lamps, vehicle engines, headlights and brushwood, leading a high false alarm rate. To overcome this difficulty, a unique descriptor, which combines both HOG and DCT features, is first used to describe extracted blobs, and then, a machine learning algorithm, SVM, is adopted to distinguish pedestrians from non-pedestrians objects. The following presents some details of the HOG and DCT descriptors and SVM.

(1) HOG descriptor: To achieve reliable and precise classification, the first step is to select appropriate features to form descriptors, which can be used to describe objects. Previous research has indicated that the HOG feature can be used to form a local region descriptor to effectively detect pedestrians in visual images. The HOG feature is based on the well-normalized local histograms of image gradient orientations in a dense grid. To extract HOG feature, the first step is to compute the gradient values and orientations of all pixel units in the image by applying the 1D centered point discrete derivative mask with the filter kernel $[-1, 0, 1]$. The second step is to create the cell histograms. In this step, the image is divided into cells, and the 1D histogram H_i is generated by binning local gradients according to the orientation of each cell. The third step is to group cells together into larger, spatially-connected blocks F_i and to normalize blocks in order to alleviate the influence of illumination and shadow. Since each block consists of a group of cells, a cell can be involved in several block normalizations for the overlapping block. Finally, the feature vector V_{HOG} is obtained by concatenating the histograms of all blocks.

The HOG feature essentially describes the local intensity of gradients and edge directions. Since the shapes or edges of the pedestrian and non-pedestrian objects are obviously different, the HOG feature is applied to characterize the difference. In our experiments, the UAV flight altitude ranges from 40 m to 60 m, and the corresponding object size approximately ranges from 35×35 pixels to 25×25 pixels. Hence, we normalize the object size by scaling the examples to 32×32 pixels. We then use 8×8 blocks with a four-pixel stride step and nine histogram bins (each bin is corresponding to $20°$ of orientation) to calculate the feature vector. Note for an image with 32×32 pixels, a total of 49 blocks can be identified (note, 49 is calculated by $\left(\dfrac{(32-8)}{4} + 1 \right)^2 = 49$). Therefore, the final HOG feature descriptor can be described by a vector V_{HOG}:

$$V_{HOG} = [F_1, F_2, \cdots, F_i, \cdots F_{49}] \tag{5}$$

where V_{HOG} is the HOG feature descriptor and F_i is the normalized block vector for the i-th block. Since each block contains four cells and each cell contains nine bins, $F_i = \left[\bar{h}_{1,i}, \bar{h}_{2,i}, \ldots, \bar{h}_{j,i}, \ldots, \bar{h}_{36,i} \right]$, where $\bar{h}_{j,i}$ is the j-th normalized value in the i-th block. Figure 7 illustrates the process of extracting the HOG descriptor.

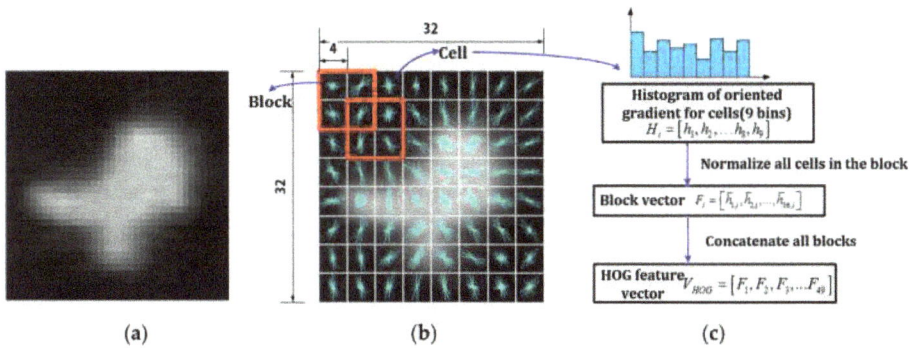

Figure 7. Illustration of HOG descriptor extraction. (**a**) Original image; (**b**) visualization of gradient orientations of all cells (white grids are cells, and red bounding boxes are block regions); and (**c**) generating the feature vector.

(2) DCT descriptor: This research further explores the potential of using the DCT feature as the descriptor. DCT has been widely applied to solve problems in the digital signal processing and face recognition fields [29,30]. In addition, the DCT feature also showed good performance for low resolution person detection in thermal images [20].

DCT essentially transforms signals (the spatial information) into frequency representation. To derive a DCT descriptor, the first step is to calculate the DCT coefficients. Typically, for an M × N pixel image, the following equation can be used to calculate the DCT coefficient matrix $G(u, v)$ at frequency component (u, v):

$$G(u,v) = \frac{2}{\sqrt{MN}} \sum_{x=0}^{M-1} \sum_{y=0}^{N-1} C(u) C(v) f(x,y) \cos\left[\frac{\pi u (2x+1)}{2M}\right] \cos\left[\frac{\pi v (2y+1)}{2N}\right] \tag{6}$$

where $f(x, y)$ is the intensity of the pixel in row x and column y in the input image. $C(u)$, $C(v)$ are defined by the following:

$$C(u), C(v) = \begin{cases} \frac{1}{2} & u = 0, v = 0 \\ 1 & otherwise \end{cases}$$

Generally speaking, using the essential DCT coefficients as the descriptors can mitigate the variation of pedestrian appearance. For most of the pedestrian thermal images, signal energy usually lies at low frequencies, which correspond to large DCT coefficient magnitudes located at the upper left corner of the DCT coefficient matrix. In our experiments, the example image is first divided into blocks of 8 × 8 pixels size (see Figure 8a), then discrete cosine transform is applied to each block to obtain the corresponding DCT coefficient matrix (see Figure 8b), and at the end, the top-left DCT coefficients are extracted via zigzag scan (see Figure 8c). Note that only the first 21 coefficients in the matrix are selected as the feature vector of the block, because more coefficients do not necessarily imply better recognition results, and adding them may actually introduce more irrelevant information [29]. Then, for each block, a final feature vector, which concatenates the 21 low-frequency coefficients, is constructed. In addition, as mentioned before, most information of the original image is stored in a small number of coefficients. This is indicated by Figure 8b, in which the value of DCT coefficients at the matrix top-left is significantly larger than the others.

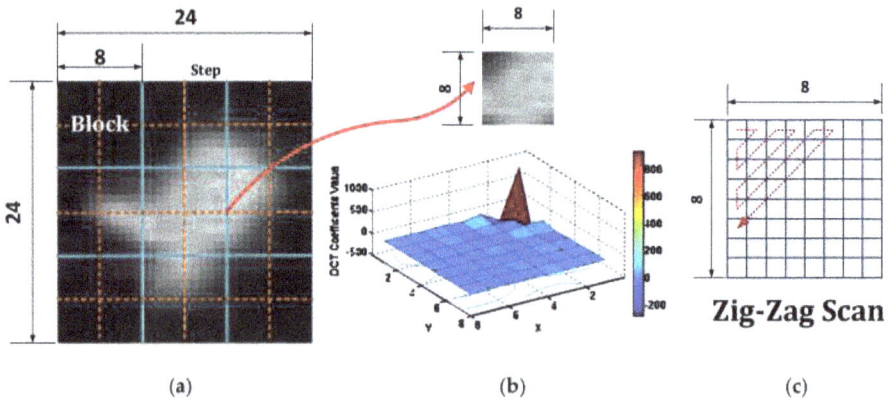

Figure 8. Illustration of the DCT descriptor extraction. (**a**) Original image division; (**b**) visualization of DCT coefficients value of block; (**c**) zigzag scan method.

(3) SVM with a combination descriptor of HOG and DCT: Unfortunately, neither HOG nor DCT shows consistently good performance in our tests. To enhance the adaptability of feature descriptors, a weighted combination descriptor of HOG and DCT is proposed here. Because HOG can help capture the local gradients and edge direction information and DCT can help address the variation of object appearance, such as the halo effects and motion blur [20], combining both features is a legitimate idea. In our experiments, the HOG feature vector V_{HOG} has been normalized by the L2-norm method. Therefore, the DCT feature vector V_{DCT} needs to be normalized, as well, due to its large magnitude. In this research, the DCT feature vector V_{DCT} is normalized using $v_i^* = v_i/\max(v_1, v_2, \cdots, v_n)$. The final combination descriptor V_F is constructed by the weighted combination of V_{HOG} and V_{DCT}, i.e., $V_F = [\alpha V_{HOG}, \beta V_{DCT}]$. The values of α and β are hyper-parameters, which are determined by cross-validation, aiming to minimize the number of false classifications in training samples.

With a hybrid descriptor, SVM is then applied to classify pedestrian and non-pedestrian objects. SVM has been proven to be a highly effective two-category classification algorithm, which looks for an optimal hyper-plane to achieve maximum separation between the object classes. SVM demonstrates good performance on high dimensional pattern recognition problems. In our case, a linear SVM trained by example images from the blob extraction stage with and without pedestrians is used to classify the pedestrian candidate regions. Since the hybrid descriptor takes advantage of both HOG and DCT features of infrared thermal objects, our testing shows great improvements of the results (see Section 4).

Note that in our practice, we only employed a simple linear kernel without any parameter selection methods, such as cross-validation and grid search, due to the following two reasons: (1) the kernel type is difficult to determine; (2) the grid search or cross-validation would cost too much time in the training stage. Our experiment has shown that it takes days to search for the optimal parameters when using cross-validation and grid-search methods. Therefore, we believe that a simple linear kernel without any parameter selection methods works best for our detection framework.

3.2. Pedestrian Tracking

After detecting the pedestrian, a tracking approach is proposed here. Our purpose is to track multi-pedestrians simultaneously once they have been detected in the detection stage and to extract the trajectory data for the subsequent motion analysis. However, due to the irregular motion of the UAV in three axes of roll, pitch and yaw [31], the video pictures rotate and drift. This brings difficulties for pedestrian tracking and trajectory extraction. Therefore, to eliminate the motion of UAV surveillance videos, an image registration technology is applied first.

3.2.1. Video Registration

Video registration aims to spatially align video frames in the same absolute coordinate system determined by a reference frame. It essentially converts the spatiotemporal information into temporal information. A typical feature-based alignment framework for video registration contains the following steps [24]:

(1) Detecting the stationary feature as the control points;
(2) Establishing correspondence between feature sets to obtain the coordinates of the same stationary point;
(3) Setting up the transformation model from the obtained point coordinate;
(4) Warping the current frame to the reference frame using the transformation model.

The registration algorithm is crucial to subsequent tracking, because errors induced by this step would impact the tracking accuracy. The proposed registration method uses the KLT feature [32], which has been proven to be a good feature for tracking. This whole process includes three steps: (1) KLT feature extraction; (2) KLT tracking; and (3) aligning using affine transformation.

KLT feature extraction is derived from the KLT tracking algorithm, which computes the displacement of the features between two small windows in consecutive frames. The image model can be described by the following equation:

$$J(x) = I(x - d) + \eta(x) \tag{7}$$

where $x = [x, y]^T$; $I(x)$, $J(x)$ represent the intensities of tracking windows centered at feature points in the current frame and the next frame, respectively; d represents the displacement vector; and $\eta(x)$ is a noise term.

The goal for tracking is to determine d that minimizes the sum of squared intensity difference between $J(x)$ and $I(x - d)$ in the tracking window, as described by:

$$\varepsilon = \iint [I(x - d) - J(x)]^2 \omega dx \tag{8}$$

where ω is a weighting function. For the small displacement, the intensity function can be approximated by Taylor series as:

$$I(x - d) = I(x) - g \cdot d \tag{9}$$

where $g = [\partial I/\partial x, \partial I/\partial y]^T$. Then, Equation (8) is differentiated with respect to d, and the result is set equal to zero. The final displacement equation can be expressed as follows:

$$Gd = e \tag{10}$$

where G is a 2 × 2 coefficient matrix:

$$G = \iint gg^T \omega dA$$

and e is the 2D vector:

$$e = \iint (I - J) g \omega dA$$

Note that to obtain a reliable solution of Equation (10), eigenvalues of coefficient matrix G must be large [33]. The KLT feature then is defined as the center of window W, where the minimum eigenvalue of G is larger than a predefined threshold. These points usually are corners or have large intensity gradient variation.

After the detection of the KLT feature in the first frame of the video, the control points for registration are selected manually to avoid finding false points on moving objects. Then, the KLT tracker is introduced to track the chosen feature points in the video sequence and to establish correspondence for computing the transformation model. We employ affine transformation to transform the current frame to the first frame, which needs four selected feature points. Mapping to the first frame avoids the error accumulation caused by registration between two consecutive frames. This KLT feature-based method can eliminate the small range motion of aerial videos. Figure 9 shows the registration example images.

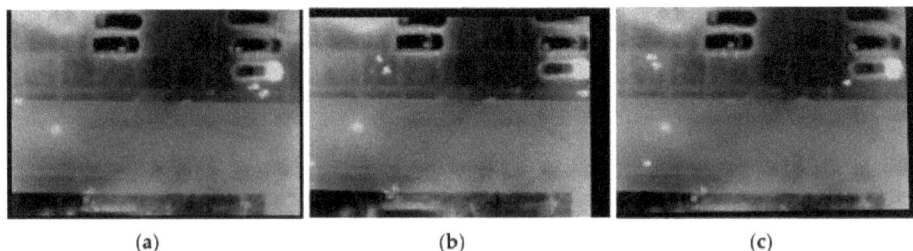

(a)   (b)   (c)

Figure 9. Illustration of video registration. (**a**) The reference frame (the first frame); (**b**,**c**) the 317th and 392nd frames mapping to (**a**), respectively. The red points are the KLT control points.

3.2.2. Pedestrian Tracking

After video registration, the KLT feature-based tracking method is applied for pedestrian tracking. Since most of the KLT features are located on the edge of the object, where the intensity difference exists, the KLT features are not accurate enough to describe the locations of pedestrians. To address this issue, the center of the detected pedestrian is determined as the beginning of tracking, and then, a pyramidal implementation of the Lucas–Kanade feature tracker is employed to accurately compute the displacement. Image pyramidal representation could handle relatively large displacement because the optical flow is calculated and propagates from the highest to lowest levels. In addition, due to the low quality of infrared imagery and motion blur caused by UAVs, the tracking points might drift away from the center during the tracking process. This is because the premise of optical flow methods assumes that the gray level of objects keeps consistent over time, but the thermal images cannot satisfy this assumption due to the high noise level and non-uniform brightness. The drift could significantly downgrade the tracking accuracy. To address this issue, a secondary detection stage in a local region followed by the feature tracking is used to correct the drift, as described in the following:

(1) Once one pedestrian object is detected in the current frame, input the center coordinate of the object into the feature tracker;
(2) Applying the pyramidal Lucas–Kanade feature tracker to estimate the displacement between consecutive frames, the estimated displacement then is used to predict the coordinate of the object in the next frame;
(3) Around the predicted coordinate, a search region is determined. In the search region, a secondary detection is conducted to accurately find the pedestrian localization.

Note that in the search region, more than one object may be detected. In this case, the minimum distance can be obtained by comparing all distances between the centers of detected objects and the predicted coordinate. If the minimum distance is less than half of the predefined blob size, the corresponding object center will be the corrected pedestrian location.

Then, the predicted point will be replaced by the detected pedestrian central point. Repeating the aforesaid process can keep the tracking continuous and accurate. Figure 10 illustrates the update process. Note that during the practice, a target may not be detected all of the time in the search region. If no pedestrians are detected in the search region, the tracking point will keep its origin, *i.e.*, the predicted point. This automatic feature tracker is crucial for providing us the coordinates of the corresponding point of the same object. In addition, to improve tracking efficiency, the detection of the whole image is only conducted every 15 frames for detecting the new pedestrian objects.

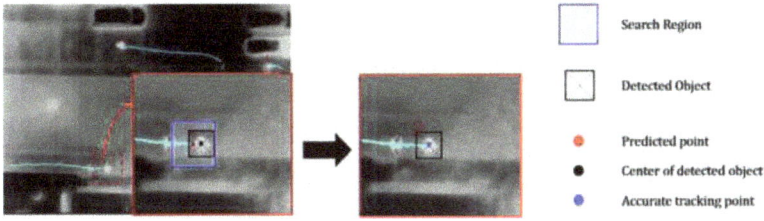

Figure 10. Illustration of the update of the secondary detection.

Overall, the proposed tracking method is similar to some tracking-by-detection methods, which are usually used in optical images and implemented using online learning. However, our detection scheme in the tracking approach utilizes the classifiers that have been trained during blob classification (see Section 3.1.2). This significantly saves time for real-time pedestrian tracking and improves the efficiency.

4. Experiments for Pedestrian Detection and Tracking

4.1. Evaluation for Pedestrian Detection

To evaluate the proposed pedestrian detection approach, experiments were conducted using aerial infrared videos captured by a thermal long-wave infrared (LWIR) camera with a 720×480 resolution mounted on a quadrotor with a flight altitude of about 40 to 60 m above the ground. Figure 11 shows the basic components of UAV platform used in this research. Five different scenes with different backgrounds, flying altitudes, recording times and outside temperatures were used for evaluation (see Table 1). The detection statistical effect was evaluated using two measurements: (1) Precision $=$ TP$/$(TP $+$ FP), which is the percentage of correctly-detected pedestrian number over the total detected pedestrians; and (2) Recall $=$ TP$/$(TP $+$ FN), which is the percentage of correctly-detected pedestrian number over the total true pedestrians. Here, TP (*i.e.*, True Positive) is the number of correctly-detected pedestrians. FP (*i.e.*, False Positive) is the number of detected objects that are not pedestrians, and FN (*i.e.*, False Negatives) is the number of misdetected pedestrians. In our evaluation, a detected blob is regard as a True Positive (TP) when it fits the following condition:

$$\frac{A_{anno} \cap A_{det}}{\min \{A_{anno}, A_{det}\}} \geqslant 0.7$$

where A_{anno} denotes the area of the annotated ground-truth rectangle surrounding an actual pedestrian and A_{det} is the area of the detected pedestrian blob.

Figure 11. Basic components of the UAV platform.

For the simple background scene, such as grassland, a playground, *etc.*, there is high contrast between pedestrians and the background (see Figure 12). For this type of scene, selecting pedestrian sample images and training the classifier are not necessary, since the pedestrian targets can be easily detected by only using blob extraction. In our test, 176 typical frames were selected from these videos to form a test dataset named "Scene 1". For this dataset, only blob extraction method was applied. The testing results are shown in Table 2. Note that the numbers in Table 2 indicate the sum of pedestrian numbers of all images in corresponding datasets. As shown in the table, the precision and recall rates are as high as 94.03% and 93.03%, respectively.

For the other four scenes, the background is much more complex due to some non-pedestrian objects, such as lamps, vehicle engines, headlights and brushwood, which are similar to true instances. The supervised classification algorithm was employed to distinguish them. To evaluate the performance of the classifier, we set up four challenging datasets composed of typical images selected from four corresponding aerial videos, which may generate a number of false positives only using blob extraction. Table 1 shows the basic information of datasets. In our experiments, a part of the images from the dataset was chosen as the training images and the remainder as the test images. The pedestrian candidate regions generated by the blob extraction method (see Section 3.1.1) in training images were labeled manually for pedestrian and non-pedestrian blobs. Then, the blobs from training images were used to train the linear SVM classifier, and then, the blobs extracted from test images were used to examine the performance of the classifier. Table 3 provides the details of images for training and testing for four datasets. Some pedestrian and non-pedestrian example blobs from four datasets are also shown in Figure 13. As shown in the figure, the appearance of pedestrians varies across the datasets because of the different flight altitudes and temperatures. For each dataset, the performances of HOG, DCT and the proposed hybrid descriptors were compared. The Receiver Operating Characteristic (ROC) curves were drawn to reflect the performance of the combination of descriptors and linear SVM.

Table 1. Basic information of datasets.

Dataset	Resolution	Flying Altitude	Scenario	Date/Time	Temperature
Scene 1	720 × 480	40–70 m	Multi-scene	April/Afternoon, Night	10 °C–15 °C
Scene 2	720 × 480	50 m	Road	20 January/20:22 p.m.	5 °C
Scene 3	720 × 480	60 m	Road	19 January/19:46 p.m.	6 °C
Scene 4	720 × 480	40 m	Road	8 April/23:12 p.m.	14 °C
Scene 5	720 × 480	50 m	Road	8 July/16:12 p.m.	28 °C

Table 2. Detection statistical results.

Dataset	# Pedestrian	# TP	# FP	# FN	Precision	Recall
Scene 1	1320	1228	78	92	94.03%	93.03%
Scene 2	9214	8216	686	998	92.30%	89.17%
Scene 3	10,029	9255	344	774	96.41%	92.28%
Scene 4	5094	4322	396	772	91.61%	84.84%
Scene 5	3175	2923	197	252	93.69%	92.06%

Table 3. The number of training and test images for datasets.

Dataset	# Images	# Training	# Test	Training Blobs # Positives/Negatives	Test Blobs # Positives/Negatives
Scene 1			No Training		
Scene 2	2783	544	2239	2347/9802	8339/35,215
Scene 3	2817	512	2305	2098/938	9237/4097
Scene 4	1446	365	1081	1560/881	4473/2817
Scene 5	1270	340	930	734/966	2984/4572

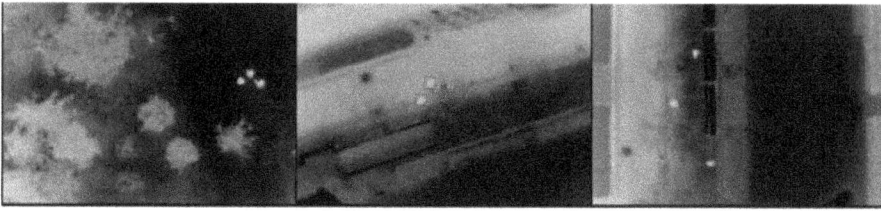

Figure 12. Pedestrian detection sample images for Scene 1 only using blob extraction.

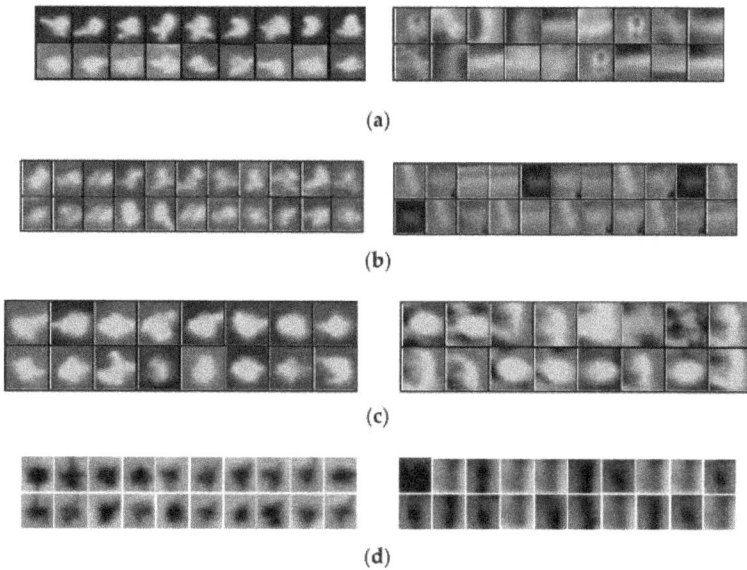

(a)

(b)

(c)

(d)

Figure 13. Pedestrian (left column) and non-pedestrian (right column) blob examples for: (**a**) dataset "Scene 2"; (**b**) "Scene 3"; (**c**) "Scene 4"; and (**d**) "Scene 5".

In dataset "Scene 2", pedestrians were easily distinguished from the background because of the high intensity (see Figure 14b); therefore, the training examples contained clear and detailed information about gradient and contour. Figure 15a illustrates the relative performance of three descriptors on this dataset. As shown, the HOG performed better than the DCT. This can be explained by the fact that the contour of pedestrian examples is clear and distinguishable between pedestrian and background examples from the salient local intensity gradient and edge direction information. The performance of the fusion descriptors is slightly better than HOG. Figure 14b shows the comparison of blob extraction results and corresponding blob classification results using the hybrid descriptor. Most wrong blobs were eliminated through classification. The combination of HOG and DCT strengthens the adaptability of feature descriptors, because the DCT descriptor can handle slight appearance changes and the halo effect. Meanwhile, the combination of both features increases the dimensionality of the feature vector, therefore containing more information for classification.

In dataset "Scene 3", pedestrians were blurry due to the higher flight altitude than in the "Scene 2" dataset. As shown in Figure 15b, all three descriptors had good performance. With a false positive rate of 5%, more than 90% of pedestrians were classified correctly. As seen in the ROC curves, both the DCT and the fusion descriptors perform well and outperform the HOG. The possible reason is that the DCT descriptor extracts the key information from the example picture (top-left DCT coefficient);

therefore, it has high discernibility for the blur appearance and imperceptible changes. The "Scene 4" dataset was captured with lower flight altitude than other datasets. Hence, the pedestrian blobs are larger and clearer, resulting in detailed gradient and edge information. However, this dataset is challenging and complex because of the interference of the road lamps and the vehicle headlights. As shown in Figure 14d (the top row), distinguishing the pedestrians and lamps is difficult. As we predicted, the DCT could not compete with the HOG descriptor due to the high similarity of true and false positive examples. As shown in Figure 15c, the true positive rate of DCT only roughly reaches half of the HOG and fusion descriptors. The fusion descriptor slightly outperformed the HOG, which was similar to "Scene 2".

The last dataset "Scene 5" includes few pedestrian objects, which are shown as black blobs because of the high environment temperature. Figure 15d indicates the high classification precision of three descriptors, because fewer non-pedestrian candidate regions were detected in the blob extraction stage. Hence, the performance of the classifier is good.

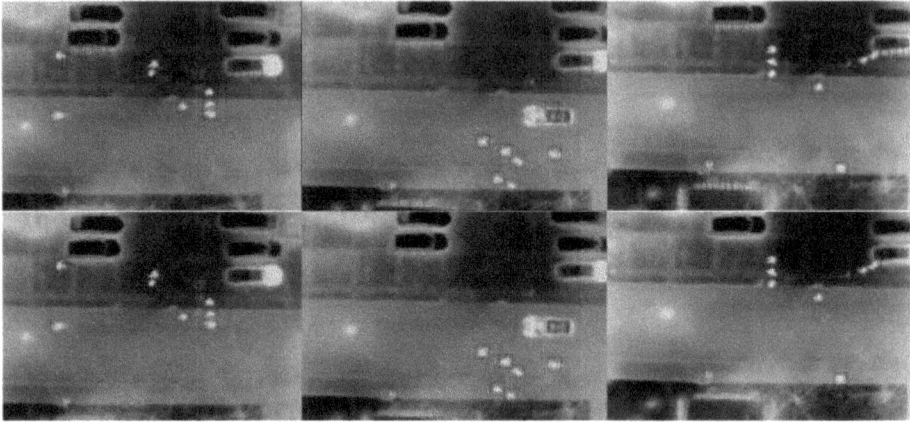

(a)

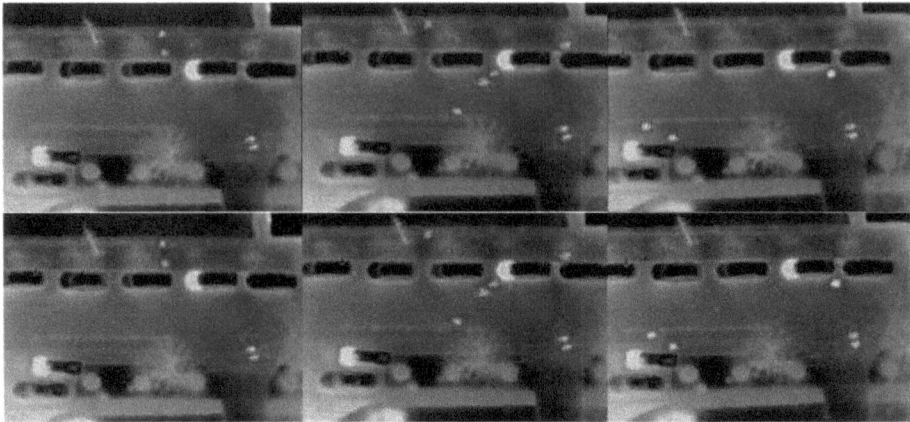

(b)

Figure 14. *Cont.*

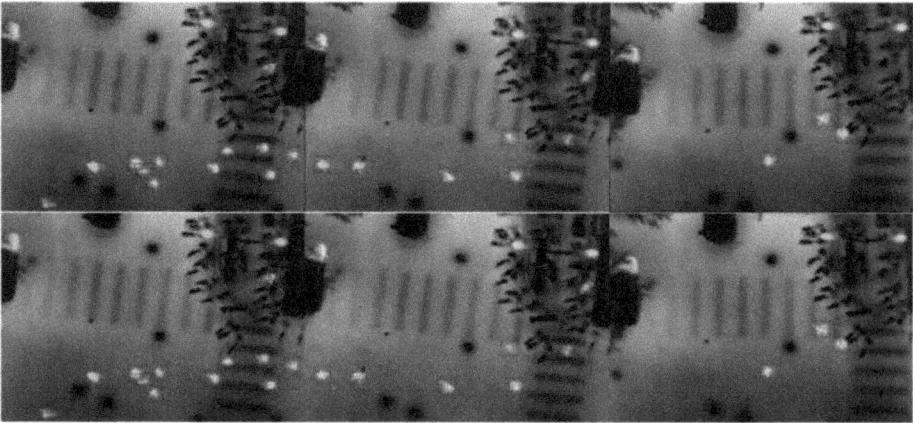

(c)

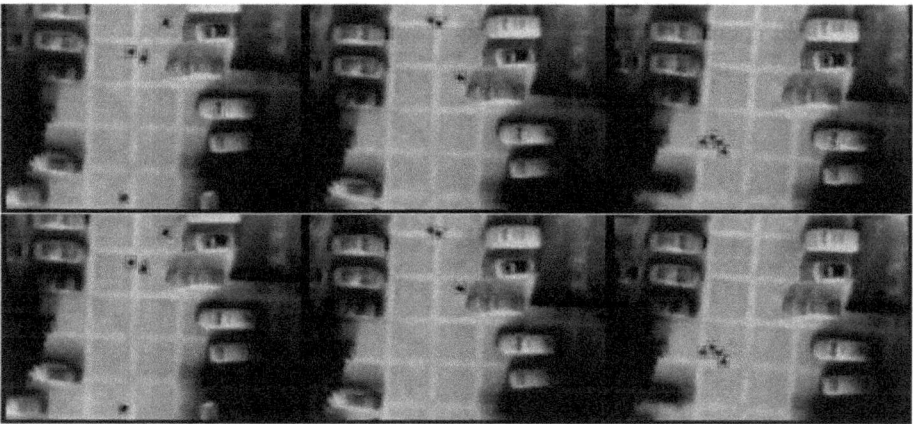

(d)

Figure 14. Pedestrian detection sample images for four scenes. (**a**) Pedestrian detection for Scene 2: the first row is the blob extraction results, and the second row is the classification results. (**b**) Pedestrian detection for Scene 3: the first row is the blob extraction results, and the second row is the classification results. (**c**) Pedestrian detection for Scene 4: the first row is the blob extraction results, and the second row is the classification results. (**d**) Pedestrian detection for Scene 4: the first row is the blob extraction results, and the second row is the classification results.

As a result, the hybrid descriptor makes full use of the advantages of both HOG and DCT in representing the pedestrian feature, promotes the performance of the standalone use of HOG or DCT and tends to the better one. As shown in Table 2, the proposed detection method achieves a high precision in all datasets above 90% and a high recall rate of 84.84%–92.28%. The overall performance is encouraging. In practice, owing to the high similarity between pedestrian and cyclist blobs in aerial thermal top views, we regard cyclists as true positives. The used hyper-parameters in our experiments are presented in Table 4. Note that the *range* could be determined by ($[k/2] - 5, [k/2] + 2$).

Table 4. Hyper-parameters used in the experiments.

Dataset	T_1	T_2	k	Range	α	β
Scene 2	3.5	3	29	(9, 16)	10	3
Scene 3	4	2	25	(7, 14)	10	3
Scene 4	3	2	35	(12, 19)	10	3
Scene 5	5	2	27	(8, 15)	10	3

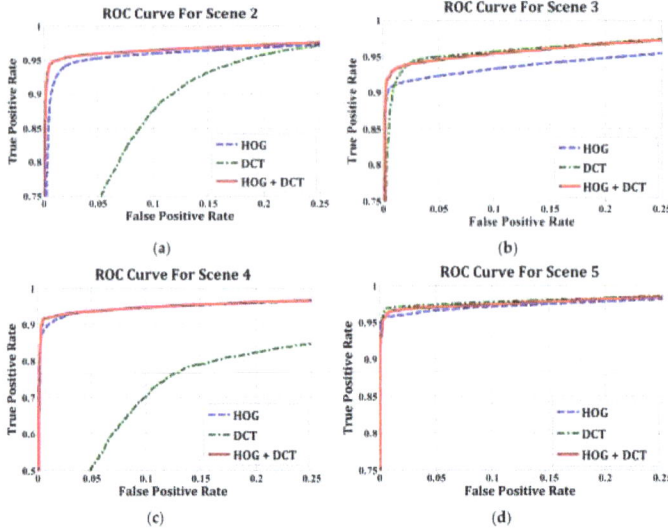

Figure 15. ROC curves for four training datasets. (**a**) scene 2, (**b**) scene 3 (**c**) scene 4 (**d**) scene 5.

4.2. Evaluation for Pedestrian Tracking

The performance of pedestrian tracking is evaluated by the Multiple Object Tracking Accuracy (MOTA) [34], as defined:

$$\text{MOTA} = 1 - \frac{\sum_t (m_t + FP_t + mme_t)}{\sum_t g_t} \quad (11)$$

where m_t, FP_t and mme_t are the number of missed pedestrians, of false positives and of mismatches, respectively, for time t, and g_t is the number of actual total pedestrians. Note that in order to apply the proposed pedestrian tracking method, images have to be registered first. However, due to the drastic shaking of videos captured by the drone without the gimbal, few video clips can be registered well. In our test, only three available registered sequences from different scenes were selected. The basic information of test sequences is presented in Table 5. The m, FP, mme and g were counted on every image from the three test sequences. The obtained results are summarized in Table 6. For comparison, the tracking results from the original KLT tracker are also presented in Table 6.

As shown in Table 6, the overall performance of the proposed method is impressive with an average MOTA of 0.88; certainly, there is a room for improvement. Furthermore, compared to the KLT tracker, the proposed method with a secondary detection in search regions ensures continuous tracking and consistent trajectories. Therefore, the performance of the proposed method is significantly better than the original KLT tracker (more than 20% improvement, as shown in Table 6).

A possible reason for the different performances between the original KLT tracker and the proposed method is that the original KLT tracker generates more tracking misses (*i.e.*, m_t) and mismatches (*i.e.*, mme_t) due to the point drift (see Table 6). Tracking misses typically occur when pedestrians fail to be detected in the tracker initialization or secondary detection. This could be due to the low contrast or pedestrians walking too close. As shown in Figure 16a, second row, the pedestrian with ID 5 was missed when three pedestrians were walking too close, and the track point drifted away from this pedestrian. This created two tracking points for the pedestrian with ID 4. Because the tracker found the wrong object, a false positive occurred in subsequent frames. On the other hand, mismatch typically occurs due to the change of the tracking ID from previous frames, also called "ID switch". In our three test sequence, mismatches are much less than tracking misses and false positives. This is partly because that the intersections between pedestrians' walking paths could be more easily avoided from the vertical top view of the aerial images and partly due to the non-congested scenes tested in this research. Figure 16 demonstrates some examples of tracking errors, which are highlighted by yellow circles. Note here that the tracking errors include the tracking misses, false tracking and mismatches.

Table 5. Basic information of sequences.

Dataset	Resolution	Flying Altitude	Scenario	Date/Time	Temperature
Sequence 1	720 × 480	50 m	Road	20 January/20:22 p.m.	5 °C
Sequence 2	720 × 480	50 m	Road	20 January/20:32 p.m.	5 °C
Sequence 3	720 × 480	45 m	Plaza	13 April/16:12 p.m.	18 °C

Table 6. Tracking statistical results.

Sequence	# Frames	#g	#m	#FP	#mme	MOTA
Sequence 1	635	3404	374	193	0	0.8334
Sequence 1 KLT	635	3404	379	956	5	0.6063
Sequence 2	992	4957	161	597	0	0.8471
Sequence 2 KLT	992	4957	198	1637	4	0.6290
Sequence 3	1186	3699	318	0	0	0.9140
Sequence 3 KLT	1186	3699	104	1367	5	0.6010

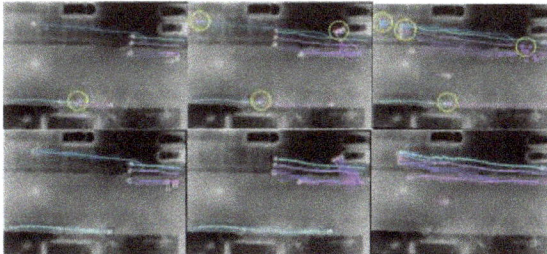

(a)

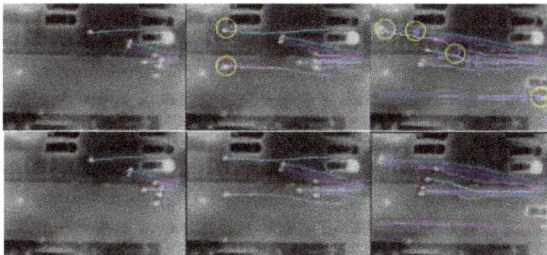

(b)

Figure 16. *Cont.*

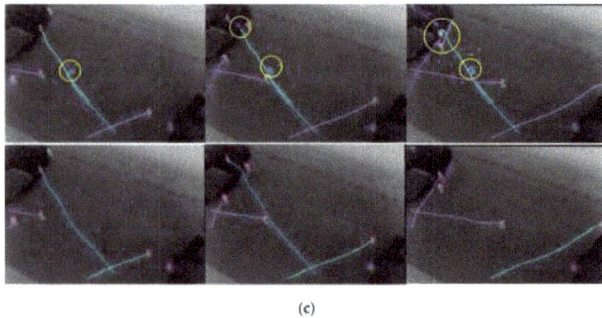

(c)

Figure 16. Pedestrian tracking sample images for three sequences. (**a**) Pedestrian tracking for Sequence 1: the first row is the KLT tracking results, and the second row is the proposed method tracking results; the yellow circles highlight the tracking errors. (**b**) Pedestrian tracking for Sequence 2: the first row is the KLT tracking results, and second row is the proposed method tracking results; the yellow circles highlight the tracking errors. (**c**) Pedestrian tracking for Sequence 3: the first row is the KLT tracking results, and the second row is the proposed method tracking results; the yellow circles highlight the tracking errors.

5. Pedestrian Tracking and Velocity Estimation

From the tracking results, the trajectories of pedestrians can be derived. Figure 17 presents some examples of pedestrian trajectories. However, from Figure 17, it is difficult to distinguish cyclists (or running men) from pedestrians, since from the top-view aerial thermal images, pedestrians and cyclists appear the same. This issue can be addressed by estimating the velocities of pedestrians and cyclists.

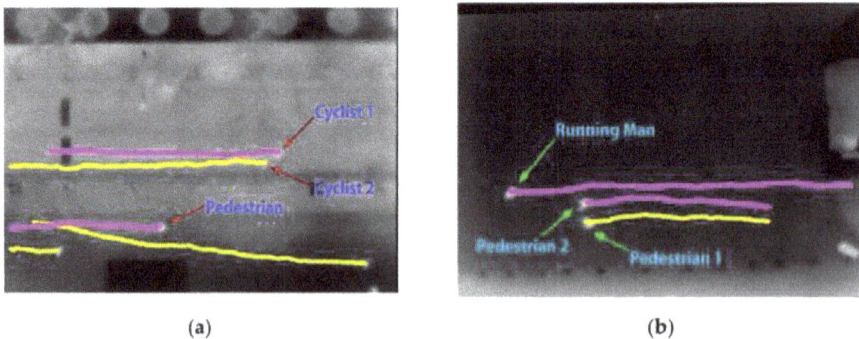

(a)   (b)

Figure 17. Cyclists and pedestrians trajectories. (**a**) Scene 1 and (**b**) Scene 2.

To estimate velocities, accurate trajectory data will be needed. Considering that the drift of track points may result in estimation errors of the true object center, the Kalman filter is used to process raw trajectory data. The Kalman filter computes the best estimate of the state vector (*i.e.*, location coordinates) in a way that minimizes the mean of the squared error according to the estimation of the past state and the measurement of the present state. Figure 18 presents the processed trajectory data, *i.e.*, time-space diagrams, of some detected objects. Since these time-space diagrams are close to straight lines, the velocity can be easily estimated by calculating the slopes of these lines. These diagrams can be used to distinguish cyclists (or running men) and walking pedestrians simply because the average velocity of cyclists (or running men) is larger than the velocity of walking pedestrians. The instantaneous velocity of an object can be estimated by the following equation:

$$velocity\,(m/s) = \frac{X\,(pixels)\,Scale\,(m/pixel) \cdot FPS\,(frames/s)}{N\,(frames)} \tag{12}$$

where X is the motion distance of pedestrians; *Scale* represents the actual length of the unit pixel; *FPS* represents frames per second; and N is the total frames of pedestrian motion.

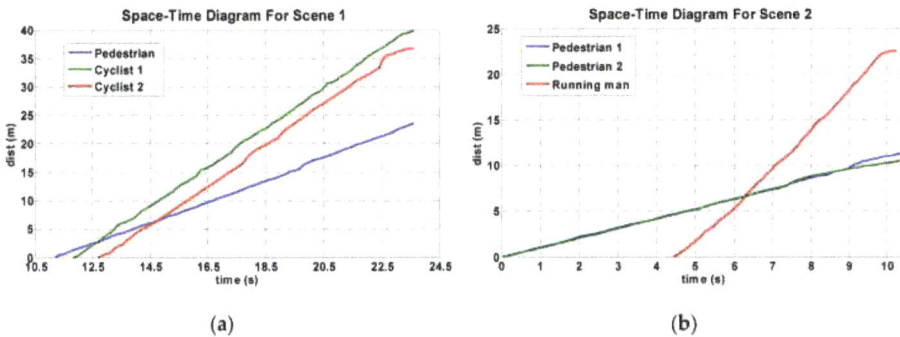

Figure 18. Pedestrian trajectory: space-time diagram. (**a**) Scene 1 and (**b**) Scene 2.

With estimated instantaneous velocity data, the frequency curves of detected objects can be derived as shown in Figure 19. These frequency curves of the instantaneous velocity can also be used to distinguish pedestrians and cyclists (or running men). As shown in Figure 19a, the maximum frequency velocity of cyclists is approximately 3 m/s, which is significantly larger than the maximum frequency velocity of pedestrian of about 2 m/s. In addition, the instantaneous velocity of cyclists ranges roughly from 1 m/s–8 m/s, but for walking pedestrians, its instantaneous velocity range is more concentrated, because a severe offset occurred for tracking fast moving targets. Note, this method cannot be used to differentiate running men and cyclists, as running men and cyclists have very similar velocity profiles.

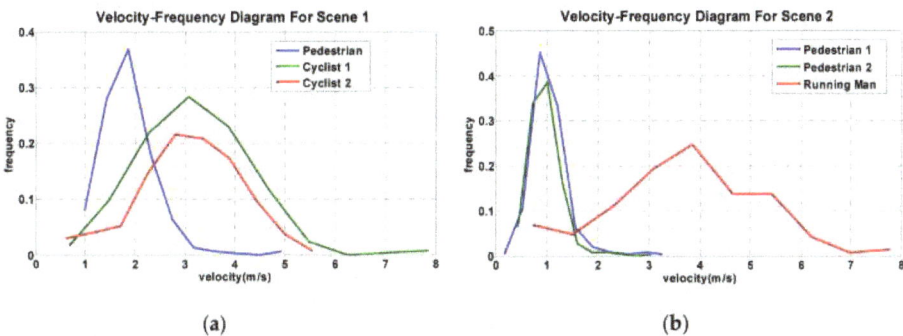

Figure 19. The frequency curve of the instantaneous velocity. (**a**) Scene 1 and (**b**) Scene 2.

6. Discussion

The proposed blob extraction and blob classification scheme shows good detection results, as shown in Table 2. However, the accuracy of final pedestrian detection is impacted by two important factors: (1) different training sets during blob classification; and (2) the accuracy of blob extraction.

Using Scenes 2 and 3 as an example, as shown in Table 1, the basic environment information (*i.e.*, temperature, flying height, *etc.*) of Scenes 2 and 3 is very similar, but the detection results are different. As shown in Table 2, the differences regarding precision and recall for Scenes 2 and 3 are 4.11% and 3.11%, respectively. Two reasons contribute to the difference of the final pedestrian detection:

(1) Different training sets during blob classifications. For different scenes, even if the environment is very similar, different training sets are used during the detection; because even a slight change of environment, such as temperature or flying height, will impact the positive and negative training samples. Therefore, one possible reason for the different detection results between Scenes 2 and 3 is because they use different training sets during blob classifications. To confirm this point, we merged the training blobs of Scenes 2 and 3 into an integrated dataset called "Scene 2 + Scene 3". We train a new model with the "Scene 2 + Scene 3" dataset and test the classification performance in the Scenes 2 and 3 dataset, respectively. As shown in Figure 20, with new training sets, the detection results show differences.

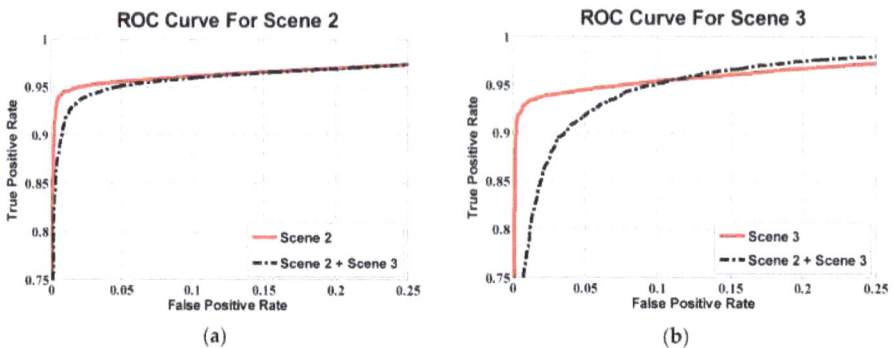

Figure 20. ROC curves for (**a**) Scene 2, (**b**) Scene 3.

(2) Accuracy of blob extraction. However, Figure 20 only shows that the new model is slightly worse than the original model. Further, referring to Figure 15a,b, the ROC curves show that the HOG + DCT classification model of Scene 2 is better than that of Scene 3, but the final detection results of Scene 2 are worse than those of Scene 3. Therefore, the training set may not be the main, at least not the only, factor. This leads to another important factor: Blob extraction.

To confirm our thought, we calculated the precision and recall of blob extraction in the Scene 2 and Scene 3 dataset. The following Table 7 presents the statistical results. As shown in Table 7, the precision and recall of Scene 3 are higher than those of Scene 2. This difference of blob extraction results is consistent with the difference of the final detection results. Thus, it is clear that one of the reasons (and maybe the main reason) why a detection difference does exist is due to the blob extraction difference.

Another indirect proof that blob extraction is one of the important reasons for the different detection results between Scenes 2 and 3 is that the Scene 3 dataset has less interference (such as lamps, vehicle headlights, *etc.*) than the Scene 2 dataset, therefore leading to the more accurate pedestrian candidate region extraction in Scene 3. Thus, there are less false pedestrian candidate regions in Scene 3, which results in higher precision and recall.

Table 7. Blob extraction statistical results.

Dataset	# Pedestrian	# TP	# FP	# FN	Precision	Recall
Scene 2	9214	8545	29,684	669	22.35%	92.74%
Scene 3	10,029	9633	3187	396	75.14%	96.05%

7. Conclusions

This paper proposes an approach to detect and track pedestrians in low-resolution aerial thermal infrared images. The proposed pedestrian detection method consists of two stages: blob extraction and blob classification. A blob extraction method based on regional gradient feature and geometric constraints filtering is first developed. This method can detect both bright and dark pedestrian objects in either daylight or at night. Furthermore, this method can work either with a static background or a moving platform. The proposed blob classification is based on a linear SVM, with a hybrid descriptor, which sophisticatedly combines the HOG and DCT descriptors. Experimental results showed the encouraging performance of the overall pedestrian detection method.

This paper further develops an effective method to track pedestrians and to derive pedestrian trajectories. The proposed method employs the feature tracker with the update of the detected pedestrian locations to accurately track pedestrian objects from the registered videos and to extract the motion trajectory data. The velocity characteristic of pedestrians and cyclists is also analyzed on the basis of velocity estimation.

We would like to point out that all methods described in this paper, individually, have been applied by other research. Some techniques, such as HOG and DCT features, SVM, video registration and KLT feature-based tracking, have been widely applied in much image processing research. We do not intend to claim the originality of these methods. In fact, the major contribution of this paper lies in how to apply these methods, as a whole, for pedestrian detection and tracking, particularly using thermal infrared images recorded from UAVs; such a topic, due to the low-resolution of imagery, platform motion, image instability and the relatively small size of the objects, has been considered as an extremely difficult challenge. The proposed comprehensive pedestrian detection and tracking system has successfully addressed this issue by nicely combining some existing classical detection and tracking methods with some relatively novel improvements, such as extracting pedestrian ROIs using a regional gradient feature and combining SVM with a hybrid HOG and DCT feature descriptor, which we have not seen in any direct applications in previous research based on our limited knowledge. In fact, the proposed blob extraction method can not only be used to extract pedestrian candidate regions in aerial infrared images, but also be used in the infrared images captured by roadside surveillance cameras mounted on low-altitude light poles with oblique views. As shown in Figure 21, using the thermal images provided by the Ohio State University (OSU) thermal public infrared dataset, our proposed blob extraction method can accurately extract pedestrian candidate regions.

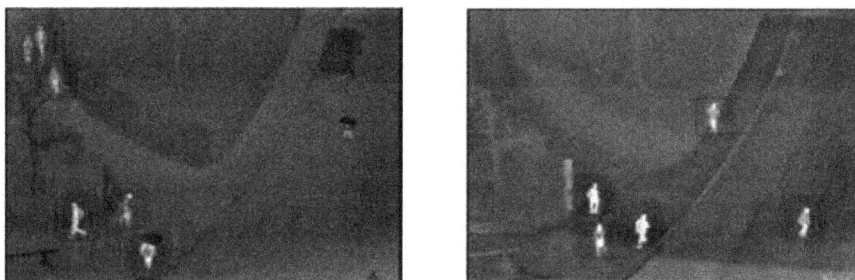

Figure 21. Blob extraction results in Ohio State University (OSU) thermal datasets.

More importantly, this research is expected to benefit many transportation applications, such as pedestrian-vehicle crash analysis, pedestrian facilities planning and traffic signal control (similar to [35]), Advanced Driver-Assistance System (ADAS) development, the multimodal performance measure [36] and pedestrian behavior study. Note that clearly the proposed pedestrian detection and tracking method can be further improved. For example, the current pedestrian detection method has difficulties dealing with the situations in which pedestrians are too close to each other, leading to the merging of pedestrian blobs caused by the halo effect. This problem brings significant missed detections. Furthermore, the employed video registration method works poorly when video pictures have drastic shaking due to wind or drone motion. Future work for addressing this issue will be considering mounting the camera gimbal on the drone to increase the stability of video pictures. More importantly, as a common drawback of all methods for object detection using thermal images, the low contrast of targets and background, especially in daylight, seriously affects the accurate detection. This issue cannot be addressed using our current method, but could be addressed by considering fusing thermal and visual images [15]. This will be a major future research topic.

Acknowledgments: This research was funded partially by the Fundamental Research Funds for the Central Universities #30403401 and by the National Science Foundation of China under Grant #61371076, #51328801, and 51278021.

Author Contributions: Yalong Ma and Xinkai Wu designed the overall system and developed the pedestrian detection and tracking algorithms. In addition, they wrote and revised the paper. Guizhen Yu and Yongzheng Xu have made significant contributions on the blob extraction and classification algorithms. Additionally, they designed and performed the experiments. Yunpeng Wang analyzed the experiment results and contributed to the paper revision.

Conflicts of Interest: The authors declare no conflict of interest.

Abbreviations

The following abbreviations are used in this manuscript:

UAVs	Unmanned Aerial Vehicles
SVM	Support Vector Machine
HOG	Histogram of Oriented Gradient
DCT	Discrete Cosine Transform
ROI	Region of Interest
CSM	Contour Saliency Map

References

1. Geronimo, D.; Lopez, A.M.; Sappa, A.D.; Graf, T. Survey of Pedestrian Detection for Advanced Driver Assistance Systems. *IEEE Trans. Pattern Anal. Mach. Intell.* **2010**, *32*, 1239–1258. [CrossRef] [PubMed]
2. Oreifej, O.; Mehran, R.; Shah, M. Human identity recognition in aerial images. In Proceedings of the 2010 IEEE Conference on Computer Vision and Pattern Recognition (CVPR), San Francisco, CA, USA, 13–18 June 2010; pp. 709–716.
3. Gaszczak, A.; Breckon, T.P.; Han, J. Real-time people and vehicle detection from UAV imagery. In Proceedings of the Intelligent Robots and Computer Vision XXVIII: Algorithms and Techniques, San Francisco, CA, USA, 24 January 2011; pp. 536–547.
4. Portmann, J.; Lynen, S.; Chli, M.; Siegwart, R. People detection and tracking from aerial thermal views. In Proceedings of the IEEE International Conference on Robotics and Automation (ICRA), Hong Kong, China, 31 May–7 June 2014; pp. 1794–1800.
5. Cortes, C.; Vapnik, V. Support-Vector Networks. *Mach. Learn.* **1995**, *20*, 273–297. [CrossRef]
6. Dalal, N.; Triggs, B. Histograms of oriented gradients for human detection. In Proceedings of the IEEE Conference on Computer Vision and Pattern Recognition (CVPR), San Diego, CA, USA, 20–26 June 2005; pp. 886–893.
7. Bouguet, J. Pyramidal Implementation of the Lucas Kanade Feature Tracker Description of the Algorithm. Available online: http://robots.stanford.edu/cs223b04/algo_tracking.png (accessed on 24 March 2016).
8. Enzweiler, M.; Gavrila, D.M. Monocular Pedestrian Detection: Survey and Experiments. *IEEE Trans. Pattern Anal. Mach. Intell.* **2009**, *31*, 2179–2195. [CrossRef] [PubMed]
9. Dollar, P.; Wojek, C.; Schiele, B.; Perona, P. Pedestrian Detection: An Evaluation of the State of the Art. *IEEE Trans. Pattern Anal. Mach. Intell.* **2012**, *34*, 743–761. [CrossRef] [PubMed]
10. Davis, J.W.; Keck, M.A. A Two-Stage Template Approach to Person Detection in Thermal Imagery. In Proceedings of the Seventh IEEE Workshops on Application of Computer Vision, Breckenridge, CO, USA, 5–7 January 2005; pp. 364–369.

11. Dai, C.; Zheng, Y.; Li, X. Pedestrian detection and tracking in infrared imagery using shape and appearance. *Comput. Vis. Image Underst.* **2007**, *106*, 288–299. [CrossRef]

12. Leykin, A.; Hammoud, R. Pedestrian tracking by fusion of thermal-visible surveillance videos. *Mach. Vis. Appl.* **2008**, *21*, 587–595. [CrossRef]

13. Miezianko, R.; Pokrajac, D. People detection in low resolution infrared videos. In Proceedings of the IEEE Computer Society Conference on Computer Vision and Pattern Recognition Workshops, Anchorage, AK, USA, 23–28 June 2008; pp. 1–6.

14. Jeon, E.S.; Choi, J.-S.; Lee, J.H.; Shin, K.Y.; Kim, Y.G.; Le, T.T.; Park, K.R. Human Detection Based on the Generation of a Background Image by Using a Far-Infrared Light Camera. *Sensors* **2015**, *15*, 6763–6788. [CrossRef] [PubMed]

15. Lee, J.H.; Choi, J.-S.; Jeon, E.S.; Kim, Y.G.; Le, T.T.; Shin, K.Y.; Lee, H.C.; Park, K.R. Robust Pedestrian Detection by Combining Visible and Thermal Infrared Cameras. *Sensors* **2015**, *15*, 10580–10615. [CrossRef] [PubMed]

16. Zhao, X.Y.; He, Z.X.; Zhang, S.Y.; Liang, D. Robust pedestrian detection in thermal infrared imagery using a shape distribution histogram feature and modified sparse representation classification. *Pattern Recognit.* **2015**, *48*, 1947–1960. [CrossRef]

17. Xu, F.; Liu, X.; Fujimura, K. Pedestrian detection and tracking with night vision. *IEEE Trans. Intell. Transp. Syst.* **2005**, *6*, 63–71. [CrossRef]

18. Ge, J.; Luo, Y.; Tei, G. Real-Time Pedestrian Detection and Tracking at Nighttime for Driver-Assistance Systems. *IEEE Trans. Intell. Transp. Syst.* **2009**, *10*, 283–298.

19. Qi, B.; John, V.; Liu, Z.; Mita, S. Use of Sparse Representation for Pedestrian Detection in Thermal Images. In Proceedings of the IEEE Conference on Computer Vision and Pattern Recognition Workshops (CVPRW), Columbus, OH, USA, 23–28 June 2014; pp. 274–280.

20. Teutsch, M.; Mueller, T.; Huber, M.; Beyerer, J. Low Resolution Person Detection with a Moving Thermal Infrared Camera by Hot Spot Classification. In Proceedings of the IEEE Conference on Computer Vision and Pattern Recognition Workshops (CVPRW), Columbus, OH, USA, 23–28 June 2014; pp. 209–216.

21. Sidla, O.; Braendle, N.; Benesova, W.; Rosner, M.; Lypetskyy, Y. Towards complex visual surveillance algorithms on smart cameras. In Proceedings of the IEEE 12th International Conference on Computer Vision Workshops (ICCV Workshops), Kyoto, Japan, 27 September–4 October 2009; pp. 847–853.

22. Benfold, B.; Reid, I. Stable multi-target tracking in real-time surveillance video. In Proceedings of the IEEE Conference on Computer Vision and Pattern Recognition (CVPR), Providence, RI, USA, 20–25 June 2011; pp. 3457–3464.

23. Xiao, J.; Yang, C.; Han, F.; Cheng, H. Vehicle and Person Tracking in Aerial Videos. In *Multimodal Technologies for Perception of Humans*; Stiefelhagen, R., Bowers, R., Fiscus, J., Eds.; Springer: Berlin/Heidelberg, Germany, 2008; pp. 203–214.

24. Bhattacharya, S.; Idrees, H.; Saleemi, I.; Ali, S.; Shah, M. Moving Object Detection and Tracking in Forward Looking Infra-Red Aerial Imagery. In *Machine Vision Beyond Visible Spectrum*; Hammoud, R., Fan, G., McMillan, R.W., Ikeuchi, K., Eds.; Springer: Berlin/Heidelberg, Germany, 2011; pp. 221–252.

25. Miller, A.; Babenko, P.; Hu, M.; Shah, M. Person Tracking in UAV Video. In *Multimodal Technologies for Perception of Humans*; Stiefelhagen, R., Bowers, R., Fiscus, J., Eds.; Springer: Berlin/Heidelberg, Germany, 2008; pp. 215–220.

26. Fernandez-Caballero, A.; Lopez, M.T.; Serrano-Cuerda, J. Thermal-Infrared Pedestrian ROI Extraction through Thermal and Motion Information Fusion. *Sensors* **2014**, *14*, 6666–6676. [CrossRef] [PubMed]

27. Viola, P.; Jones, M. Rapid object detection using a boosted cascade of simple features. In Proceedings of the 2001 IEEE Computer Society Conference on Computer Vision and Pattern Recognition, Kauai, HI, USA, 8–14 December 2001; pp. I-511–I-518.

28. A Threshold Selection Method from Gray-Level Histograms. *IEEE Trans. Syst. Man Cybern.* **1979**, *9*, 62–66.

29. Hafed, Z.; Levine, M. Face Recognition Using the Discrete Cosine Transform. *Int. J. Comput. Vis.* **2001**, *43*, 167–188. [CrossRef]

30. Vaidehi, V.; Babu, N.T.N.; Avinash, H.; Vimal, M.D.; Sumitra, A.; Balmuralidhar, P.; Chandra, G. Face recognition using discrete cosine transform and fisher linear discriminant. In Proceedings of the 11th International Conference on Control Automation Robotics & Vision (ICARCV), Singapore, 7–10 December 2010; pp. 1157–1160.

31. Shastry, A.C.; Schowengerdt, R.A. Airborne video registration and traffic-flow parameter estimation. *IEEE Trans. Intell. Transp. Syst.* **2005**, *6*, 391–405. [CrossRef]

32. Shi, J.; Tomasi, C. Good Features To Track. In Proceedings of the IEEE Conference on Computer Vision and Pattern Recognition (CVPR), Seattle, WA, USA, 21–23 June 1994; pp. 593–600.

33. Tomasi, C. Detection and Tracking of Point Features. *Tech. Rep.* **1991**, *9*, 9795–9802.

34. Bernardin, K.; Stiefelhagen, R. Evaluating multiple object tracking performance: The CLEAR MOT metrics. *J. Image Video Process.* **2008**, *2008*. [CrossRef]

35. Sun, W.; Wu, X.; Wang, Y.; Yu, G. A continuous-flow-intersection-lite design and traffic control for oversaturated bottleneck intersections. *Transp. Res. Part C Emerg. Technol.* **2015**, *56*, 18–33. [CrossRef]

36. Ma, X.; Wu, Y.J.; Wang, Y.; Chen, F.; Liu, J. Mining smart card data for transit riders' travel patterns. *Transp. Res. Part C Emerg. Technol.* **2013**, *36*, 1–12. [CrossRef]

sensors

MDPI

Article

Autonomous Aeromagnetic Surveys Using a Fluxgate Magnetometer

Douglas G. Macharet [1,*], Héctor I. A. Perez-Imaz [1], Paulo A. F. Rezeck [1], Guilherme A. Potje [1], Luiz C. C. Benyosef [2], André Wiermann [2], Gustavo M. Freitas [3], Luis G. U. Garcia [3] and Mario F. M. Campos [1]

[1] Department of Computer Science, Universidade Federal de Minas Gerais, Belo Horizonte 31270-901, Brazil; hector.azpurua@dcc.ufmg.br (H.I.A.P.-I.); rezeck@dcc.ufmg.br (P.A.F.R.); guipotje@dcc.ufmg.br (G.A.P.); mario@dcc.ufmg.br (M.F.M.C.)

[2] Observatório Nacional/MCTI, Coordenação de Geofísica, Rio de Janeiro 20921-400, Brazil; benyosef@on.br (L.C.C.B.); andrew@on.br (A.W.)

[3] Vale Institute of Technology, Ouro Preto 35400-000, Brazil; gustavo.medeiros.freitas@itv.org (G.M.F.); luis.uzeda@itv.org (L.G.U.G.)

* Correspondence: doug@dcc.ufmg.br; Tel.: +55-31-3409-6579

Academic Editors: Felipe Gonzalez Toro and Antonios Tsourdos
Received: 30 September 2016; Accepted: 5 December 2016; Published: 17 December 2016

Abstract: Recent advances in the research of autonomous vehicles have showed a vast range of applications, such as exploration, surveillance and environmental monitoring. Considering the mining industry, it is possible to use such vehicles in the prospection of minerals of commercial interest beneath the ground. However, tasks such as geophysical surveys are highly dependent on specific sensors, which mostly are not designed to be used in these new range of autonomous vehicles. In this work, we propose a novel magnetic survey pipeline that aims to increase versatility, speed and robustness by using autonomous rotary-wing Unmanned Aerial Vehicles (UAVs). We also discuss the development of a state-of-the-art three-axis fluxgate, where our goal in this work was to refine and adjust the sensor topology and coupled electronics specifically for this type of vehicle and application. The sensor was built with two ring-cores using a specially developed stress-annealed CoFeSiB amorphous ribbon, in order to get sufficient resolution to detect concentrations of small ferrous minerals. Finally, we report on the results of experiments performed with a real UAV in an outdoor environment, showing the efficacy of the methodology in detecting an artificial ferrous anomaly.

Keywords: geophysical surveys; fluxgate magnetometer; autonomous vehicles; rotary-wing UAVs; drones in mining applications

1. Introduction

Robotics has received notoriety in our society mainly due to the wide use of manipulators. However, recent advances on the research of autonomous vehicles have shown a much more vast range of applications, such as exploration, surveillance and environmental monitoring. In this sense, the use of aerial robots have a clear advantage over other kind of robots, since they are able to navigate over large areas faster and with a privileged view from above.

Autonomous mapping in Robotics could be defined as the task of acquiring models of the environment using autonomous mobile robots [1]. One of the most versatile type of mapping is the aerial one. Aerial surveys can use a great variety of sensors such as cameras, lasers, radio receptors, barometers, magnetometers, etc. Currently, most aerial surveys are created using a manned aircraft with expensive and special equipment, and therefore it is a high cost and time consuming task, as it generally must be done on large areas. In this context, autonomous aerial robots can bring many benefits to the workers and companies. Mapping with autonomous robots can reduce the risk of endangering human lives, and it can also reduce the economic costs of the task.

In this context, geophysical maps such as magnetic, gravitational and textured digital elevation maps can be used in different situations. Some practical applications that benefit from such maps are: (i) in geology it is possible to use a coverage strategy to detect minerals of commercial interest beneath the ground; (ii) in the military, to find unexploded ordnance (UXOs), sunken ships and submarines at sea; (iii) in archeology, they are used to create maps of subsurface buried artifacts; among others.

One of the main aspects that must be considered before executing a mapping is which type of information must be acquired, and consequently the sensors that will be used. In geophysical surveys, magnetometers are among the most widely used ones, allowing for the evaluation of the magnetic field at a certain region.

The fluxgate magnetometer is a traditional and well known instrument that have been used in geophysical exploration for decades, mostly due to its small size, good sensitivity, low noise and relatively low power consumption. This low noise characteristic is however directly related to a strong limit on its frequency response.

Fluxgate sensors are built using materials of high magnetic permeability that are excited up to the saturation limits by a source with periodic current. A significant part of this noise is due to magnetic properties of the sensor core. The right choice of materials and special thermal treatment can produce sensors with very low noise figures. The characteristics of the core material are also fundamental to improve the intensity of the induced signal in the sense coil [2]. There are several new amorphous and crystaline (nanocrystaline) materials available [3–5], and deeper investigation would be necessary to see if this class of material has the potential to be used in fluxgate applications.

In most uses as in geomagnetic stations, the fluxgate integration filter is set to frequencies usually below 1 Hz, with 0.1 Hz being typical [6], which is not adequate for moving platform applications. To increase the sensor frequency response, a higher frequency is also necessary for the core excitation, consequently affecting the core material choice. Cobalt based amorphous alloys are among the best for fluxgate sensor due to its lower intrinsic noise and lower conductivity which result in smaller eddy currents, favoring its use at higher frequencies [7].

In this work, we discuss the development of a state-of-the-art three-axis fluxgate magnetometer to be used in aeromagnetic surveys. The proposed sensor is specially designed for small rotary-wing UAVs, considering possible applications in the mining industry such as prospection of ferromagnetic minerals beneath the ground.

The magnetometer uses two ring-cores with CoFeSiB amorphous ribbon developed for the detection of small ferrous minerals concentration. Due to UAV restricted resources, the electronic design aims to reduce its size and power consumption. The sensor housing is assembled far from the electronics, in order to allow enough distance from the UAV's motors, a substantial source of magnetic noise.

The remainder of this paper is structured as follows: Section 2 presents the details regarding the magnetometer development. The coverage path planning strategy for autonomous mapping is proposed in Section 3, followed by Section 4 in which we describe the methodology for data acquisition and map generation. The experiments and analysis about the results are presented in Section 5 and finally in Section 6 we draw the conclusions and discuss paths for future investigation.

2. Magnetometer Development

Detailed geophysical surveys need accurate sensor information. These mapping executions with aerial techniques are very restricted. In most cases, such missions are done manually using a manned aircraft, which increases the survey duration, as well as the economical costs, whilst the quality of data is not necessarily improved. Other types of magnetic surveys can be done by foot or by ground vehicles, thus augmenting the survey time.

On the other hand, mapping with autonomous robots can reduce risks and costs related to the task. Due to the UAV's potential to increase productivity while mitigating safety hazards, mining companies already benefit from such equipment integrated into their operations.

The sensor described here is designed specifically for mining applications, focusing on aeromagnetic surveys for prospection of ferromagnetic minerals beneath the ground. The instrument is capable to detect small ferrous minerals concentration, and suitable for installation in moving platforms. Considering the UAV restrictions, the developed magnetometer presents particular characteristics, including reduced weight, dimensions, and power consumption. We also decrease costs greatly by considering the use of small and inexpensive robots. The proposed sensor is at par with other fluxgate magnetometers on the market with a similar scale range and sensitivity [8,9] but with an increased data rate of $\approx$4–5 Hz and smaller size.

Table 1 presents a summary of the three-axis fluxgate magnetometer proposed to be used in autonomous aeromagnetic surveys.

Table 1. Developed magnetometer specifications summary.

Full Scale Range	$\pm$ 60 μT
Resolution	<0.5 nT
Frequency Response	DC to 4 Hz (-1 dB)
Overall Linearity	0.02%
Precision	Better than 10 nT
Power consumption	$\approx$0.9 W @ 12 VDC
Weight	$\approx$89 g (head) + 174 g (body)
Size Head	5.5 cm $\times$ 4 cm $\times$ 4 cm

Figure 1 shows the complete sensor mounting.

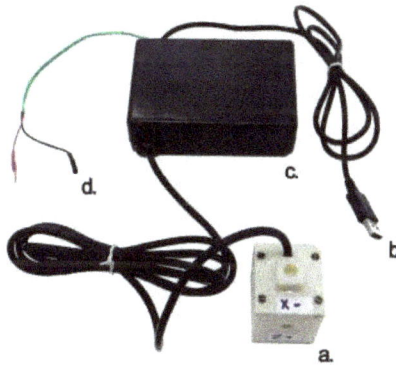

Figure 1. The developed sensor exposing the connectors. (**a**) The sensor core; (**b**) serial connector to acquire the data; (**c**) the Printed Circuit Board (PCB) protected by a small plastic case; and (**d**) power connectors.

2.1. Materials

In this work, we used a $Co_{67.5}Fe_{3.5}Si_{17.4}B_{11.6}$ composition melted by melt-spinner method [10] and after submitted to a convenient stress-annealing technique [11] increasing its sensibility and reducing its noise level.

As shown in the literature, there is a dependence between the magnetization process of the amorphous ribbon and the induced effects when an external load ("stress") is applied on the ribbon during the adjustment process in the core [12]. So to reduce them is convenient to use the ring-core geometry as applied in this experiment.

The ribbon was improved using an adequate stress-annealing that is fundamental to induce the appearance of transverse magnetic anisotropy causing rotation of the spontaneous domains of magnetization. A domain study of magnetization processes in a stress-annealed metallic glass ribbon for fluxgate sensors [12] confirms the importance of the longitudinal axis of the amorphous ribbon to be an easy magnetization axis, as well as the need of a convenient rotation of the domain walls, to obtain low levels of noises in fluxgate sensors. When the ribbon is accurately thermally annealed, it makes possible to obtain a fluxgate sensor with low level of noise, with a performance superior to the commercial crystalline materials [13].

2.2. Instrument Construction

The electronic circuit devised for this application is based on the traditional topology of second harmonic synchronous demodulation with current feedback, with the sensor operating in a null field condition [14]. This arrangement allows for a large dynamic range and great stability, reducing the circuit sensitivity to core parameter variations. A simplified block diagram for the sensor processing circuitry is shown on Figure 2.

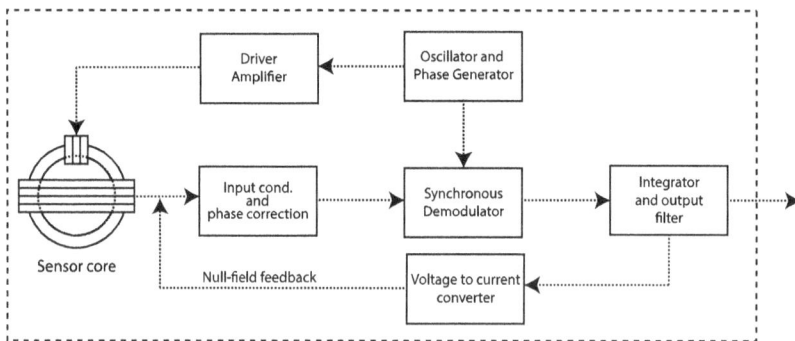

Figure 2. Block diagram of the sensor conditioning electronic circuit.

Although based on a traditional configuration, some minor changes were made to assure a good performance under fast changing field conditions as faced on a moving platform. No resonant filter was used at the sensor secondary winding and a special capacitive sampling and hold acted as a synchronous demodulator. All control pulses were produced by a small micro controller embedded in the oscillator and phase generation stage, conferring also great flexibility during the adjustment and calibration tasks. In order to reduce overall sensor size, two axes (X, Y) were wounded over one single core. This arrangement additionally saves 30% of drive power when compared to traditional three ring cores topology.

After demodulation and integration, the analog signal from each axis is carried to a three channel 22 bit sigma-delta A/D converter, controlled by a small micro-controller responsible for taking the measurements, apply calibration coefficients and transmit the resultant information throughout a serial communication port as in Figure 3.

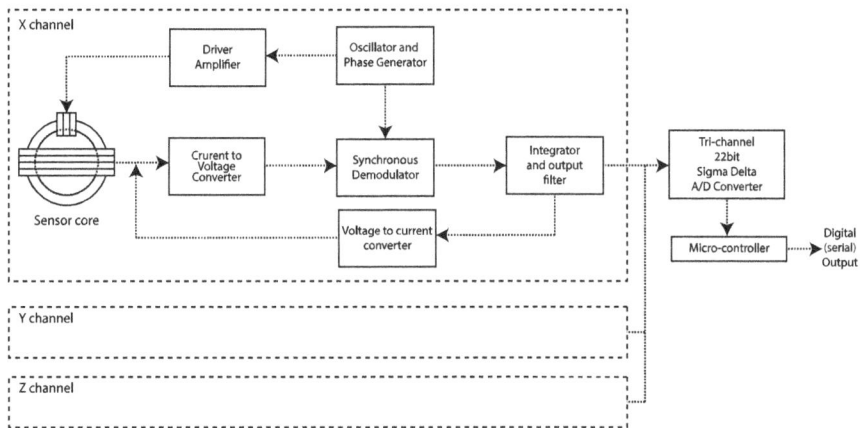

Figure 3. Complete magnetometer block diagram showing three channels and digital control.

3. Autonomous Mapping

3.1. Related Work

Path planning is a fundamental task for any kind of autonomous mobile robot. This problem in the context of area coverage is a popular research topic in recent years due to its varied uses. It can be assumed that the quality of the mapping task (e.g., resolution, energy consumption, execution time, etc.) is very dependent of an efficient coverage, so defining efficient routes is mandatory in many circumstances.

To perform a correct coverage of a region with a robot, some requirements need to be fulfilled: the robot needs to be able to reach all the points conforming the area, traversing the region using simple and/or optimal paths without overlapping or repeating them, and avoiding any obstacles for a safe navigation [15]. In addition, there are other constraints related to the particular characteristics of the robots and the environment such as autonomy optimization and availability of prior knowledge about the region. Execution time and vehicle autonomy can be critical in defining what type of trajectory will be generated, as small unmanned robots may have very limited autonomy.

In the literature, several coverage patterns have been studied, and in between the most efficient ones there is the lawnmower pattern [16–18]. This pattern is simple to implement as it is composed by several parallel straight lines with a space between them.

Other works adapt ground coverage solutions for aerial coverage. In [19] the authors proposed the use Boustrophedon cell decomposition [20] on the region of interest to then generate a Euler circuit with all the connected cells. Finally a lawnmower path is used to comb the cells. A photometric mapping system with quad rotors was developed in [21]. The authors proposed a wavefront planner that uses square cell decomposition with a simple breadth-first search (BFS) on the resulting graph to generate the routes.

A very important aspect when dealing with large areas to cover and small robots is cooperation. Robotic cooperation can be defined as a group of robots working together to perform one or several tasks in order to increase the efficiency and robustness of the task.

Aerial cooperative coverage has been a great research area over the last years due the advances and miniaturization of components and sensors. In this context, the work of [22] is one of the first to propose a cooperative coverage approach using such vehicles. It uses a polygonal area decomposition with heuristics to assign the routes between the robots considering their autonomy limitations as shown in Figure 4. Next, it is generated a collection of straight lines with the maximum length possible with the assumption that robots do not waste time on curves and stay most of the time at full speed.

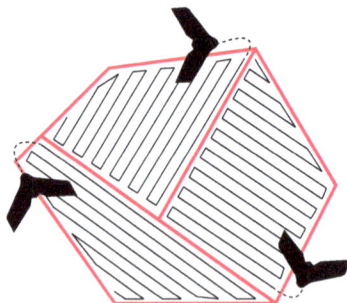

Figure 4. Example illustrating the method proposed in [22]. The area is decomposed using the robots autonomy and inside every cell the coverage pattern is optimized to decrease unnecessary turns.

In [23] the authors developed a path planning algorithm for multiple Unmanned Aerial Vehicles (UAVs) also based on cell decomposition. The allocation of the cells depends on the sensor footprint of every aircraft. In [24] the region of interest is subdivided with an greedy algorithm which minimizes the number of turns inside every sub-area. The authors also proposed the generation of sweep lines with the biggest length possible, and then the visit order of the cells is calculated by using an undirected graph.

In [25] it is proposed an algorithm which does not consider the environment segmentation. Initially, a collection of waypoints is assigned to the boundary of the region of interest, next it is created a graph composed by these waypoints, finally the original problem is formulated as the Vehicle Routing Problem (VRP) [26]. The main contribution is the incorporation of specific features that are relevant in a real-world deployment, such as a setup time. However, they do not consider cases where the number of vehicles or their autonomy are not sufficient given the size of the area to be covered.

In this sense, this work proposes a novel magnetic survey pipeline, that aims to increase versatility, speed and robustness of surveys by autonomous mapping using UAVs. The proposed approach is angle configurable to match the magnetic north, and allows configure several other parameters to match different mapping conditions. Quality of the magnetic data is increased due the stationary acquisition on every survey point.

3.2. Methodology

For the particular task of finding and recognizing magnetic anomalies we propose the use of modified version of the Maza et al. [22] path planning algorithm. The path planning algorithm is based on the lawnmower coverage pattern, that is one of the most efficient ways to cover a region [16–18]. The coverage process is divided in 3 steps once a region of interest is defined: (i) Parameter definition; (ii) Path generation and optimization; and (iii) Feasibility verification. A region of interest is defined as area bordered by the vertexes list $V = v1, v2, v3, ..., vn$ forming a polygon P, which we assume is convex.

3.2.1. Parameter Definition

To allow the aircraft be able to fulfill the tight constrains of a conventional magnetic survey some parameters had to be defined before generating the coverage path.

- Coverage angle, since it is recommended to keep the sensor aligned with the magnetic north of the region, this parameter allows the aircraft to define a coverage angle and/or specific yaw when acquiring the magnetic data on hover.
- Separation between survey waypoints, depending of the survey resolution needed the distance between survey waypoints need to be greater or smaller. For small anomalies, this distance can vary from 1 m to 3 m. For greater anomalies the distance could be several hundred meters [27].
- Flight height, the quality of the magnetic sensing decay to the cube of the distance from the anomaly [27], so this parameter could make the entire survey campaign useless is settled very high. For a balance between great resolution and safety we fly at approx 15 m off the ground.
- Waypoint hover time, since the fluxgate magnetometers have a low acquisition rates, the robot will need to hover every waypoint for a specified time to be able to acquire information with less noise possible. We empirically choose to hover 5 s at every survey point due that it was enough time for the UAV to stabilize.

3.2.2. Path Generation and Optimization

Once a region of interest and a set of parameters is defined, the survey path can be generated. The developed fluxgate sensor require the minimum movement possible to acquire low noise measurements. Quadrotors, unlike other aircraft such as fixed wing airplanes, can fly or "hover" over a specific point indefinitely. The survey path proposed subdivides the total survey path in small sections (generally 1 to 5 m) to allow the aircraft hover over every acquisition point, pointing the magnetic north an specified amount of time to collect enough readings to recreate that point the more accurately possible.

Algorithm 1 Survey path (P, α, d)

1: $P \leftarrow rotateCW(P, \alpha)$

2: $(h_{min}, h_{max}, v_{min}, v_{max}) \leftarrow getBoundaries(P)$

3: $L \leftarrow getVerticalLines(v_{min}, v_{max}, d)$

4: $R \leftarrow$

5: $i = 0$

6: **while** $i < |L|$ **do**

7:      $p = L_i$

8:      **if** $i + 1 < |L|$ **then**

9:          $p \leftarrow p \cup L_{i+1}$

10:          $i+ = 2$

11:      **else**

12:          $i+ = 1$

13:      **for** $j \in \{0, ..., |p|\}$ **do**

14:          $K \leftarrow intersection(h_{min}, h_{max}, p_j)$

15:          $D \leftarrow subdivide(k, d)$

16:          **if** $(i + j) \% 2 == 0$ **then**

17:              $D \leftarrow reverse(D)$

18:          $R \leftarrow R \cup D$

19:      $R \leftarrow rotateCCW(R, \alpha)$

20: **return** R

The algorithm for survey path generation can be seen in Algorithm 1 and a result of the generated path can be seen in Figure 5. The input of Algorithm 1 is the P polygon denoting the region of interest, α coverage angle and ω width separation between survey points. The algorithm first rotate P to be aligned horizontally, and then it calculates the position of the survey lines. Then it defines the size of the survey line and it subdivides it in small ω segments. Finally it again rotates back to the desired α angle and return the R path with the survey points.

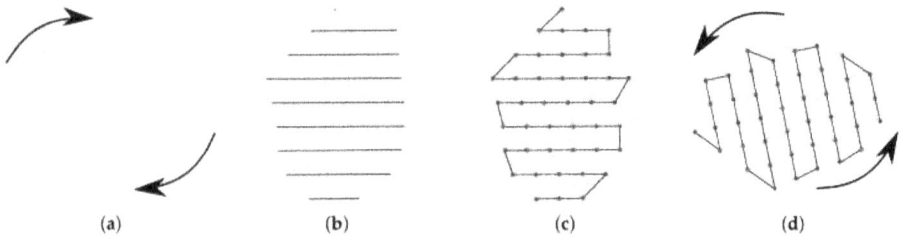

| (a) | (b) | (c) | (d) |

Figure 5. Survey path. (**a**) A region of interest is defined P and then rotated to be parallel to the X axis; (**b**) P is segmented in sections with an d distance; (**c**) Every line segment is subdivided in smaller d size segments and the visiting sequence is defined. Every survey visit point is denoted in red; (**d**) The final path is then rotated back to the desired angle.

In special cases where the survey require to have a specially low distance between survey points, a particular localization problem arises with small aerial robots. The GPS sensor on many of those small and expendable robots have an accuracy of 1 to 2.5 m [28], therefore setting such small distance between points will make the robot to skip the points inside the error radius of their on-board GPS sensor.

To tackle this problem we propose an algorithm that reorganize the visiting points allowing the aircraft to visit them with a minimum distance threshold to avoid those common GPS problems. The method can be observed in Algorithm 2 (Figure 6), it receives as input the R survey points and a β distance treshold. The algorithm first generate a graph where the vertexes are the survey points and the edges are created only for survey points that are inside a distance threshold. After that we used the DSATUR vertex coloring algorithm [29] to label the neighbor vertexes. The new generated group of labeled vertexes will not share any direct neighbors inside the distance threshold. Every set of vertexes is then optimized using a TSP (Traveling Salesman Problem) solver to obtain the visiting order that minimizes distance. Finally all the paths are concatenated: if there is any points closer than the distance threshold the process is repeated, else the new survey point sequence is returned.

Algorithm 2 Spaced survey points (R, β)

  1: $G \leftarrow getGraphEdges(R, \beta)$
  2: $S \leftarrow graphColoring(G)$
  3: $R \leftarrow \{\}$
  4: $f \leftarrow False$
  5: **while** f is *False* **do**
  6:     **for** $i \in \{1, ..., |S|\}$ **do**
  7:        $s \leftarrow S_i$
  8:        $s \leftarrow TSPSolver(s)$
  9:        $R \leftarrow R \cup s$
 10:     $f \leftarrow checkThreshold(R)$
 11: **return** R

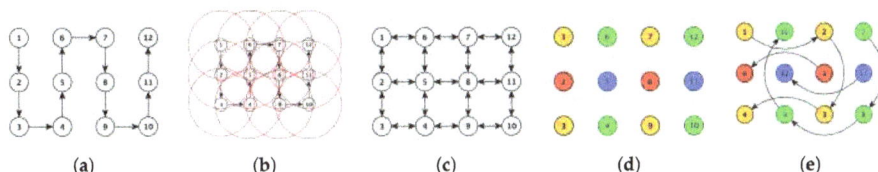

Figure 6. Spaced visiting sequence; (**a**) Normal survey path *R*; (**b**) A new graph is created, the red circles are the distance threshold to connect to neighbors; (**c**) Neighbors connected *G*; (**d**) Edges labeled by color; (**e**) Final path.

3.2.3. Feasibility Verification

Finally, in order to verify if the route can be completed by the robot, a time estimation is generated using the following equation:

$$t_{total} = (|R| \times t_h) + (|R| \times d)/v,$$

where t_h is the hovering time, d the distance between survey points and v the robot speed (m/s). Therefore, if t_{total} is greater than the actual flight time of the robot, the path cannot be done with the specified robot.

4. Magnetic Map Generation

4.1. Data Acquisition Strategy

Once a path is generated by the algorithm described in Section 3, the aerial vehicle will obtain the mean of 10 reading samples from the magnetometer for each point in an approximate regular grid within the desired area of coverage. The grid is said to be approximately regular mainly because of the GPS errors (less than 2 m) and the vehicle position instability.

The GPS navigation proven to be superior in time and with a relative low error in comparison with other techniques such as markers. GPS navigation allow the robot to be deployed fast and reach regions where terrestrial locomotion is difficult or dangerous. GPS error position can be also improved to less than 5 cm error with the use of differential GPS systems.

Generally, the final results of a interpolation depend on the spatial distribution of the points to be interpolated. A regular grid profile [30] is a good choice for this scenario in order to avoid artifacts and distortions on the final map. An irregular profile caused by disproportional spacing between points, will not accurately represent the actual signal that originated the data. The sampling value was empirically selected to increase reading accuracy, and decrease the effects of the UAV displacement over the plane and vibrations.

The average euclidean norm value *T*, which is also known as the total field, is computed from 10 samples obtained for each point in order to minimize noise influence. The unit used for *T* is nanotesla. Figure 7 shows an example of the generated grid for an arbitrary mapping profile to be followed by the drone. We follow the traditional approach of calculating the euclidean norm value *T* (total field) for each point, which represents the intensity/strength of the magnetic field of a point in the 3D space, as described in [27]. Other methods may use the gradient of *T* (gradiometers), but all of these methods consider the magnitude of the signal to generate a raster of values for each pixel on the final map.

Figure 7. Example of generated path following magnetic north, with 5 m spacing between samples. The pink area is the region of interest, the blue lines are the expected path to be followed and the red circles are the discrete waypoints to visit.

4.2. Plotting and Interpolation

Potential fields such as the magnetic field are continuous in space and devoid of abrupt changes and discontinuity. Therefore, an interpolator can be used to obtain a smooth continuous curve that fits on the observed points, and can fairly approximate to the real function that generated the original signal if there is sufficient known points [27]. The most used methods for interpolating geographic data are the Kriging [31], Splining [32] and Inverse Distance Weighting (IDW) interpolation [33].

By using the GPS coordinates and the mean total field T calculated for each position of the grid collected from the survey, we first generate a sparse 2D map, using a color map for the values of T.

To obtain the final map, we use a multiquadratic Radial Basis Function (mRBF) interpolator to estimate a smooth curve that passes through the scattered sample points. The choice of mRBF interpolator was made by experimenting with several different interpolation methods, including the most used ones. To evaluate the interpolators' accuracy and also optimize their parameters, we used a two-fold cross-validation method that consisted in splitting the dataset into train and test set by randomly selecting points uniformly, using the mean squared error as metric. We concluded that that the mRBF interpolation is able to provide better results, specially when tested on noisy data. In our case, it is very important that the interpolator is able to handle noisy data due to the sensor intrinsic noise and other sources of dynamic interference that may be caused by the aerial vehicle itself or external sources.

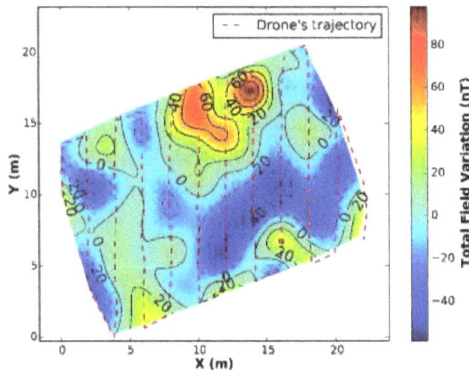

Figure 8. Illustration of the interpolated data represented by a coloured contour map.

Finally, we generate a regular grid of a desired resolution that can be visualized through different kind of maps, e.g., shaded reliefs, color maps or contour maps. Figure 8 depicts the resulting maps generated from a campaign.

5. Experimental Evaluation

Additional to the standard testing and calibration protocols, the magnetometer was subject to a few characterization procedures in order to confirm its specifications as required for this unique application.

5.1. Noise and Sensitivity

While typical commercial sensor cores present noise level of 100 pT rms in the interval of 0.1 Hz to 10 Hz, the sensor used in this experiment presents—in the same level—noise value below 47.5 pT rms (Figure 9). The sensor output (before data conversion) was measured by a digital Spectrum Analyzer (SR850, Stanford Research Systems/CA-USA). During this analysis the sensor elements were kept inside a demagnetized mu-metal shield with five layers.

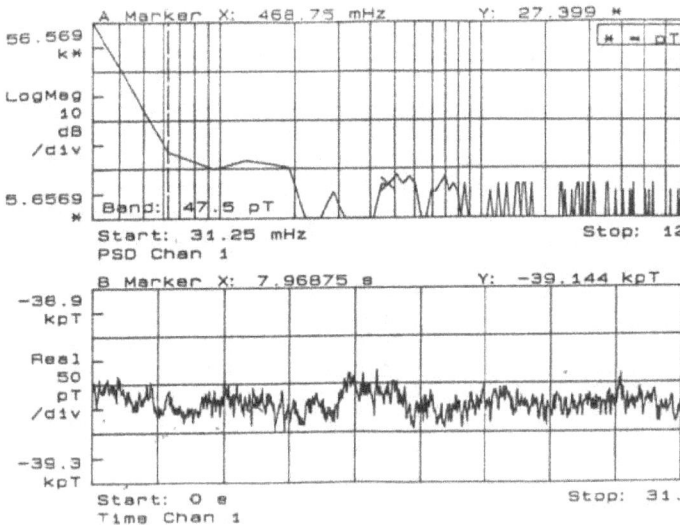

Figure 9. Noise level for selected CoFeSiB fluxgate sensor.

For sensitivity characterization a 0.1 Hz square wave current driving a 100 cm wide triaxial Helmholtz coil was used. The current square wave was obtained from a computer controlled source calibrator (2400 SourceMeter, Keithley Tektronix/OR-USA) and a lab current amplifier with current sensors inline. A set of three 6 1/2 digit voltmeters (8846A 6.5 Digit Precision Multimeter, Fluke/WA-USA) were used to compare each channel reading. Signals as low as 0.5 nT were detected showing excellent sensitivity for the instrument.

5.2. Frequency Response

To test the magnetometer frequency response, a sinusoidal current sweep ranging from a few milihertz to 10 Hz was applied to the Helmholtz coil (single axis at a time). For these low frequencies direct AC rms readings are not possible due to voltmeter low band limitations. A fast DC sampling with instantaneous normalization was used instead. After some adjustments the magnetometer achieved a −3 dB cutting just above 7 Hz as shown in the Figure 10.

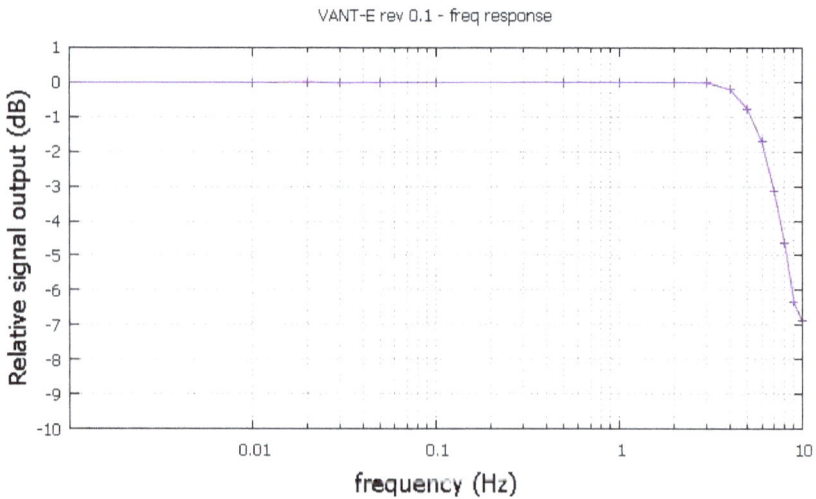

Figure 10. The sensor frequency response.

5.3. Sensor Behavior Characterization

Using a magnetometer sensor in proximity with an electrically motorized drone requires special considerations. The motor iron components and its strong operational current represent a significant source of noise, capable to disturb and even impair the intended magnetic survey. Under normal operation, the drone internal computer supplies the motor with a wide range of power levels, depending on the flight profile and attitude.

In order to empirically estimate the UAV intrinsic magnetic noise and avoid as much as possible its disturbance, a set of measures were taken around the vehicle with the help of a GSM-19T proton magnetometer. The GSM-19T is an instrument of regular use for geophysical surveys, showing great sensitivity and precision more than adequate for our noise investigation. To map all possible situations, we tested the drone at different distances from the GSM-19T exercising motor current levels ranging from off state to full power, searching for a minimum possible distance for the magnetometer to be installed where no interference could be detected. Since this is a much more sensitive base magnetometer, we found safe distances of aproximately 3 m.

We have also conducted an experiment with the developed fluxgate magnetometer. The experiment is composed by three tests, each one consisted in a mapping profile containing a single line of 6 m long aligned to the magnetic north, at a place free of sources of strong magnetic interference. Each measurement was obtained every 0.5 m along the line, where for each measurement we collected the mean norm of 10 readings from the sensor, in order to remove the high frequency noise. In all tests, the drone was positioned in the middle of the line.

The goal for this analysis was to reduce as much as possible the distance between drone and sensor, considering that a sensor hanging in a long cable can pose limitations on flight control and generate safety issues. Below a distance of one meter the disturbance was such that in some events the magnetometer was unable to provide any readings even for the lowest power levels applied. We found that distances starting from 2 m can be used depending on the drone size and motor power required for a given flight profile. Our conclusion is that beyond 3 m of separation between sensor and drone no significant change on readings was present at any power level.

In the first test, the drone was turned off and the battery cable was disconnected. The second test consisted in powering the drone, excepting the motors. Finally, in the third test, the motors were rotating at moderate speed, without taking off the ground.

By analyzing the results, we observed a peak at approximately the 1.0 m mark. So we safely assume the distance of the next stable measurement, as a minimum distance to locate the sensor. Finally we decided to add 10 cm extra for cautiousness, ending up with the sensor being located at 1.1 m of the drone. Based on this simple (and conservative) approach, our observations have shown that the impact of the drone's airframe, electronics and power levels were negligible at the chosen distance and position. Others works in the literature, for example [34], report on the use of booms of similar size (considering the distance to the motors).

This distance is very important because the magnetic field of the robot will change depending on payload and thrust applied to motors, so incorrectly setting this distance will interfere directly with the measurements in a non-predictable way. We have not conducted an exhaustive characterization of the setup, as at this moment the focus was to find a compromise solution that would guarantee the system operation, allowing us to demonstrate the feasibility of the concept. Nevertheless, measuring the sensitivity of the complete setup is part of a future work.

Finally, the last evaluation applied was a small survey over a controlled site submitting the sensor to real environment conditions of local noise and gradients. A 12 m profile where a ferromagnetic container of half cubic meters and 300 kg was placed over the magnetometer path as depicted on Figure 11. The results for this survey (Figure 12) exhibit an anomaly having an amplitude of approximately 2000 nT next to the position the container was placed.

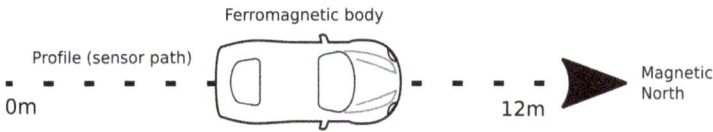

Figure 11. Profile used for testing a basic survey with the drone magnetometer.

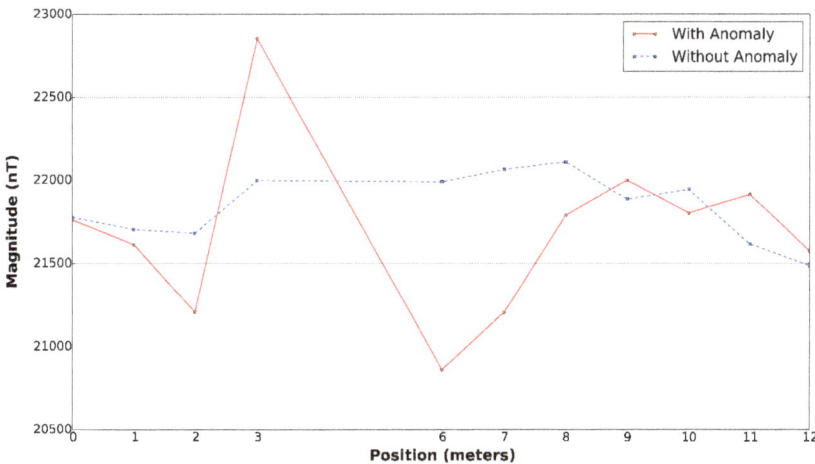

Figure 12. Survey along a test profile aligned to the magnetic north. Two survey has been carried, with and without a ferromagnetic test body containing approximately 300 kg of low carbon.

5.4. UAV Sensor Mounting

Small aerial robots have serious payload limitations, and their small size also creates a challenge to accommodate a large number of sensors and payload. The developed magnetometer has a weight of 89 g (head) and 174 g (PCB with case). Due the light weight nature of this sensor, typical pendulum carrying wasn't possible: small wind perturbations greatly alter the sensor position and angle, thus decreasing sensor accuracy.

The way the sensor is mounted on the drone is key to having correct measurements. We tested different mountings such as pendulum and vertical mountings. Despite the pendulum mounting don't had any serious impact in aircraft mobility, wind and even the normal movement of the drone generated a unpredictable motions on the sensor, thus generating extra noise. A vertical mounting do not prove to be robust enough on this type of drone due the expected distance between the sensor and the drone body.

Given those limitations we proposed a "boom" mounting. A 1.1 m lightweight 100% aluminum bar pointing at the front of the robot was the more stable mounting. The magnetic permeability of aluminum is known to be close to μ_0, making it a good material for the magnetometer mounting. This mounting can be seen on Figure 13 on a X8+ drone from 3DR. Also a companion computer and extra batteries needed to be added to acquire and process the sensor data accurately. A Raspberry Pi 2 computer was connected to the *Pixhawk* control board to process sensor information when a survey point was reached.

All the weight on the aircraft needed to be balanced before flight to avoid misbehavior and control problems during flight. This balancing was done on the X and Y axes of the aircraft, carefully moving all components of the system until the drone stays still on his center of gravity.

Figure 13. Close up of chosen Unmanned Aerial Vehicle (UAV): X8+ from 3DR. Important components are highlighted on the picture. (**a**) Sensor head with an 1.1 m aluminium boom; (**b**) compainion computer: Raspberry Pi 2; (**c**) sensor body.

5.5. UAV Field Experiment

In order to validate our technique, we performed a complete survey using the proposed methodology. All the steps required for the mapping and data processing phases were performed automatically by the robot and processing software with little to no human interaction.

The objective of this experiment was to generate consistent maps, where one is able to identify an artificial anomaly caused by a metallic structure placed at a known position. A 2.10 m × 3.00 m car was used as an anomaly. The chosen location for the survey was the middle of a soccer field, far away from other sources of magnetic interference such as power lines or big metal structures. Figure 14 depicts the experiment setup. A video illustrating the complete execution of a survey campaign is available at [35].

(a)

(b)

Figure 14. Field experiment scene. (**a**) The sensor components are being carried by the octarotor, and its core is supported by a pole of 1.1 m, in order to avoid as much as possible the magnetic interferences caused by the drone itself. In the background, the car used as the magnetic anomaly originator; (**b**) UAV flying a mission.

The mapping profile created for this experiment has a 2 m of separation between survey points, and 7 s hover time over an area of 381 m^2. The survey length was 181 m. At every survey point the robot faced the sensor to the magnetic north. The planned survey path and the executed one can be seen on Figure 15.

We executed the same profile configuration for 4 different surveys at 3, 5, 7 and 20 m of height above the ground. Figure 16 presents the results of the aeromagnetic maps.

By comparing the interpolated maps from all the surveys, it is clear that the magnetometer was able to detect the variation of the magnetic field in the vehicle's surroundings, leaving an expected dipolar magnetic signature on the generated map, specially for the first case (flight height = 3 m). However, as expected, with a greater distance between the sensor and the anomaly (flight height = 20 m), the map looses quality/resolution and the anomaly is undetectable.

(a)

Figure 15. *Cont.*

(b)

Figure 15. Field experiment survey path. (a) The planned path with a 2 m separation between survey points on an area of 381 m². The red dots are the survey points where the robot hovers ($t_h = 7$ s), and the red rectangle the position of the artificial anomaly; (b) The real executed trajectory of the robot on red.

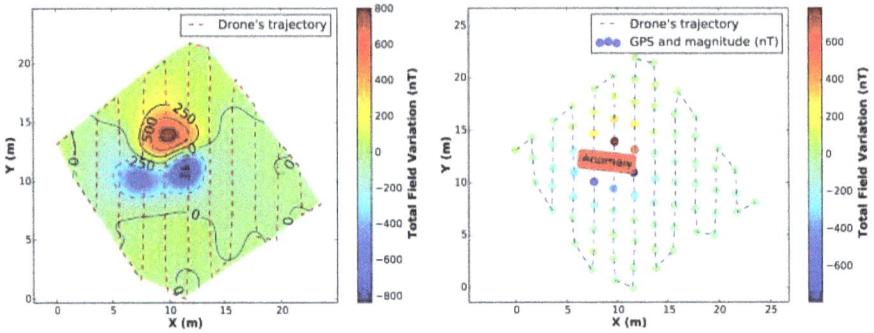

(a) Flight height = 3 m.

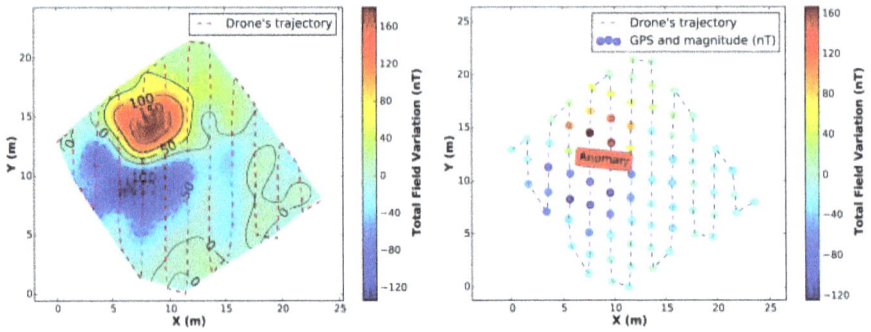

(b) Flight height = 5 m.

Figure 16. *Cont.*

(c) Flight height = 7 m.

(d) Flight height = 20 m.

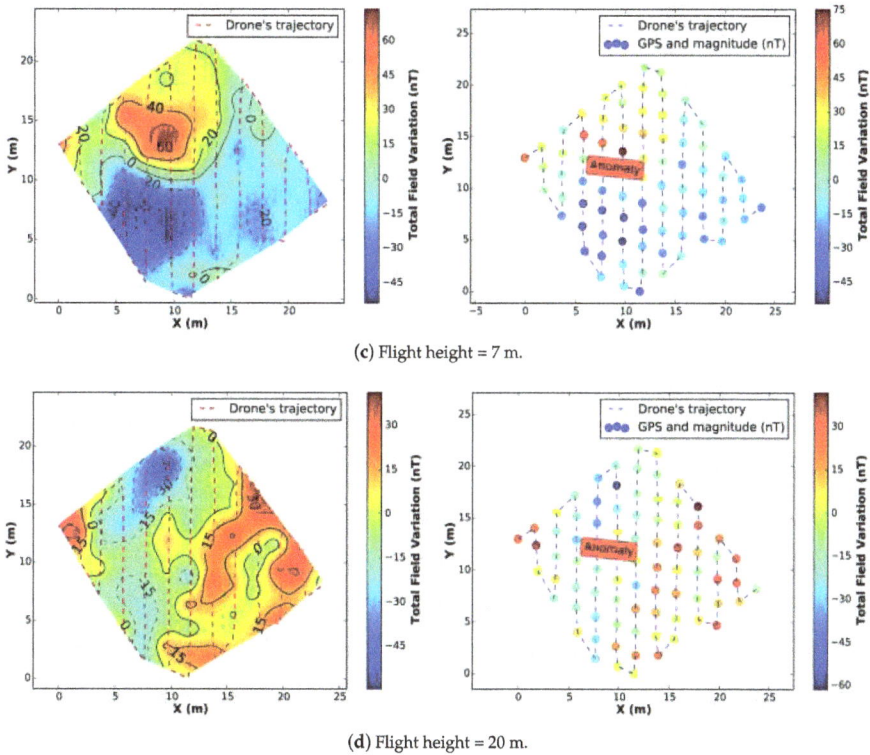

Figure 16. Generated maps from the complete survey at different heights with a artificial anomaly placed in the hightlighted position on the right (sparse) map.

6. Conclusions and Future Work

6.1. Conclusions

The interest and research in UAVs has been increasingly growing, specially due to the decrease in cost, weight, size and performance of actuators, sensors and processors. These type of vehicles clearly have their niche of applications, being one of their main advantages a privileged view from above. Considering the mining industry, it is possible to use such vehicles in the prospection of minerals of commercial interest beneath the ground.

In this context, we discussed the development of a state-of-the-art three-axis fluxgate magnetometer to be used in autonomous aeromagnetic surveys. The electronic was built aiming to reduce its size and power consumption, and the sensor housing was assembled far from the electronics in order to allow enough distance from important sources of magnetic noise, such as the vehicle's motors. Due to the sensor small size and weight, it is very sensitive to noise generated from fast rotational movements and vibration. We proposed the use of rotary-wing autonomous aerial robots that can hover over survey points. This limitation makes it impractical for mounting this sensor on other types of aircraft such as fixed-wing aircrafts.

For the particular task of aeromagnetic surveys, we have proposed the use of modified version of the Maza et al. [22] path planning algorithm, which is based on the lawnmower coverage pattern. It was modified to subdivide the path on several small size segments and to customize survey angle to

Sensors **2016**, *16*, 2169

match the magnetic north or another arbitrary angle. Also we proposed a method to overcome GPS accuracy on surveys, changing the visiting sequence of those survey points.

Through an extensive set of experiments performed to determine the mapping profile parameters, map generation methodology and adequate vehicle's mounting configuration, we were able to successfully generate meaningful magnetic maps. Our method is highly automated, and the system is easily reproducible and scalable to multiple drones with small addition in costs.

6.2. Future Work

The work has several future directions that can be pursued in order to improve the methodology, which includes the extension of the proposed technique to multi-robot ensembles and to environments with obstacles, both providing better representations of real-world scenarios.

Furthermore, every survey campaign has unique characteristics, in this sense, on future work we aim to improve the parameter selection for surveys such as hover time, sample size and distance between survey points. We also intend to evaluate how the influence of different sensor mounting positions affect accuracy of data acquisition.

Future work will also involve the application of the proposed methodology on a prospective mining scenario, comparing results to known previous generated maps with conventional methods.

Acknowledgments: This work was developed with the support of Vale Institute of Technology (ITV).

Author Contributions: Douglas G. Macharet, Héctor I. A. Perez-Imaz, Paulo A. F. Rezeck, Guilherme A. Potje were responsible for the design and development of the path planning algorithm and map generation methodology. They also carried out the UAV flight experiments. Gustavo M. Freitas and Luis G. U. Garcia have contributed in the data analysis, writing and revision of the paper. Luiz C. C. Benyosef and André Wiermann have conceived and built the fluxgate magnetometer. Mario F. M. Campos has supervised the work.

Conflicts of Interest: The authors declare no conflict of interest.

References

1. Thrun, S. Robotic Mapping: A Survey. In *Exploring Artificial Intelligence in the New Millennium*; Morgan Kaufmann Publishers Inc.: San Francisco, CA, USA, 2003; pp. 1–35.
2. Benyosef, L.C.C.; Teodósio, J.R.; Taranichev, V.E.; Jalnin, B.V. Improvements on CoFeSiB amorphous ribbon for fluxgate sensor cores. *Scr. Metall. Mater.* **1995**, *33*, 1451–1454.
3. Kennedy, J.; Leveneur, J.; Turner, J.; Futter, J.; Williams, G.V. Applications of nanoparticle-based fluxgate magnetometers for positioning and location. In Proceedings of the 2014 IEEE Sensors Applications Symposium (SAS), Queenstown, New Zealand, 18–20 February 2014; pp. 228–232.
4. Kennedy, J.V.; Leveneur, J.; Williams, G.V.M. Wide Dynamic Range Magnetometer. US Patent 20,150,323,615, 12 November 2015.
5. Leveneur, J.; Kennedy, J.; Williams, G.; Metson, J.; Markwitz, A. Large room temperature magnetoresistance in ion beam synthesized surface Fe nanoclusters on SiO_2. *Appl. Phys. Lett.* **2011**, *98*, 053111.
6. Jankowski, J.; Sucksdorff, C.; Geomagnetism, I.A. *IAGA Guide for Magnetic Measurements and Observatory Practice*; Boulder, CO: Warsaw, Poland, 1996.
7. Shirae, K. Noise in amorphous magnetic materials. *IEEE Trans. Magn.* **1984**, *20*, 1299–1301.
8. Bartington. MAG03 Fluxgate Three Axis Magnetometer. 2016. Available online: http://www.bartington.com/presentation/mag-03-three-axis-magnetic-field-sensor (accessed on 25 September 2016).
9. Stefan-Mayer. FLC3-70 3-Axis Magnetic Field Sensor. 2016. Available online: http://www.stefan-mayer.com/en/products/magnetometers-and-sensors/magnetic-field-sensor-flc3-70.html (accessed on 25 September 2016).
10. Budhani, R.C.; Goel, T.C.; Chopra, K.L. Melt-spinning technique for preparation of metallic glasses. *Bull. Mater. Sci.* **1982**, *4*, 549–561.
11. Benyosef, L.C.D.C.; Stael, G.C.; Bochner, M.A.C. Optimization of the magnetic properties of materials for fluxgate sensors. *Mater. Res.* **2008**, *11*, 145–149.
12. Záveta, K.; Nielsen, O.V.; Jurek, K. A domain study of magnetization processes in a stress-annealed metallic glass ribbon for fluxgate sensors. *J. Magn. Magn. Mater.* **1992**, *117*, 61–68.

13. Benyosef, L.C.C. Effect of Stresses on the Magnetic Properties of Amorphous Ribbons for Fluxgate magnetometers. *J. Adv. Mater.* **1996**, *3*, 140–143.
14. Primdahl, F. The fluxgate magnetometer. *J. Phys. E Sci. Instrum.* **1979**, *12*, 241–253.
15. Cao, Z.L.; Huang, Y.; Hall, E.L. Region filling operations with random obstacle avoidance for mobile robots. *J. Robot. Syst.* **1988**, *5*, 87–102.
16. Lumelsky, V.J.; Mukhopadhyay, S.; Kang, S. Dynamic path planning in sensor-based terrain acquisition. *IEEE Trans. Robot. Autom.* **1990**, *6*, 462–472.
17. Acar, E.U.; Choset, H.; Rizzi, A.A.; Atkar, P.N.; Hull, D. Morse decompositions for coverage tasks. *Int. J. Robot. Res.* **2002**, *21*, 331–344.
18. Peless, E.; Abramson, S.; Friedman, R.; Peleg, I. Area Coverage with an Autonomous Robot. US Patent 6,615,108, 2 September 2013.
19. Xu, A.; Viriyasuthee, C.; Rekleitis, I. Optimal complete terrain coverage using an unmanned aerial vehicle. In Proceedings of the 2011 IEEE International Conference on Robotics and Automation (ICRA), Shanghai, China, 9–13 May 2011; pp. 2513–2519.
20. Choset, H.; Pignon, P. Coverage path planning: The boustrophedon cellular decomposition. In *Field and Service Robotics*; Springer: Berlin, Germany, 1998; pp. 203–209.
21. Barrientos, A.; Colorado, J.; Cerro, J.D.; Martinez, A.; Rossi, C.; Sanz, D.; Valente, J. Aerial remote sensing in agriculture: A practical approach to area coverage and path planning for fleets of mini aerial robots. *J. Field Robot.* **2011**, *28*, 667–689.
22. Maza, I.; Ollero, A. Multiple UAV cooperative searching operation using polygon area decomposition and efficient coverage algorithms. In *Distributed Autonomous Robotic Systems 6*; Springer: Berlin, Germany, 2007; pp. 221–230.
23. Santamaria, E.; Segor, F.; Tchouchenkov, I. Rapid Aerial Mapping with Multiple Heterogeneous Unmanned Vehicles. In Proceedings of the 10th International ISCRAM Conference, Kristiansand, Norway, 24–27 May 2015.
24. Jiao, Y.S.; Wang, X.M.; Chen, H.; Li, Y. Research on the coverage path planning of uavs for polygon areas. In Proceedings of the 2010 the 5th IEEE Conference on Industrial Electronics and Applications (ICIEA), Taichung, Taiwan, 15–17 June 2010; pp. 1467–1472.
25. Avellar, G.S.C.; Pereira, G.A.S.; Pimenta, L.C.A.; Iscold, P. Multi-UAV Routing for Area Coverage and Remote Sensing with Minimum Time. *Sensors* **2015**, *15*, 27783.
26. Toth, P.; Vigo, D. (Eds.) *The Vehicle Routing Problem*; Society for Industrial and Applied Mathematics: Philadelphia, PA, USA, 2001.
27. Reeves, C. *Aeromagnetic Surveys: Principles, Practice & Interpretation*; Geosoft: Toronto, ON, Canada, 2005.
28. Ublox. NEO-7 Series. 2016. Available online: https://www.u-blox.com/en/product/neo-7-series (accessed on 25 September 2016).
29. Brélaz, D. New methods to color the vertices of a graph. *Commun. ACM* **1979**, *22*, 251–256.
30. Haxby, W.; Karner, G.; LaBrecque, J.; Weissel, J. Digital images of combined oceanic and continental data sets and their use in tectonic studies. *Eos Trans. Am. Geophys. Union* **1983**, *64*, 995–1004.
31. Stein, M.L. *Interpolation of Spatial Data: Some Theory for Kriging*; Springer: New York, NY, USA, 2012.
32. De Boor, C. *A Practical Guide to Splines*; Applied Mathematical Sciences; Springer: New York, NY, USA, 1978; Volume 27.
33. Zimmerman, D.; Pavlik, C.; Ruggles, A.; Armstrong, M.P. An experimental comparison of ordinary and universal kriging and inverse distance weighting. *Math. Geol.* **1999**, *31*, 375–390.
34. McKay, M.D.; Anderson, M.O. Development of Autonomous Magnetometer Rotorcraft For Wide Area Assessment. In Proceedings of the 13th Robotics & Remote Systems for Hazardous Environments and 11th Emergency Preparedness & Response, LaGrange Park, IL, USA, 2011.
35. Macharet, D.G.; Perez-Imaz, H.I.A.; Rezeck, P.A.F.; Potje, G.A.; Benyosef, L.C.C.; Wiermann, A.; Freitas, G.M.; Garcia, L.G.U.; Campos, M.F.M. Autonomous Aeromagnetic Surveys Using a Fluxgate Magnetometer. Available online: https://youtu.be/Z64kTn6kIs8 (accessed on 5 December 2016).

MDPI

St. Alban-Anlage 66

4052 Basel, Switzerland

Tel. +41 61 683 77 34

Fax +41 61 302 89 18

http://www.mdpi.com

Sensors Editorial Office

E-mail: sensors@mdpi.com

http://www.mdpi.com/journal/sensors

www.ingramcontent.com/pod-product-compliance
Lightning Source LLC
Chambersburg PA
CBHW051707210326
41597CB00032B/5404